Intelligent Vehicle Technologies

Intelligent Vehicle Technologies

Theory and Applications

Edited by

Ljubo Vlacic
Griffith University, Brisbane, Australia

Michel Parent
INRIA, France

Fumio Harashima
Tokyo Metropolitan Institute of Technology, Japan

OXFORD AUCKLAND BOSTON JOHANNESBURG MELBOURNE NEW DELHI

Butterworth-Heinemann
Linacre House, Jordan Hill, Oxford OX2 8DP
225 Wildwood Avenue, Woburn, MA 01801-2041
A division of Reed Educational and Professional Publishing Ltd

A member of the Reed Elsevier plc group

First published 2001

British Library Cataloguing in Publication Data
Intelligent vehicle technologies: theory and applications
1 Motor vehicles – Electronic equipment
I Vlacic, Ljubo. II Parent, Michel. III Harashima, Fumio.
629.2′549

Library of Congress Cataloguing in Publication Data
Intelligent vehicle technologies/edited by Ljubo Vlacic, Michel Parent, Fumio Harashima.
p. cm.
Includes index.
ISBN 0 7506 5093 1
1 Automobiles – Automatic control. I Vlacic, Ljubo. II Parent, Michel.
III Harashima, Fumio.
TL152.8.I573 2001
629.2′3–dc21 2001025276

ISBN 0 7506 5093 1

Typeset in 10/12pt Times Roman by Laser Words, Madras, India
Printed and bound in Malta

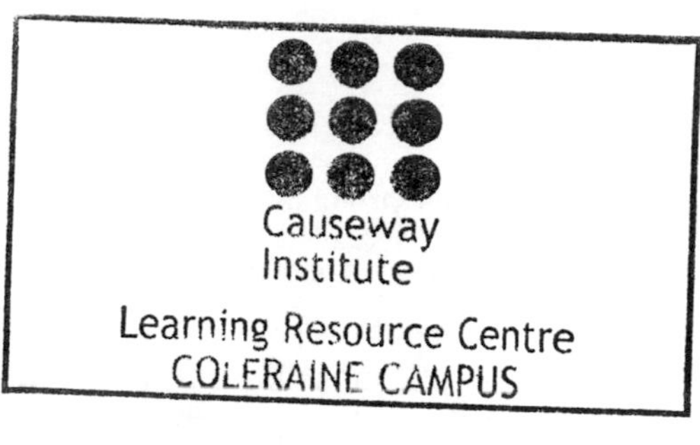

Contents

Preface

Intelligent vehicle technologies are rapidly growing worldwide. They have been recognized as technologies that are enabling enhancement in road safety, road transport operational efficiency, and increasing driving pleasure. Intelligent vehicle technologies are being deployed under the umbrella of the 'driver assistance concept' (to assist human beings while driving the vehicle) or the 'autonomous driving/intelligent vehicle concept' (to enable a vehicle to drive independently (autonomously) along the road, with no or limited assistance from humans).

This book is aimed at giving a thorough introduction to the many aspects of these technologies whether they are to support only the driver assistance concept, the autonomous driving concept, or both. However, the latter was selected as the focal point of this book bearing in mind that the driver assistance concept is a subset of the intelligent vehicle concept.

Written by the leading experts in the field – practising professionals from both industry and academia – this book provides an insight into intelligent vehicle technologies. It will, we strongly believe, serve as the text in various pedagogical settings as well as a reference book for both educators and practitioners, not just in automotive engineering but in all other, intelligent vehicle related disciplines from electronics and computing to control and communications. We also believe the book will be of immense value to our students and will greatly assist them in exploring the synergy between a number of the technologies that contribute to the exciting and fast growing development of intelligent vehicles, and related research disciplines and industry.

If intelligent vehicles are to operate autonomously, they should then be able to: (a) sense their own status as well as the environment in which they are in; (b) process the sensed information and communicate with the immediate environment; and (c) undertake a decision about the most appropriate (i.e. the safest) manoeuvre that should be performed and execute that manoeuvre. This simple classification of the intelligent vehicle's abilities has influenced the composition of this book which is as follows.

Part One: Introduction

Chapter 1 addresses some of the beauties and nuisances that are introduced into society by vehicles, discusses the needs of implementing intelligent vehicle technologies and related concepts, and advocates the deployment of the intelligent vehicle concept, particularly in dense urban areas where further investment in new road infrastructure is hardly justifiable, almost impossible.

Part Two: Intelligent Vehicle Sensor Technologies

In Chapter 2, the basic concept and the functional principles of the CAN (controller area network) data bus system are described along with its main features and technical advantages. An effort was made to depict CAN implementations not only for in-vehicle applications (for interconnection of in-vehicle electronic sensors and devices) but elsewhere in engineering.

Chapter 3 gives the reader an idea of micro-controllers and microelectronic technologies that are behind the intelligent vehicle's ability to communicate with the environment they are in (i.e. people, roads, other vehicles and drivers).

The sensors that are currently in use by intelligent vehicles are explained in Chapters 4–8. In particular, Chapter 4 introduces the vehicle detection sensor (a sensor that detects the distance and direction of a vehicle running ahead) using a technology for road environment sensing (road surface condition recognition). Chapter 5 elaborates on vision systems and their application to vehicle environment sensing. It discusses applications of vision sensors in increasing the driver's safety as well as an implementation of a multi-sensor based vision system. This allows complete autonomous control of a vehicle and assists a vehicle to perform a tight autonomous following. A broad implementation of the vision system in a road traffic environment, including lane and traffic sign recognition, and shape-based road object recognition (cars, pedestrians, etc.) is discussed in Chapter 6.

Radio communication technologies for vehicle-to-vehicle and road-to-vehicle communication concepts are described in Chapter 7. Optical and millimetre-wave devices, the key technologies behind these communication concepts, are also explained as well as the currently available vehicle information and communication systems.

The global positioning system (GPS), the technology that provides the ability to accurately determine a vehicle's position on demand, and to deliver customized location-based services to the vehicles, is explained in Chapter 8. It also provides an overview of the components and solutions which are available for use in GPS-enabled applications.

Part Three: Intelligent Vehicle Decision and Control Technologies

Chapter 9 brings the essence of adaptive control, the essential techniques for the moving vehicle's control problems. It demonstrates that the adaptive control techniques are powerful tools in implementing a variety of intelligent vehicle control tasks such as intelligent cruise control, inter-vehicle distance warning and control, computer controlled brake systems, platooning, lateral control, vehicle path following and vehicle collision avoidance, etc.

The solution to the distance and tracking control – the two very important control tasks of intelligent vehicles – is explained in great detail in Chapter 10. The chapter also describes fuzzy control, one of the most popular and widely used intelligent control techniques as well as the design steps of the fuzzy control system design process.

Chapter 11 discusses the decision and control architectures for motion autonomy, particularly those architectures that are relevant to intelligent vehicle operations such as the sensor-based manoeuvre concept. It describes both the theoretical and experimental considerations for a static obstacle avoidance, trajectory tracking, lane changing, parallel parking and platooning manoeuvres of the experimental intelligent vehicle – Cycab.

Chapter 12 addresses another important control problem, that of the intelligent vehicle's automatic brake control. This control problem is becoming increasingly important with the recent development of the adaptive cruise control concept and its specific case, stop-and-go adaptive cruise control. These adaptive control tasks are specifically discussed in Chapter 13, which brings results from experiments with the three adaptive cruise control techniques. To date, these techniques have been developed as a typical driver assistance system where a driver retains the authority over, and the responsibility for, the vehicle at all times. Intelligent, driver-less, vehicles would require an automatic cruise control unit capable of taking over both the stop-and-go manoeuvre and collision prevention at all times. This development is being considered by many researchers.

Part Four: Case Study

This concluding section of the book is devoted to the performance review of the operations of the ARGO intelligent prototype vehicle which currently performs obstacle avoidance, lane detection, vehicle detection and pedestrian detection manoeuvres. The chapter also provides concluding remarks on the experimental results collected during the 2000 km operational testing of the ARGO prototype vehicle performance.

Ljubo Vlacic
Michel Parent
Fumio Harashima

Acknowledgements

The book is a result of contributions submitted by leading experts in the field – practising professionals from both industry and academia. Without their enthusiasm and dedication, the manuscript for the book would not have been completed nor its high quality achieved. Our profound gratitude goes to all chapter authors and to the many research colleagues and students around the globe. They have inspired our thinking in this emerging field of intelligent vehicle technologies and created warm, productive and enjoyable dialogues that have enriched the substance of this book.

The editors are most grateful to numerous people for their help in converting the manuscripts into this book. Without their support this volume may not have come to fruition. In particular, the editors are thankful to Ms Siân Jones, Commissioning Editor at Arnold, who approached the editors with an initiative to prepare this title for publishing and the editor of then Arnold's Automotive Engineering Series, Professor David A. Crolla of the University of Leeds, UK, who provided valuable guidance in the early stages of this book as well as constructive feedback to the authors through his extensive review process. This was prior to the consequent transfer of this title from Arnold to Butterworth-Heinemann.

After its transfer to Butterworth-Heinemann, the title was under the responsibility of Ms Renata Corbani, Desk Editor, who professionally guided it throughout and brought it to timely completion. Her patience and expert guidance were instrumental in converting the manuscript into a book. For this, we are sincerely appreciative.

Finally, we are indebted to our wives and families for setting the wonderful supportive home environment that assists us in exploring knowledge.

The Editors

Contributors

Khurshid Alam, Griffith University, Australia
Massimo Bertozzi, University of Parma, Italy
Alberto Broggi, University of Pavia, Italy
Gianni Conte, University of Parma, Italy
Stéphane Donikian, IRISA-CNRS, France
Thierry Fraichard, INRIA Rhône-Alpes and Gravir, France
Alessandra Fascioli, University of Parma, Italy
Uwe Franke, DaimlerChrysler AG, Germany
Masayuki Fujise, Ministry of Posts and Telecommunications, Japan
Zoran Gajic, Rutgers University, USA
Dariu Gavrila, DaimlerChrysler AG, Germany
Axel Gern, DaimlerChrysler AG, Germany
Steffen Görzig, DaimlerChrysler AG, Germany
Fumio Harashima, Tokyo Metropolitan Institute of Technology, Japan
Patrick Herron, Motorola Inc., USA
Tetsuo Horimatsu, Fujitsu Limited, Japan
Mark Hitchings, Griffith University, Australia
Reinhard Janssen, DaimlerChrysler AG, Germany
Vojislav Kecman, The University of Auckland, New Zealand
Muhidin Lelic, Corning Incorporated, USA
Christian Laugier, INRIA Rhône-Alpes and Gravir, France
Nan Liang, Siemens AG, Germany
Dragos B. Maciuca, BMW of North America, LLC., USA
David Maurel, Renault, France
Shingo Ohmori, Ministry of Posts and Telecommunications, Japan
Frank Paetzold, DaimlerChrysler AG, Germany
Michel Parent, INRIA, France
Dobrivoje Popovic, University of Bremen, Germany
Chuck Powers, Motorola Inc., USA
Yuichi Shinmoto, OMRON Corporation, Japan
Michael Solomon, Motorola Inc., USA
Christoph Stiller, Robert Bosch GmbH, Germany
Kiyohito Tokuda, Oki Electric Industry Co., Ltd, Japan
Ljubo Vlacic, Griffith University, Australia
Christian Wöhler, DaimlerChrysler AG, Germany

Part One Introduction

1

The car of the future

Michel Parent
INRIA, France

1.1 Such a wonderful product...

At the beginning of this twenty-first century, nobody can ignore the tremendous impact the development of the car has had on our lives and on society in general. In just a bit more than a century, this product has become at the same time something now indispensable for the daily living of billions of inhabitants, and also a major nuisance for many of them and a threat to the planet.

Over less than 100 years, the private car has changed from an eccentric product to a mass market commodity of the highest technology and at an incredibly low cost. If you consider the number of parts, the weight or the service rendered, it is certainly the product with the best performance/price ratio. Its performances are simply amazing and have been constantly improved over the last century. Not just the top speed which has been 'sufficient' for most usage (and possibly even too high) for quite some time, but in terms of road handling, braking, comfort, safety and, most of all, in terms of reliability.

The car has changed our society in a profound way. It has allowed many of us to live in the place of our choice, often far away from our work place. It has made possible the dream of many families to live in a single detached house with a large garden, creating in consequence the suburban life with its shopping centres, two-car families and city highways (see Figures 1.12–1.14). It has allowed many to live a more fruitful life with many more opportunities to work, to meet people, to shop, and for leisure, and education, etc. Among the most important steps in the life of a young adult are acquiring the ability to drive and then the ownership of a car. The car, with its enormous choice of styles and performances, is also a prime status symbol and at the same time a very exciting product to use for many drivers who perceive the driving task as a 'sport' which can be performed daily on any road.

The car has also created a huge industry which has often been the driving force of the economies of many countries. Not just for the manufacturing of the product but also for its maintenance, for the exploitation, processing and distribution of oil products, for highway construction and maintenance, for tourism, for real-estate, etc.

So, in short, the car is such a wonderful product that it become the preferred transportation means for almost anyone. This is true for trips from a few hundred metres to several hundred kilometres (with the same product making it the most flexible of all

Fig. 1.1 Quadricycle De Dion Bouton (1895–1901).

Fig. 1.2 Ford Model T (1908–1927).

Fig. 1.3 VW Beetle (1946–).

transportation means), whether you are alone, two or with the entire family, with or without all sorts of cargo. It is not only a transportation means but also an extension of your home where you can feel and behave as you wish, whether you are driving or not. At the end of the twentieth century, there were around 800 million vehicles on the roads, mostly in industrial countries with densities around 500 cars for 1000 inhabitants and an average distance driven of 15 000 km per car, per year. And of course, in all the countries where the densities are much lower, such as in China, Africa or India, the desire of the individuals (for their own freedom) as well as the desire of the states (for their economy) is to catch up with these figures. So, some economists predict that the number of vehicles on the road could double in the next 20 years (mostly due to the newly developed countries), while the miles travelled by each car are also on the increase because of new infrastructures and better use of these infrastructures.

1.2 Difficulties now and ahead

The success of the car has created a number of problems which will certainly increase over time if strong steps are not taken in the next decade. These problems cannot be ranked by order of importance because they are not affected by the same variables (it would be like comparing apples and oranges), but it is difficult not to place safety

Fig. 1.4 Citroën DS 19 (1955–1965).

Fig. 1.5 Audi A8 (1994–).

Fig. 1.6 Drive-in theatre.

as the top concern of the population and of governments. Just in Europe (the same is true in North America), there is an average of 150 deaths due to car accidents each single day – the equivalent of an aeroplane crash. The economic impact of these accidents (not just death but also injuries which often last a lifetime) is just enormous and not fully supported by the drivers through their insurance. Fortunately, the safety

Fig. 1.7 Old service station.

Fig. 1.8 Highway.

has improved enormously in the last decades of the twentieth century in industrial countries through improvements of both the car and of the infrastructure. The car has much better active and passive safety and through the construction of motorways and the elimination of many dangerous spots, the road is safer. However, it seems that in many countries, a plateau has been reached and safety cannot be improved without

Fig. 1.9 Car park.

Fig. 1.10 Underground car park.

taking some drastic steps such as reduced speed limits and better control of drivers' abilities. It is well known that almost all accidents are due to human errors, mostly reaction time too long and inability to control the vehicle in emergency situations. It has been observed through analyses of videos on motorways, that in case of sudden emergencies, about 30 per cent of all drivers take improper action and often lose completely the control of their vehicles.

Of great concern is in particular the license to drive of older persons. It is a fact that with age, their abilities can decrease drastically, so with the fast ageing of the population (and in particular when the 'baby boom' will reach this critical age which is quite soon) the politicians will have to face the fact that many should not be allowed to drive an ordinary car. Should it be no car at all, a 'sub-car' with minimal performances such as those allowed in Europe for drivers without a licence, or an advanced, safer, car?. . . .

Fig. 1.11 Highway interchange.

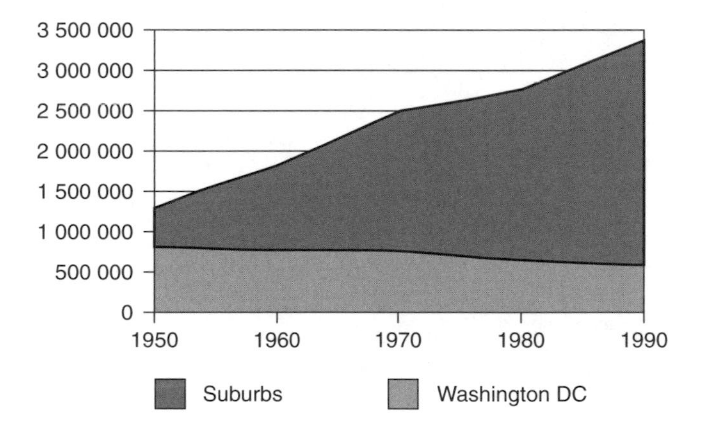

Fig. 1.12 Urbanized area growth: 1950–90.

The second concern in the list of governments (who issue the rules) and manufacturers (who must comply with them) is the energy consumption, and most precisely the burning of fossil fuel. Although there is still some debate about the warming of the planet, most industrial nations have agreed to reduce the production of CO_2 in the years to come. In many countries, the biggest contributor is road transport. Although some improvements have been made in the efficiency of engines through better control of ignition, of injection, of valve timing, of transmissions, the fact remains that consumers buy bigger and bigger cars (especially in the USA with the success of SUVs – sports and utility vehicles), with more comfort (air-conditioning is now a must in Europe) and most of all, they drive more and more. The solutions are in two directions: drive less (move less or use other transportation modes which do not burn fossil fuels), or develop another type of power plant for the car.

The third concern is with local pollutants, in particular in cities. Here the technology has been more successful. All pollutants have been drastically diminished in the last 20 years, sometimes by a factor of almost 100. Here again these reductions have been

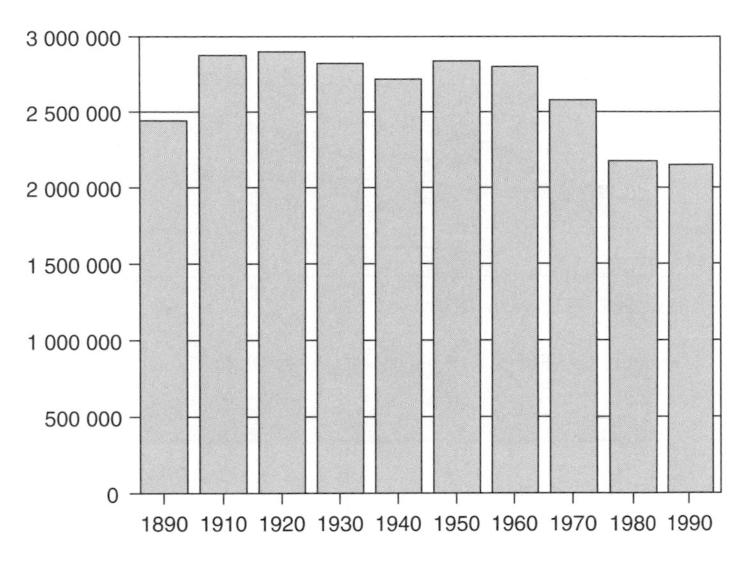

Fig. 1.13 Central city population from 1890: Paris.

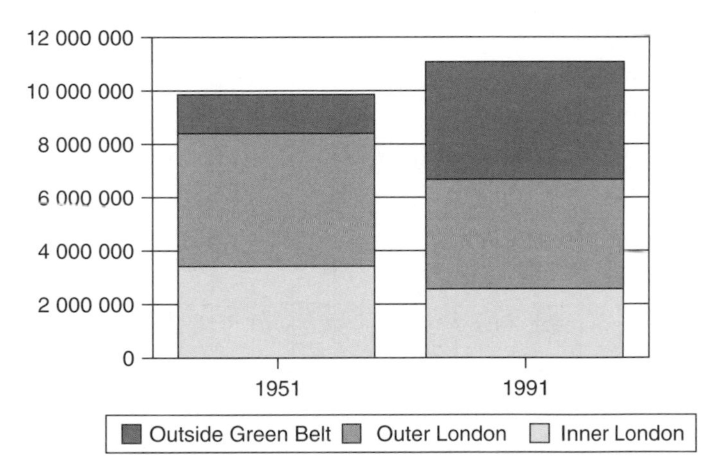

Fig. 1.14 London population: 1951–91.

obtained through the use of the best control technologies in terms of timing, injection, treatment of the exhaust gases and through new advanced sensors. However, cities are still faced with old cars (the 'clunkers') and also old buses in face of growing concerns with health. So, many cities are experiencing with various ways to curb the usage of cars, in general or just on days of peak pollution.

The next (in the list, not in importance) concern, is all the nuisances the automobile has brought to cities. Although it may be nice to live in a detached house with a large garden and a five-car garage (a must in new California developments), it is more and more difficult to go downtown for shopping or pleasure because 'downtown' cannot expand as much as the suburbs (which have no limits). There is just not enough space for all the cars that would want to go there, not enough freeways, not enough parking.

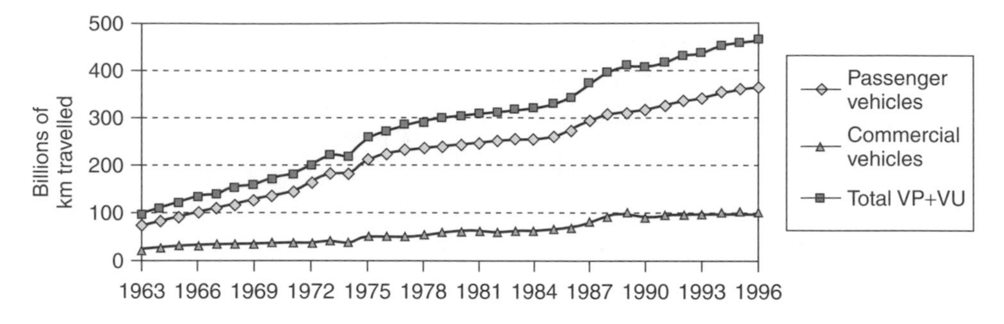

Fig. 1.15 Annual kilometres travelled for French registered vehicles (total).

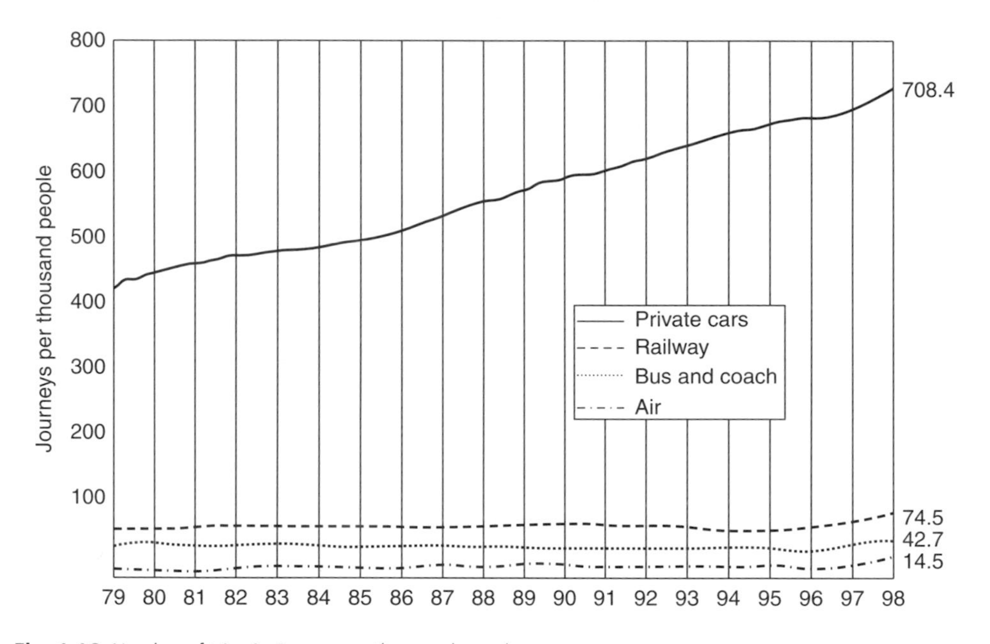

Fig. 1.16 Number of trips in France per thousand people.

And the demand exists since the roads are filled with cars that make it unpleasant for both the drivers and for the residents. Among the nuisances mentioned most often are the noise, the pollution and the lack of space for pedestrians. In short, a poor quality of life for both residents and for visitors. The solutions here are coming slowly, with more and more restrictions on the use of private cars: better parking control (and more expensive parking), pedestrian zones, city pricing (in Singapore), highway pricing and at the same time, alternatives to the car: better mass transit, park and ride, car-pooling, car-sharing, station cars. . . . Europe seems to be leading the way to protect the cultural heritage of its historic centres, although there are several interesting rehabilitation developments in North America where the cars are 'hidden'. In Europe, the driving force is high-end tourism and in particular cultural tourism. City officials now recognize that a car-free environment is very attractive to visitors who often come by high speed train or by plane and who stay longer in a more friendly environment. As a bonus, old

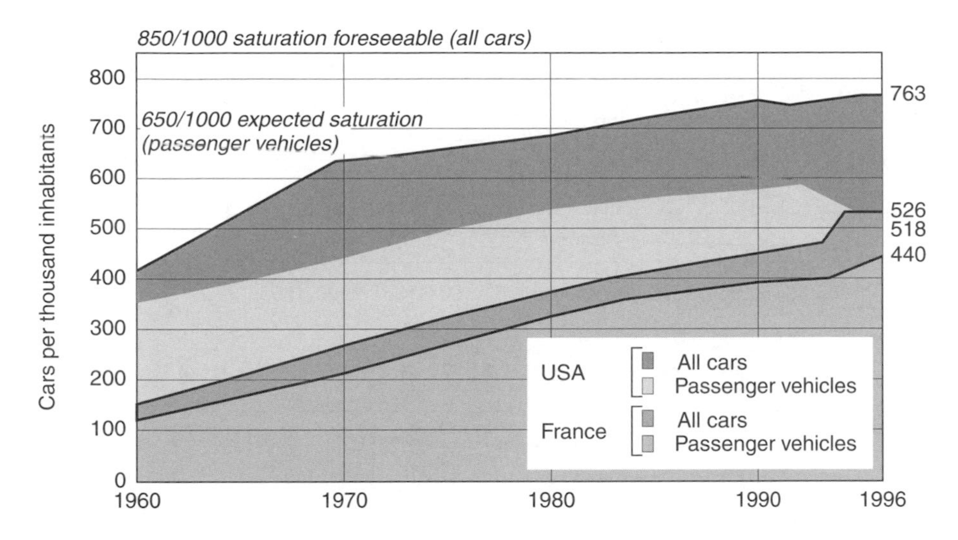

Fig. 1.17 Increase in the number of cars in USA and France.
Sources: Fédération routière internationale (IRF) et Comité des constructeurs français d'automobiles.

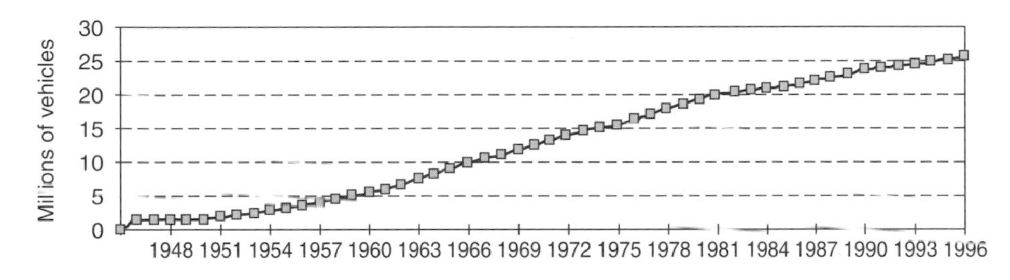

Fig. 1.18 Total number of privately registered vehicles.

buildings do not deteriorate as fast... However, one must be careful not to 'kill' a city by prohibiting all forms of transportation. So, some vehicles must still be allowed but under close control and probably new types of vehicles, well adapted to the city and well controlled, must be offered (with drivers or on a car-sharing basis) for those who cannot walk long distances.

The last item of concern is the dependency on the car our society in industrial countries have allowed to happen. If the car has created a new form of urbanism with suburban life, this form of life is totally dependent on the car. Without a car, life cannot be supported. So, in the same fashion that the car has created the ghettos in the cities of North America by allowing residents to move out to suburbs, the reverse trend may happen if these residents cannot have access to a car because of whatever restriction that may occur. The biggest threat is with the ageing population – who might become dissatisfied with the suburban way of life or simply who may not be allowed to drive anymore.

There is already some tendency in the upper classes of North America to favour the city life where many conveniences are close by and where it is quite possible to live without owning a car. This has always been the case in Europe where the most

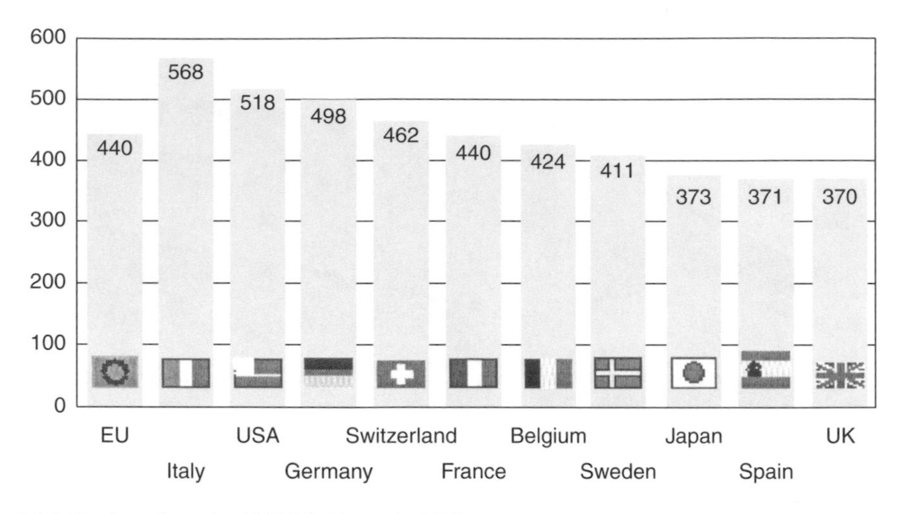

Fig. 1.19 Number of cars for 1000 inhabitants in 1996.

Fig. 1.20 Congestion.

desirable places to live are in the historic centres of cities. In North America, some developers have started to propose 'close-knit communities' with higher densities and where the car is down-played and pedestrian travel is favoured. In such environments, the ownership of a car is not so important and various forms of 'car-sharing' are in the development.

Fig. 1.21 Crash.

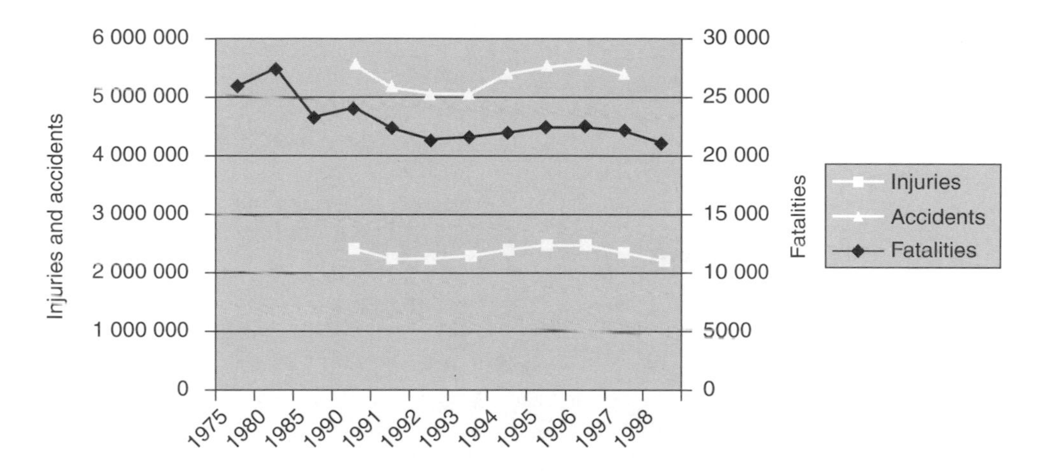

Fig. 1.22 Number of injuries, fatalities and accidents by car (USA)
Source: Bureau of Transportation Statistics.

1.3 Emerging technologies

The recent improvements to the car in order to alleviate many of the problems mentioned above are, for the most part, issued from computer and control technologies inside it. In this book we will describe most of these technologies and their future. However, new developments now concentrate on a system approach with the advent of ITS (intelligent transportation systems), with advances also in computers and on control but also on telecommunications and not only at the level of the vehicle but also on infrastructures. These new developments will allow a better usage of the infrastructures which have reached a saturation level in many countries. It is indeed more and more difficult, especially in dense urban areas where most problems facing car use are concentrated, to justify new road infrastructure. At the moment, ITS are mostly used for information purposes but control techniques are also being tested

Fig. 1.23 French 'sub-car'.

Fig. 1.24 Pollution.

to regulate the flow of traffic through access control, variable speed limits and road pricing.

For the moment, the interaction between the infrastructure and the car is only through the driver who gets information through message signs, traffic lights, radio, the Internet, etc. and acts in view of this information. In the future, the infrastructure may interact directly with the vehicle, for example through speed control or lateral guidance.

In fact, we might see in the next decade a progressive disconnection between the driver and his/her car with the possibility (which already exists in some very dedicated

Fig. 1.25 Toyota IMTS.

Fig. 1.26 Toyota Crayon.

vehicles) to leave the driving entirely to machines. This represents a change of paradigm in the usage of the automobile which, up to now, is under the total control of the driver, even if reckless actions are taken (such as driving in the wrong way on a motorway). In the future, such manoeuvres may be impossible to execute and driving may become more and more under automatic control. This trend is clearly under way with brake control (first, several generations of anti-lock braking system (ABS), then with electronic stability programme (ESP), with torque control (which prevents the Smart from flipping backwards) and now with adaptive cruise control (ACC). The normal end of this trend is the full automation of the driving, not just for comfort but for safety and for better management of the infrastructures.

For improving the safety of their vehicles, the manufacturers have improved drastically the vulnerability of the automobiles and the passenger restraints. However, it is better to develop techniques to prevent the crash. These techniques could cover driver warning and vehicle control. Curiously, while vehicle control seems more difficult to develop and to implement inside the vehicle, it is now widely available in many vehicles. The first such system is of course the ABS which takes over some control from the driver in order to limit slippage of wheels during braking. Now this principle has been extended with traction slip control and electronic stability

Fig. 1.27 People-mover (Frog).

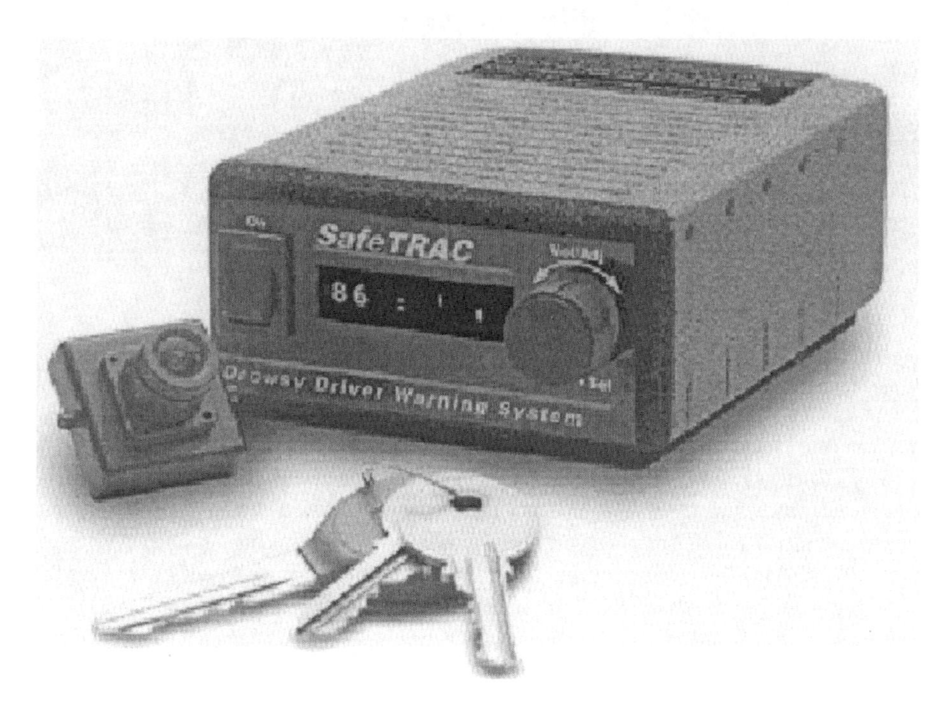

Fig. 1.28 SafeTRAC System.

programmes (ESP), which can prevent a car from spinning in a turn. The next level of control comes with ACC which can regulate the speed of the vehicle depending on the relative distance and speed of the vehicle ahead. In the years to come we will see lateral guidance to keep the vehicle in its lane and obstacle avoidance. More and more people now accept the possibility that in the not so distant future the car may be driven

Fig. 1.29 ControLaser System.

under full automatic mode without any supervision from the driver. This technology is already available in people-movers which drive at low speed (up to 30 km/h).

In the meantime, on the ordinary roads, we will have to do with assistive technologies such as ABS but at the same time we will see the arrival of driver monitoring. Indeed, the major source of accidents is human error. So, and in particular for professional drivers, drowsiness monitoring is now close to being installed in some vehicles. Several techniques are still in competition such as eye blinking, lane keeping, steering monitoring, collision warning, but these products are now being tested and the potential benefits (in particular for commercial fleet operators) are so great that it is only a matter of a few years before they will become widely used. Such systems, coupled with precise maps and accurate localization, could also warn the driver if he/she is not driving at a safe speed.

In the various chapters of this book, we will present all these existing and emerging technologies. For the moment, these technologies are being introduced by the car manufacturers (but often developed by equipment manufacturers) to satisfy legal constraints (such as those concerning pollution) or to improve the value perceived by the customer. These improvements essentially concern safety and comfort but they must come at a very marginal cost; otherwise they will not be accepted.

Part Two Intelligent vehicle sensor technologies

2

The CAN bus

Nan Liang
Siemens AG, Germany

and

Dobrivoje Popovic
University of Bremen, Germany

2.1 Introduction

During the last two decades or so, the most convenient interfaces used for information passing between computers and the attached controllers have been, apart from RS 232 and RS 448 standards, *local area networks* that have been developed for automation in the processing industries and in manufacturing. The most typical local area networks have been defined by the FIELDBUS standard for automation of industrial plants, and by the MAP/TOP standard primarily developed for applications in manufacturing and production industry, such as in the car making industry where even the MAP/TOP standard has been elaborated by a special task group on the initiative of General Motors. However, none of the local area networks standardized in the past has been seen as appropriate enough for application in the car itself for inter-networking various sensors, actuators and controllers. This was due to the complexity of network architecture and communication protocols, due to the relatively low network reliability and their high costs. The situation changed in the early 1990s as Robert Bosch GmbH (Bosch, 1991) decided to standardize CAN (controller area network), a bus-oriented communication system specially tailored for applications in automotive facilities, such as cars, trucks, buses, trains, and other vehicles. The rationale was drastically to reduce the wiring of complex data acquisition and processing systems present in contemporary vehicles containing multiple microcomputer- and microcontroller-based instrumentation and control systems for engine management, suspension control, ABS, etc.

Initially, Robert Bosch GmbH and Intel jointly worked on CAN bus protocol refinement and on its implementation in IC technology. In the meantime, the CAN bus has gained a widespread acceptance as a local area network not only by the automotive industry but also by many other users in other industrial sectors. This was reason enough for ISO (the International Standardization Organization) and SAE (the Society of Automotive Engineers) to standardize the CAN specifications internationally. Nowadays, by scanning the list of CAN bus producers we can identify the names of well-known computer and controller vendors, such as Alcatel, Fujitsu, Hitachi, Intel, Intermetall, Mitsubishi, Motorola, National Semiconductor, NEC, Philips, SGC-Thompson, Siemens, Texas Instruments, Toshiba, and others. Beside the bus, they also produce the compatible microchips and devices, particularly microcontrollers, transceivers, microprocessors, interfaces, etc.

In this chapter the basic concept and the functional principles of the CAN bus system are described, along with its main features and technical advantages. The attention will particularly be focused on hierarchical systems structure and the internal data exchange protocols, error detection and error handling provisions and some additional specific features of the bus. Furthermore, effort will be made to depict the standard system implementations and their applications in vehicles and elsewhere in engineering. Reference to the existing standards of low-speed and high-speed CAN systems and of its components will close the chapter.

2.1.1 What is CAN?

CAN, the controller area network, is a hierarchically organized distributed communication system for serial data transfer in real-time applications. The data transfer protocol of the system strictly relies on a simplified OSI (open system interconnection) model of ISO (IEEE, 1983), mainly containing the physical layer and the data link layer of the model. In addition, depending on the field of application, the CAN standard also incorporates a number of application protocols.

The principal features of the CAN system are the high reliability, availability and robustness that are required in safe real-time communication systems operating in extremely harsh environments. Moreover, the excellent error detection and confinement capabilities of the system increase its reliability in noise-critical environments. Finally, the achievable data transfer rate of 1 Mbps makes the CAN system comfortable for applications in relatively high-speed real-time control.

In the nodes of the CAN bus the *node stations* are situated to communicate over the bus using the multi-master principle of bus arbitration. Theoretically, a single CAN network is capable of interconnecting up to 2032 devices via a single twisted pair of wires. However, due to the practical limitation of the hardware (transceivers), it can practically link many fewer devices, for instance with the Philips chip 85C250 up to 110 devices.

Message frames of CAN generally contain a 0–8 bytes long data field and an 11 or 29 bits long identifier that determines the bus transmission priority and the destination node of the message.

Typical applications for CAN are motor vehicles, utility vehicles and industrial automation. Other applications for CAN are trains, medical equipment, building automation, household appliances and office automation. Due to high-volume production in the automotive and industrial markets, low-cost protocol devices are available.

Since the very beginning, the CAN bus system has been applied in advanced vehicles such as the S-class of Mercedes, to be followed by BMW, Porsche and Jaguar. This promised a good future that came soon with its acceptance by Volkswagen, Fiat, Renault, etc., as well as by various manufacturing industries that decided to use the CAN bus system for inter-networking of programmable controllers, intelligent sensors and actuators, and other instrumentation elements in automatic control of material handling automata and robots. The automotive application of the CAN bus has favourably been extended to trucks and buses, agricultural and forestry vehicles, passenger and cargo trains, etc.

2.1.2 How does it work?

CAN, as an asynchronous serial bus system, relies on an open logical line with a linear bus structure and with two or more nodes. The configuration of the CAN bus can be changed on-line by adding or deleting the nodes without disturbing the communication of other nodes. This is of great advantage when the modification of system functions, error recovery, or bus monitoring requires it. For instance, additional receiver nodes can be attached to the network without requiring any changes in the bus hardware or software.

The bus logic corresponds to a 'wired – AND' mechanism, 'recessive' bits (mostly, but not necessarily equivalent to the logic level 1) are overwritten by 'dominant' bits (logic level mostly 0). As long as no bus node is sending a dominant bit, the bus line is in the recessive state, but a dominant bit from any bus node generates the dominant bus state. Therefore, the CAN bus line as the data transfer medium must be chosen so that transfer of the two possible bit states, the 'dominant' and the 'recessive', is enabled. In the majority of cases the twisted wire pair is chosen as the low-cost solution.

The CAN bus lines to be connected directly or via a connector to the nodes are usually called 'CAN_H' and 'CAN_L', whereby there is no strictly standardized CAN connector type. The maximum bus speed of 1 Mbaud can be achieved with a bus length of up to 40 m. For bus lengths longer than 40 m the bus speed must be reduced (a 1000 m bus can be realized with a 40 Kbaud bus speed). For a bus length above 1000 m, special drivers have to be used.

To reduce the electromagnetic emission of the bus at high baud rates, the shielded bus lines can be chosen. Erroneous messages are automatically retransmitted. Temporary errors are recovered. Permanent errors are followed by automatic discovery of defective nodes.

For encoding of binary information the NRZ (non-return-to-zero) code is used with low level for the 'dominant' state and with high level for the 'recessive' state. To ensure a high operational synchronization of all bus nodes the bit-stuffing technique is used, i.e. during the transmission of a message a maximum of five consecutive bits may have the same polarity. Otherwise, whenever five consecutive bits of the same polarity have been transmitted, the transmitter, before transmitting further bits, inserts a stuff-bit of the opposite polarity into the bit stream. The receiver also checks the number of bits with the same polarity and removes the added stuff-bits from the bit stream (this is known as de-stuffing).

The messages transmitted over the bus do not include the addresses of source or destination nodes. Instead, they contain the identifier that is common throughout the bus and specifies the message priority. All nodes within the bus receive all transferred messages and identify, by an individual test, the acceptance of each message received. The non-accepted messages are ignored.

For optimal arbitration of transmission medium, a combination of the CSMA/CD (carrier sense multiple access with collision detection) principle and of a non-destructive bitwise arbitration is used, that enables the maximum use of CAN bus data transfer capability by determining the priority of messages, and provides the collision resolution. This is particularly important when more than one node intends to transfer urgent data at high transfer rate. Moreover, non-destructive bitwise arbitration is essential for resolving the potential conflicts on the bus in accordance with the 'wired – AND' mechanism that in conflict cases a *dominant* state overwrites a *recessive* state. This is possible because, during the planning phase of the bus, a priority is assigned to each message by assigning the priority to its identifier.

2.1.3 Main features of the CAN bus

The most essential features of a CAN bus system are directly determined by their field of application, which requires highly-reliable real-time control of distributed systems. Such distributed systems are, in addition to process and production plants, the modern vehicles that are increasingly becoming equipped with more and more sensors, actuators, microcontrollers, alarm annunciators, and terrestrial orientation aids distributed within the vehicle. This, in the sequence, defines the major design requirements of the bus system and the features to be implemented. The requirements could briefly be summarized as:

- efficient transfer of relatively short messages
- simple and transparent communication principles
- short response time to enable high-speed control
- low implementation costs.

To meet such high-performance requirements the CAN bus communication concept has been developed. This includes the following features:

- communication medium accessibility through message priority
- bus arbitration that includes conflicts resolution
- address-free data transfer through distribution of all messages to all nodes, each node being able to identify the messages destined to it
- extensive error check strategies jointly employed by each active node
- safe data consistency: a message is accepted by all active nodes or by no node
- availability of various bus management approaches
- availability of various higher layer protocols for different applications
- automatic handling of data transfer functions through powerful suites of communication management features that include:
  - priority-based bus contention resolution
  - data transmission error detection
  - automatic re-transmission of failed messages
  - message delivery acknowledgement
  - automatic shutdown of failed nodes.

The implemented features enable the bus to manage the following essential problems:

- availability of the entire system through multi-master system architecture capable of detecting and localizing a failed node, retaining the communication system availability

- system fail-safety and on-line re-configuration
- event-driven data transfer enabling each node to initiate a data transfer whenever required
- broadcast communication through the principle that all nodes receive all messages
- content-oriented message transfer through their identifiers.

2.1.4 Advantages of CAN

Once implemented in the CAN bus system, the features listed in Section 2.1.3 render the following functional advantages in real-time applications:

- high-speed data transfer of 1 Mbps at bus length of up to 40 m.
- predictable maximum latency time of messages: a trigger message with the empty data field and the highest priority has a maximum latency time of 54 μs at 1 Mbps transfer rate
- extremely high robustness
- reliable and powerful error detection and error handling technique: the total residual error probability of 8.5×10^{-14} for one corrupted message in 1000 transmissions
- automatic re-transmitting of faulty data
- non-destructive collision detection by bitwise arbitration
- remote message support
- capability of immediate transmission of latest available data by functional units upon request from any other unit
- automatic disconnect of nodes suspected to be physically faulty
- Functional Addressing of node units by 'broadcasting', and acceptability test of all active units on the bus
- firm priority assignment to all messages on the bus
- possibility of alternative message transmission in point-to-point mode or by broadcasting or multicasting
- commercial availability of low cost CAN controllers and the controller ICs from Intel, Motorola, Philips, Siemens, NEC, etc. like 82032 (Standard CAN) or 536.870.912 (Extended CAN).

2.1.5 Application fields

Although originally developed for automotive purposes, CAN soon proved to be very useful as a low-cost networking option, for various industrial applications, for instance as a general purpose sensor/actuator bus system for distributed, real-time microcontroller-based industrial automation. In Germany, such applications are promoted by a users and manufacturers group CiA (CAN in Automation), already joined by a number of EC companies.

Examples of non-automotive uses of CAN are:

- transportation systems, such as public traffic and pneumatic post control systems, mail and package sorting systems, etc.
- navigation and marine systems

- production line packaging systems
- paper and textile manufacturing systems
- agricultural machinery
- lift and escalator, heating and air conditioning control systems and general building automation systems
- measurement systems
- robot control systems
- medical laboratory systems for data collection from X-ray and tomography sources, etc.
- PC-controlled manufacturing systems.

Nowadays, CAN is largely used in the above application areas. According to an earlier estimate, there have been more than 20 million CAN nodes in use world-wide in 1997. By the end of 2000 the number of nodes is estimated to be around 150 million. According to a CiA (CAN in Autimation, a CAN user group at Karlsruhe, Germany, http://can-cia.de) survey, around 100 million CAN nodes were sold in 1998, the overwhelming majority (around 80 per cent) of them being installed in automotive objects. It is expected that by the end of 2002 around 180 million controllers will be installed world-wide.

2.2 Functional concepts

2.2.1 Data exchange concept

As already pointed out, for data transfer no CAN bus node is directly addressed. Messages transmitting the data contain the address of neither the transmitting nor the receiving node. The identifier of the message bears the address information and indicates the priority of the message based on the principle: the lower the binary value of the identifier the higher the message priority.

For bus arbitration the CSMA/CD (carrier sense multiple access/collision detection) medium access method is used with NDA (non-destructive arbitration) is used (see Section 2.3.3). The method operates in the following way.

When the bus node ***A*** wants to transmit a message across the network, it first checks that the bus is 'idle', i.e. free; this corresponds to the 'Carrier Sense' part of the method. If this is the case, and no other node intends to start a transmission at the same moment, the node ***A*** becomes the bus master and sends its message. All other nodes switch to receive mode during the first transmitted bit (start of frame bit). After correct reception of the message (which is acknowledged by each node) each bus node checks the message identifier and stores the message if it is required by the node, otherwise the node discards the message. If, however, two or more bus nodes start their transmission at the same time, this corresponds to 'multiple access' and their collision is avoided by bitwise arbitration. The bus access is handled via the advanced serial communications protocol carrier sense multiple access/collision detection with non-destructive arbitration. Each node first sends the bits of its message identifier (MSB) and monitors the bus level. A node that sends a 'recessive' identifier bit but reads back a 'dominant' one loses bus arbitration and switches to receive mode. This condition occurs when the message identifier of a competing node has a lower binary value (i.e. the 'dominant' state or logic 0) and therefore the competing node is sending a message with a higher priority. In this way, the bus node with the highest priority message wins the bus arbitration without losing time by having to repeat the message. All other nodes automatically try to repeat their transmission intention once the bus returns to the 'idle' state. It is not permitted for different nodes to send messages with the same identifier because in that case the bus arbitration could collapse, which would lead to collisions and errors.

The original CAN specifications, i.e. the Versions 1.0, 1.2, and 2.0A (Etschberger *et al*., 1994; Lawrenz *et al*., 1994a) define the message identifier as having a length of 11 bits and giving the possibility of 2048 different message identifiers. The specification has been updated as Version 2.0B to remove the limitation. This specification allows message identifier lengths of 11 and/or 29 bits to be used (an identifier length of 29 bits allows over 536 million message identifiers).

The core part of CAN message frame is the identification and data field, available to the user for programming. The remaining fields and bits are not accessible by the user. While being involved in bus arbitration, the identifier field corresponds in fact to the bus arbitration field. The available 29 (for standard format 11) bits could be also used to transmit data, and vice versa, some of the 'data' bits could be used to 'identify' the data or the transmitting node. This is because although the CAN rigidly specifies the field format, there is still considerable flexibility in how these fields are used. However, within a CAN frame there is a limited number of bits for transferring data (8 bytes if we stick to the defined data fields, and only a few more if we steal some of the identification field bits). Certainly, there will be data that exceeds the limits of single CAN frame.

Version 2.0B CAN is often referred to as 'Extended CAN', the earlier versions (1.0, 1.2 and 2.0A) being referred to as 'Standard CAN' (see Section 2.6.1 for further details).

2.2.2 Message and frame formats

CAN defines four types of message packets:

- data frame, used to send some data to other devices and having up to 8 bytes of data
- remote frame, having no data and used to request data from a remote device
- error frame, used to notify all devices on the network when an error has been detected
- overload frame, used to signal that a device is not ready to receive another frame.

Data frame

A data frame is generated by a node when the node wishes to transmit data. The standard CAN data frame is shown in Figure 2.1. In common with all other frames, the frame begins with a start of frame bit (SOF as 'dominant' state) for hard synchronization of all nodes. The SOF is followed by the arbitration field, consisting of 12 bits, the 11-bit identifier (reflecting the contents and priority of the message) and the RTR (remote transmission request) bit used to distinguish a data frame (when RTR is 'dominant') from a remote frame (see Section 2.3.2). The next field is the control field, consisting of 6 bits, the first bit being the IDE (identifier extension) bit and which is at 'dominant' state, specifying that the frame is a standard frame. The bit next to this is

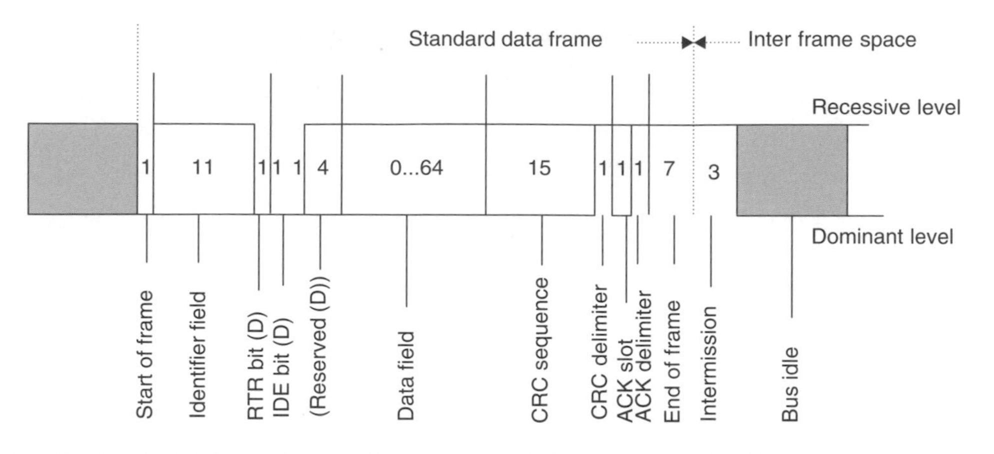

Fig. 2.1 Standard CAN data frame. RTR, remote transmission request; IDE identifier extension; CRC cyclic redundancy check; ACK acknowledgement.

reserved and defined as a 'dominant' bit. The remaining four bits of the control field, i.e. the data length code (DLC), specify the number of bytes of data to be sent, stored in the neighbouring data field (0–8 bytes).

The CRC (cyclic redundancy code) field is used to detect possible transmission errors. It consists of a 15-bit CRC sequence and is completed by the recessive CRC delimiter bit. Thereafter, the acknowledge field follows, used by the transmitting node to send a 'recessive' bit. The nodes that have received an error-free frame acknowledge this by sending back a 'dominant' bit (regardless of whether the node is configured to accept that specific message or not). This demonstrates that CAN belongs to the 'in-bit-response' group of protocols.

The 'recessive' acknowledge delimiter completes the acknowledge slot and may not be overwritten by a dominant bit. Seven recessive bits (end of frame) end the data frame. Between any two frames the bus must remain in the 'recessive' state for at least 3 bits of times that is called bus intermission. If thereafter no node wishes to transmit then the bus remains in the 'recessive', i.e. bus idles, state.

Extended data frame

In order to enable standard and extended data frames to be sent across a shared network, it is necessary to split the 29-bit extended message identifier into an 11 most-significant-bits section and a 18 least-significant bits section. This ensures the identifier extension bit (IDE) to remain, both in standard and extended frames, at the same bit position.

In the extended data frame the start of frame bit (SOF) is followed by the 32-bit arbitration field, the first 11 bits being the most significant bits of the 29 bits long identifier ('base-ID'). The bits are followed by the substitute remote request bit (SRR) which is transmitted as a 'recessive' bit. The SRR is followed by the IDE bit, which is also 'recessive' and denotes that it is an extended frame. If after transmission of the first 11 bits of the identifier, the conflict situation is not resolved, and one of the nodes involved in arbitration is sending a standard data frame with the 11-bits identifier, the standard data frame will, due to the assertion of a dominant IDE bit, win the arbitration. Further to this, the SRR bit in an extended data frame must be 'recessive' in order to allow the 'assertion' of a dominant RTR bit by a node that is sending a standard remote frame.

In the frame, the SRR and IDE bits are followed by the remaining 18 bits of the identifier ('ID-extension'), and the remote transmission request bit (with the 'dominant' RTR bit for a data frame). The next field is the control field, consisting of 6 bits, the first 2 bits being reserved and at 'dominant' state. The remaining 4 bits are the data length code (DLC) bits specifying the number of data bytes in the data field (like in the standard data frame).

The remaining bit string of the frame (data field, CRC field, acknowledge field, end of frame and intermission) is constructed in the same way as for a standard data frame.

Remote frame

The format of a standard remote frame is shown in Figure 2.2.

Normally, data transmission within the CAN bus is performed on an autonomous basis with the data source node (e.g. a sensor) sending a data frame. It is, however, possible that a destination node requests the data from the source instead. For this purpose the destination node sends a 'Remote Frame' with an identifier that matches the identifier of the required data frame. The appropriate data source node will then send a data frame as a response to this remote request.

Remote frame, as compared to the data frame, has the RTR-bit at the 'recessive' state and contains no data field. In the case that both a data frame and a remote frame with the same identifier are simultaneously transmitted, the data frame – due to the dominant RTR bit following the identifier – wins arbitration. In this way the node transmitting the remote frame receives the desired data immediately.

Error frames

Error frames are generated by the nodes that have detected a bus error. The frame consists of an

- error flag field, which bit pattern depends on the 'error status' of the node that detects the error, followed by an

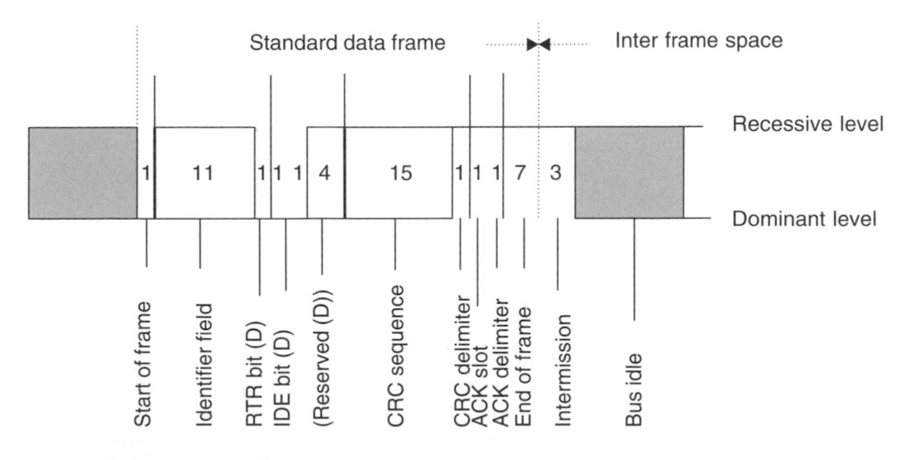

Fig. 2.2 Standard CAN remote frame.

- error delimiter field, consisting of 8 'recessive' bits and allowing the bus nodes to re-start bus communications cleanly after an error.

If an 'error-active' node detects a bus error the node interrupts transmission of the current message by generating an 'active error flag', composed of 6 consecutive 'dominant bits'. This bit sequence actively violates the bit stuffing rule. The node stations recognize the resulting bit stuffing error and in turn generate error frames themselves. The error flag field therefore consists of 6 to 12 consecutive 'dominant' bits, generated by one or more nodes.

After completion of the error frame, bus activity returns to normal and the interrupted node attempts to re-send the aborted message.

If an 'error passive' node detects a bus error, the node transmits an 'error passive flag', also followed by the error delimiter field. The flag consists of 6 consecutive 'recessive' bits, and the error frame of the 'error passive' node of 14 'recessive' bits. Thus, unless the bus error is detected by the node that is actually transmitting and being the bus master, the transmission of an error frame by an 'error passive' node will not affect any other node on the network. If the bus master node generates an 'error passive flag', this may cause other nodes, due to the resulting bit stuffing violation, to generate error frames. After transmission of an error frame an 'error passive' node must wait for 6 consecutive 'recessive' bits on the bus before attempting to rejoin the bus communications.

Overload frame

Overload frames have the same format as an 'active' error frame generated by an 'error active' node. The frames, however, can only be generated during the inter-frame space. This distinguishes an overload frame from an error frame that is sent during the message transmission.

Overload frames consist of two fields:

- the overload flag, consisting of 6 dominant bits followed by overload flags generated by other nodes giving, as for the 'active error flag', a maximum of 12 'dominant' bits
- the overload delimiter, consisting of 8 'recessive bits'.

Overload frames can be generated by a node when the node detects a 'dominant' bit during the inter-frame space or if the node, due to internal conditions, is not yet able to start the reception of the next message. A node may generate a maximum of two successive overload frames to delay the start of the next message.

2.2.3 Error detection and error handling

The CAN protocol provides sophisticated error detection mechanisms. The following errors can be detected:

- *Cyclic redundancy check error:* With the cyclic redundancy check (see Section 2.3.2 for details) the transmitter calculates the CRC bits for the bit sequence on the partial length of the frame to be transmitted, from the start of a frame until the end of the data field. The receiving node also calculates the CRC sequence using the same formula and compares it with the received bit sequence. If an error has occurred, the mismatch of CRC bits will be detected and an error frame generated. The transmitted message is repeated.

- *Acknowledge error:* In the acknowledge field of a message the transmitter checks if the acknowledge slot, sent out as a 'recessive' bit, contains a 'dominant' bit. If not, no other node has received the frame correctly, an acknowledge error has occurred and the message is repeated. No error frame is generated though.
- *Form error:* If a transmitter detects a dominant bit in one of the four segments end of frame, interframe space, acknowledge delimiter, or CRC delimiter, a form error has occurred and an error frame is generated. The erroneous message is repeated.
- *Bit error:* A bit error occurs if a transmitter sends a 'dominant' bit and detects, when monitoring the actual bus level and comparing it with the just transmitted bit, a 'recessive' bit or if it sends a 'recessive' bit and detects a 'dominant' bit. In the later case no error occurs during the arbitration field and the acknowledge slot.
- *Stuff error:* If between the start of frame and the CRC delimiter 6 consecutive bits of the same polarity are detected, the bit stuffing rule has been violated. A stuff error occurs and an error frame is generated. The erroneous message is repeated.

Detected errors are made public to all nodes via error frames, the transmission of the erroneous message aborted, and the frame as soon as possible repeated. Furthermore, depending on the value of its internal error counter, each CAN node is in one of the three error states: 'error active', 'error passive', or 'bus off'. The 'error-active' state is the usual state in which the bus node can transmit without any restriction messages and active error frames, made of 'dominant' bits. In the 'error-passive' state the messages and the passive error frames, made out of 'recessive' bits, may be transmitted. The 'bus-off' state temporarily disables the station to participate in the bus communication, so that during this state messages can neither be received nor transmitted.

2.3 Hierarchical organization

The need for integrated interconnections of individual sensors and actuators in technical systems is required for sharing the common system resources and for direct access to various data files within the system. When used in vehicles, such interconnections also reduce the cabling installation costs and increase the reliability of data transfer. For this purpose, the ISO has worked on a number of network standards strongly related to automotive applications, such as on ISO 11519 and ISO 11898 International Standards for Low Speed CAN Bus and High Speed CAN Bus.

CAN bus was designed as a *multi-master communication system* based on the OSI seven-layer model of ISO. This model defines the *network architecture* by specifying the basic network functions and the relationships between them, as well as the related interfaces and the data transfer protocols required for implementation of network services. It represents an international standard on what was named *open systems interconnection* in which a layered protocol concept was proposed appropriate for complete description of any data communication system of whatever complexity (Figure 2.3). The concept decomposes the relatively complex network structure into a number of small and functionally independent sub-structures, called *layers*. For specification of individual layer functions a description was proposed that is independent

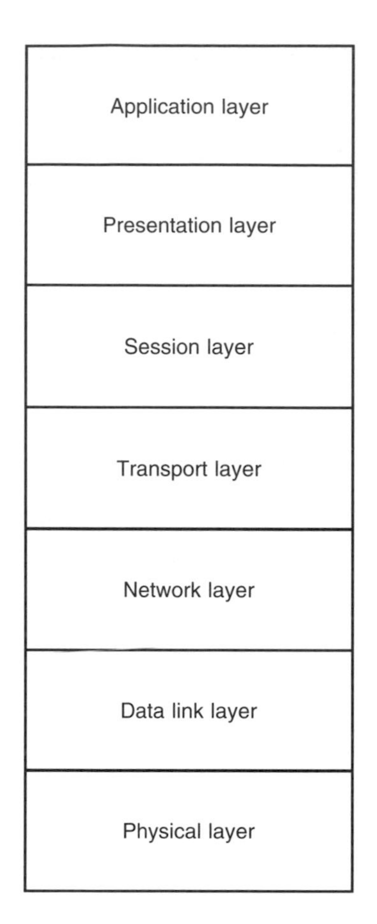

Fig. 2.3 OSI reference model.

from their hardware implementation. Consequently, the entire system specification is independent from its hardware or software implementation. The description merely specifies *how* the implementation should conceptually look and *what* the individual parts of it should perform. The term 'open' in the model name should only indicate the possibility of two implementations meeting the standard requirements to be easily internetworked, i.e. physically and functionally interconnected with each other.

OSI standardization defines the number of layers, the services provided by each layer, and the information to be exchanged between the layers, required for exchange of data between the network users. To each layer a sub-net of functions is assigned, required for the layer performance and for communication with the neighbouring layers because each layer provides services to the next higher layer, based on the services provided to it by the next lower layer. The chain of services within each layer is independent of the layers really implemented in a given communication system, this in the sense that the removal of an existing layer or the insertion of a new layer does not require functional changes in other layers of the system. Hence, the individual layers are mutually independent functional modules. This has enabled the structuring of the CAN bus system out of some selected layers, as will be shown in Section 2.3.1.

When implemented as a full version, the OSI reference model of ISO should have seven hierarchically organized levels, numbered in the sequential order starting with the lowest layer as represented below.

1. *The physical layer*, the lowest layer of the hierarchical model, is required for direct control of data along the transfer medium. It defines the *physical* (i.e. electrical and mechanical) *functional* and *procedural* standard recommendations for *accessing* and *using* the data transfer medium, without specifying the type and the parameters of the cable to be used as such medium. The layer primarily helps *establish* and *release* the communication links, supports the *synchronization* and *framing* in transmission process, controls the *data flow*, detects and corrects the *transmission errors* etc. The physical layer mainly handles electrical voltages, pulses, and pulse lengths, detects the collisions during the transmission (when CSMA/CD medium access control technique is applied), and serves as an interface to the transmission medium. Typical standards for the layer are RS 232-C, RS 449/422A/423A and X.21.
2. *The data link layer* provides the medium access control technique and analyses the incoming data strings in order to identify the flags and other characteristic bit patterns in them. It also provides the outgoing data strings with flags, error checking, and other bit patterns, and provides the reliable transfer of data along the communication link. The layer is generally responsible for transfer of data over the channel by providing the synchronization and safety aids for error-free transfer of data to the addressed destination. An example of a standard related to this is the HDLC protocol.
3. *The network layer* is responsible for establishing the transmission paths and direct communication links within the network by specifying individual links along the communication path that connect the terminal devices intending to exchange the data. It sets up the *routes* for data under transfer, builds the data packets for sequential transport, and reassembles them at the receiving end in order to restore the completed data sent by the transmitting equipment, and provides inter-networking services and multiplexing procedures. The best-known network layer standard is the X.25 protocol.
4. *The transport layer* provides the interface between the data communication network and the upper three layers of the open system interconnection model. It determines a reliable, cost-effective, transparent data transfer between any two network participants, optimizes the service of network layer, increases their reliability by end-to-end error detection and recovery, monitors the service quality, and interprets the session layer messages. The layer specifies *how* the logic transport connections between the corresponding terminal devices should be established, supervised, released, etc. An example of a transport layer protocol is the IP/TCP (internet protocol/transmission control protocol) protocol.
5. *The session layer* provides the mechanism for control of dialogue between the session and the transport layer, for code and character translation services, modification of data frames, and for syntax selection. It supports the data compression, encryption, and conversion between specific terminal characteristic of application programs. The layer opens, structures, controls and terminates a *session* and relates the addresses to the names, controls the dialogues, determines the art of dialogues (*half-duplex or full duplex*), and the duration of transmission. Finally, it provides a

checkpoint mechanism for detecting and recovering the failures occurring between the points.

6. *The presentation layer* provides the *syntax* of data presentation in the network. It does not check the *content* of data, or *interprets* their meaning, but rather accepts unconditionally the data types generated by the application layer and uses a pre-selected syntax for their presentation. In this way the layer provides the independence of user data from their presentation form. This makes the data transfer independent of user's device. In addition, presentation layer *adapts* the user application to the entire network communication protocols by uniquely presenting each information to be transferred. For this purpose the layer translates the user information to be transmitted into a language appropriate for transmitting.
7. *The application layer* generally supports specific user applications related to the communication system. It is equipped with the necessary management functions and manipulating mechanisms for support of *distributed applications*. Within the layer, two basic application service elements are present:
   - *common application service elements (CASE)*, such as *password check, log-in, commitment, concurrency, recovery*, etc.
   - *specific application service elements (SASE)*, such as *message handling, file transfer, file manipulation, database access and data transfer, system management*, etc.

Generally, the layering concept of computer network architecture, in which the network functions and protocols are hierarchically ordered, provides for:

- transparent decomposition of complex communication systems
- use of standard hardware and software interfaces as boundaries between network functions
- use of standard tools for system description, unique for network designer, vendor and user
- easy integration of user devices and application programs originating from different vendors.

The system layering concept was introduced primarily in order to provide for what is called *peer-to-peer information exchange* within the communicating systems and to enable a modular, mutually independent configuration and reconfiguration of a system. This enables individual layers to be independently deleted from the system or a new layer inserted into the system without disturbing the system itself. For instance, many real-time communication systems, such as the CAN bus or *field bus* system, have mainly the physical, data link and the application layer because the other layers are here of no use.

The mutual independence of individual layers has been achieved through the definition of *services*, provided by each layer to the layer above it, so that the layering technique used in the model can be understood in the following way: each layer transfers its own messages, based on its services, to which the messages of all lower layers will be added, based on their messages. This enables the peer-to-peer communication to be implemented as follows (Figure 2.4). The information generated by the User A will be *top-to-bottom* transported through all layers of the *source system* A and, after being transferred by the physical medium, it will again be *bottom-to-top* transported through all layers of the *destination system* B, up to the user device B. On the transport way, each layer of the source system adds its parts to the *transfer protocol*, which will be

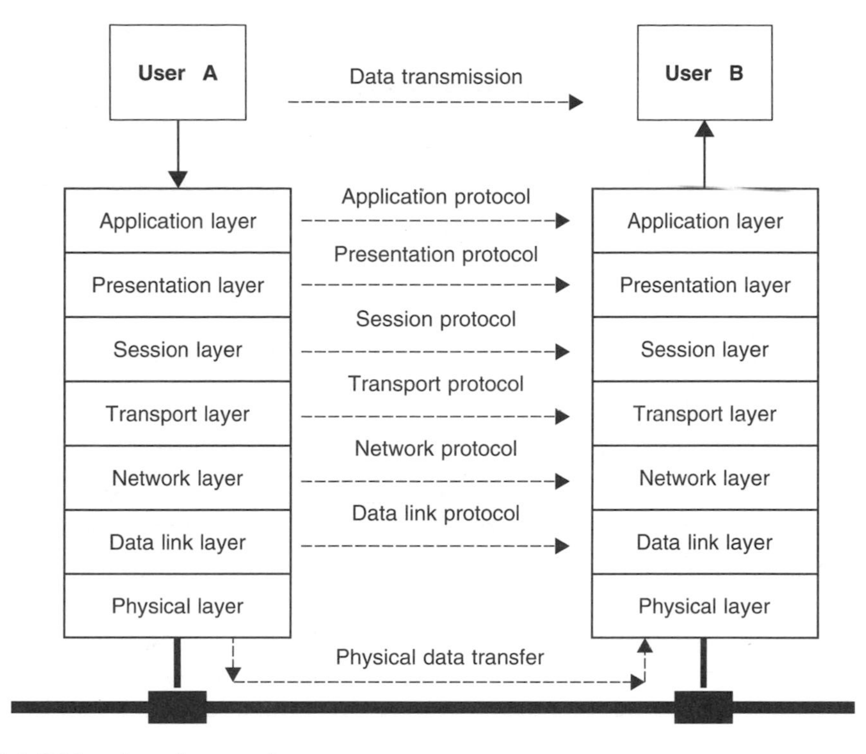

Fig. 2.4 OSI-based transfer protocol.

taken away and processed by the corresponding layer of the destination system. This implements a 'direct' transfer of protocols of individual layers, i.e. the peer-to-peer exchange of layer protocols, as shown in Figure 2.4.

2.3.1 Layering concept

The CAN bus, in view of the OSI model of ISO, is an open architecture communication system consisting of only three hierarchical layers:

- application layer
- data link layer
- physical layer

The application layer is generally a direct *interface layer* between the OSI styled communication system and the user terminal devices or sensors and actuators. It is the *top layer* of OSI model and has only one neighbouring layer – depending on the character of the communication system, the presentation, session, transport, or network layer. In CAN bus systems the application layer, in the absence of higher communication layers, directly communicates with the data link layer (Figure 2.5).

The application layer handles the sending and receiving of the messages and helps access the resources adequate for the given application. It also provides services to the application processes, which include:

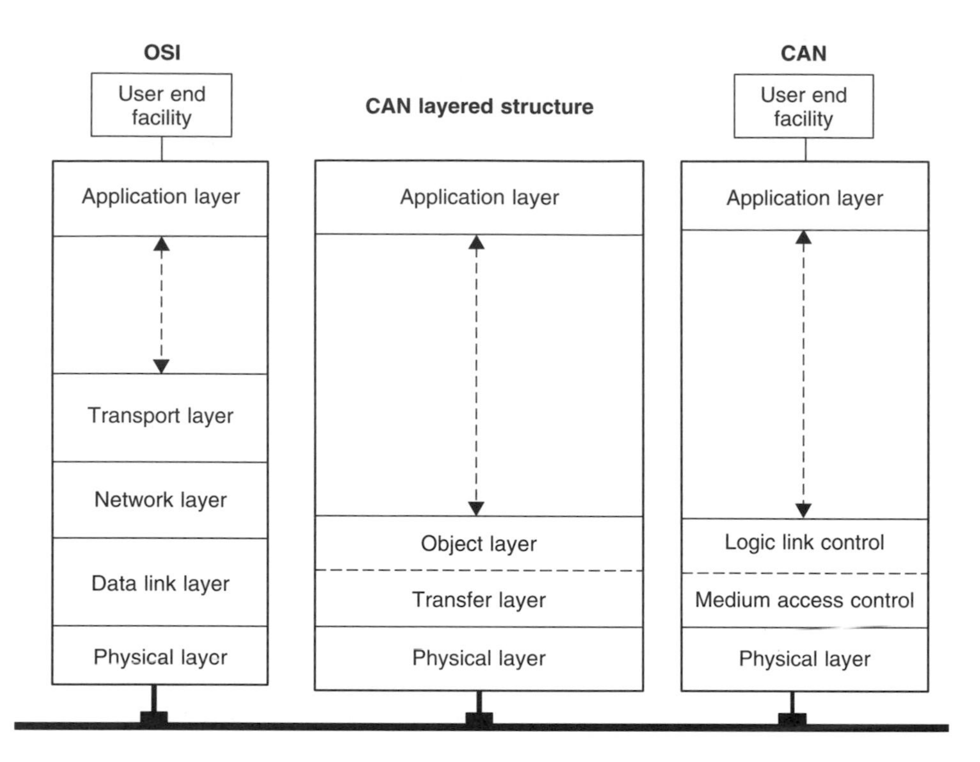

Fig. 2.5 Interconnection of OSI and CAN application layer.

- data transfer
- identification and status check of communication partners
- synchronization of communication dialogue
- initiation and release of dialogue conventions
- localization of error recovery and identification of responsibilities
- selection of control strategies for insurance of data integrity.

The data link layer basically implements the services of:

- logical link control (LLC), responsible for acceptance filtering, overload notification, and recovery management, and
- medium access control (MAC), responsible for data encapsulation/de-capsulation, frame coding, medium access management, error detection, error signalling, acknowledgement, and serialization/de-serialization.

In the CAN bus standard the data link layer is understood to contain two operative sub-layers:

- the object sub-layer, corresponding to the logic link control, and
- the transfer sub-layer, corresponding to the medium access control.

The object sub-layer provides the interface to the application layer and is in charge of

- message filtering, and
- status handling.

The transfer sub-layer provides the interface to the physical layer and mainly implements the transfer protocol. It controls the:

- message framing and validation
- error detection and signalling
- fault confinement
- bus arbitration, etc.

The physical layer basically operates at signal level and is responsible for:

- bit representation (encoding and decoding)
- signal transmission (Bit Timing and Synchronization)
- electrical parameter specification of transmitted signals.

2.3.2 Physical layer

The physical layer is the basic layer of a communication system, placed on its physical boundaries. The layer specifies the physical *interface* between the devices, i.e. sensors and actuators attached to the network, by specifying:

- *the mechanical parameters* related to the connector plugs to be used, such as the most frequently used 25-pin connector RS-232-C
- *the electrical parameters* related to the voltage levels, timing of voltage transitions, pulse rates, etc.
- *the functions* assigned to the signals and the related pins of the connector, like terminal ready, data set ready, transmitted data, received data, request to send, clear to send, receiver line signal detected, signal quality detected, etc.
- *the procedures*, i.e. the required sequences of events needed for transmission.

The specifications enable implementation of basic services at physical level, such as

- *physical connections* of individual terminal devices via a common transmission medium
- *bit sequencing*, required for *bit delivery* at destination point in the order they have been submitted at the source point
- *data circuit identification*, consisting in identification of physical communication path required or determined by higher layers
- *service quality specification* by assigning specific values to the quality parameters of transmitting path
- *fault detection* of bit strings received from data link layer.

The services are required for

- *activation* and *deactivation* of physical connections
- *activities management* within the physical layer
- *synchronous* and/or *asynchronous bit transmission*.

The physical layer of the CAN bus is divided into three sub-layers:

1. Physical signalling (PLS) sub-layer, implemented in the CAN controller chip, that encompasses:

- bit encoding and decoding
- bit timing
- synchronization

2. Physical medium attachment (PMA) sub-layer, mainly defining the transceiver characteristics
3. Medium dependent interface (MDI), that specifies the cable and connector characteristics.

The last two sub-layers are in various ways recommended by national, international and industrial standards. By selection of physical connections, however, it should be taken into consideration that to achieve the data transfer rate of 1 Mbps, a wide frequency band should be available. In the CAN users' community the use of driver circuits recommended by the ISO 11898 standard and various cables and connectors proposed by CiA (CAN in Automation, an international users and manufacturers group) is preferred for this purpose. Also, for implementation of physical medium attachment various recommendations are available, related to the ISO standards such as:

- ISO 11519-1 standard for low-speed CAN implementation
- ISO 11898-2 standard for high-speed CAN implementation
- ISO TC22 SC3 standard draft for fault-tolerant transceivers for car body electronics
- ISO 11992 standard for truck/trailer connections, also with the fault tolerant capabilities.

The transmission medium, the physical path between transmitters and receivers of the communication system, conveys data between the data nodes. The media predominantly used within the CAN bus are the *coax cables* with, according to ISO 11898-2, the nominal impedance of 120 Ω, the length-related resistance of 70 mΩ/m, and a specific line delay of nominal 5 ns/m in order to implement the transmission rate of 1 Mbit/s. The cable should be terminated with a 120 Ω resistor.

In the low-cost communication links, such as DeviceNet (designed for interconnection of limit switches, various sensors, valve manifolds, motor starters, panel displays and operator interfaces), the use of transmission medium is open, so that here for interconnections of elements in the node also the twisted pairs are used, i.e. the pairs of usually coated copper or steel wires, several of which are wrapped with a single outer sheath. Without a single repeater, the lines can be used in point-to-point connections for digital data transfer at a distance of up to 1 km. Taped twisted-pairs have a lower transmission domain, but they remain a suitable communication medium.

The use of coax cables is preferred because their noise immunity is superior to that of twisted pairs, especially in the higher frequency range. However, the coax cables are considerably more costly than the twisted-wire pairs.

In the CAN bus network, the maximum bus length is specified by:

- the delay of the bus line and of the intermediate bus nodes
- the difference in the bit time quantum length caused by oscillator tolerances between the individual nodes
- the amplitude drop of transferred signals due to the cable resistance and the resistancc of individual nodes.

For instance, the bus length of 30 m is achievable with a bit rate of 1 Mbps and with the nominal bit-time of 1 μs, of 100 m with a bit rate of 500 kbps and the nominal bit-time of 2 μs, and of 5.000 m with a bit rate of 10 kbps and with the nominal bit-time of 100 μs.

In the CAN bus, both *baseband* and *broadband* data transmission techniques are used. Signals in a baseband network are represented using two voltage levels: the *dominant-bit level d (0 V)* representing the 0-bits of the binary signal and the *recessive bit-level r (5 V)* representing the 1-bits of the binary signal. In such networks messages are transmitted as a series of direct current pulses, representing the coding elements of the information content of the message being transferred. This restricts the transmission over the line to only one message at a time, i.e. the transmission medium is *time-shared* among the participants attached to the network. For this reason the transmission capacity of the network as a communication channel is relatively low. This can be improved using symmetric circuits technology and trapezoidal pulse forms.

For the baseband data transmission the 50 *Omega* coaxes are used, enabling a maximum transfer rate of 10 Mbps. When using the twisted pairs, only a transfer rate of 1 Mbps is achievable, and this appears to be satisfactory for low-cost networks. In both cases one needs the signal repeaters at each 1 km of distance.

The advantages of the baseband system are mainly:

- low investment and installation costs, and
- easy installation, maintenance and extension

The disadvantages are single transmission channels, lower channel capacity, shorter transmission distances, and grounding difficulties.

In *broadband transmission*, messages are transmitted as a series of modulated pulses, which increases the network transfer capacity by enabling many transmissions to take place simultaneously, each transmission using its own carrier *frequency*, i.e. its *own transmission channel*. Here, the number of possible channels basically depends on the frequency characteristics of the transmission medium used. For representation of signals to be transmitted *amplitude, frequency* and *phase modulation* is used, and at the receiver end demodulated for extraction of DC component that carries the transmitted information. At both ends of the transmission path modems are used for preparing of signals to be transmitted and for interpretation of the received signals.

Bit encoding and decoding

Bit encoding used in the CAN bus is the *non-return-to-zero* (NRZ) encoding, i.e. the pulses of equal width and equal amplitude, but of different polarity, i.e. 0 and 1 or *low* and *high*, are used for information coding. Thus, in the coded message not every bit within the bit string contains a rising or a falling edge, which creates two substantial difficulties:

- identification of starting and ending point of individual bits in the bit string of the same polarity is difficult and creates some synchronization problems
- due to the unbalanced number of different polarities a DC component is present in the signal.

To improve the identification of bits the bit-stuffing technique is used, which limits the number of consecutive bits of the same polarity to five (Figure 2.6). This requests the transmitted bit sequence to be stuffed at the transmitting end and to be de-stuffed at

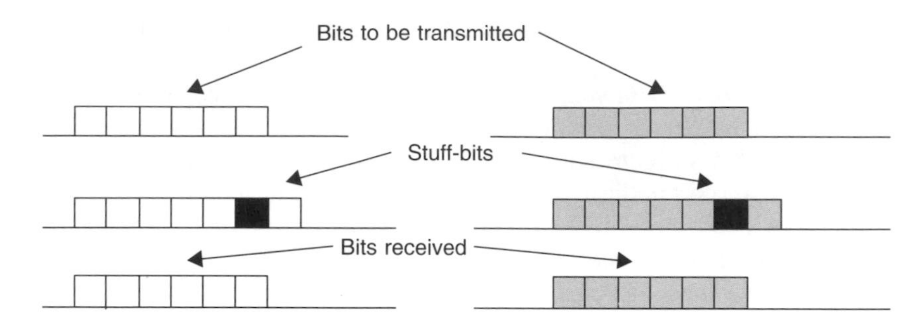

Fig. 2.6 Bit-stuffing technique.

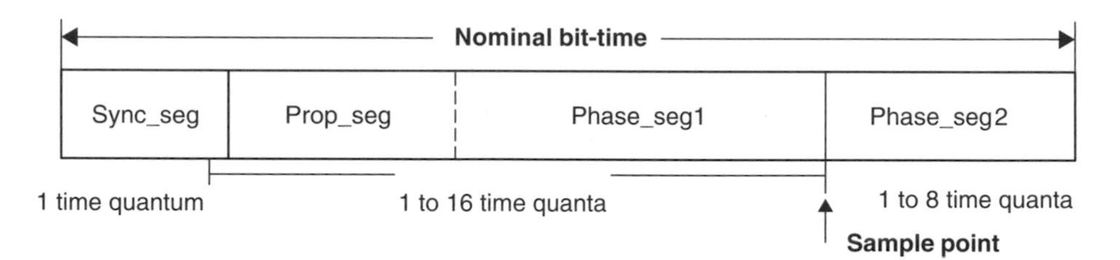

Fig. 2.7 Sub-bit segments.

the receiving end, which as a consequence leads to a message bits to stuff bits ratio of 5:1 or, in the worst case, to a ratio $(n - 1)/4$, n being the number of message bits.

Bit timing

In the CAN bus the *bit timing* is so organized that each bit period is thought to be subdivided into four *time segments*, each time segment consisting of a number of *time quanta* (Figure 2.7), a time quantum being a fixed unit of time defined by the oscillator frequency.

- SYNC_SEG, the 1-quantum-long nodes synchronization segment, synchronizes the individual bus nodes so that within the quantum an edge is supposed to lie
- PROP_SEG, the 1 to 8-time-quanta-long propagation time segment, compensates for the physical delay times of signals across the network
- PHASE_SEG1, the 1 to 8-time-quanta-long phase buffer segment 1, compensates for edge phase errors and can be lengthened during the re-synchronization
- PHASE_SEG2, maximum 8-time-quanta-long phase buffer Segment 2, also compensates for edge phase errors and can be shortened during re-synchronization.

This results in a total number of time quanta of 8 to 25, whereby the information processing time is equal or less than two time quanta.

The *sample point* is the instant of time at which the bus level is read and the real bit value identified. Its selection helps optimize the *bit timing*.

Synchronization

Synchronization is a serious problem of data transfer. A terminal device intending to communicate with another device has first to identify its communication capability and its communication readiness. Thereafter the device can initiate the communication process.

Synchronization and re-synchronization are essential mechanisms for orderly communication along the bus and for compensation of phase shifts within the transmitted frames. Phase shifts can occur in different bus nodes because each node is clocked by its own oscillator. This is removed by the soft synchronization or re-synchronization process during the frame receiving.

Hard synchronization, again, causes the internal bit time to restart with the synchronization segment and compels the hard synchronization edge to stay within the synchronization segment of the restarted bit time. The synchronization is always applied when a 'dominant' edge precedes a 'recessive' edge when the bus is idle. It could only be alternatively applied to the hard synchronization.

2.3.3 Data link layer

The data link layer is defined as a layer that provides functional and procedural tools for establishing, maintaining, and releasing data-link connections between the terminal devices. The layer has the capacity to implement:

- *establishment, splitting* and *release* data-link connection
- *delimiting* and *synchronization*
- *sequence and flow control*
- *error detection and recovery*
- *control* of physical interconnections.

Unlike the physical layer, which provides only raw bit stream services, the data link layer is expected to guarantee a reliable physical link and to manage it during the completed process of data transfer. The protocol of the layer enables:

- *transparent (code-independent) data transfer*
- flexible transmission path configuration
- various operational modes (*simplex, duplex*, etc.)
- minimal overhead length (number of non-user data bits)
- high transfer reliability.

The data link layer contains two sub-layers known as:

- the LLC (logic link control) sub-layer, responsible for
  - acceptance filtering
  - overload notification
  - recovery management
- MAC (medium access control) sub-layer, responsible for
  - data encapsulation and decapsulation
  - frame coding
  - medium access management
  - error detection and signalling.

In the CAN bus the services of data link layer are provided by:

- the object layer, responsible for
  - message filtering and
  - message and status handling, and

- the transfer layer, responsible for
  - message framing and validation
  - error detection and signalling
  - fault confinement
  - arbitration
  - transfer rate timing
  - acknowledgement.

Logic link control

Logic link control (LLC), the sublayer of the data link layer, copes with the problems in data transmission using the LLC protocol. The sublayer is in charge of:

- *data encapsulation* and *data decapsulation*
- *framing*
- *addressing the data blocks.*

Data encapsulation consists in forming the blocks to be transmitted by packing together the data to be transmitted and the control information, such as:

- *address* of the sending and/or the receiving device
- *error detection field*
- *protocol control field.*

Data decapsulation is the process in which the user data is extracted from the information block received. It is followed by the *data reassembling*, a process in which the user data fields of sequentially transferred data blocks, or messages, are chained to give the original data string decomposed for transmission purposes. As a result, the meaningful messages are extracted in the ordering as decomposed and sent by the transmitting device.

Framing is the operation of finding the boundaries between the successive frames, i.e. location of the positions where the previous frames stop and the following frames start. In the framing process also the idle fill between two frames, such as intermitted synchronous bit pipe, is separated.

Framing becomes a problem when the transmission error occurs and the receiving device cannot identify the end of the frame. Here, depending on the circumstance, the starting flag of the next frame could be identified using CRC approach or parity bits check.

Addressing is an operation similar to the *framing* in which the incoming data blocks are analysed with the objective of identifying, at each node, *how* to forward the received information to the destination device. The problem can efficiently be solved using, in the header of each block, the address of the sending and the receiving device, along with the *session identification number* to which the data transfer process belongs. The session number is essential when different blocks of the same session are transmitted over different paths in the network.

The data link and physical layers are the most essential layers of data communication networks. This is because communication networks such as buses and rings do not have a complex structure. Using the two lowest layers and the application layer, such communication networks can be built relying most frequently on bit-oriented protocol *HDLC (high level data link control)* with the transmission frame containing

- 8 bits FLAG for transmission synchronization
- 8 bits ADDRESS
- 8 bits CONTROL field for functions specification
- *n* bits DATA field
- 16 bits CRC field
- 8 bits of trailing FLAG.

The protocol provides

- *transparency* (code-independence) of operation
- *simplex* and *duplex* mode of data transfer
- *high-channel efficiency* using a minimum of overhead bits
- *high transmission reliability*
- *flow control* by frame acknowledgement
- *error control.*

Medium access control

Keeping in mind that the bus topology network uses a single communication path for interconnection of all attached terminal equipment, the question arises, 'How can individual connections be established in competitive circumstances, i.e. how can individual terminals access the common communication medium, the bus or ring network?'

The simplest way was found in direct initiation of a data transmission to a destination terminal at any time and in 'listening', i.e. in monitoring whether the initiated transmission was successful or was disturbed by another simultaneous transmission. If so, the disturbed data transmission is started again and the same monitoring procedure carried out. This can be repeated until the data transfer procedure has been successfully terminated. This principle has originally been used in the ALOHA medium access approach, known as *listen-while-talking protocol*. The protocol was used in the original version of *Ethernet*.

However, the ALOHA approach obviously suffers from extreme communication inefficiency of the network and from low transmission rate. It happens that the communication medium is largely in conflicting communication state and can not meet the high-speed data transfer requirements of computer-based terminal equipment attached to the network. Thus, to meet such requirements, a more efficient approach of medium access had to be found that includes some better monitoring steps of the operational state of communication medium. The approach should at least check for the *free* and/or *busy* state of the communication medium, or assign the sequence of permitted activities to the attached terminals. The solution was found in the introduction of *CSMA/CD (carrier sense multiple access/collision detection) medium access control*. It is a medium access approach based on the extension of the ALOHA concept so that the terminals, intending to initiate a data transfer, monitor the transmission medium *before* initiating the data transfer as well as *during* its execution. This is done by sensing the carrier frequency of the data transfer possibly taking place along the medium. The concept requests the terminal device intending to send the data to first 'listen' into the network in order to check whether the transmission path is free. If it is not free, the device has to wait until it becomes free and then start its own transmission process. It could, however, happen that two devices, listening at the same time for the free transmission path, after discovering that the path is free, both start transmission of data. This will cause mutual *communication collision*, which, according to the proposed standard, should be detected by *listening while transmitting*. In this case both devices interrupt their transmissions and wait for a period of time before they try again to monitor for the free transmission path. The waiting time here is specified for each device, or it is randomly generated in each device. This at least minimizes the new *communication collision probability* between the same devices and generally increases the traffic volume of the network to over 90 per cent, which makes the CSMA/CD approach appropriate for use in high-performance computer networks.

The advantage of the technique is that it does not require specific knowledge about the internal structure of the data packets to be transferred, and that it can easily be implemented as hardware, i.e. at signal level. For transmission purposes, the waveform of the analogue signal in the transmission channel can directly be monitored. It is detected, due to the superposition of two carrier signals generated by two different transmitting devices, as over-voltage.

The MAC protocol of the CAN bus supports the services of its transfer layer containing:

- message framing and validation
- error detection and signalling
- fault confinement
- arbitration
- transfer rate timing
- acknowledgement.

Message framing and validation For efficient and reliable transmission the data to be transferred are grouped in a number of packages using the *framing principle*. The CAN bus standard supports two frame structures having two different lengths of the identifier. In Figure 2.1 the structure of *message frame* CAN bus standard specification 2.0A is shown with the

- *arbitration field*, that includes the identifier and the RTR bit
- *control field*, that includes the DIE and the reserved bits
- *data field*, 0 to 64 bits long
- CRC *field*, 15 bits long
- *acknowledge field*, 1 bit long
- *end of frame*, 7 bits long.

In Figure 2.1, the acronyms are used:

- RTR for remote transmission request
- IDE for identifier extension
- CRC for cyclic redundancy check
- ACK for acknowledgement.

The start of frame consists of a single 'dominant' bit marking the beginning of a frame.

The arbitration field contains the identifier bits and the remote transmission request (RTR) bit indicating whether the frame is a data frame (when the RTR bit is 'dominant') or a remote frame (when the RTR bit is 'recessive').

The next field, the control field, contains the identifier extension bit IDE, two reserved bits, and the 4-bit data length code, representing the binary count of data bytes in the data field.

The data field contains the data (up to 8 bytes) to be transferred during the frame transmission.

The next field, the CRC field, is used as a frame security check by detection of transmission errors and its neighbouring ACK field, containing a slot and a delimiter bit, is used for acknowledgement of the received *valid message* by the receiver to the transmitter. Two 'recessive' bits are used for this purpose.

The end of frame field terminates the frame by a flag sequence consisting of seven 'recessive' bits. Thereafter the idle bus period follows, during which new transmission can start.

The CRC field, which represents the *error check field* or *error control field*, checks whether the user data field transferred has reached the receiver in the ordering as sent by the sender. It includes the CRC sequence and the CRC delimiter, consisting of a single 'recessive' bit.

In the CAN bus message transfer the following frames are standardized (see Section 2.2.2):

- data frame, used for data transfer
- remote frame, used for transmission request
- error frame, transmitted by detection of a bus error
- overload frame, used for frame transfer delay.

The *validity of the message* for the transmitter is given when there has been no error observed within the start of frame and end of frame, and for the receiver when there was no error observed up to the last but one bit of end of frame.

Error detection and signalling To increase the reliability and the efficiency of data transfer, both *error detection and error correction* (i.e. *recovery from error*) are required. Introduction of a certain redundancy into the frame to be transmitted and the redundancy check at the receiving end of the transmission path helps detect the transmission error.

The CAN bus protocol identifies, in addition to the CRC error that occurs when the calculated CRC result at the receiving end is different from that of sending end, the following error types:

- bit error, detected by receiving a different value of a monitored bit from that sent
- stuff error, occurring when six consecutive bits of the same level are detected while using the bit stuffing technique
- form error, indicating that a standard form bit field contains illegal bits
- acknowledgement error, stated by the transmitter when the expected acknowledgement message is not correct.

The detected errors are signalled by transmitting an error flag.

For error detection at message frame level, the CAN bus protocol combines the following approaches:

- cyclic redundancy check, which evaluates the information in the frame and inserts the redundancy check bit string at the transmission end, which is checked at the receiving end

- frame check, which checks the frame format and its individual fields to discover possible format errors
- acknowledgement check, which verifies that the ACK message has followed a successful transmission.

In the CAN bus the cycle redundancy check is used for error detection in signal transmission, based on *cyclic codes*. Here, a bit string is represented in form of a special *cyclic code polynomial*. For instance, the bit string 11001101 is represented by $1 + x + x^4 + x^5 + x^7$, with 1 belonging to the *least significant bit* (LSB) and x^7 to the *most significant bit* (MSB).

The *check key* for error detection is generated by a *cycle algorithm*, both at sending and the receiving station, this after a *block check character* has been inserted into the message to be transmitted. The related operations of multiplication and division of polynomials can easily be implemented using the appropriately designed hardware (shift registers).

A cyclic code is defined by a *generator polynomial* G(x), say of the degree k, and a *message polynomial* M(x), say of the order n, e.g. in the above example:

$$\mathrm{M}(x) = 1 + x + x^4 + x^5 + x^7$$

with n = 7. In order to encode M(x) using G(x), we consider

$$x^{\mathrm{k}}\mathrm{M}(x) = \mathrm{Q}(x)\mathrm{G}(x) + \mathrm{R}(x)$$

where Q(x) is the quotient of division of $x^{\mathrm{k}} \times \mathrm{M}(x)$ by G(x) and R(x) is the corresponding remainder. For instance, choosing

$$\mathrm{G}(x) = 1 + x^2 + x^4 + x^5$$

for encoding of $\mathrm{M}(x) = 1 + x^2 + x^5 + x^9$ we get

$$x^5 \times \mathrm{M}(x) = (1 + x + x^2 + x^3 + x^7 + x^8 + x^9)\mathrm{G}(x) + (1 + x)$$

The code polynomial to be transmitted is now built as

$$\mathrm{P}(x) = \mathrm{R}(x) + x^{\mathrm{k}}\mathrm{M}(x)$$

In our case it is

$$\mathrm{P}(x) = (1 + x) + x^5 \times (1 + x^2 + x^5 + x^9) = 1 + x + x^5 + x^7 + x^{10} + x^{14}$$

that is equivalent to 110001010010001.

It is to be pointed out that the special polynomials considered above obey the laws of algebra, with the exception of addition operation which is *modulo 2*, i.e. the addition and the subtraction give here the same results. Hardware implementation of cyclic encoding and decoding relies on application of *shift registers* for implementation of division and multiplication of polynomials.

Error detection of a cycle encoded message relies on the fact that, in case of erroneous data transfer, the transmitted polynomial P(x) will have an error component at the receiving end, say the polynomial E(x), so that at this end the modulo 2 polynomial sum

$$Q(x) = F(x) + E(x)$$

is received. The transmission error has occurred if Q(x) is not divisible by F(x). This of course, will not hold if also E(x) is divisible by P(x), so that the choice of P(x) is a dominant problem to be solved with respect to the expected (most probable) error pattern of the transmission.

An additional advantage of cyclic codes is that, as a code capable of detecting a double error, it is also capable of *correcting* a single error. This relies on the fact that in case of a single error, no error will appear if the erroneous bit is corrected, but in case of a double error, the error will appear if one of the erroneous bit is corrected.

In the CAN bus as the generator polynomial the polynomial

$$X^{15} + X^{14} + X^{10} + X^8 + X^7 + X^4 + X^3 + 1$$

is used by which the polynomial is divided, the coefficients of which are defined by the de-stuffed bit stream that includes the

- start of frame
- arbitration field
- control field, and
- data field (when present)

bits and the 15 lowest coefficients of which are equal to zero. The remainder of the division is then transmitted within the frame as the CRC sequence.

Error signalling is used to inform all bus stations that an error condition was discovered. The station that has detected such a condition – for instance a bit error, stuff error, a form error, a CRC error, or an acknowledgement error – announces this by transmitting an error flag that can be an active error flag or a passive error flag.

Fault confinement The individual stations attached to the bus can be in one of the following states:

- error active state, participating in data communication and sending an active error flag upon detecting an error
- error passive state, participating in data communication and sending a passive error flag upon detecting an error
- bus off state, not participating in data communication because it is switched off, and not sending any error flag.

Each station is provided with two counts: the transmit error count and the receive error count for fault confinement. For this purpose they follow a number of rules, such as:

- when a receiver detects an error, the receive error count will be increased by 1
- when a receiver, after sending an error flag, detects a dominant bit as the first bit, the receiver error count is increased by 8
- when a transmitter sends an error flag the transmit error count is increased by 8
- when a transmitter, while sending an active error flag or an overload flag, detects a bit error the transmit error count is increased by 8

Bus arbitration The arbitration of the CAN bus is a bus access method particularly tailored to meet the requirements of real-time data transfer with 1 Mbps and with swift bus allocation among many stations intending to send the messages. The method takes into account not only the announcing time of bus requirement of stations but also their priority, which is specified by the related message identifier. This is because the urgency of data transfer can be very different. For instance, the messages concerning the car speed or brake-force should be transferred more frequently than the message concerning the cabin temperature. The priority assignment to the messages is, of course, the decision of the system designer.

CAN bus allocation is based on station demand: the bus is allocated only to the stations waiting for message transfer, and this uniquely to only one station meeting the priority requirements, which is achieved by bitwise arbitration based on identifiers of the messages to be transmitted. This increases bus capacity and avoids bus overload because in this way only useful messages are transmitted. In addition, due to the decentralized bus control multiply implemented within the system, its reliability and availability are highly increased.

An alternative, message-oriented method of CAN bus arbitration is carrier sense multiple access/collision detection (CSMA/CD) which also avoids message collision if two or more bus participants, after discovering that the bus is idle, would like to transmit at the same time. The alternative method, however, guarantees a lower bus transfer capacity and could, by bus overload through higher transfer requirements of its participants, lead to the collapse of the whole transmission system.

2.3.4 Application layer

The application layer is a direct *interface layer* between the OSI styled communication system and the user terminal devices (CiA, 1994; Honeywell, 1996; Lawrenz, 1994b). It is the *top layer* of the OSI model and has only one neighbouring layer – depending on the character of the communication system, the presentation, session, transport, or network layer. In the distributed computer systems and the related local area networks the application layer, in the absence of higher communication layers, also directly communicates with the data link layer (Figure 2.5).

The principal objective of the layer is to provide services to the user application processes outside of the OSI layers, which include:

- data transfer
- identification and status check of communication partner
- communication cost allocation
- selection of quality of services
- synchronization of communication dialogue
- initiation and release of dialogue conventions
- localization of error recovery and identification of responsibilities
- selection of control strategies for insurance of data integrity.

The *applications* are the collections of information processing functions defined by the *user*, rather then by the user terminal device. In modern automation systems the applications are even distributed and multi-user defined.

The protocols of the layer

- support the file transfer, sending and receiving messages, etc.
- ensure that the agreed semantic is used by the cooperating partner
- help the user access the resources adequate for the given application.

The CAN bus application layer is described in Section 2.3.5 discussing the CANopen system.

2.3.5 CANopen communication profile

CANopen is a serial bus system particularly appropriate for motion-oriented systems, such as vehicles, transport and handling systems, off-road vehicles, etc. Its profile family is maintained by CiA and is freely available there (CiA, 1999). The profile family includes the CiA specifications:

- DS-301, application layer and communication profile
- DSP-302, framework for communication profile
- DRP-303–1, recommendations for cables and connectors.

The CANopen transceivers and controllers, however, are specified by ISO 11898 and its internal structure relies on OSI model of ISO, having

- the physical layer, mainly managing the bit timing and the connector pin assignment
- the data link layer, to which also the high-speed transceiver control is included
- the application layer, with the options like device profile, interface profile, application profile, 'manager' profile, etc.

Inter-layer communication between the bus participants corresponds to the peer-to-peer communication within the OSI model of ISO (Figure 2.4).

Physical medium recommendations of CANopen follow ISO 11898, which standardizes the differentially driven two-wire bus line with common return. The pin assignment, however, is specified by DRP-303-1 of CiA for the 9-pin D-sub connector according to the DIN 41652 Standard but also some other connectors are admitted, particularly the mini style connector and the open style connectors.

CiA divides the CANopen devices into three main parts:

- object dictionary, which describes the data types, communication objects, and application objects used in the device
- communication interface and protocol software, which provide services for exchanging the communication objects along the bus
- process interface and application program, which define the interface to the hardware and the internal control object dictionary function.

Data types permitted in CANopen include, apart from the usual Boolean, integer, and floating values, also the date and time intervals. Most typical here is the data type *visible string*, related to *visible car, array of visible car*, etc.

Object dictionary is the key part of CANopen devices in which devices accessible on the bus using a pre-defined approach are listed and hexadecimally encoded. Description of objects' entries to the dictionary has to follow firm instructions concerning the entry category, data type, access type, PDO mapping, value range, and default and substitute value.

Generally, objects could be

- communication objects, which could be
  - process data object (PDO), for real-time data transfer
  - service data object (SDO), related to the object dictionary
  - synchronization object (SYNC)
  - time-stamp object
  - emergency object
  - network management objects (NMT), for bus initialization and error and device status control.
- application objects that describe the CANopen application profiles.

The PDO communication profile enables the following message triggering modes:

- event- or timer-driven triggering, where the message transfer is initialized by the occurrence of an event specified in the device profile or after a pre-defined time has elapsed
- remotely requested triggering, where the asynchronous transmission is initiated on a remote request
- synchronous transmission triggering, where the synchronous transmission is triggered on reception of a SYNC object after a prescribed transmission interval.

The SDO supports the access to entries of device object dictionary using the client/server command specifier containing information such as *download/upload, request/response, segment/block, number of data blocks*, etc. SDO also enables the peer-to-peer communication between two devices to exchange data of any size.

The synchronization object generates the synchronization signal for sync-consumer that start their synchronous tasks upon receiving the synchronization signal. For a timely access of the transmission medium the synchronization object is assigned a very high priority identifier. This is vitally important for transmission of process data objects. However, synchronous signals are also important for implementation of cyclically transmitted messages.

The time stamp object provides the bus participants with the absolute time within the day with a resolution of 20 ms and the emergency object organizes the emergency service, which includes automatic triggering of emergency messages in case a fatal error was identified in a device to inform other devices about the emergency event. Special emergency error codes help identify the type of error present in the failed device.

Using the communication objects described, the following communication models can be implemented:

- The producer/consumer model, with the broadcast communication features based on acceptance filtering. The node stations listen to all messages transmitted along the bus and decide on acceptance of individual messages received. Moreover, the model supports the transmission of messages as well as the requests for messages.
- The client/server model enables transfer of longer data files in segments on the peer-to-peer communication principle using the same identifier.

- The master/slave model, in which the slaves are permitted to use the transmission medium for data transfer or for transmission request only on respective master command.

The master/slave model is used in CANopen node-oriented network management, which provides:

- module control services that initialize the NTM slaves intending to communicate
- error control services that control the nodes and the system status
- configuration control services that upload and download the configuration data on request.

The competence of the NMT master is stretched over the all system nodes, whereby every NMT slave is competent for its node only.

In addition, CANopen network management provides the command specifier based services:

- start remote node
- stop remote node
- enter pre-operational
- reset node
- reset communication.

Network management of the CANopen is provided with the node and life-guarding feature, which considerably increases the system reliability. This is achieved in the following way:

- the NMT master creates a database in which the expected status of each node slave is stored for node guarding, i.e. for cyclic checking for the absent devices that do not regularly send the process data objects
- the NMT slaves themselves iteratively check the NMT master for its node guarding activity based on their life-guarding feature.

For correct and reliable operating of data exchange services within the entire system the CANopen manager has been defined, and is made up of

- network manager master
- service data objects manager
- configuration manager

The application objects of the CANopen define the control and execution functions of data exchange with the system environment where the application activity takes place. Constituent parts here are:

- device profiles, which include the
  - detailed specification of device type object
  - PDO transmission and mapping parameters
  - additional emergency codes and data types
  - application object description
  - CANopen interface profiles, mainly related to the IEC 1131 interfaces
  - CANopen application profiles, such IBIS-CAN specifications for omnibus applications.

Examples of device profiles are:

- I/O module profiles, such as digital input, digital output, analogue input, and analogue output object
- drive and motion control profiles
- device profiles for HMI, etc.

2.4 Implementations

Although there is not a stringent specification for implementation of CAN systems, two CAN implementations are still dominant:

- basic CAN, used in low-price standalone CAN controller systems or in mixed controller systems that also include microcontrollers, and
- full CAN, used in high performance mixed controller systems.

The two implementations mainly differ in:

- message filtering and evaluation
- answering mode to the remote frames
- message buffering.

Basic CAN controllers usually have two receive buffers, in FIFO configuration, and one transmit buffer each. FIFO configuration enables a new message to be received into one buffer while reading the message from the other buffer. By buffers overload the oldest message is saved and the most recent one lost.

Message filtering is based on identifier information using two registers for storing the filtered message accepted. Typical for the Basic CAN is that the receiving nodes (controllers) have no support for answering the remote frames. This is done by the application concerned.

Full CAN controllers, unlike the Basic CAN controllers, have a set of buffers, called mailboxes, for receiving and/or transmitting messages depending on their pre-programming. The acceptance of a received message is decided by comparing the message identifier with the identifier codes previously stored in mailboxes for test purposes. Messages that do not match any of the identifier codes stored are rejected. Similarly, the identifier of a remote message is checked against the identifier in the transmit mailbox. In addition, Full CAN controllers have support for answering remote frames. They, however, keep the most recent message and lose the previous one when overloaded.

In the meantime a number of semiconductor producers have launched silicon implementations of ISO 11898 standardized CAN modules for various hierarchical layers and protocols, most of them relying on the Bosch reference model.

This concerns the implementation of standalone and integrated CAN controllers, single-chip CAN nodes, as well as the CAN I/O expander chip SLIO, etc.

Two alternative versions of the CAN reference model with the identical functions have been elaborated by Bosch:

- C reference CAN model
- VHDL reference CAN model.

The model-related specifications represent the guide for CAN protocol implementation for the completed controller area network.

2.4.1 General layout

General layout of a CAN system includes the CAN protocol controller, for message management on the bus through arbitration, synchronization, error handling, acceptance filtering, message storage, etc. In the CAN nodes the following is implemented:

- transceiver functions, including the signalling and bus failure management
- CAN protocol controller with its message filter, message buffer, and the CPU interface
- higher layer protocols and application functions.

For applications in vehicles a CPU scalable in performance and functionality is used, capable of managing up to 15 000 interrupts per second.

2.4.2 CAN controller

The general architecture of a CAN controller is shown in Figure 2.8. The main components of the controller are

- a protocol controller that manages the
  - data encapsulation
  - frame coding
  - bit encoding
  - error detection
  - synchronization
  - acknowledgement
- hardware acceptance filter to release the controller from message filtering
- receive message buffers, the number of which is application dependent
- transmit message buffers, the number of which is also application dependent
- CPU interface.

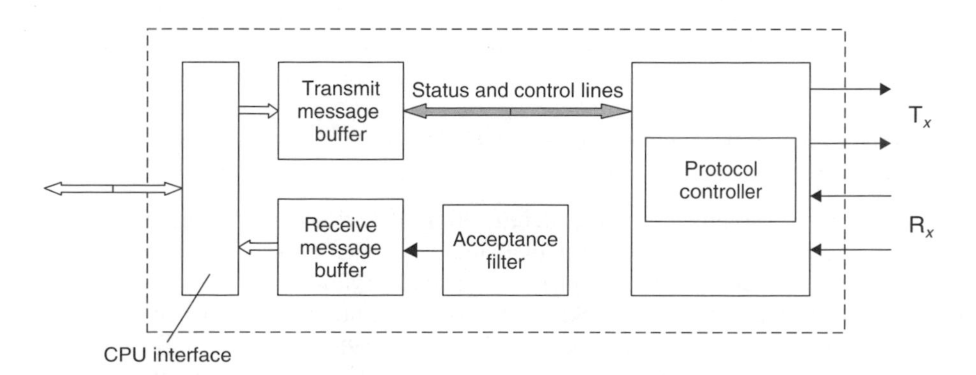

Fig. 2.8 Architecture of a CAN controller.

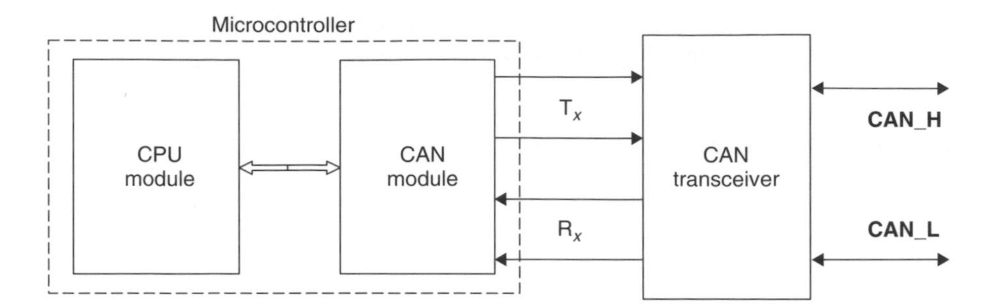

Fig. 2.9 Integrated CAN controller.

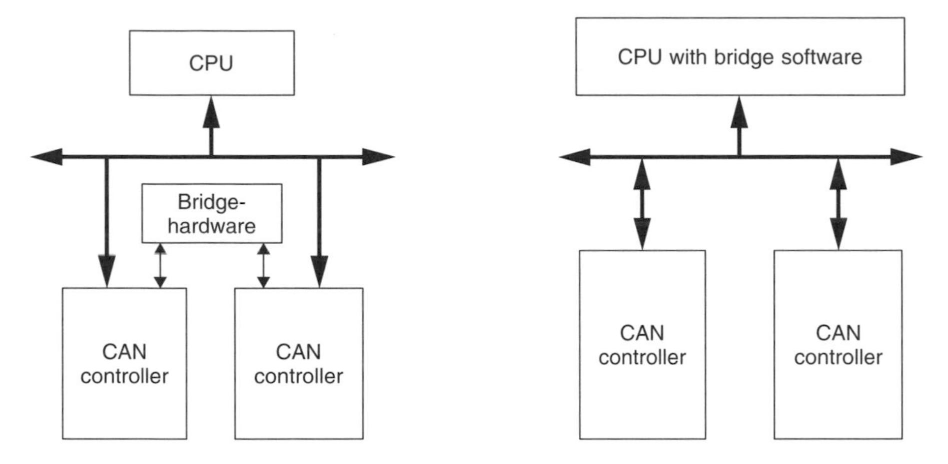

Fig. 2.10 CAN dual controller.

The architecture of integrated CAN controllers (Figure 2.9), however, includes the:

- CAN transceiver, required for signalling and bus failure management and the
- microcontroller containing the CPU module and a CAN module.

The architecture of a standalone CAN controller is similar to that of an integrated CAN controller, the only difference being that the CPU module is replaced by a microcontroller. Available on the market are also double CAN controller implementations on one microcontroller with the hardware as well the software bridge implementation (Figure 2.10).

Some commercially available advanced CAN controllers are equipped with additional features, such as:

- programmable warning limits for identification of various error types
- interrupt request generation in particular situations, such as warning limits proximity, new message reception, successful message transmission, etc.
- arbitration lost capture for checking the communication behaviour in conjunction with the single shot option, as specified by ISO 11898
- time stamp capability and frame counting using multiple counters, required for real-time applications

- readable error counters, useful for diagnostic purposes
- sleep mode for power reduction.

2.4.3 CAN products

Currently, there are more than two dozen manufacturers who have launched their CAN-related products on the marketplace, among them the world famous ones like Alcatel, Fujitsu, Hitachi, Intel, Intermetall, Mitsubishi, Motorola, National Semiconductors, NEC, Philips, Siemens, Texas Instruments, Toshiba and, of course, Bosch. Alcatel, in co-operation with WABCO Westinghouse Fahrzeugbremsen GmbH, has developed the MTC-3054 Transceiver and Dallas Semiconductor dual CAN 2.0B compatible controllers. The double CAN controller has also been developed by Fujitsu, Hitachi, and Mitsubishi.

Intel has developed the first CAN controller AN82526, soon replaced by AN82527, a standalone controller compatible with the V2.0A and V2.0B standard. A standalone controller with the same compatibility was also developed by Siemens that meets the FullCAN specifications.

Remarkable progress in elaborating fault tolerant CAN interface devices has been made by Motorola, Philips and Siemens and in elaborating low-cost one-time programmable (OTP) microcontrollers by National Semiconductors, Philips and Siemens.

Finally, serial linked I/O (SLIO) devices have been launched by National Semiconductors and Philips.

2.4.4 HECAN

HECAN is a CAN controller belonging to the HECAN TIM-40 modules of a modular scalable system that supports both Standard CAN and Extended CAN protocol specifications. A HECAN-based system is a CAN bus oriented distributed system integrating an Intel 82627 CAN controller and various sensors and controllers, supervised by a TMS320C40 digital signal processor.

2.5 CAN applications

2.5.1 General overview

CAN bus systems are used as communication multi-microcontroller systems, as well as open communication systems for intelligent devices. For instance, the CAN serial bus system, originally developed for use in automobiles, has increasingly been used in industrial and other automation systems. This is because the requirements for industrial field bus systems are similar to those for passenger car networks, above all as regards the installation and maintenance costs, reliability, availability, robustness and the ability to function in a difficult electrical environment, high levels of real-time performance and ease of use. For example, in the field of medical engineering some users opted for CAN because they have to meet particularly stringent safety requirements. Similar problems are faced by manufacturers of machine equipment, robots, and transportation systems, that must meet very high operational requirements.

A considerable advantage of CAN application is its scalability in the sense of implementation using Basic CAN controllers in building a low-price multi-node communication link, as well as using the Full CAN controllers as high-performance elements of the multi-controller system. When used in data communication systems, Basic CAN controllers require, in order to benefit from the real-time capabilities of the host and to implement an adequate response time to the external events, a better and more intensive service from the host. Full CAN, again, requires a minimal service from the host. Both type of controllers are widely available on the market, such as:

- Basic CAN controllers MCAN-Modul of Motorola, COP684BC and COP884BC of National Semiconductors, and 82C200, 8xCE592 and 8xCE598 microcontroller from Philips
- Full/Basic CAN controllers 8xC196CA and 8xC196CB of Intel, TOUCAN Microcontroller of Motorola, and 81C90/81C91 of Siemens.

2.5.2 Applications in vehicles

The need for serial communication link in vehicles

Contemporary vehicles already have a large number of elements of automotive electronics, including smart electronic modules for different automation purposes. This is the result partly of the customer's wish for better comfort and greater safety and partly of the authorities' regulations for improved emission control and reduced fuel consumption. Control devices that meet these requirements have been in use in the area of engine timing, gearbox and carburettor throttle control, anti-lock braking systems (ABS), acceleration skid control (ASC), etc.

The high number of operational, monitoring, alarming, and control functions in present vehicles, implemented using automotive electronics, requires exchange of data between them. The conventional way of data exchange by means of point-to-point data transfer is becoming complex and requires very high installation and maintenance costs. This is particularly the case when the control functions become ever more complex, particularly in the sophisticated complex control systems, such as in Motronic, where the number of connections is enormous. In the meantime, a number of systems have been developed which implement functions covering more than one control device. For instance, ASC reduces the torque when drive wheel slippage occurs, based on the interplay of the engine timing and the carburettor control. Another example of functions spanning more than one control unit is the electronic gearbox control where ease of gear-changing can be improved by a brief adjustment to the ignition timing.

Considering the future developments aimed at overall vehicle automation, it becomes obvious that the inter-networking problems of installed electronic devices have to be resolved. This can be done flexibly by inter-networking the system components using a serial data bus system. It was for this reason that Bosch developed the controller area network (CAN), which has since been internationally standardized as ISO 11898 and implemented in silicon by several semiconductor manufacturers. Using the CAN bus system, the bus node stations such as sensors, actuators and controllers are easily interconnected via the bus data transfer medium, which can be a symmetric or asymmetric two wires serial bus, screened or unscreened, specified by ISO 11898. For the bus, suitable driver elements are available as chips from a number of manufacturers.

The CAN protocol corresponding to the data link layer of the OSI reference model of ISO meets the real-time requirements of automotive applications. Unlike cable trees, the network protocol detects and corrects various transmission errors, including those caused by electromagnetic interference, and the data communication system is easily on-line re-configurable. Thus, the major advantage of using CAN in vehicles is the possibility that every node at any station of the bus may communicate directly with any other node station.

Use of CAN in vehicles

There are many applications for serial communication in vehicles, each application having different requirements. As already mentioned, Daimler-Benz was the first car producer that used the serial bus system CAN in engine management. Presently, most of the European passenger car manufacturers use the system as a standard communication link not only for engine management, but also for multiplexing of body electronic ECUs and entertainment devices. Recently, CAN oriented automatic car test pilot systems have come into use as diagnostic elements. Here, the 'Diagnostics on CAN' ISO 15765 standard was published, in which the physical, data link and application layer is defined, and which should become mandatory in Europe.

The automotive industry (in its effort to balance the human preference for personal comfort and security, public regulations and community appeals for reducing the environmental pollution and traffic congestion), has long been concerned with the increased integration of the various functions to be used by drivers for different purposes. They largely extend the basic function of starting, driving and stopping. For instance, it is almost usual in the modern car to have various controllers for engine timing, transmission, and brakes, as well as various provisions for lighting control, air-conditioning, central locking, and seat and mirror adjustment. In higher-class vehicles some additional advanced functions are integrated, such as stability and suspension enhancement, passenger protection, energy management, steering strategy improvement, sophisticated anti-block breaking and interactive vehicle dynamics control. In the near future, smart collision avoidance functions will be integrated that will include the enhanced night vision, obstacles and human road user detection, driver condition monitoring. Furthermore, the present communication and entertainment services will be extended to include the multimedia features, Internet linkage, navigation and route guidance and the like. For instance, some more advanced functions to be integrated are defined in the Prometheus project, such as vehicle-to-vehicle and vehicle-to-infrastructure communication. All this, evidently, requires a high-performance inter-networking system, or a combination of separate communication links like CAN.

As regards the general problem of cabling and inter-networking within the vehicle, the SAE (Society of Automotive Engineers) has elaborated a document in which the following application classes are defined:

- A-application class, for interconnections in the area of chassis and power elements, such as conventional switches, front, rear and brake lights, flash signals, seat and mirror adjustment, window manipulation, door locking, etc. The actions accompanying the use of such services are real and of short duration. They can be carried out

using a low-speed communication link, say with the transfer rate of 10 kbps, with a single wire and the chassis as data transfer medium.
- B-application class, for interconnection of more complex and more smart elements or equipment, such as information panels, air-conditioning facility, etc. Here the data transfer rate of up to 40 kbps is required.
- C-application class, for high-speed real-time transfer of data and for critical responses to the signals, say within 1 to 10 milliseconds. The signals are related to the engine, gears and brake management functions and require a transfer rate of 250 kbps to 1 Mbps.
- D-application class, for inter-networking of most advanced information and communication facilities within the cabin, such as sound and vision equipment, monitoring and navigation guidance aids, computers and databases, etc. For this, the data transfer rate of 1 to 10 Mbps is recommended.

When using the CAN bus system to manage the above tasks, its major function will be to interconnect the available sensors and actuators in order to build the required control loops and in this way implement the real-time operational functions. The most common sensors are the:

- acceleration sensor
- fuel level sensor
- fuel tank pressure sensor
- wheel speed sensor
- tyre pressure sensor
- temperature sensing sensors
- humidity sensor
- load/weight sensors
- coolant level/temperature sensor
- oil pressure sensor, etc.

The rapid increase of number of sensing elements, controllers, and other smart equipment installed in the vehicle has not only made their originally used point-to-point interconnection installation and maintenance expensive, but also conceptually obsolete because the individual terminal elements and equipment are expected to mutually exchange the information. This is particularly critical in high-performance control such as the engine torque control that is based on high-speed sampled values of *throttle position, ignition instant* and *fuel injection flow*. In this way instantaneous engine power control and optimal fuel consumption control are possible.

In the S-Class Daimler-Benz car catalyst ageing protection is remarkable. It includes the initial catalyst heating before it is fully used and prevention of semi-exhausted gases to spoil the catalyst, which is achieved by avoidance of engine misfires.

2.5.3 Other CAN applications

The similarity of requirements for vehicles and industrial bus systems – such as low installation and maintenance costs, operational simplicity, robustness, and high-speed real-time capabilities – has made the CAN bus system highly attractive for application in industrial automation. The successful application of CAN standard in S-Class

Mercedes cars and the adoption of the CAN standard by US car manufacturers has made a number of other industrial users alert, like the manufacturers of mobile and stationary agricultural and nautical machinery, as well as the manufacturers of medical apparatus, textile machines, special-purpose machinery, and elevator controls, that have already selected the CAN bus as a particularly well-suited communication link for inter-networking of smart I/O devices, sensors, actuators, and controllers within the machines or plants.

The first CAN bus applications 'out of car' have been recorded in the utility-vehicle-making industry, where the fork-lift trucks, excavators, cranes, tank lorries, and various agricultural and forestry machines have used the CAN bus to implement the leverage, pump, and driving power control, as well as engine state monitoring by acquisition of engine revolution speed, vehicle driving speed, and of related parameters like pressures and temperatures. Also, within a tank lorry, acquisition of delivery and tank state data is included and the corresponding bills printed for the customer. This makes the vehicle a mobile book-keeping office.

The CAN bus has also benefited public transport such as tram and underground rail lines, where the control, signalling, and alarming electronics have been inter-networked using the CAN bus system. Here, data are exchanged related to the driving and braking conditions, door state monitoring and failure detection and isolation, overall vehicle status analysis, etc. In Germany, a *board information and control system* was developed by Kiepe Elektronik in which two CAN bus systems are integrated via a central arbitrating unit, one bus being in charge of data exchange within the vehicle, the other one of data exchange between the vehicles. The data transfer rate used in the buses is 200 kbps and 20 kbps respectively. Based on data collected from the train and from its individual vehicles, a detailed analysis of the entire system and generation of extensive status reports is possible. Also, on-line diagnostics of available electronic units is possible by their periodical preventive tests. Finally, the record of a vehicle's long-term operation can be archived.

The next application step 'away from the passenger transport facilities' would be CAN application in heavy rolling machines, excavators, and cranes, etc. Similar to the local traffic system described in the proceeding paragraph, the CAN buses here can also be installed in individual vehicles and inter-networked via a central bus that enables the current vehicle's status overview and efficiency of their use, individual task distribution, and optimal use of available vehicle fleet, and the like. The same holds for the machines available in the industrial sector, where the CAN bus has been accepted as a standard inter-networking system, and this not only in the machine tools industry and manufacturing but also in textile and paper making industries for production control, printing and packaging, etc.

The textile machinery industry is one of the pioneers of CAN industrial application. The very beginning was made in 1990 as a manufacturer equipped his looms with modular control systems communicating in real-time via the CAN networks. In the meantime several textile machinery manufacturers have joined together to form the CAN Textile Users Group, which has also joined CiA. Bus requirements similar to those of the textile machinery are to be found in packaging machinery and machinery for paper manufacture and processing.

In the USA a number of enterprises are using the CAN bus in production lines and machine tools as an internal bus system for networking sensors and actuators within the line or machine. Some users, for instance in the medical engineering sector, decided in favour of CAN because they had particularly stringent safety requirements. Similar problems are faced by other manufacturers of machinery and equipment with particular requirements with respect to safety, such as for robots and transportation systems automation. Here, apart from the high message transmission reliability and availability, the low connection costs per station are a further decisive argument for CAN. In applications where price is critical, it is essential that CAN chips are available from a variety of manufacturers. The compactness of the controller chips is also an important argument, for instance in the field of low-voltage switchgear.

In 1997, a carton packaging machine was presented by Wepamat in Germany in which the CANopen bus controlled the available I/O devices, microcontrollers and PCs. Similar machines have been presented by EDF, Island Beverages, Soudronic, and TetraPak, as well as a printing machine by Ferag at Heidelberg. Still, a wide application field of CAN bus systems is industrial robotics, where Bosch, ABB, Engel and Kuka have done the pioneer work. Finally, a number of CAN bus applications exist in conveyer and lift monitoring and control, as well as in general automation of buildings for air-conditioning and safety assurance.

This application survey of the CAN bus should not be closed before mentioning in brief its increasing applications in medical engineering and dentistry, mainly for reliable patient data collection in computer tomography, radiology, fluoroscopy, and in other instrumentation sectors. Some early steps here have been made by Philips Medical Systems, followed by Combat Diagnostics, Dräger, Fresenius, GE Medical Systems and Siemens.

2.6 CAN-related standards

CAN was originally developed, on the initiative of Mercedes, by Robert Bosch GmbH, Germany, as a distributed communication system for interfacing the electronic control units in vehicles. The prototype of Bosch development was fabricated by Intel in 1987. Thus, the first CAN standard specification was elaborated by Bosch, the Version 2.0 of which was documented as

- Standard CAN, Version 2.0A and
- Extended CAN, Version 2.0B.

The two versions differ in the length of the identifier (i.e. 11/29 bits) and consequently in the message frame format. The CAN protocol specification document that describes the entire network functions the CAN 2.0 Addendum also represents a guide for CAN implementation, as described in Section 2.4.

Later on, two international standards have been worked out by ISO TC22/SC3:

- ISO 11898 Standard (Data Link Layer and Transceivers) for automotive and general purposes, and
- ISO 11519 (CAN Low Speed).

In addition, a number of CAN-related standards for vehicles and for general and dedicated control applications have been launched by various national organizations and companies. Among such organizations, outstanding standardization efforts in higher

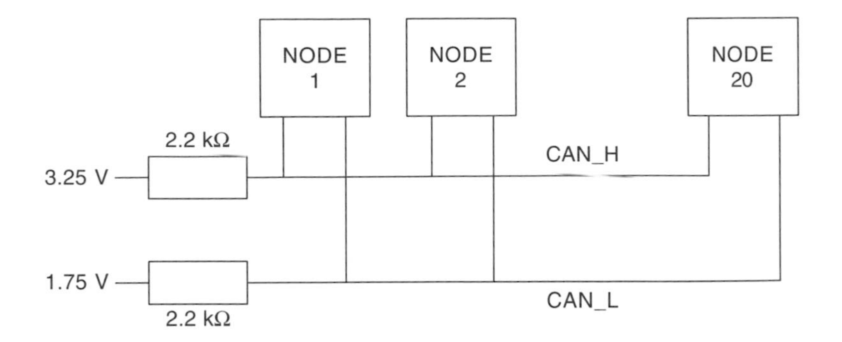

Fig. 2.11 CAN bus configuration according to ISO 11519.

layer protocols have been made by CiA (CAN in Automation), an international users and manufacturers non-profit group that supported the development of CAN Application Layer, CANopen, DeviceNet, and CAN Kingdom.

2.6.1 ISO 11519: Low speed CAN

The ISO standard for low speed CAN (ISO, 1994) defines the nominal voltage for

- 'dominant' state: 4.0 V for CAN_H and 1.0 V for CAN_L, and
- 'recessive' state: 1.75 V for CAN_H and 3.25 V for CAN_L

ISO 11519 does not prescribe the terminating resistors because the maximal transfer rate of 125 kbps does not suffer under strong wave reflections (Figure 2.11). In addition, only 20 nodes are permitted per bus configuration.

2.6.2 ISO 11898: High speed CAN

ISO 11898 defines (ISO, 1993) the nominal signal levels of high speed CAN for

- 'dominant' state: 3.5 V for CAN_H and 1.5 V for CAN_L state, and
- 'recessive' state: 2.5 V for CAN_H and 1.5 V for CAN_L state.

For the 'recessive' state the high and low signal level is the same in order to decrease the power drawn by terminating loads. Besides, for ISO 11898-compatible bus devices the maximum bus length is limited to 40 m for transfer rate of 1 Mbps. For lower transfer rate the total bus length can be prolonged to 100 m (for 500 kbps), 200 m (for 250 kbps), 500 m (for 125 kbps), etc. Up to 30 nodes can be implemented along the bus that should be terminated at both ends with the 125 Ω impedance (Figure 2.12).

ISO-11898 compatible CAN transceivers considerably simplify the CAN bus implementation, particularly in automotive applications where they provide the additional features like thermal overload and short-cut protection, energy saving in stand-by mode, immunity against bus disturbances, etc.

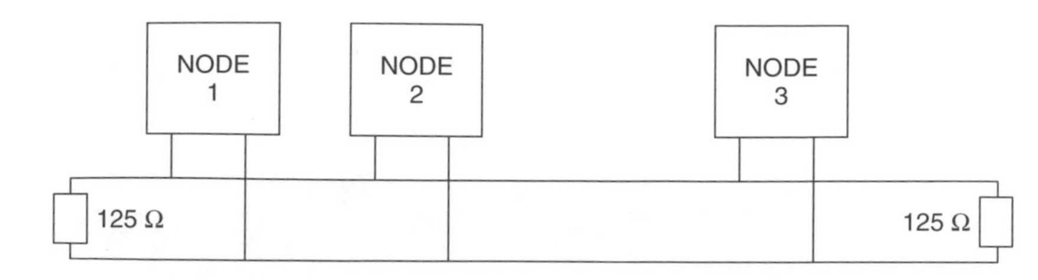

Fig. 2.12 CAN bus configuration according to ISO 11898.

2.6.3 Vehicle-related components standards

ISO TC22 SC3 as well as a number of national organizations of standardization and companies' standardization departments have worked out various CAN device standards, CDs and work documents for application in vehicle automation. Some examples of such standards are listed below:

- ISO TC22 SC3:
  - IS 11992: Transceiver cum Application Level
  - CD 15031: OBDII Diagnosis
  - CD 15031: Diagnosis on CAN
  - CD 16844: CAN for Tachograph Systems
- ISO TC23 SC19:
  - WD 11783: Serial Control and Communications
- ISO TC173 SC1:
  - CD 717617: M3S Network
- CiA:
  - WDP-4XX: CANopen Device Profiles for Off-Road Vehicles
  - WDP-4XX: CANopen Device Profiles for Railways
  - WDP-407: CANopen Application Profile for Public Transportation
- DIN
  - DIN 9684: Agricultural Bus System
- SAE:
  - J1939: Serial Vehicle Network for Trucks and Buses
- Kvaser:
  - CAN Kingdom: Application Layer and Profiles.

2.7 The future of CAN

Currently, most of the CAN interfaces are applied in the automotive business, but in recent years more and more industrial applications have been reported. Due to the increased CAN bus applications in the automotive and process industries, the sales of CAN chips increases steadily. Today, 10 to 20 million chips per year are sold, and by the end of the decade, sales of 40 million CAN interfaces per year and more are anticipated.

The forecast of the evolutionary trend of the CAN bus indicates that there will be more and more differently performing chips available concerning the processor interface, addressing some data link layer issues, such as increased quantity and higher 'intelligent' mailboxes. Real-time related application issues will be more and more on the agenda. In the meantime, specific mailbox organizations are on the market to address especially the issue of timeliness. Numerous efforts have also been made related to the higher layers interface specification. For example, a variety of application layer specifications have been worked out. The development trend will even cross the borders of the application layer and enter the area of standardized operating systems for automotive and industrial purposes that incorporate various communication drivers, etc. Again, the important issue will be to address the real-time multitasking features of operating systems in conjunction with communication through networks. As automotive industries typically require high-volume, low-cost, high-performance solutions, consuming very few controller resources such as ROM, RAM, CPU time, etc., this trend will be very beneficial for the industrial users as well.

Current development trends in CAN technology indicate that in the future system interfaces are to be expected that will make the communication simpler, more efficient, and with predictable delay time, that is important for real-time applications. Integrating the communication layers into operating systems will unify today's variety of existing solutions or at least will make a distinction between the more accepted solutions and the rest. This is expected to make it easier for the designers looking for a powerful application support, providing a basis for easy and efficient integration of multi-vendor solutions into one system. On the other side, the integration process will generally continue in semiconductor industries. More and more transistor functions will be offered at the same price each year. Semiconductor industries will follow the integration path, integrating on the same piece of silicon augmented CAN solutions together with a microcontroller, power supply regulator, and an intelligent line transceiver module, which is capable of performing line diagnoses and reacting to faults as well as being able to cope with EMC requirements and constraints. The industry will also work on integration of higher layer communication drivers and parts of future operating systems. This tendency can be recognized from the past years' experiences: along with the next-generation chips, the silicon suppliers typically integrate those functions that have been provided by 'higher layer' software drivers and which are desired by a majority of applications and thus had become a *de facto* standard. As a consequence, only a small variety of solutions will be proliferated, raising the interest of the silicon industries to provide related standard products at attractive prices.

Today's communication solutions are split up between hardware and software. The overall system performance is characterized by the throughput time or the access time. Over the time, the higher degree of integration will result in a decrease of the access time, as some parts of the time consuming software drivers will migrate into hardware, providing a better throughput time factor. Unfortunately, over time, the demands for increased functionality will rise. Therefore, the decrease in throughput time will be more moderate than expected, if the functionality were constant over time. Nevertheless, the overall system throughput time will be reduced as the degree of parallelism increases over time. Thus, communication in general and CAN-based communication particularly will play a major role in future systems solution.

References

Robert Bosch GmbH, (1991). *Bosch CAN Specification Version 2.0.* Stuttgart.

Cassidy, F. (1997). *NMEA 2000 & the Controller Area Network (CAN).*

CiA, (1994). *CAN Application Layer – CAL.* CAN in Automation.

Cia, (1999). *CANopen.* CAN in Automation.

Etschberger, K. *et al.*, 1994. *CAN Controller-Area-Network*, Carl Hanser Verlag, München (in German).

Honeywell, (1996). *Smart Distributed System – Application Layer Protocol Version 2.0.* Honeywell Inc., Micro Switch Division, Freeport, IL, USA.

Institute of Electrical and Electronics Engineers (USA), (1983). Special Issue on ISO Reference Model. *Proc. IEEE*, Vol. 71, No. 12.

International Standards Organisation, (1994). ISO 11519-2, *Road vehicles – Low speed serial data communication – Part 2: Low speed controller area network (CAN).*

International Standards Organisation, (1993). ISO 11898 *Road vehicles – Interchange of digital information – Controller Area Network (CAN) for high speed communication.*

Lawrenz, W. *et al.* 1994a. *CAN Controller Area Network.* Hüting Verlag, Heidelberg (in German).

Lawrenz, W. (1994b). Network Application Layer. Paper 940141. SAE Detroit.

Lawrenz, W. (1997). CAN system engineering – From theory to practical applications. Springer, ISBN 0–387–94939–9.

Müller, H. (1998). CAN – Kommunikationsdieste und Bus – Arbitrierung, CAN in der Automatisierungstechnik; Tagung Langen, 19./20. Mai 1998/VDI/VDE – Gesellschaft Mess- und Automatisierungstechnik (in German).

Zeltwanger, H. (1998). Fehlererkennung und -eingrenzung in CAN-Netzwerken, CAN in der Automatisierungstechnik; Tagung Langen, 19./20. Mai 1998/VDI/VDE – Gesellschaft Mess und Automatisierungstechnik (in German).

Web sites

http://www.omegas.co.uk/CAN/index.html

http://141.44.61.248/NT/CAN/Welcome.html

http://www.can.bosch.com

http://can-cia.de

http://www.mitsubishichips.com

http://nec.de

Microcontrollers and micro-electronic technology

Khurshid Alam and Ljubo Vlacic
School of Microelectronic Engineering, Griffith University, Australia

3.1 Introduction

Intelligent vehicles are to operate autonomously or with no intervention from humans. This ability requires the intelligent vehicles to: (a) sense their own status and the environment they are in; (b) process the sensed information; and (c) compute the required course of action to accomplish the set task. To perform these processing and decision-making tasks, intelligent vehicles need assistance from sensors and microcontrollers. This chapter presents a brief description of the microcontrollers. The sensors that are in use by intelligent vehicles are explained in Chapters 4–8.

These processors and sensors are manufactured using the same microelectronic technologies as developed and used for the fabrication of integrated circuits. The micro-electronic technologies refer to the technologies used in producing very fine patterns in semiconductors, mostly silicon, to create active and passive electrical components. These are connected together to form any level of complex circuits, from microcontrollers and digital signal processors to the powerful microprocessors such as Pentium IV*. Those readers who wish to become familiar with the main idea of the development of the micro-electronic technologies and their use in the fabrication of the most common circuit element, the transistor, may wish to read the second part of this chapter. The purpose of this chapter is to give the reader an idea of the micro-electronic technologies without the level of detail needed for the actual practice of the technologies.

3.2 Microcontrollers – an embedded microcomputer

The architecture of today's modern microcontrollers can vary from one design to another. However, all designs contain the same basic elements – a central processing

* Pentium IV is a trademark by INTEL.

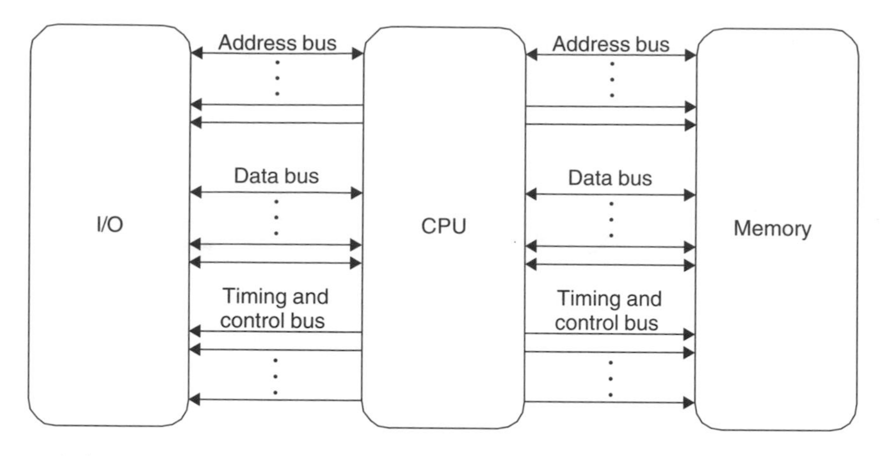

Fig. 3.1 The basic architecture.

unit, memory, input/output circuitry and the address, data and control buses – as it is depicted in Figure 3.1. More details of a microcontroller's architecture are depicted in Figure 3.2. The detailed description of a design of the microcontroller-based products can be found elsewhere (for example, Lilly and Vlacic, 1999). The References at the end of this chapter list a number of other sources that describe the properties of the particular microcontrollers.

3.2.1 Address bus, data bus and control bus

The common function of the address bus, data bus and control bus is to allow communications within a microcontroller, i.e. among different sections of its architecture. The address bus is used to set the unique memory location of the data that it wishes to access. This is typically performed by the central processing unit (CPU) enabling it to retrieve and store values in memory or peripheral devices. The amount of memory that a CPU can access depends directly on the size of the address bus. For example, a 16-bit address bus can typically access 64 kbytes of data.

The data bus is used to transfer or copy data from one address location to another as 'pointed' to by the address bus. A 16-bit processor has an internal data bus that is 16 bits wide.

The control bus contains the control signals or control lines that determine the type of transfer that is going to take place. It is usually made up of signals to enable read and write access, and a clock signal that determines when these events happen. The address bus and data bus are typically clocked at the same rate (which is typically in the MHz region).

3.2.2 Central processing unit

The central processing unit (CPU) is the core element of the microcontroller's architecture. This is where instructions are fetched, decoded, administered and executed. The

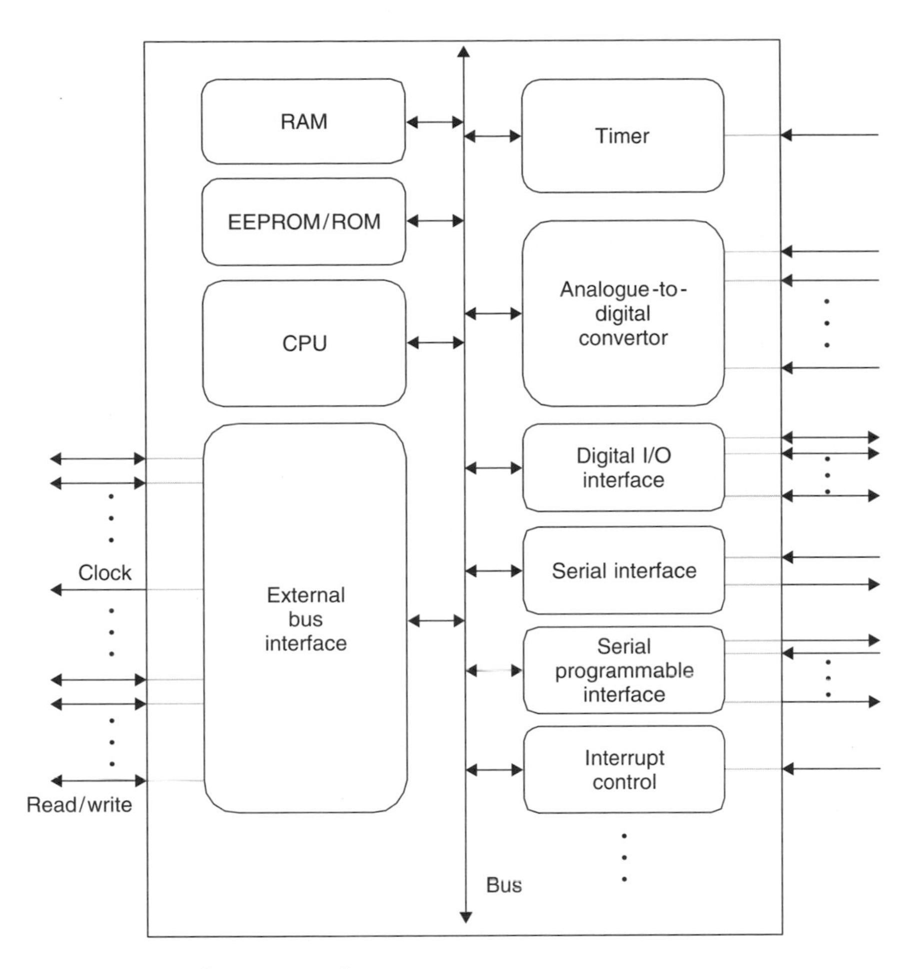

Fig. 3.2 Block diagram of a micro-controller.

CPU is divided into three main sections: the arithmetic logic unit (ALU), the control unit and the registers.

The registers in a CPU are fast memory locations used for temporary storage of data and status flags of the CPU. General purpose registers are also used at the input of the ALU to hold data ready for execution and/or manipulation. The common registers found in a CPU are:

ACC	Accumulator
PC	Program counter
IR	Instruction register
SP	Stack pointer
R0-R7	General registers
SF	Status flag register
EAR	Effective address register

The arithmetic logic unit is where all the mathematical and logic operations are executed. It takes the data out of the general registers, performs the required operation and then puts the result back into another or the same register. After execution, a number of status flags in the status flag register are set, cleared or left unchanged depending on the instruction executed. For example, the zero flag will be set if the result from the ALU is a zero.

The control unit controls the fetching of an instruction or op-code from the memory and decodes it. It decodes the instruction into a sequence (steps) which explains the events that have to occur in order to execute that instruction.

Instructions may contain more complex steps or simple ones such as moving data from one address location to another. Once a series of instructions are sequentially added together, we call it a *program*. A program is written for a particular task and can contain millions of different variations of instructions to perform its goal.

3.2.3 Memory

Memory contains the programs to be executed in the CPU. Different types of memory are in use with microcontrollers:

- ROM-based memory (masked ROM) is hard-coded into a microcontroller during manufacture and is contained in the same circuit as the CPU. As the name suggests, it can only be written once but read many times.
- RAM can be written to and read from many times. It can be found in the same circuit with the CPU or externally to the microcontroller. This type of memory can be found in a variety of different speeds (access times). High-speed RAM is used internally (within a microcontroller) for various functions including registers. Slower-speed RAM is used externally to the microcontroller to hold data such as temporary variables and for temporary storage of programs during prototyping.
- EPROM is the most common memory type for long-term storage of programs. The EPROM is erased when it is placed under UV light of a particular wavelength for a period of time.

E^2PROM or EEPROM is used for the long-term storage of data such as the system configuration of a circuit or for logging events of a monitoring system. E^2PROM does not require UV light to erase it. Instead it can be erased electrically. This memory is not used as RAM as writing the data is very slow. It is used where long-term storage of data is needed and a fast write time is not a major concern.

An alternative solution which has a relatively fast write time is the flash E^2PROM. Flash E^2PROM can be used in place of EPROM for the storage of programs and has the facility of being programmed while contained in the circuit.

3.2.4 Input/output (I/O) interface

In general terms, the I/O interface allows a microcontroller to communicate to the outside world. Two kind of such devices can be used, namely: (a) monitors, printers, LCD screens, status indicators, etc., which display messages to the user; and (b) keyboards, switches, push buttons and key pads, which allow the user to instruct the microcontroller-based system to perform the required operations.

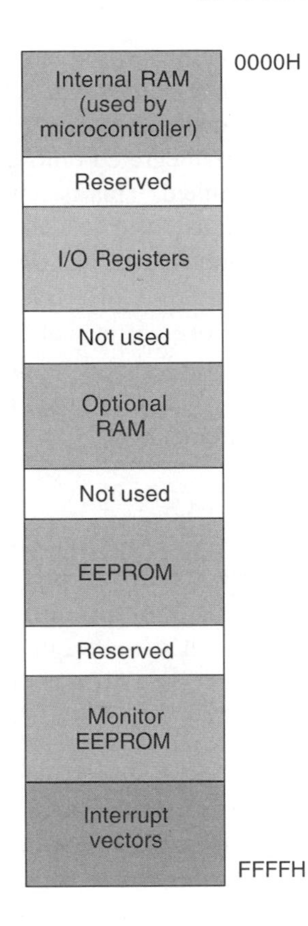

Fig. 3.3 A typical memory map.

Speaking in terms of electronic circuits, I/O interface may consist of digital parallel ports, serial communication ports, analogue-to-digital (ADC) converters and digital-to-anaologue converters (DAC). These types of interfaces are used to display warning signals, measure sensor inputs, control motor driving circuits, and so on.

A memory map describes the position of various memory devices, I/O interfaces and registers in terms of their memory locations – see Figure 3.3. The purpose of the memory map is to designate the physical memory address location where each of the devices is located so that they do not overlap each other and cause conflicts. For example, the interrupt vector table occupies a memory segment at the top of the memory map, usually between FFCOhex and FFFFhex.

After deciding what the system contains and the devices that are needed, a memory map has to be designed so that all components addressable by the memory of the system are accessible. This step is the transition between pure hardware design and the software or programs that access the hardware.

3.3 Microprocessor or microcontroller?

When only the CPU with some interrupt control and timers are found on the chip, such a configuration is usually named a *microprocessor*. In this case, all memory, input/output and other peripheral devices are wired externally to the chip. However, if the application is required to be small or compact, a *microcontroller* is to be used. It has a high degree of integration – as depicted in Figure 3.2 – so that all the elements of the architecture from Figure 3.2 are integrated onto the chip. However, the number of bytes of ROM and RAM is still limited. This is not considered a disadvantage as most of the microcontrollers allow the use of external memory.

Microcontrollers also utilize comprehensive and diverse interrupt procedures that are in-built in to the hardware/software basis of a microcontroller. This is useful for applications where a product needs to operate in real-time conditions, i.e. to measure sensor information and produce a control signal at defined time slots. This makes microcontrollers useful for intelligent vehicle applications where real-time sensoring and control are the most typical requirements.

3.4 Product design using a microcontroller

This section lists the main design steps from an initial design concept to a working prototype. A more detailed discussion on how to design a microcontroller-based product can be found in Lilly and Vlacic (1999), and Valvano (2000) for example. The main issues to be addressed during the product design process are as follows:

- Description of the function that the final system must perform; this should also show the interactions the system must have with the outside world.
- Selection of a microcontroller that is suited to the application; the hardest decision in a design can be what microcontroller to use. It is vitally important since such a decision influences the rest of the design; design/specification influences include the following (not in any particular order):
  - processing power – a speed must be fast enough to execute the software in real time
  - on-board memory – the size must be big enough to hold all parameters, instructions and variables
  - amount and type of I/O devices
  - ADC resolution (8/10 bit etc.)
  - number and size of internal counters
  - enough parallel ports to interface with all digital I/O signals
  - enough serial ports to interface with other computers
  - PWM signal if the application is in control
  - cost
  - expandability (memory)
  - ease of design
  - power requirements (particularly if the system is to be battery operated)
  - second source availability
  - compatibility with already existing microcontroller-based systems.
- If an application requires the device to operate at very high or very low temperatures, its operating temperature will also be a deciding factor. Most standard devices have a temperature range of −40°C to +80°C. However, military devices are available which extend the operating range to +125°C.

- The final and probably the most decisive factor is cost. The cost of the device itself is not always the best benchmark. To develop the system using the tools available is generally the deciding factor. This is why many companies utilize the tools they already have and develop products based around only a handful of different devices. A decision can also be made on the ease with which the device can be designed and programmed as this will save expensive development time. This may be one of the main influences in choosing a certain chip.

No one microcontroller can be utilized for all applications. Microcontrollers are application specific and the choice can change the design of the rest of the system. Consultation of information (such as from Table 3.1) is vital throughout the design so that a designer can focus on the purpose of the system. Table 3.1 is adopted from *Electronics Engineering*, 2000, where a more detailed review of currently available microcontrollers is discussed. This reference guide has been designed by *Australian Electronic Engineering* (published by Reed Business Information Pty Ltd) to provide a starting point when looking for microprocessors/microcontrollers. The reference guide, in its full size (covering more than fifty microprocessors and microcontrollers), can be found in Vol. 36, No. 6 of the *Australian Electronics Engineering* magazine, pp. 14–24, 2000.

3.5 Microtechnology

3.5.1 Introduction – brief outline of circuit integration

The roots of microelectronic technologies lie in the early efforts to reduce the size of electrical circuits through process innovation. The trend of circuit miniaturization that started about four decades ago is still going on, and there does not seem to be an end to this trend of circuit miniaturization, at least not in the near future. The following paragraphs attempt to outline the microelectronic technologies that are utilized to achieve the continued high-level integration of electronic circuits and their miniaturization.

Circuit integration is possibly the most significant factor in technological development in recent history that has changed the way modern society works. The powers of computations and communications that are available today are the fruits of the ability to build staggering amounts of circuitry in tiny chips. This trend of circuit integration is not a recent thing. In fact, as early as the first heyday of electronics in the 1940s and 1950s, when vacuum tubes were the mainstream of active devices, and bulky and hot at that, miniature tubes were developed to reduce the space taken by circuits in radios, radars and televisions. A mainframe computer of modest processing power using old technology used to occupy a whole room almost filling it from floor to ceiling. Things changed dramatically when the transistor was invented in 1948 and its commercial use became widespread in the 1960s. The use of the transistor in place of the vacuum tube in itself was a big step in circuit miniaturization. But with this came a change in the way the circuit components were connected together. In the vacuum tube equipment, all components were connected with flying leads, individual wires soldered between components. Each tube would typically need five or six connections: two for the heater filament, a cathode sometimes isolated from the filament, two grids and an anode. All this would add to the size of the circuit. The three-terminal transistor obviously needed

Table 3.1 Microprocessor/microcontroller reference guide (courtesy of Reed Business Information Pty Ltd)

Series name	Toshiba 47E	Sony SPC	Hitachi H8/300L	Rockwell R6502P	Microchip PIC	Microchip PIC17Cxxx	TI TMS320C24	NEC 78 K/IV	Zilog3 Z180	Motorola MC683xx	NEC V850	Motorola MPC860
Distributor	Arrow	Fairmont	Insight	Rs Comp	Avnet, Rs Comp	Arrow	Arrow	Soanar	REC	Arrow, Avnet	Soanar	Arrow
Architecture	4 bit	4–8 bit	8 bit	8 bit	8 bit	8bit	16 bit	16 'bit RISC	16 bit	32 bit	32 bit RISC	32 bit RISC
Clock frequency	up to 8 MHz	10–16 MHz		1–3 MHz	4–20 MHz	up to 33 MHz	30 MHz	32 MHz	33 MHz	16–25 MHz	33 MHz	up to 100 MHz
MIPS	1		1		1–10		30			2.5–4	62	up to 100
Program memory	up to 16 K	1–60 K	8–60 K		0.5–16 K	up to 16K × 16	32K × 16	256 K		up to 4 K	512 K	to 5 K
Memory type*	A,EE,M, O	A,E,O	F,M,O		A,E,EE, F,M,O	E,O,R,S	F,R	F,L,M,O, R,S	X	A	A,F,L,M, R	A
I/O pins	up to 56		64–100	0	6–68	up to 66	26–32	86	32		135	up to 120
Timers	1		up to 16	0	1–4	5	12	8/16 bit	4	1	16 bit	4
A/D converters	yes	4–8	yes		yes	up to 16	2	yes	no	yes	10 bit	no
UARTs	1	1	yes		yes	USART	1	multiple	2	1	multiple	6

*Memory type: A = RAM, E = EPROM, EE = EPROM, F = Flash, L = ROMless, M = Mask, O = OTP, P = PROM, R = ROM, S = SRAM.

about half the number of the tube connections. The transistor could also be mounted close to other components because it dissipated little heat compared with the vacuum tubes. These characteristics of the transistor circuits lend it to compactness, that later led to monolithic integration of circuits on very large and ultra large scales.

The development of alternative methods to connect components together, other than flying leads, was the starting point in circuit integration. One of the earlier methods of circuit connection was the *thick film hybrid* method. The major development was that now interconnecting conductors were screen printed on ceramic substrates, and circuit components like resistors, capacitors and inductors, along with active devices were mounted on the substrate. Later, resistors and capacitors were also deposited directly on the substrate using the screen printing technology. Formation of a simple circuit in the thick film hybrid technology is illustrated below.

A diode-transistor logic (DTL, now obsolete) inverter circuit is shown Figure 3.4, with three diodes, three resistors and a transistor. This circuit performs the inversion of the input digital signal that is applied at V_{in}. The inverted output signal appears at the collector of the transistor, labeled V_{out}. The two levels or states of the digital signal are low, or 0, and high, or 1. The input voltage is 0 for a low state, and it is equal to the supply voltage for a high state. When the input is in low state (i.e. 0 volts, close to ground potential), then the diode D1 (silicon device) is clamped at one diode drop potential equal to 0.7 volt. Two diode drops across diodes D2 and D3 bring the base of the transistor to one diode drop below ground potential. At this potential the transistor does not conduct. The output voltage V_{out} is high, equal to the supply voltage connected through the $2\,k\Omega$ resistor R2. On the other hand, when the input is high, or close to the supply voltage, then the diode D1 is reverse biased and the transistor is biased in saturation through the $2\,k\Omega$ resistor R1 and the diodes D2 and D3. The transistor conducts heavily and the collector is close to the ground potential. This means that the V_{out} is in low state. Thus the V_{in} and V_{out} are always in opposite states and the circuit performs the function of an inverter.

The layout of the circuit in Figure 3.4 implemented in thick film hybrid technology by Colclaser (1980) is shown in Figure 3.5. All conductors and resistors are screen printed and the semiconductor devices are bonded on to the substrate on appropriate bonding pads. In the screen printing process, an ink in the form of a paste is used that is squeezed on to the substrate through a masking screen with the appropriate

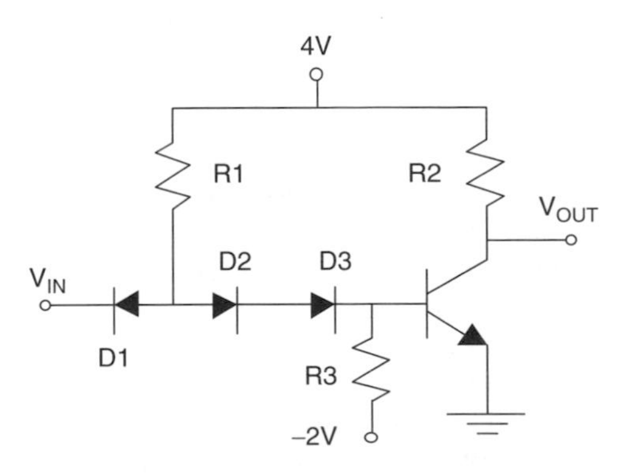

Fig. 3.4 A DTL inverter.

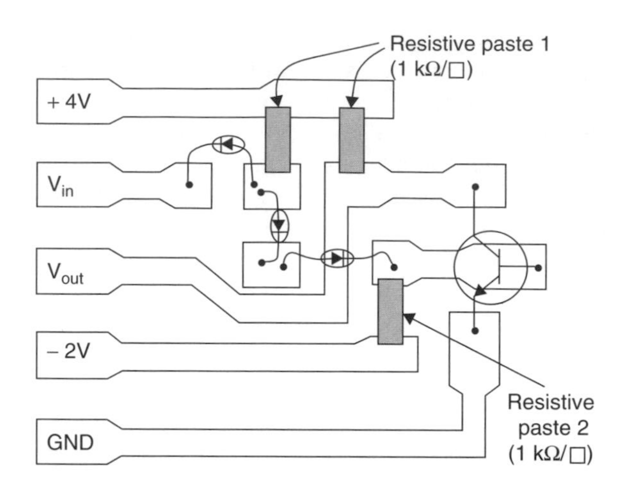

Fig. 3.5 Hybrid thick film DTL microcircuit.

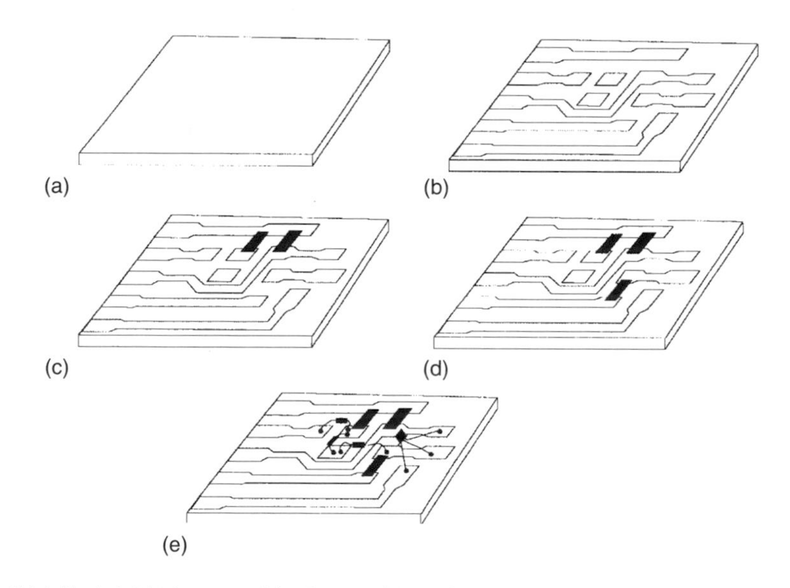

Fig. 3.6 Thick film hybrid fabrication: (a) substrate, (b) conductor tracks, (c) resistor print (paste 1), (d) resistor print (paste 2), (e) solder add-on components.

pattern. The printed pattern is then fired or dried at about 500°C to remove the liquid solvent and leave a firm film of the solids of the paste. The composition of the paste determines the nature of the printed tracks. The tracks can be highly conductive for use as inter-connects of components. For resistors, the deposited layers can have different resistivity to accommodate resistance values of different magnitude in the limited space that is available on the substrate. So, resistors R1 and R2 in this circuit will be printed with one paste and resistor R3 will be printed with another paste.

Figure 3.6, by Colclaser (1980), shows the fabrication steps involved in the thick film hybrid. The starting substrate is shown in Figure 3.6(a). Figure 3.6(b) shows the conductor tracks. After the conductor print, the ink is dried and fired at high temperature. Then the resistors R1 and R2 are printed and dried, as shown in Figure 3.6(c). Resistor R3 is next printed and dried, as in Figure 3.6(d). All resistors are fired at the same time. Figure 3.6(e) shows the semiconductor components soldered on to complete the circuit.

In the evolution of circuit integration, the next development was to deposit layers of conducting or resistive films and then use photographic procedures (known as photolithography in micro-electronic technology parlance) to define small and fine geometry interconnects and resistors. This technology, known as *thin film hybrid* technology, has reduced the circuit dimensions by a factor of at least two and the overall size by a factor of about five compared to the thick film technology.

The thin film hybrid technology is utilized to implement the emitter-coupled logic (ECL) inverter circuit shown in Figure 3.7. The layout by Colclaser (1980) is shown in Figure 3.8. The fabrication process is outlined in Figure 3.9. Figure 3.9(a) shows the starting substrate. Figure 3.9(b) shows the substrate coated with a thin layer of tantalum nitride. This is a high resistivity material and resistors will be made in this layer. Figure 3.9(c) shows the substrate coated with a thin titanium layer and another layer of gold on the previously deposited layer. Titanium is interposed between tantalum nitride and gold to overcome the difficulty of gold adhering well to tantalum nitride. The conductor pattern to be used in forming the circuit is transferred to the substrate by photolithographic process. The process consists of coating the substrate using a spinner, with a thin layer of photoresist. The photoresist is dried and baked and is ready for the transfer of the pattern to be etched on it. A mask with the desired pattern is used to expose the photoresist as in photography. Ultraviolet light is normally used for this process. The exposed substrate is developed in a developing solution, then washed and dried. This process leaves the substrate with a photoresist pattern the same as that to be retained on the surface. Other areas of the substrate are exposed bare. In the next step, the layer is etched in an active solution. The bare areas of the layer are removed and after stripping of the resist the transferred pattern is left on the substrate. Figure 3.9(d) shows the transferred conductor pattern. The tantalum nitride is present

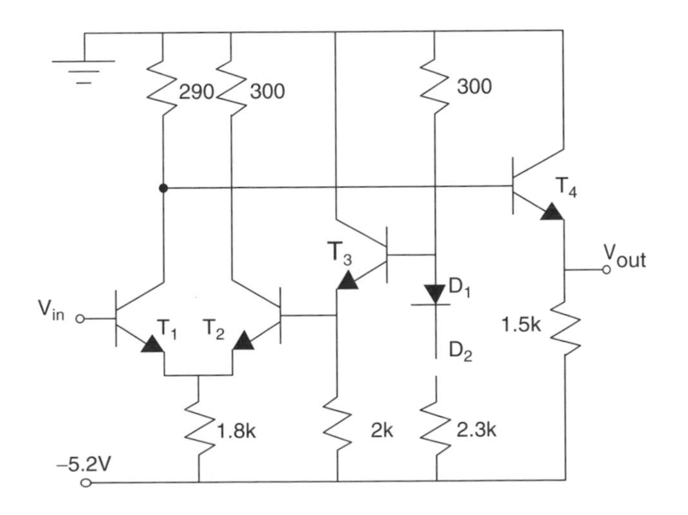

Fig. 3.7 An emitter coupled logic inverter.

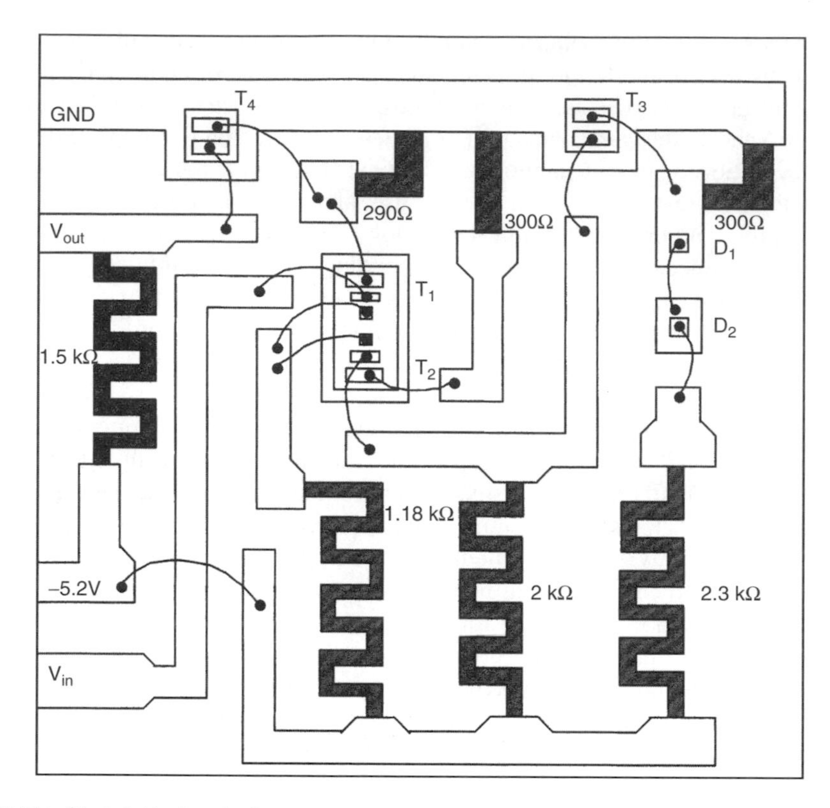

Fig. 3.8 Thin film hybrid microcircuit.

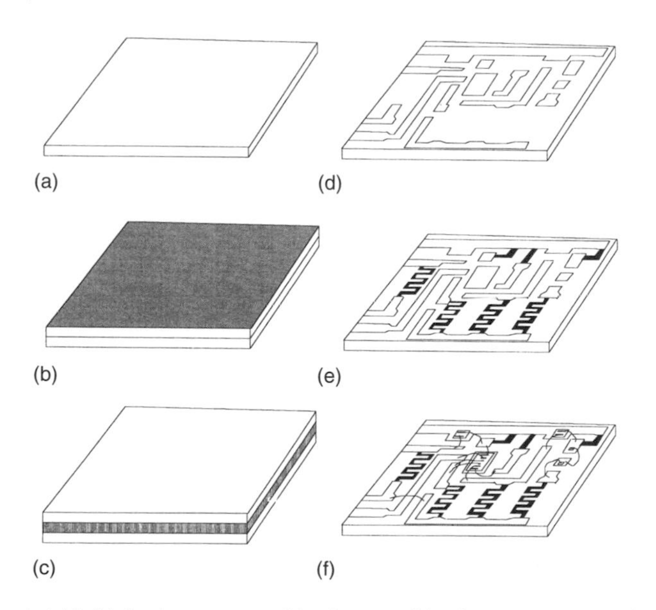

Fig. 3.9 Thin film hybrid fabrication process: (a) substrate, (b) substrate coated with tantalum nitride, (c) composite tantalum, titanium and gold layers, (d) etched conductor pattern, (e) etched resistor pattern, (f) chip bonded complete circuit.

under the pattern and everywhere else. In the next step, the resistive layer is etched to leave zigzag lines between appropriate points. This again uses a photolithographic step as in the conductor definition process. The selective etch leaves the resist material under the conductor pattern but it does not interfere with the circuit operation as it is covered with a highly conducting gold layer. The appearance of the substrate after this step is shown in Figure 3.9(e). Lastly, semiconductor components in chip form are wire bonded to complete the circuit. This is shown in Figure 3.9(f).

The monolithic (single block) process started with the concept of planar technology in 1958. This was a considerable shift from the prevalent method used for manufacturing discrete transistors. Transistors were fabricated using grown or alloyed junctions. In the grown junction technology, crystals were drawn from a melt in which the dopant was changed at the appropriate time so that alternate n-p-n or p-n-p regions were formed. Slicing and dicing would then produce discrete transistors. In the alloyed junction technology, an n-type silicon or germanium blank was the starting material. The blank was thinned where the transistor was to be formed. Aluminium was deposited on both sides of the thinned region and then heated to incorporate it in the semiconductor, giving the desired p-n-p structure of the transistor.

The thickness of the base width layer is critical in transistor operation. The technology used for grown and alloyed junction has a poor control of the base width. In comparison, the planar technology in which devices are made in a thin layer of a wafer offers precise control of the base width. The technology depends on the ability to selectively dope any region of the semiconductor wafer by solid state diffusion of impurities. The idea was first developed by Kilby at Texas Instruments in 1958. He fabricated transistors, resistors and capacitors in the planar technology. At about the same time, Noyce at Fairchild was working on monolithic integrated circuits. He developed the method for isolating devices made on the same substrate using reverse biased diodes, so that many transistors and resistors could be made at the same time and interconnected using conducting layers. It took only a few years for this technology to see commercial use and integrated circuits were available commercially by 1961.

The growth of integrated circuits (ICs) has been phenomenal. Early ICs were digital circuits using bipolar transistors. Their speed and power consumption was of the essence in wide spread acceptance for circuit design and application. Innovative circuit configurations were developed to improve the power–speed figure. Two of the well known logic circuit configurations are the emitter-coupled logic (ECL) and the integrated injection logic (IIL or I^2L). IBM Laboratories, in West Germany, introduced the IIL technology in 1972, but they termed it merged transistor logic. Philips Research Labs in the Netherlands independently developed the same technique and termed it I^2L by which name it came to be known in the market. The famous mainstream transistor-transistor logic (TTL) 7400 series was introduced by Texas Instruments in 1964. The military version of this, series 5400, has a wider operating temperature range. Many popular anaologue circuits were also available by the early 1960s. The power consumption of bipolar transistors was a hindrance in making very large-scale integrated circuits. The development of metal-oxide semiconductor (MOS) transistors, which became available in 1965, was a turning point in semiconductor technology and hastened the growth of integrated circuits. This fact is evident by looking at the chronology of integrated circuit component count through the years:

1951: Discrete transistors commercially available.
1960: Small-scale integration (SSI), less than 100 components per chip.
1966: Medium-scale integration (MSI), more than 100 components but less than 1000 components per chip.
1969: Large-scale integration (LSI), more than 1000 components but less than 10 000 components per chip.
1975: Very large-scale integration (VLSI), more than 10 000 components but less than half a million components per chip.
1985: Ultra-large-scale integration (ULSI), more than half a million components per chip.
1990: Giant-scale integration (GSI), more than 10 million components per chip.

Gordon Moore of Intel Corporation presented a paper at the IEDM of IEEE in 1975 entitled 'Progress in Digital Integrated Electronics', in which he showed that the device count per chip has followed a geometric progression, doubling every year since 1959 (Moore, 1975). He predicted that this trend would continue till about 1980, after which the component count would increase at the rate of doubling about every two years. This prediction has been surpassed as the DRAM chips have continued to grow with amazing speed. Demand for larger and larger memories has kept the growth rate close to that originally established. Deep sub-micron technology being developed these days points to the fact that bigger and bigger circuits with larger and larger component counts will remain the flavour of the day for a long time.

3.5.2 The transistor – the active element of integrated circuits

The active element of all integrated circuits, including the microprocessors and the microcontrollers, is the transistor. It is not necessary to understand the operation of the transistor or the internal operation of the IC to use ICs, but a brief description of transistors is included here for the benefit of those readers who may like to know a little more on the subject.

There are a few types of transistors available for circuit implementation in ICs. The two common types are the bipolar junction transistor (BJT) and the metal-oxide semiconductor field effect transistor (MOSFET).

The bipolar junction transistor

The structure of a BJT consists of three layers of semiconducting materials. The layers are known as emitter, base and collector because of their specific roles. The two outer layers, the emitter and the collector, have opposite conductivity to that of the internal layer, the base. So, the structure can either be p-n-p or n-p-n, these being the two types of BJTs. The general structure of a p-n-p bipolar transistor is shown in Figure 3.10. The circuit symbol of the p-n-p transistor and its structure in the planar technology is also shown in the figure. The emitter is indicated on the symbol by an arrow. It points inwards towards the base for a p-n-p transistor and outwards for an n-p-n transistor. The operation of the two types of bipolar transistors is the same, the difference being that in describing the two, the n- and p-types are swapped between the two structures. For the simple structure shown, the emitter and the collector can be interchanged; but when actual transistors are fabricated, the fabrication process mostly dictates that a

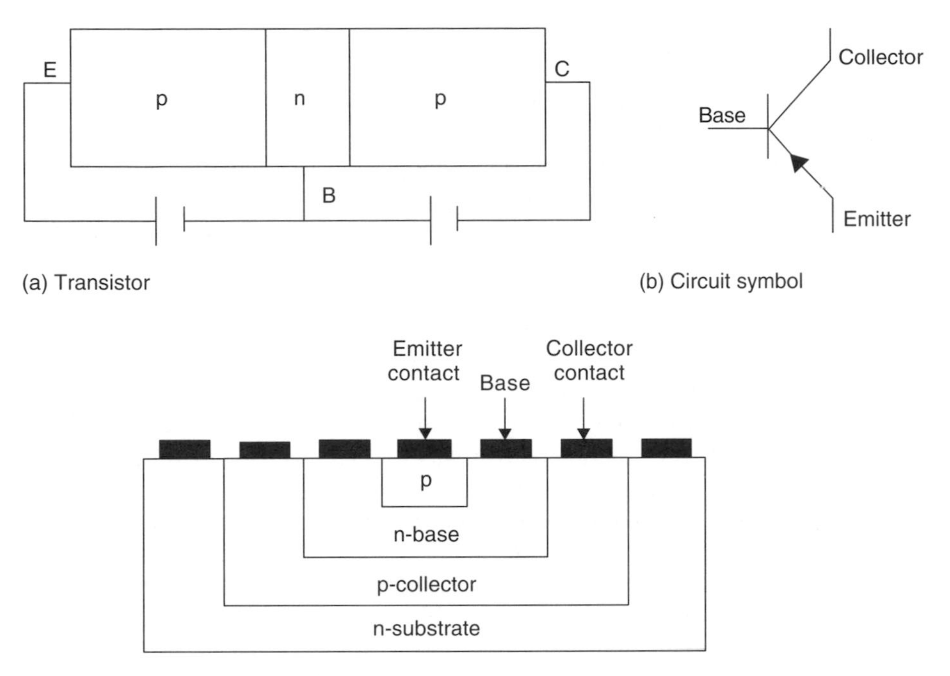

Fig. 3.10 p-n-p transistor, circuit symbol and planar structure.

particular layer be used as the emitter or the collector. There is also a fundamental requirement on the width of the base layer, which states that this width be very much smaller than a characteristic length of the base material known as the diffusion length. No transistor action will occur if the base width is larger than a diffusion length. The reason will become apparent as the operation of the transistor is explained below.

In normal operation, a voltage is applied between the emitter and the base, so that it is forward biased. A voltage is also applied between the collector and the base, so that this junction is reverse biased. Under the influence of the forward emitter voltage, holes will be injected from the emitter to the base (at the same time electrons are injected from the base to the emitter, but their magnitude is much smaller as the doping of the emitter is much higher than that of the base). The injected holes in the base region will diffuse towards the collector junction as there are more holes near the emitter junction due to injection. There is a continuous recombination of the injected holes in the base region as they are moving towards the collector. All of the injected holes are expected to recombine within a diffusion length of the emitter junction. So, if the base width is large, few holes will reach the collector. In the worst case, no holes will reach the collector and there will be no transistor action. But when the base width is very small, most of the injected holes will reach the collector junction. There, an electric field due to the collector voltage is present with such polarity that it propels the holes through the junction to the collector region. A corresponding current will flow in the collector circuit even in the presence of a large value resistance. If the emitter current is made to fluctuate in response to an input signal, then the collector current will also fluctuate in a similar manner. This will generate a voltage which can be many times larger than the input signal, as the output current in the collector circuit flows through a large resistance. The name transistor comes from the fact that there is a transformation of resistance (trans-resistance) from the input to the output circuit.

The configuration described above is called the common base amplifier configuration, in which the base is common to both the input and the output circuit. A more commonly used configuration for voltage amplification is the common emitter configuration in which the emitter is the common terminal between the input and the output. Yet again, the common collector configuration is used in output stages of power amplifiers.

The actual structure of the bipolar transistor is technology dependent. The earlier technologies used for the fabrication of discrete transistors were the alloy junction and the grown junction technologies. The diffused planar technology was developed for VLSI and later epitaxial layer transistors have been developed for devices of enhanced performance. It may be mentioned that in modern-day integrated circuit processing, ion implantation has become an alternative way of introducing impurities in place of diffusion of impurities in semiconductors.

The diffused planar technology depends on a controlled diffusion of impurities in selected regions on the surface of a semiconductor wafer. This is done through the use of a mask that stops impurities penetrating the masked regions and reaching the semiconductor. Silicon-dioxide, a stable oxide of silicon, has been found to have excellent masking properties for common impurities, and is used extensively in silicon planar technology. For the p-n-p structure in one variation of this technology, the starting material is a wafer of silicon with n-type impurity of thickness 250–380 μm. Masking oxide of sufficient thickness is grown in a furnace, which is calculated to stop diffusion of impurities dependent on time and temperature of the diffusion cycle. Photolithography is used to define a window in the oxide. P-type impurities are diffused in this region which will become the collector of the transistor. After this diffusion, the masking oxide is stripped and a second masking oxide is grown. In this oxide a window is opened that is aligned inside the first diffusion. N-type impurities are diffused this time through the window in the p-type impurity of the previous step to create the base of the transistor. The incorporated impurity level in the window is of sufficient magnitude to compensate the previous impurity and give a net n-type impurity. A further p-type diffusion is done in the base region to create the emitter of the transistor. This creates the p-n-p structure of the transistor. Next, another oxide layer is grown on the surface. Windows are cut in the oxide and metal contacts are made to the base, emitter and collector of the transistor. The cross-section of the final structure is shown in Figure 3.10(c).

MOS field effect transistor (MOSFET)

The metal-oxide semiconductor field effect transistor works on a different principle compared with the bipolar transistors. To understand the operation of the MOS transistor, first consider the structure of a MOS capacitor shown in Figure 3.11(a). It consists of a p-type (or n-type) semiconductor on which there is a thin layer of oxide, and then a layer of metal completes the structure. This is a parallel plate capacitor: the top metal forms one plate and the semiconductor (through a bottom metal contact) forms the other plate.

What happens at the semiconductor surface near the oxide when a potential is applied to this capacitor is the basis of transistor action in MOSFETs. For the p-type MOS

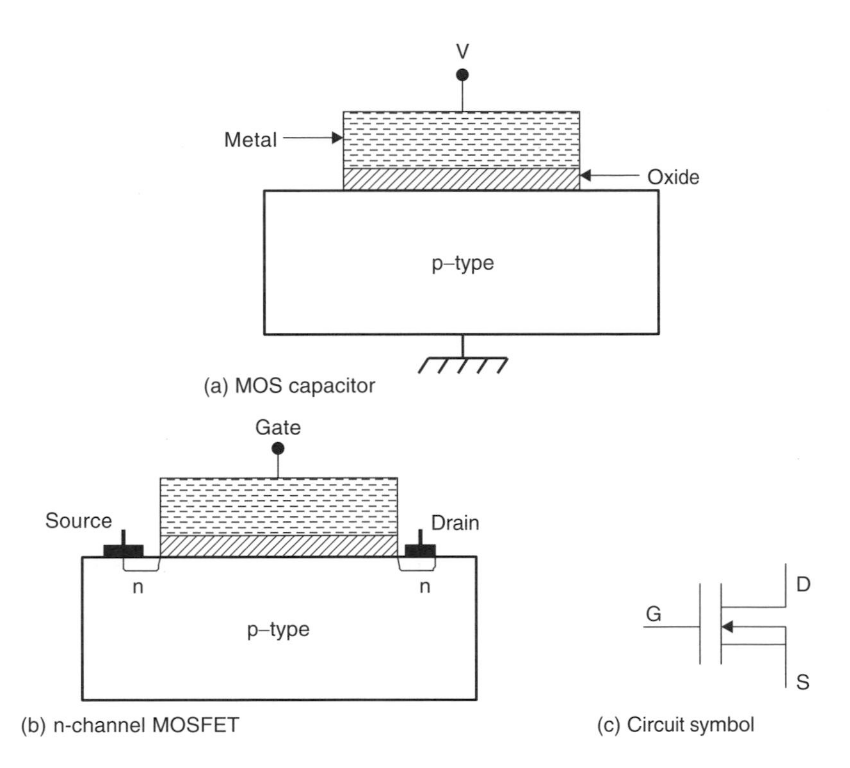

Fig. 3.11 MOS capacitor and a MOSFET.

capacitor, when a small positive potential is applied to the metal plate with respect to the bottom contact, positive charge builds on the metal plate and a corresponding negative charge builds at the semiconductor surface. This happens through the removal of holes in the surface layer and the semiconductor is said to be in depletion. When the applied potential is raised, the depletion region increases. There are no mobile carriers in the depletion region so no current can flow in the surface layer of a depleted semiconductor. When the potential is increased further, the depletion width reaches a maximum and the charge balance is obtained by accumulation of electrons in the surface region. This condition of formation of a layer of opposite type carriers in the semiconductor is termed inversion. For sufficiently high positive voltage on the metal plate or the gate, the semiconductor goes into strong inversion. In this condition there is a ready supply of carriers in the inversion layer for current conduction. The carrier density in the inversion layer depends on the gate potential. The current flow in the inversion layer can be controlled by controlling the potential on the gate. The structure to achieve this control is the MOS transistor shown in Figure 3.11(b). Here, two additional contacts have been added, the source through which current carriers enter the transistor, and the drain through which the carriers exit the transistor.

There are a variety of MOS transistors depending on the material and technology used. The transistor structure described above is called an n-channel device because the current channel is made of electrons. Use of n-type material would give a p-channel device. Each type can again be either an enhancement type device or a depletion type device. If there is no initial channel in the device in absence of a gate bias and the channel has to be formed for current conduction, then the device is known to be working in the enhancement mode as the one described before. For the case when a channel is initially present in absence of a gate voltage, then the control on current flow is achieved by reducing or depleting the carriers in the channel. This type of working is known as the depletion mode operation. Most integrated circuits are implemented using the enhancement mode transistors.

There are a number of MOS technologies available for circuit integration. N-MOS technology uses n-channel transistors for circuit implementation. In p-MOS technology, the circuit elements are p-channel transistors. Simultaneous use of both n-type and p-type devices in a circuit is termed complementary MOS or CMOS technology. CMOS technology is the current technology for digital circuits because, the way circuits are configured, it has a very small current drain from the power supply. A recent trend in circuit design is to include both anaologue and digital circuits on the same chip. This is known as BiCMOS technology.

3.5.3 Technologies for monolithic chip integration of circuits

As is apparent from the previous few sections, there are a number of technologies that are required to fabricate monolithic integrated circuits. The use of these technologies in the fabrication of a CMOS inverter is given below. This serves the dual purpose of describing the technology and also outlining the steps involved in producing a CMOS integrated circuit. It should be realized that what follows is a general introduction to the fabrication steps. The actual process followed at any fabrication house will vary depending on the facilities and experience available there.

The circuit of a CMOS inverter is shown in Figure 3.12(a). It consists of an n-channel and a p-channel transistor connected in series. The source of the n-channel device is connected to the V_{SS} supply, and the source of the p-channel transistor is connected to the V_{DD} supply. The drains of the two transistors are tied together. The gates of the two transistors that are connected together receive the input signal. The output is derived from the common drains. The circuit operation is as follows. When the input signal is close to the V_{SS} supply, the n-transistor is turned off while the p-transistor is turned on, and the drain is near the V_{DD} potential. As the gate voltage rises towards V_{DD}, the state of the transistors changes. The p-transistor is turned off while the n-transistor turns on. In this state the drain potential is low close to V_{SS}. Thus the output is always of the opposite state to that of the gate and the circuit performs the 'inverter' function.

The CMOS technology uses both n-channel and p-channel transistors on the same substrate. This is handled by making one type of transistor in the starting substrate and the other type in a 'well' formed by converting the conductivity of the well region using deep diffusion. The well region is used as the substrate for the complementary type transistors. In a p-well process the starting substrate is n-type, whereas in the n-well process the starting material is p-type. The p-well process is used in the following description.

Efficient use of the chip area for implementing the circuit requires that a compact layout of the circuit be developed. This layout is the plan of the circuit detailing what elements of the circuit will be located where. Design rule restrictions like feature

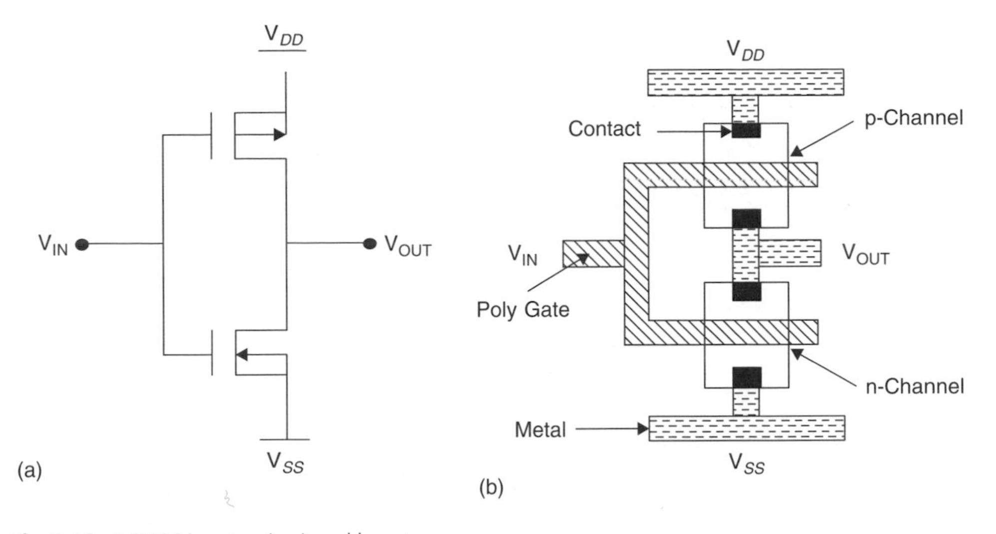

Fig. 3.12 A CMOS inverter circuit and layout.

size, placement and separation of the circuit elements are the key considerations in developing the layout. CAD tools are generally used for layout generation where design rule violations are checked and flagged for correction as the layout develops. Masks are then generated that define the regions for a series of processes needed to implement the circuit element. A typical layout of the inverter is shown in Figure 3.12(b). Many variations of the layout are possible and are used by different circuit designers.

The processing steps for the CMOS inverter is depicted in Figure 3.13. The starting material for the chip is an n-type wafer, Figure 3.13(a). The p-well where the n-type transistor will be made is created first. To do this, a masking silicon dioxide (SiO_2) is grown on the wafer in pure oxygen ambient in a quartz-tube furnace operating at about 1100°C. Photolithography and p-well mask are used to open the p-well area in the photoresist. The exposed oxide is then etched in a diluted solution of hydrofluoric acid. The photoresist is stripped in a solvent and the wafer is cleaned in de-ionized water and dried with compressed air or nitrogen, Figure 3.13(c).

The p-well is created by solid state diffusion of a p-type impurity in the unmasked area of the wafer. The process is similar to the oxidation step except that the gas ambient now contains the diffusion impurities. At the end of the process the wafer cross-section is as in Figure 3.13(d). The masking oxide is stripped and a thick oxide is grown on the wafer to define the area for the next processing step.

The current generation of integrated circuits uses a polysilicon for gate contacts in MOS devices compared to the metal gates used in the old technology. The use of a polysilicon (or poly) gate makes processing simpler and cheaper. This technology is the normal process for all fabrication houses. The added advantage of this technology is that the source and the drain can be diffused or implanted with the gate as a mask giving self-aligned source and drain. The fabrication technology has greatly benefited from this innovation.

To continue the process in the polygate technology, the active areas where the n- and p-transistors are to be made are now defined using the thin oxide mask. The thick oxide is etched there and a thin oxide (a few tens of nanometers depending on the

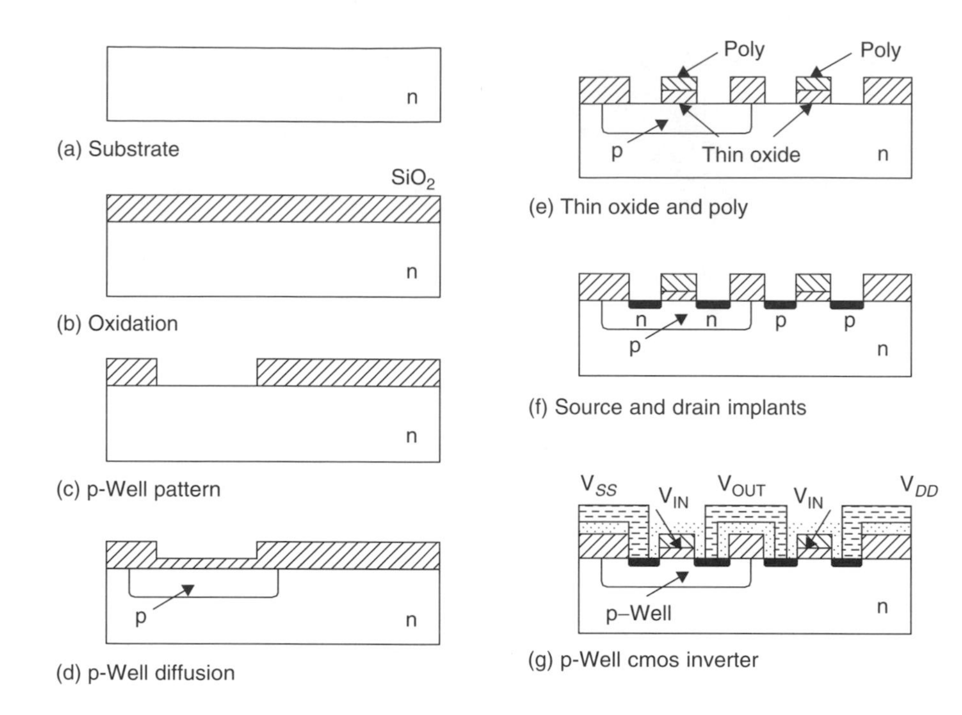

Fig. 3.13 Processing steps for CMOS inverter.

technology) is grown as in the oxidation step. Polysilicon is next deposited (and doped) on the wafer. A gate mask is used to define the polysilicon pattern for the gates and any other tracks used as interconnects. Plasma processing is the common method for poly etching. The thin oxide in areas not covered by poly is also removed. The status of the wafer is shown in Figure 3.13(e).

Next, a p-mask is used to open the area for ion implantation of p-type species for the self-aligned source and drain of the p-type transistor. In the ion implantation process, ions of the desired species are accelerated in a high electric field and made to impinge on the wafer, thus embedding them in the surface layer. The process is repeated with the next mask for implantation of n-type species for the source and drain of the n-type transistor. The implanted wafer is heated to a high temperature to activate the implanted species. The state of the wafer after this process is shown in Figure 3.13(f).

All devices have now been fabricated. What is left is to provide the interconnects to complete the circuit. This is done through the contact and metal masks. The wafer is prepared for this by depositing an oxide layer on the wafer. The contact mask is used to open cuts in the oxide to reach the silicon or the poly so that the sources, drains, gates and the substrate can be appropriately connected. Next, metal is deposited on the wafer using high vacuum evaporation. In this process, strands of metal, normally aluminium, are heated to a high temperature in a high vacuum chamber using tungsten filament or boat so that the metal evaporates and deposits on the wafer, leaving a thin uniform layer of the metal. Photolithography is again used to transfer the metal mask pattern on the wafer. Aluminium is wet etched in phosphoric acid or dry etched in a plasma reactor. A sintering step at about 400°C may also follow to make good contacts. The state of the wafer is shown in Figure 3.13(g).

Normally an over-glass layer is deposited for long-term protection of the metal. Another mask is then needed to cut holes in the over-glass to access contact pads that are used to connect the chip circuitry to the outside world. Depending on the wafer size and the circuit complexity, hundreds to thousands of copies or pieces of a circuit can be present on the wafer, not all of which may be functional. To mark the defective pieces the wafer normally goes through a test stage. After this, the wafer is scribed and broken into individual pieces. The good pieces are mounted on carriers. Contact pads are connected to carrier pins using thin gold wires. Automated bonding machines are used for this task. The carrier is then sealed and, possibly after a final test, the IC is ready for use.

3.6 Conclusion

Microelectronic technologies have evolved and developed over a long period of time as innovations were devised and challenges met in the fabrication and miniaturization of integrated circuits. The precision, control and repeatability required of these technologies demand such devotion and good practices from those who use them that the technologies are practised as an art form. Theory and detail of the technologies are available in many books, but the art of the technologies is learned only through practice.

Successful development of microprocessors and microcontrollers is the direct result of the recent advances in electronic semiconductor technology developments. This chapter has exposed the basic underlying concepts that can be applied later in practice in conjunction with the development of microcontroller-based products such as those that are being implemented onto intelligent vehicles. While microcontrollers are different, they operate on the same principles, the core of which has been presented in this chapter. For the interested reader, a list of books related to both topics, microcontrollers and microelectronics, is given in the References.

References

Selected references on microcontrollers and microprocessors

Alexandridis, N. (1993). *Design of Microprocessor-Based Systems*. Prentice-Hall, Inc.

Australian Electronics Engineering, (2000). *Microprocessor Reference Guide*. Reed Business Information, Vol. 33, No. 6, June.

Lilly, B. and Vlacic, L. (1999). Microcomputer Technology. In D. Popovic and L. Vlacic (eds) *Mechatronics in Engineering Design and Product Development*. Marcel Dekker, Inc.

Barnett, R.H. (1995). *The 8051 Family of Microcontrollers*. Prentice-Hall, Inc.

Huang, H.-W. (1996). *MC68HC11: An Introduction – Software and Hardware Interfacing*. West Publishing Comp.

Rafiquzzaman, M. (1992). *Microprocessors*. New Jersey: Prentice-Hall, Inc.

Valvano, J.W. (2000). *Embedded Microcomputer Systems: Real-Time Interfacing*. Brooks/Cole.

Web sites

http://www.hitachi.com/
http://www.intel.com/
http://www.microchip.com/
http://www.motorola.com/
http://www.nex.com/
http://www.philips.com/
http://www.siemens.com/

Selected references on micro-electronics

Colclaser, R.A. (1980). *Microelectronics: Processing and Device Design*. John Wiley & Sons.

Hamilton, D.J. and Howard, W.G. (1975). *Basic Integrated Circuit Engineering*. New York, McGraw-Hill Book Company.

Millman, J. (1979). *Micro-Electronics – Digital and Analog Circuits and Systems*. New York, McGraw-Hill Book Company.

Moore, G.E. (1975). Progress in Digital Integrated Electronics. In: *Technical Digest of the International Electron Device Meeting of IEEE*, 1975, pp. 11–13.

O'Mara, W.C., Herring, R.B. and Hunt, L.P. (1990). *Handbook of Semiconductor Silicon Technology*. Park Ridge, NJ, Noyes Publications.

Weste, N. and Eshraghian, K. (1988). *Principles of CMOS VLSI Design–A System Perspective.* Addison-Wesley Publishing Company.

4

Vehicle optical sensor

Yuichi Shinmoto
OMRON Corporation, Japan

4.1 Introduction

Sensors are a very important technology for controlling automated systems. The role and importance of sensor technology is ever increasing in advanced systems which are supported by the remarkable advancements in information/data transmission systems.

In order to realize the driving support system for a vehicle, many kinds of sensors have been developed. In this chapter we shall introduce three vehicle-installed optical sensor models that detect the road environment, namely laser radar, ground speed sensor and GVS (ground view sensor). GVS can distinguish road surface conditions by using an optical spatial filter.

On the other hand ITS (intelligent transportation system) is a new system that reduces traffic accidents, traffic jams and environmental problems. For example electronic toll collection (ETC) systems relieve traffic congestion at tollgates, improve driver convenience and reduce the labour costs of toll collection. In Japan, ETC was developed to serve these purposes.

In order to realize ITS including ETC, vehicle detection and road environment sensing are very important. This chapter also introduces the vehicle detection sensors that will be employed in ETC along with the features of two sensor models, the compact vehicle sensor and the laser-employed axle sensor.

4.2 Laser radar

Currently, as a measure to decrease traffic accidents, the development of the driving support systems is well underway. The role of the driving support system is to function as a supplement to the driver's effort to steer clear of a foreseeable and anticipated dangerous traffic situation. For this system, a sensor that detects the distance and direction of a vehicle running ahead is important. We shall introduce one of such laser radar system in this chapter.

4.2.1 The basic principles

Laser radar utilizes a laser beam to determine the distance to the measuring object.

Pulse, modular and interferometer systems are generally known as distance measurement methods utilizing a laser. The pulse method is used more frequently in low-priced, compact systems. As shown in Figure 4.1, the pulse method measures distances by calculating the reciprocating time of light pulses between the objects. It can be expressed in the following formula of $R = C \times \tau/2$, where C represents the speed of light, τ represents the reciprocating time and R represents the distance.

It is understood that the quantity of reflected light and the performance of the receiver determines the maximum detectable distance. This quantity of reflected light can be expressed as shown below, where the performance capability of a reflector is used as the parameter.

In this case, horizontal parting of emitted light width is less than or equal to the width of reflector:

$$Pr = \frac{KK \times At \times H \times T^2 \times Pt}{\pi^2 \times R^3 \times (Qv/4) \times (\phi/2)^2}$$

Here, horizontal parting of emitted light width is greater than the width of reflector:

$$Pr = \frac{KK \times Ar \times At \times T^2 \times Pt}{\pi^2 \times R^4 \times (Qv \times Qh/4) \times (\phi/2)^2}$$

where: Pr = Quantity of light received (W)
KK = Reflector reflexibility
ϕ = Reflector radius (rad)
H = Longitudinal width of reflector (m)
Ar = Reflector area (m^2)
T = Atmospheric transmission factor
Qv = Longitudinal spread radius of emission (rad)
Qh = Transversal spread radius of emission (rad)
At = Area of receiver lens (m^2)
Pt = LD power (W)

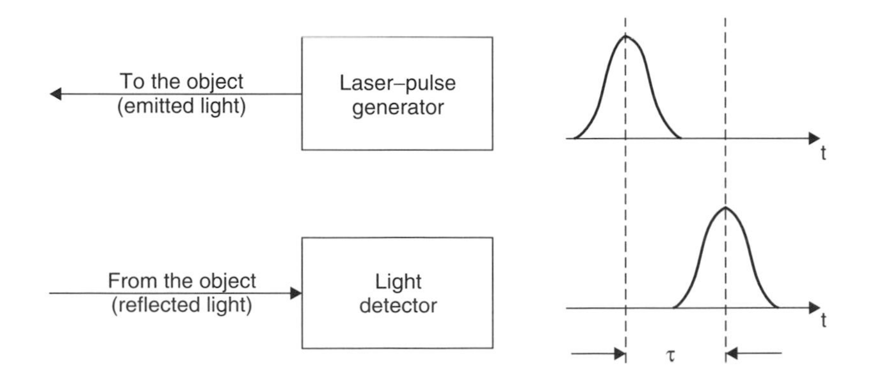

Fig. 4.1 Principle of the pulse method.

Table 4.1 The features of respective methods. XX = very bad; X = bad; √ = good

	Optical	Millimetric wave	Image
Distance capability	X	√	X
Directional resolution	√	XX	√
Signal processing capability	√	√	X
Price	√	X	XX

Table 4.2 Major specifications of the laser radar

Description	Specification
External dimensions	119 mm (W) × 60 mm (H) × 68.5 mm (D)
Weight	450 g
Distance capability	Clear day: over 100 m Rain: over 55 m
Distance precision	±1.2 m
Distance resolution	0.15 m
Perimeter radius	Horizontal: 210 mrad Vertical: 57.5 mrad
Output interval	100 mS
Directional resolution	2.5 mrad
Transversal spread radius of beam	1 mrad

As the distance from the radar increases, the value of *Pr* diminishes and reaches the point of equilibrium with the noise output power of the receiver. This point is regarded as the maximum detectable distance.

For directional measurements, a scanning mechanism that scans the laser beam horizontally is utilized. The scanning mechanism is made from mirror and actuator, such as galvanic-motor, DC motor, stepping motor, etc. The former methods tend to have higher resolution but the costs are comparatively high as well.

Also, besides the optical method, the millimetric wave method and image method are available as detection equipment. Their general features are shown in Table 4.1.

4.2.2 Example of laser radar

As an example of a laser radar, we have listed the major specifications (Table 4.2) and external appearance photograph (Figure 4.2) of a sensor developed by OMRON. The sensor is installed in front of the vehicle (Figure 4.3), and the galvanic motor is adapted for scanning in this sensor so as to acquire improved horizontal resolution.

Figure 4.4 shows the schematic diagram. The laser radar is composed of the following four blocks: data processing, light receiver, emission and scanner. On instruction from the CPU the timing circuit of the gate-allay operates. When the LD starts to emit using the light emission circuit, the counter starts simultaneously. The light is scanned by mirror and is emitted forward. The light reflected from the object is collimated through a receiver lens onto a PD. After having been photo-electrically exchanged and amplified, and depending on the condition of background light, if the signal received is above that of the optimized threshold value, the operation of the counter is terminated. The CPU computes the distance from this counter value.

Fig. 4.2 External appearance of the laser radar.

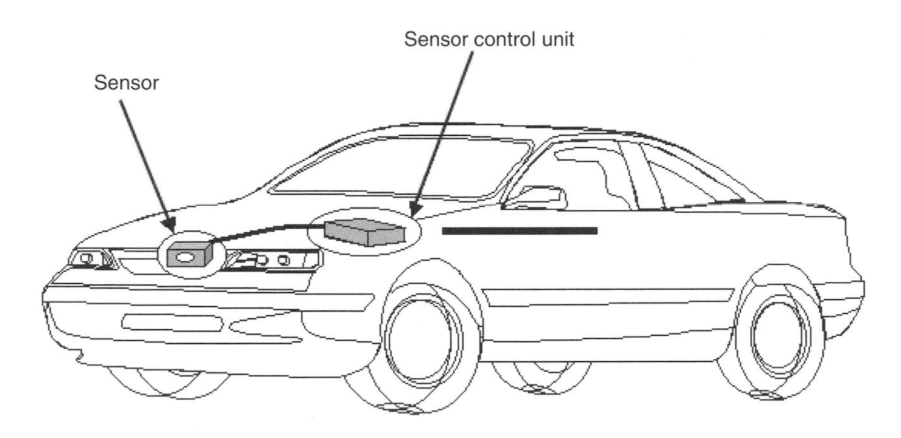

Fig. 4.3 Installation position of the sensor.

Also, this CPU obtains the present directional data from the radius detection circuit attached to the motor and its feedback controls the directional deviation of the object by means of installed software.

The CPU infers a vehicle ahead based on the above distance and directional data. As one example, Figure 4.5 illustrates vehicle recognition through curved-inference processing. The curved line is a hypothetical vehicle line calculated by curved-inference processing. A large circle represents a vehicle ahead on the same lane and smaller circles represent vehicles ahead in another lane. You can see that the system correctly recognizes its own lane and the vehicle ahead even at a curve.

4.3 Non-contact ground velocity detecting sensor

4.3.1 Comparison of respective methods

It is important to detect vehicle velocity in order to enhance advancement of vehicle control systems such as ABS (anti-lock braking system).

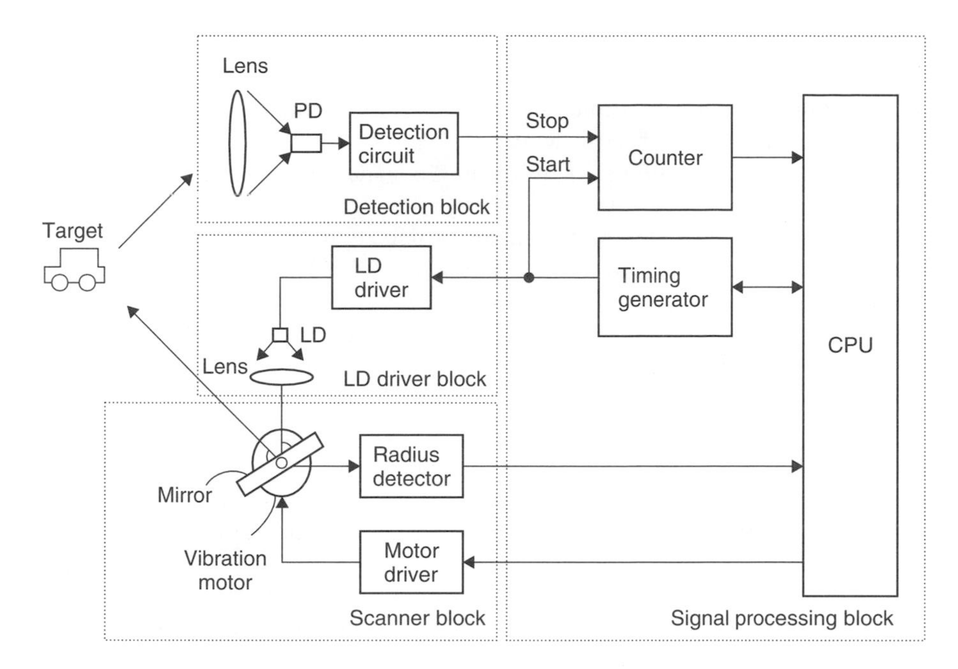

Fig. 4.4 Laser radar block diagram.

Fig. 4.5 Vehicle recognition through curved-inference processing.

A generally accepted method of detecting vehicle velocity is to calculate the speed from the rotation of a wheel. However, in this method of using wheel rotation as a parameter for calculation, a large computation deviation is perceivable in the case where the wheel slips or races. In order to cope with such a problem, it is necessary to calculate ground velocity by means of a non-contact system. The methods below are generally available:

1. **Spatial filter method**: Extracts a particular cycle from surface pattern and measures frequency.
2. **Radio doppler method**: Emits radio beam on ground surface and measures frequency shifts of the reflective wave.
3. **Supersonic doppler method**: Measures in the same manner as 2.

In these methods, depending on the angles of detection objects (such as the sensor) against ground surface, the calculation deviation of detected ground velocity fluctuates. For instance, in Figure 4.6, the spatial filter method operates perpendicular to the ground surface but the radio and supersonic Doppler methods have to be in oblique positions. Therefore, if the affixed angle differs by about 1°, the following deviations in detection are generated:

1. Spatial filter deviation of about 0.02 per cent (affixing angle, perpendicular)
2. Radio doppler deviation of about 1.7 per cent (affixing angle, 45°)
3. Supersonic doppler deviation of about 1.7 per cent (affixing angle, 45°)

When vehicle vibration is taken into consideration, the spatial filter method is at an advantage in terms of detection precision. As reference, the performance capability comparisons of these methods are shown in Table 4.3.

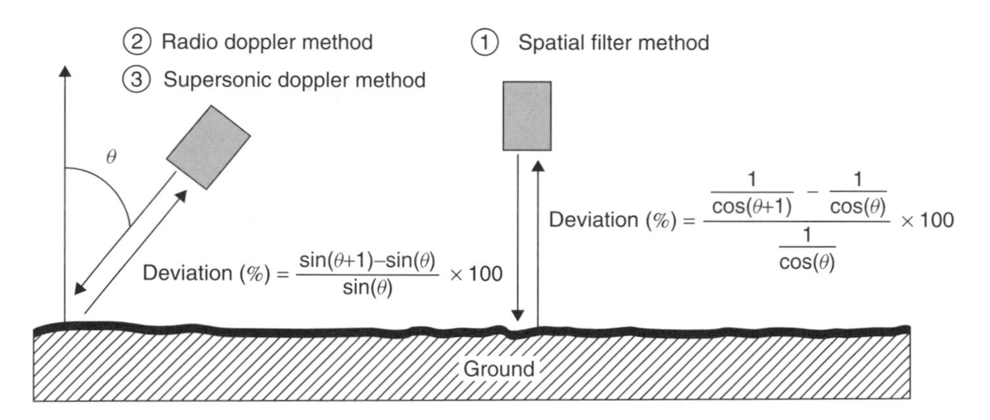

Fig. 4.6 Installation method for the ground velocity sensor and theoretical detection variances.

Table 4.3 Performance comparison of respective methods. XX = very bad; X = bad; √ = good; √√ = very good

	Deviation by wheel racing	Deviation by vibration	Cost
Wheel rotation	XX	√√	√√
Spatial filter	√	√	X
Radio doppler	√	X	XX
Supersonic doppler	√	X	√

4.3.2 The principles of the spatial filter method

The basic construction of spatial filters is shown in Figure 4.7. The illumination shows a uniform distribution of light quantity on the object (ground surface); the reflexivity distribution and undulations of the ground surface produce bright and dark patterns. These bright and dark patterns are formed on the spatial filter (lattices composed of slit-arrays) through lens. The light that passed through the slit-array is detected by a photodiode (PD).

The changes in light intensity of bright and dark patterns are obtained when it passes through the spatial filter and repeats its stress dynamism, showing a sine-wave like variation as depicted in Figure 4.8. The frequency of this sine-wave signal is dependent on the Cycle P of the slit-array, also as the transit speed increases the bright and dark patterns move faster and the detected signal frequency increases. Inversely, when the transit speed decreases the detected signal frequency also decreases. In a

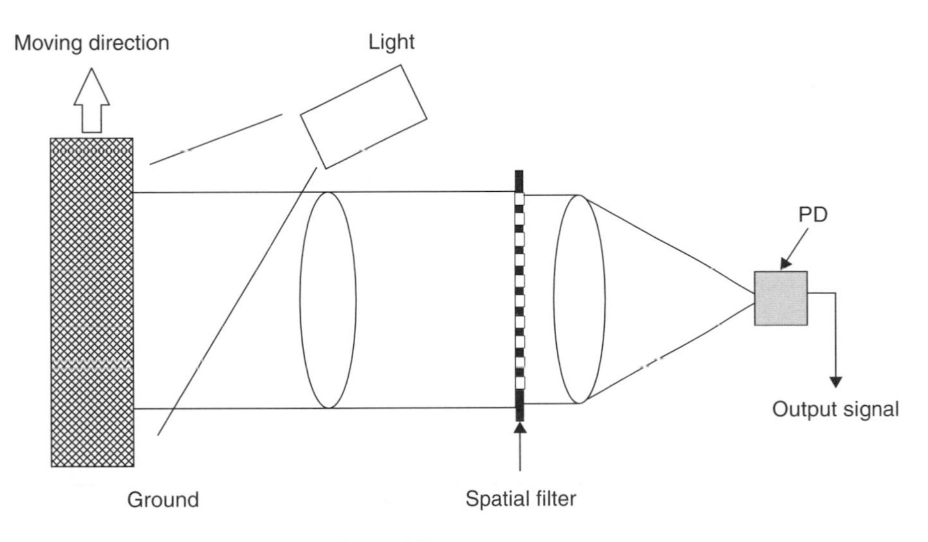

Fig. 4.7 The basic optical construction of spatial filters.

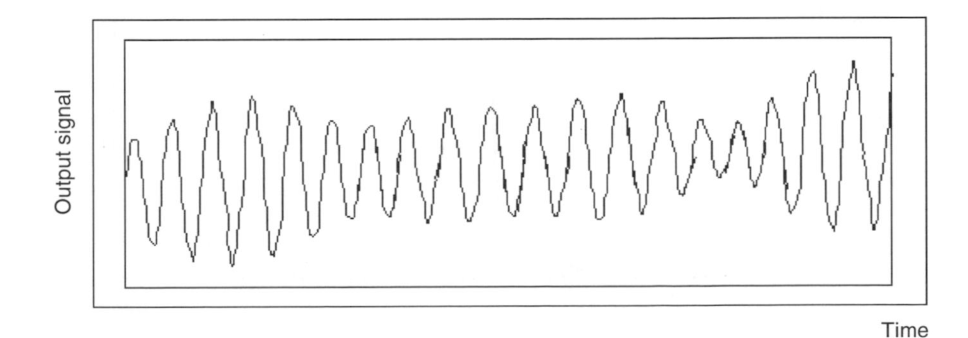

Fig. 4.8 Example of wave pattern form.

stationary state, there will be no variation and it will become a direct signal. The relationship between the detected signal frequency and transit speed becomes linear, as shown in Figure 4.9. The transit speed or ground velocity V is expressed in the equation: $V = P \times f$.

The Cycle P of the slit-array is determined when the spatial filter is designed; non-contact ground velocity can be detected by measuring the frequency of the signals detected by the spatial filter (Aizu and Asakura, 1987).

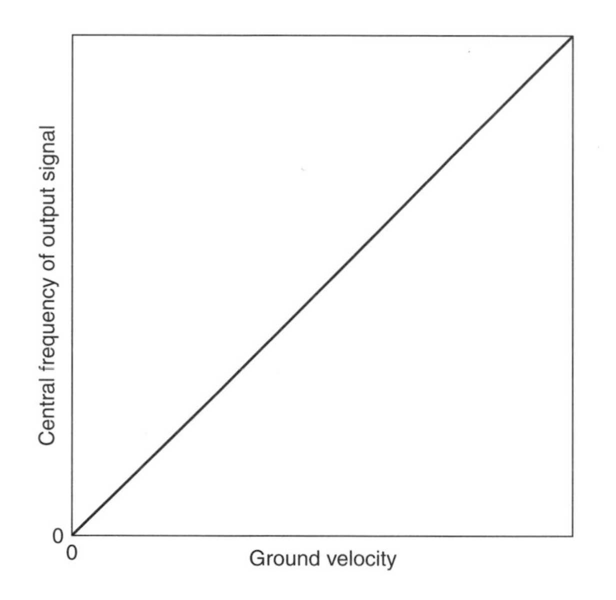

Fig. 4.9 Output property of ground velocity detection.

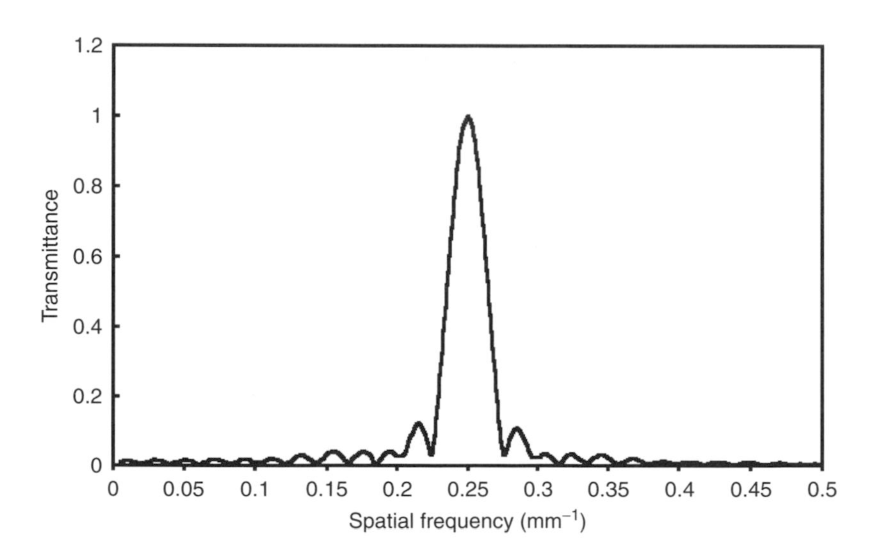

Fig. 4.10 Spatial filter transmission property.

4.3.3 Ground velocity sensors for vehicles

When applying the spatial filter principle to the ground velocity sensor for a vehicle, detection performance must be free from the influence of the operating environment. It is especially important that it is not influenced by sunlight, an external disturbance. In order to stay clear from the influence of external disturbing light, it is recommended that the spatial filter is constructed in such a way as to cut off the low-frequency domain. By using two photodiodes and amplifying the alternate output, it is possible to design the spatial filter in such a way as to cut off the low-frequency domain (Naito *et al.*, 1968). The structure of the spatial filter designed for ground velocity sensors for vehicle use and its transmission frequency properties are shown in Figure 4.10. As the centre of spatial frequency is $0.25\,\text{mm}^{-1}$, this filter detects the 4 mm pitch pattern on the object. Surfaces from asphalt to gravel road have actually been measured, and as a result, spatial frequencies that enable detection against any road surface are being developed. Also, due to the alternate operation structure and the property that does not allow the low-frequency domain of spatial frequency to pass through, external disturbance from uniform light such as sunlight is eliminated.

In velocity detection, the achievement of measurement precision is important. Since velocity deviation ΔV can be expressed as $\Delta V = \Delta F \times P$, the pitch P of the spatial

Fig. 4.11 Velocimeter exterior photograph.

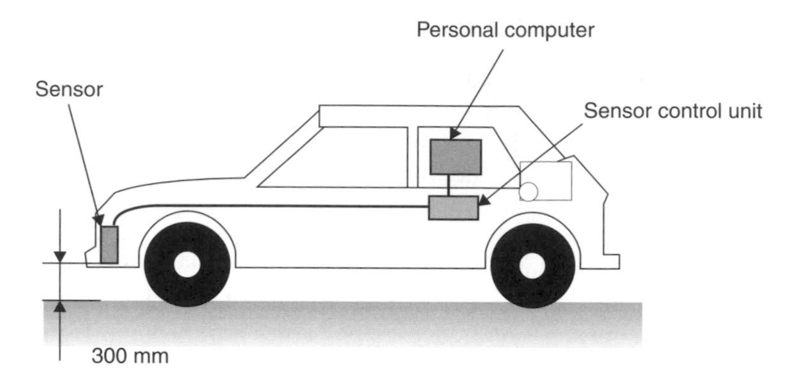

Fig. 4.12 Installation position of the sensor.

Table 4.4 Velocimeter specification

Description	Specification
Detectable velocity range	0.4–400 km/h
Velocity measurement precision	±1%
Working distance	300–500 mm
Wave length of emitted LED	850 nm
External sensor dimensions	200 × 260 × 143 mm

filter should be made smaller or measurement deviation ΔF should be made smaller to increase velocity detection accuracy. It is possible to make ΔF smaller by increasing the number of slits that make up the spatial filter. An exterior example of a velocimeter is shown in Figure 4.11 and the installation position of the sensor is shown in Figure 4.12. The major specifications are shown in Table 4.4.

4.4 Road surface recognition sensor

For the sake of traffic safety, it is important to know road surface conditions during the winter time. This because of the coefficient of sliding friction changes in accordance with the road surface conditions.

Measurement of surface reflexibility is the generally accepted method for optical detection of the road surface condition. In this method, the surface condition is recognized through the utilization of the known fact that the diffused reflection factor dominates in a dry climate and the reverberating reflection factor dominates in a damp climate. However, in this method, the influence of vehicle body vibration becomes quite substantial. Recently, new measurement methods utilizing microwaves (Rudolf *et al.*, 1997) and spatial filter (Shinmoto *et al.*, 1997) sensors were suggested as methods that are relatively resistant to vibration. We would like to introduce the one method that measures reflexibility (Section 4.4.1) and another one that utilizes spatial filters (Section 4.4.2).

4.4.1 Measuring reflexibility

In the electromagnetic theory, the mirror reflexibility of a plane wave is expressed by the equation:

$$\rho_h(\theta) = \left| \frac{\cos\theta - \sqrt{n^2 - \sin^2\theta}}{\cos\theta + \sqrt{n^2 - \sin^2\theta}} \right|^2$$

$$\rho_v(\theta) = \left| \frac{n^2\cos\theta - \sqrt{n^2 - \sin^2\theta}}{n^2\cos\theta + \sqrt{n^2 - \sin^2\theta}} \right|^2$$

where: θ = Light incident radius
n = Relative refractive index
$\rho_h(\theta)$ = Parallel polarization factor of reflection ρ
$\rho_v(\theta)$ = Perpendicular polarization factor of reflection ρ

$\rho_v(\theta) = 0$ is the equation in this instance and there exists a light incident radius θ_b, which is generally referred to as the Brewster radius.

On the other hand, Polarization P that indicates the polarization characteristic of the reflection in terms of quantity is defined by the equation:

$$P(\theta) = \left| \frac{\rho_h(\theta) - \rho_h(\theta)}{\rho_h(\theta) + \rho_h(\theta)} \right|$$

When the light incident radius is that of the Brewster radius and when the reflecting surface is mirror-like, then the equation is $P(\theta_b) = 1$. When the reflecting surface is that of the perfect diffused reflection then, regardless of light incident radius, the equation is expressed as $\rho_v = \rho_h$, consequently, $P(\theta) = 0$. As shown in Figure 4.13, since a dry surface produces diffused reflection, the equation becomes $P = 0$. Since the damp surface generates mirror reflection, the equation therefore will be $P = 1$. In other words, by measuring the surface polarization P it becomes possible to measure whether the reflection is generated by that of mirror reflection in a damp climate or not. By measuring the temperature at the same time, it is possible to determine whether the

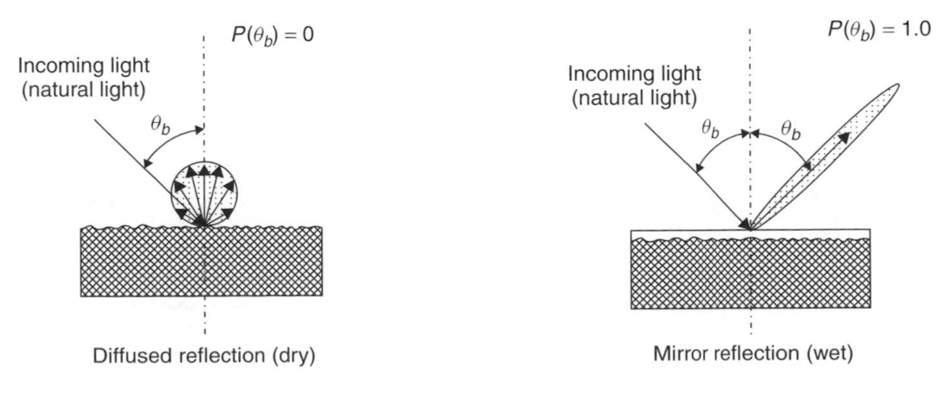

Fig. 4.13 Diffused reflection and mirror reflection.

surface is frozen or wet and damp. It is possible to construct a sensor in a relatively simple way, but it is not possible to fix the light incident radius at that of the Brewster radius when vehicle body vibrates. In such a case, a mechanism that compensates vibration to obtain stable operation becomes necessary.

4.4.2 Surface recognition by the spatial filter method

It is also possible to capture the road surface condition by means of spatial filters as introduced in Section 4.3–Non-contact ground velocity detecting sensor.

For explanation purposes, the measured results of representative road surfaces captured through the spatial filters are illustrated in Figure 4.14 and Figure 4.15. Figure 4.14 is signal detected from dry asphalt and is detected as an arc sine-like wave signal corresponding to the velocity. Figure 4.15 is that of the snow-covered surface and an arc sine-like wave signal compounded by low-frequency fluctuation is detectable. These two situations show that the spatial frequencies of the reflecting objects (road surface) comprising the road surface have different distribution characteristics. The differences are attributable to the varying sizes of particles, brightness, surface irregularities and undulation. In order to analyse such output signals, the wave form is converted to the spatial frequency distributions by using FFT (Fast Fourier Transform). The spatial frequency distributions of dry asphalt and snow are shown in Figure 4.16. As is clear from Figure 4.16, we can observe marked differences in

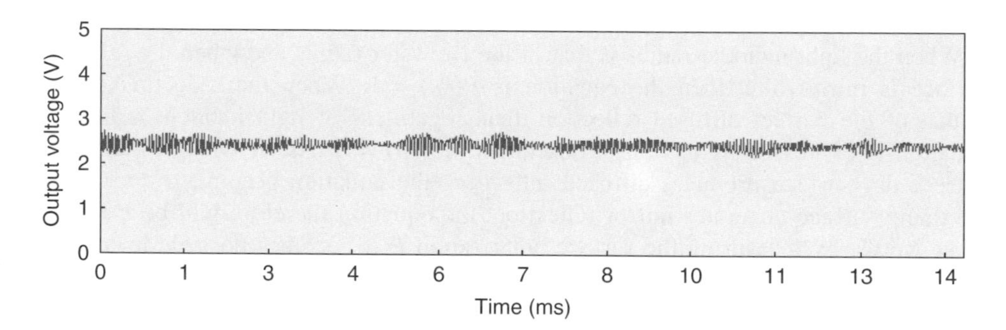

Fig. 4.14 Output signal of dry asphalt.

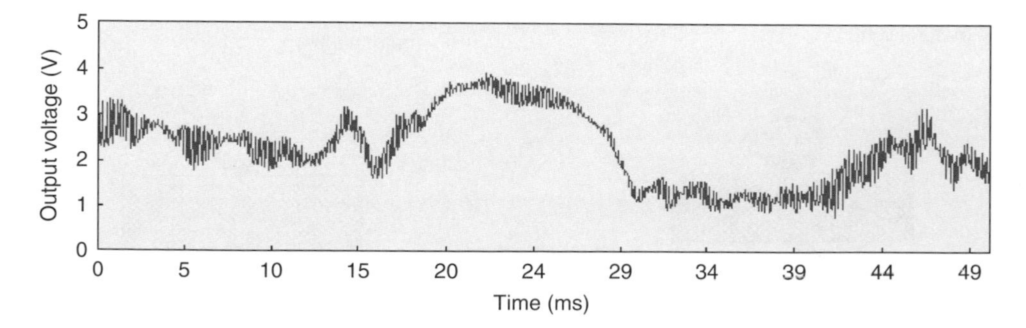

Fig. 4.15 Output signal of snow.

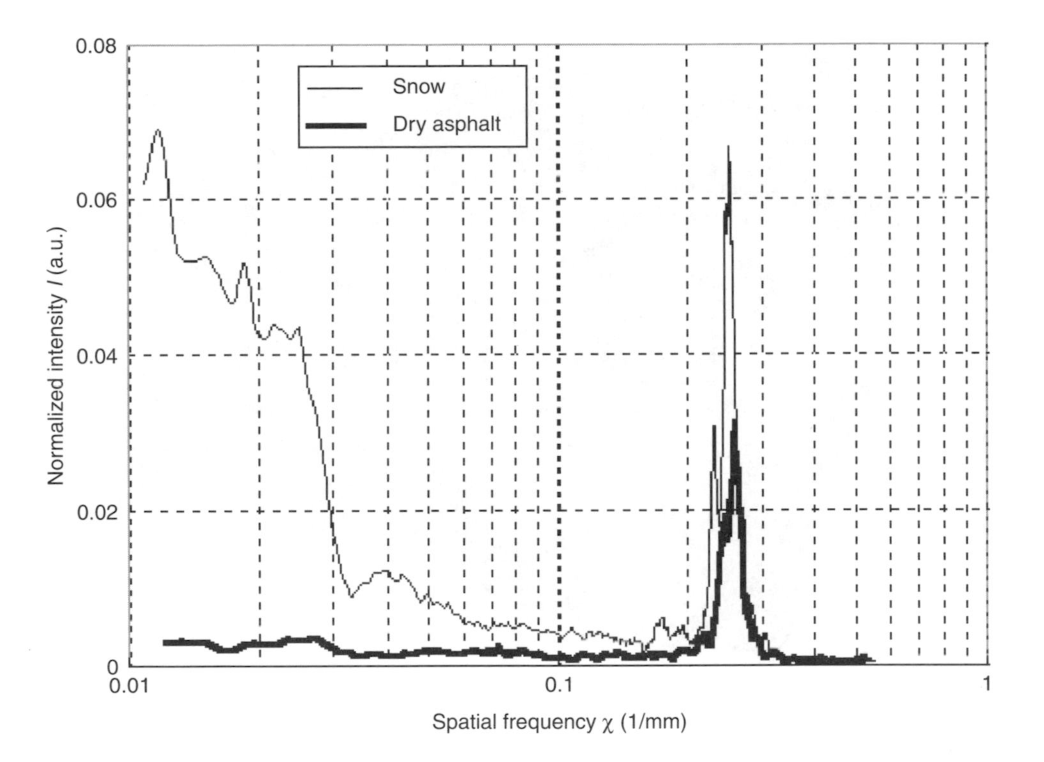

Fig. 4.16 Spatial frequency distribution.

low-frequency factors. As described above, by analysing spatial frequency and by paying attention to the differences in the distribution, the road surface condition can be captured.

4.4.3 GVS (ground view sensor)

We would like to introduce the GVS (ground view sensor) developed by OMRON (Shinmoto *et al.*, 1997). GVS is a road surface distinction sensor that utilizes a spatial filter. Figure 4.17 shows the external photograph of the GVS and the major specifications are listed in Table 4.5.

In order to recognize road surface conditions using spatial filters, it is necessary to convert road surface data obtained as a function of time into a function of space; stable velocity detection is therefore a necessity. As to detect velocity, a stable detection is possible by utilizing diffused reflection light, which has low brightness fluctuation. For instance, if reverberating reflecting light is used, the quantity of reflected light from a water-covered road surface becomes so large the light receiving photodiode gets saturated and it becomes impossible to correctly detect the velocity.

On the other hand, the naked human eye tends to look for surface brightness when trying to recognize the road surface condition. This means that the recognition is done through the use of reverberating reflection light. The same principle applies to sensors. A more effective road surface recognition becomes possible by utilizing

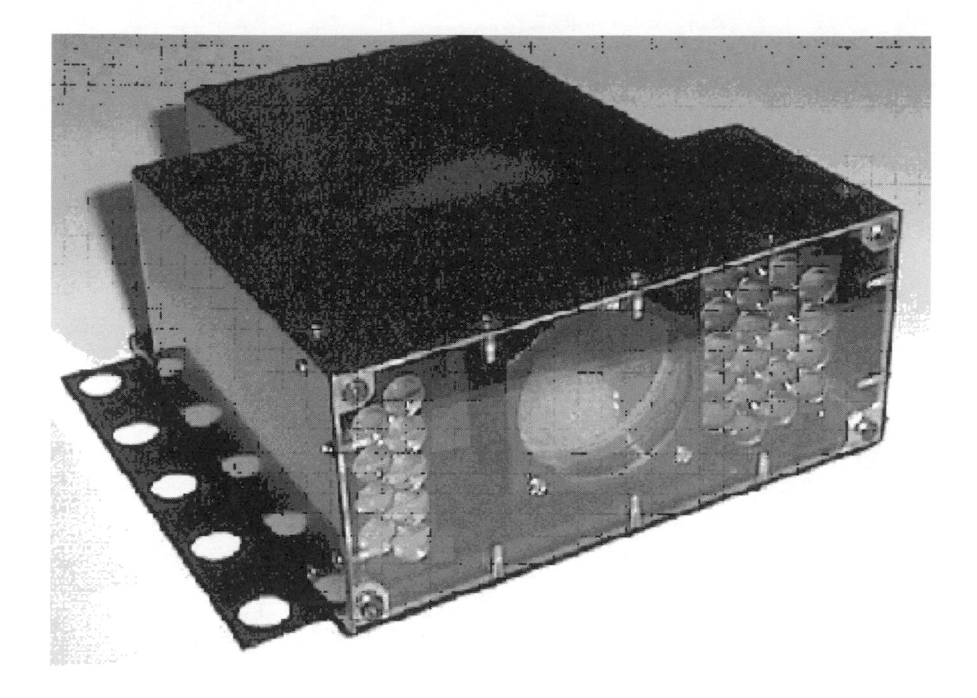

Fig. 4.17 Exterior photograph of GVS.

Table 4.5 GVS major specifications

No.	Description	Specification	Remarks
1	Types of surface recognition	6 types	Dry, Wet, New snow, Trampled snow, Ice, Black ice
2	Recognition ratio	Over 90%	
3	Response time	0.06 S	
4	Recognition accommodating velocity	5 to 100 km/h	

reverberating light refection with larger quantity of fluctuation instead of using diffused light reflection.

From what has been described so far, to achieve a stable condition for sensor to operation, it is most adequate to use diffused light for velocity detection and reverberating light reflection for road surface recognition. For this purpose, as shown in Figure 4.18, the GVS employs a structure that simultaneously detects both kinds of reflected light. A uniform road surface illumination is done by 20 infrared LED units (LED 1) to acquire diffused reflection and 9 infrared LED units (LED 2) to acquire reverberating reflection. The respective reflected light passes through spatial filters composed of slit-allay and prism-allay and is captured by two photo-diodes (PD). To obtain the selected spatial frequency of spatial filters to be 0.25 mm^{-1} on the surface, the space between image magnification of receiver lens and slit-allay has been adjusted. With this arrangement, a pattern with 4 mm frequency on road surface is detectable.

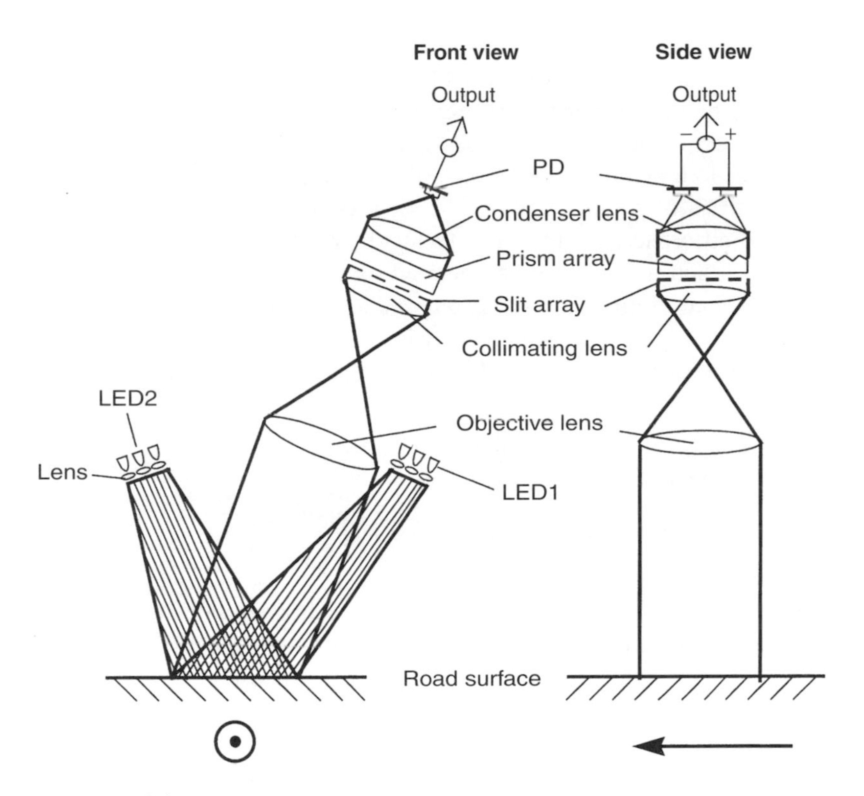

Fig. 4.18 GVS optical design.

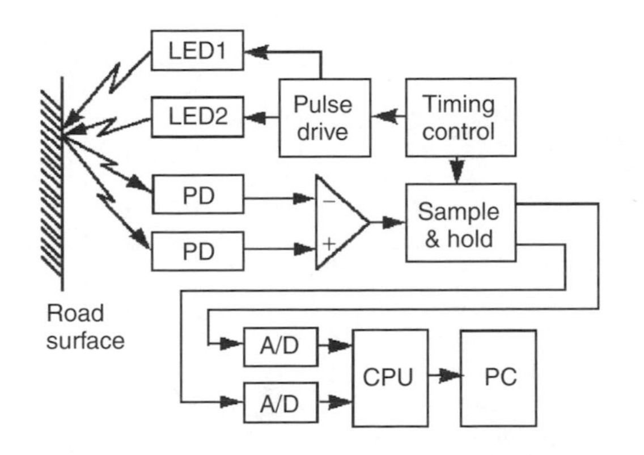

Fig. 4.19 GVS block diagram.

The schematic block diagram of this sensor is shown in Figure 4.19. Against one receiver unit, instead of alternately emitting diffused reflection light (LED 1) and reverberating reflection light (LED 2) and regenerating the analogue signal respectively in synchronization, it acquires the analogue signals of both reverberating and diffused reflection light from the road surface. The two analogue signals are converted into a digital signal and then transmitted to the CPU. CPU performs Fast Fourier Transform (FFT), and the road surface condition is recognized by spatial frequency analysis. FFT analysis is performed in real time using data from the last 10 m of road surface, and new output is provided every 1 m of vehicle movement.

As mentioned before, through a frequency analysis (FFT) of the wave patterns of reverberating reflection in Figure 4.14 and Figure 4.15, the results shown in Figure 4.16 are obtained. The longitudinal axis is standardized making the peak value equal 1 when a frequency analysis of a sine wave of 1 V is performed.

To clarify the differences of frequency factors by road surface, the two characteristic indices of Index 1 and Index 2 classify the road surface. The sum total of spatial filter's low-frequency domain signal intensity is referred to as Index 1. The sum total of spatial filter's central-frequency area signal intensity is referred to as Index 2. The computation equations for low-frequency factor and central-frequency factor are as follows:

$$\text{Low-frequency factor} = \int_{0}^{0.125} I(\chi)\,\mathrm{d}x$$

$$\text{Central-frequency factor} = \int_{0.125}^{0.375} I(\chi)\,\mathrm{d}x$$

The results of actual road travel and plotting in accordance with the two indices are shown in Figure 4.20. The intensity from dry asphalt is normalized to 1. After having travelled on various road conditions, it has been confirmed that classifications of dry surface, damp surface and winter road surface (new snow, trampled snow, ice and black ice) are possible. Road distinction parameters derived from Figure 4.20 are shown in Table 4.6. The results of the field test using these two indices show that the ratio of

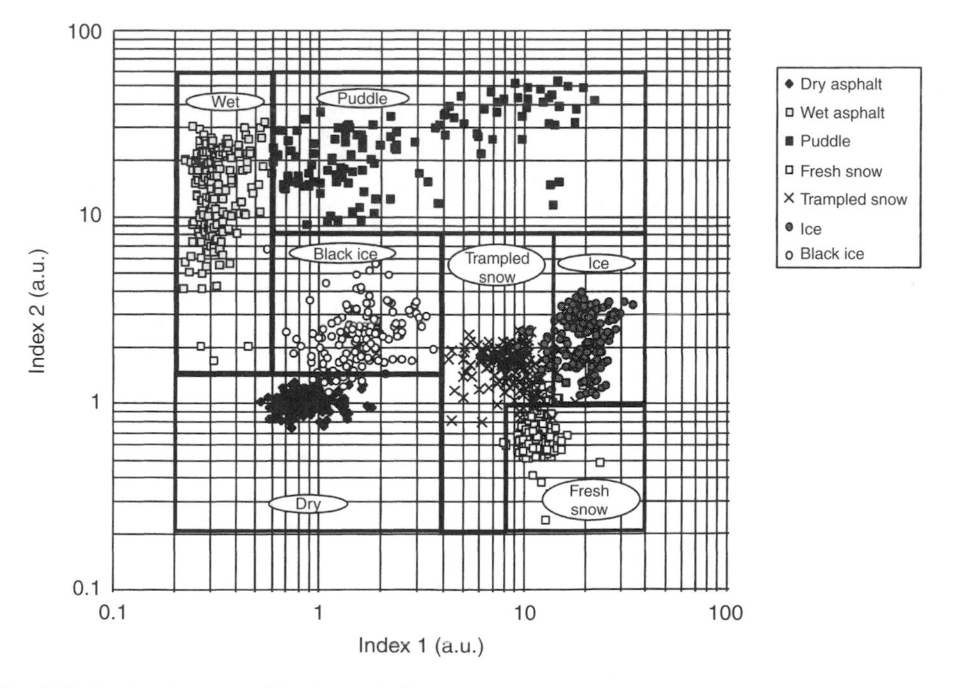

Fig. 4.20 Road surface recognition by two indices.

Table 4.6 Road surface distinction parameters

Surface condition		Index 1	Index 2
Dry surface		0.2–0.4	0.2–1.3
Damp surface	Wet	0.2–0.6	1.3–60
	Water	0.6–40	8.0–60
Winter surface	Fresh snow	8.0–40	0.2–1.0
	Trampled snow	4.0–8.0 8.0–12	0.2–8.0 1.0–8.0
	Ice	12–40	1.0–8.0
	Black ice	0.6–4.0	1.8–8.0

Fig. 4.21 Photograph of the installed sensor.

road recognition was over 90 per cent. The sensor is installed in front of the vehicle and the photograph of the installed sensor is shown in Figure 4.21.

4.5 Vehicle sensors for ETC systems

Currently, development work on electronic toll collection (ETC) systems are well underway (Iida *et al.*, 1995; Yasui *et al.*, 1995).

It is absolutely necessary for these systems to be equipped with devices that separately detect vehicle types for toll collection. It is understandable that the most difficult job of sensors in such systems is to distinguish between two separate vehicles and a

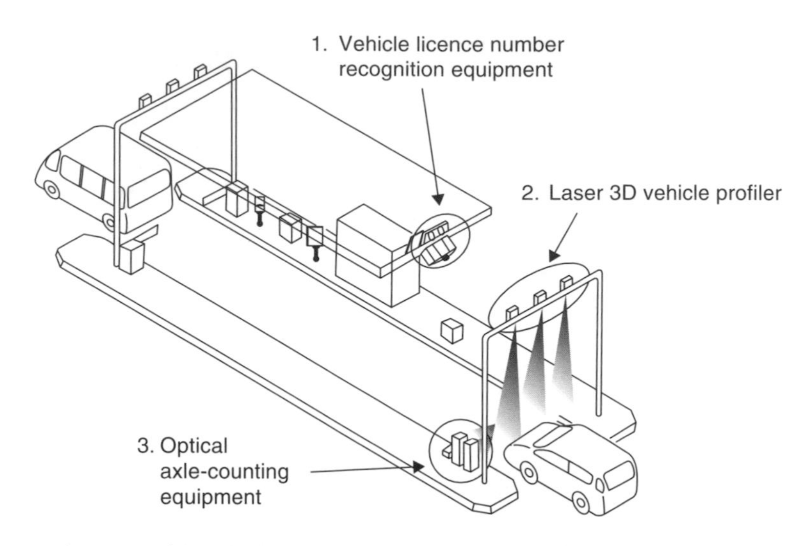

Fig. 4.22 Configuration of the vehicle classification system.

trailer unit where one carriage is pulled by a trailer head. Conventional vehicle-contact-detection sensor models like axle weight sensors and inductive loop vehicle sensors cannot carry out this particular job (Tyburski, 1989).

Optical sensors are currently preferred for the above distinction. Furthermore, by narrowing the light beam, it is possible to achieve higher resolution, which allows detection of a trailer pole with a small diameter such as $\phi 50$ mm.

On the other hand, vehicle classification systems are also very important for ETC in order to classify vehicles in accordance with the traffic fee table and to prevent dishonest and improper entry into highways. In order to realize the said classification, we have devised an instrument that utilizes the three following non-contact type sensor elements, which are also shown in Figure 4.22 (Ueda *et al.*, 1996):

1. Vehicle licence number recognition equipment using a CCD camera.
2. Laser 3D vehicle profiler.
3. Optical axle-counting equipment.

The system that includes these devices was designed so as effectively to detect the characteristics of vehicles travelling at 60 km/h or more.

The compact vehicle sensor and the axle-counting sensor are described in Sections 4.5.1 and 4.5.2.

4.5.1 Compact vehicle sensor

In order to realize ETC, it is important to distinguish between two tailgating vehicles and a trailer. However, this is considered to be the most difficult task and situation for vehicle detection and separation sensors (Figure 4.23).

Generally speaking, optical sensor methods like the opposed method (*Traffic Technology International*, 1996) and diffused method (Hussain *et al.*, 1993; Mita and Imazu, 1995; Naito *et al.*, 1995; Gustavson and Gribben, 1996; Kreeger and

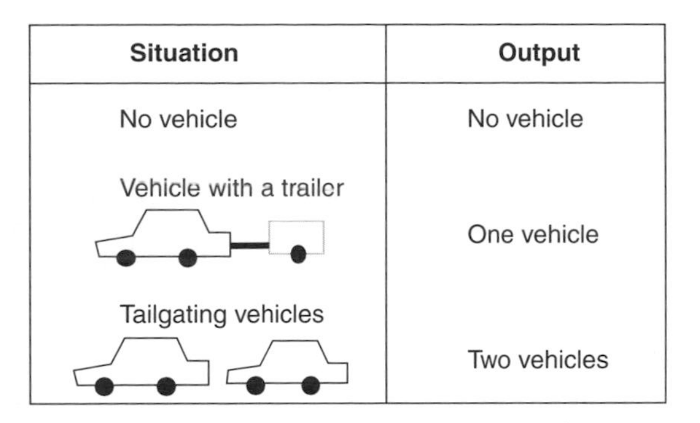

Situation	Output
No vehicle	No vehicle
Vehicle with a trailer	One vehicle
Tailgating vehicles	Two vehicles

Fig. 4.23 Function of vehicle sensor for ETC system.

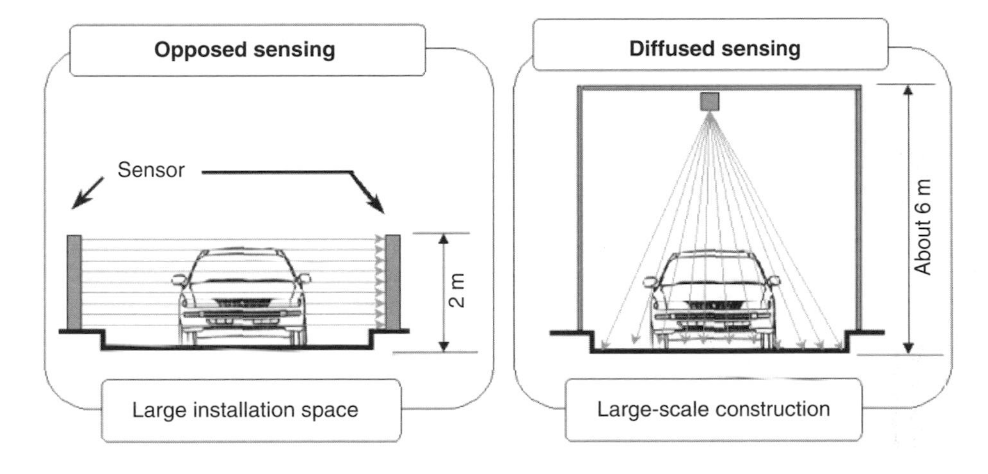

Fig. 4.24 Optical vehicle sensor.

McConnell, 1996) are currently preferred to make said distinction (Figure 4.24). However, since the opposed mode requires a pair of large towers mounted with a lot of light emitters and receivers, certain restrictions are imposed on the installation at the tollgate. Furthermore, in order to detect a trailer pole up to the smallest diameter of ϕ 50 mm, it is necessary to make a narrow light beam so that the corresponding spatial resolution can be increased. As a result, the number of light emitters and receivers becomes extremely large.

In the case of the diffused sensing method, a large construction such as a gantry is required. In addition, the reflectivity of the vehicle body surface dramatically influences the accuracy of the vehicle detection.

In order to overcome these problems, a compact vehicle detection and separation sensor that is easily installed and detects small objects has been developed for tollgate application. This sensor system consists of a sensor and a retroreflector and employs a retroreflective optical system. The sensor and the retroreflector are positioned opposite to each other on the roadside, as shown in Figure 4.25 and Figure 4.26.

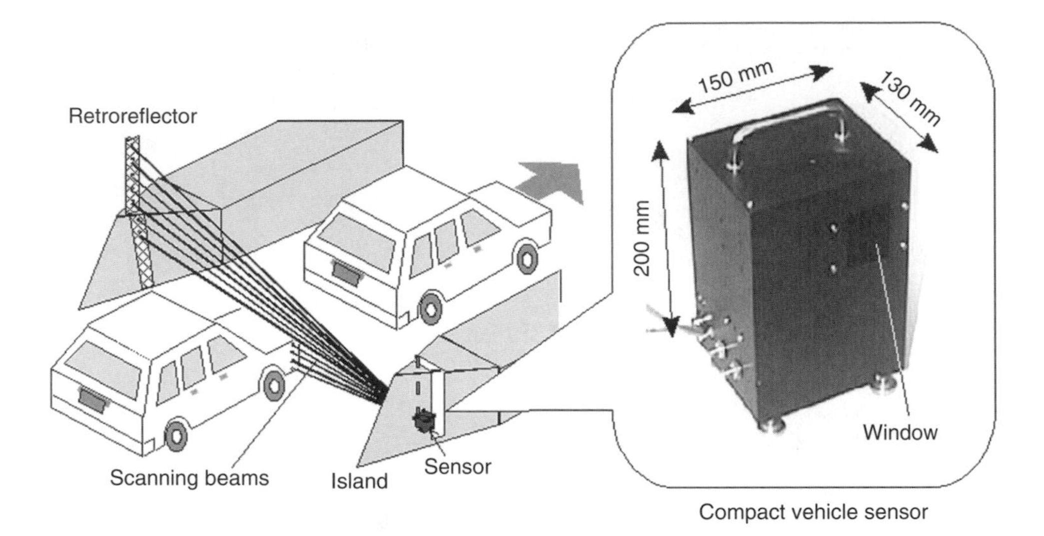

Fig. 4.25 Conceptual image of compact vehicle sensor.

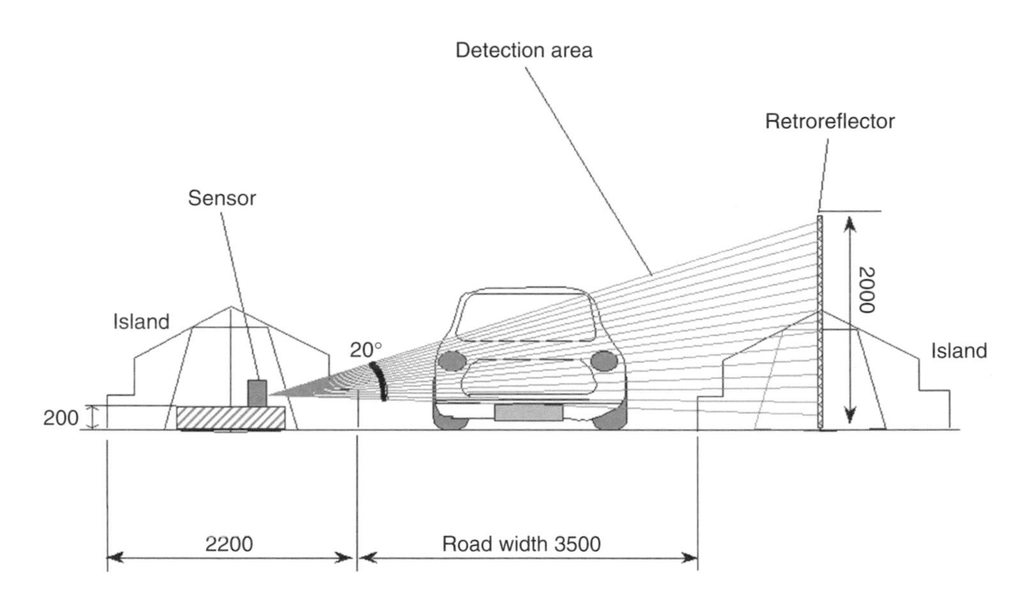

Fig. 4.26 Detection area (all units in mm).

Figure 4.27 shows the operation and optical design of the compact vehicle sensor. The sensor is coupled with an optical scanner and a galvanometer scanner emits a scanning beam toward the retroreflector at 56 cycles/sec. The emitter propagates a scanning light beam and it is transmitted to the receiver through the retroreflector. In order to block reflected light from a glossy object such as a shiny vehicle body, two polarizing filters are placed in front of both the emitter and receiver lens and their polarizing direction is rotated 90° relative to each other (Garwood, 1995). Since the retroreflector rotates the light beam's direction of polarization by 90°, the receiver is

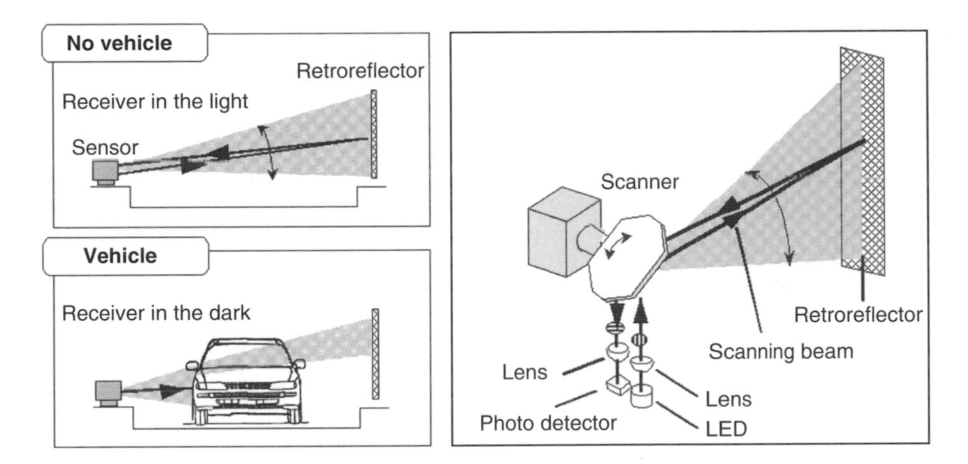

Fig. 4.27 Operation and optical design of the compact vehicle sensor.

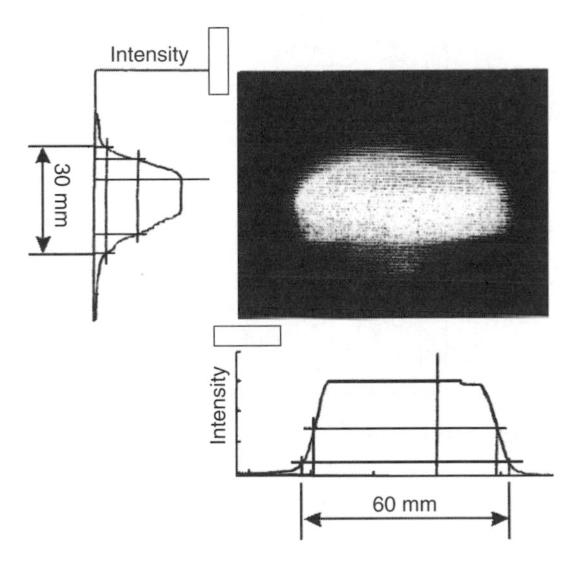

Fig. 4.28 Beam profile.

able to detect the light beam through the polarizing filter. On the other hand, when a vehicle obstructs the beam and the vehicle body reflects the beam, the reflected light's direction of polarization is not rotated and, henceforth, the receiver does not detect the light. Consequently, the presence of a vehicle can be detected when a vehicle obstructs the beam (Figure 4.27).

In order to detect a trailer pole with a minimum of ϕ50 mm, it is necessary to emit a narrow light beam. On the contrary, a wide light beam is required for preventing beam disruptions caused by interference such as snowflakes and raindrops. Therefore, the cross-sectional profile of the light beam is rectangular as shown in Figure 4.28. Laser diodes are conventionally used as light sources for vehicle detection above without Kreeger and McConnell. Because the sensor emits the light beam directly at the vehicle

Table 4.7 Light beam attenuation characteristics

Environment	Attenuation
Rain (rainfall 300 mm/h)	0.19 dB
Snow (snowfall 30 mm/h)	0.9 dB
Fog, smog (visual range 50 m)	2.6 dB
Dirty window (the worst sample)	3.0 dB

Table 4.8 Compact vehicle sensor specifications

Item	Specification
Minimum detectable object	50 mm diameter
Tailgating vehicles	300 mm tail to nose (at 60 km/h)
Maximum lane width	6500 mm

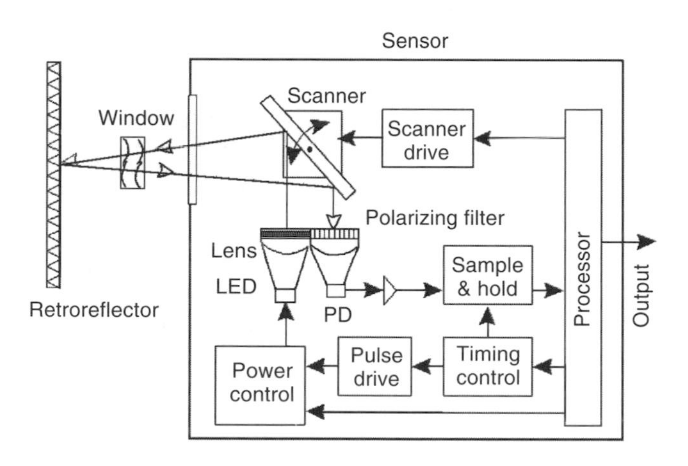

Fig. 4.29 Compact vehicle sensor block diagram.

from the roadside, passengers in the vehicle could be exposed to the beam. In order to protect passengers' eyes from the beam, a light-emitting diode (LED) is preferred as the light source.

Light beam attenuation properties in propagation through the atmosphere and a dirty window are estimated in the Morita report (Morita and Yoshida, 1969) and the results of our experiments are shown in Table 4.7. With consideration to these attenuations, the optical power system was designed to maintain an adequate signal-to-noise ratio in any environment.

The specifications and the block diagram of the newly developed sensor are shown in Table 4.8 and Figure 4.29.

4.5.2 Optical axle-counting sensor

A laser-scanning range finder is a device (Gustavson and Gribben, 1996; Chpelle and Gills, 1995; Mita and Imazu, 1995; Rioux, 1984) used to actualize some non-contact vehicle detecting systems. As an example of a system that adopted this technique, I would like to introduce the optical axle-counting sensor.

The sensor is placed on the roadside, and irradiates a laser beam on the road surface. This laser beam is collimated by a lens, scanned across the lane by a polygon scanner, and produces numerous laser beam illuminated spots on the road surface. By using these illuminated spots, the sensor detects wheels on the road and counts the vehicle axles. Figure 4.30 shows an example of the illuminated spots on the road surface or vehicle.

When a vehicle enters into the laser-scanned area, the sensor irradiates a laser on the side of a vehicle and detects the range image. Figure 4.31 shows three example range images obtained by one scan. As you can see from the figure, the pattern of the range image can be classified because the pattern make-up depends on the part of vehicle body scanned.

In Figure 4.31, the dots indicate laser-illuminated spots. Because the road surface is even, the range image obtained by gathering these dots is naturally flat. In the case of a wheel on the road, the range images will be a combination of road surface and a wheel, where it contacts the road at a right angle. In the case of a vehicle body, the range image is also perpendicular to the road. However, since the scanning is done only between the road surface and the bottom of the vehicle, there is nothing above the road in the range image. Therefore, the wheel model that is capable of making

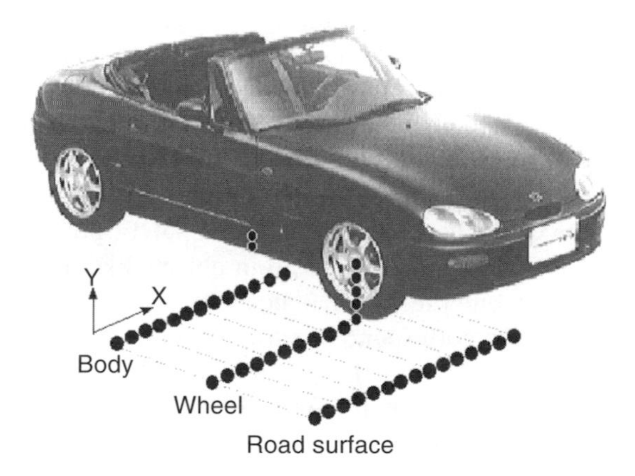

Fig. 4.30 Example of the laser illuminated spots.

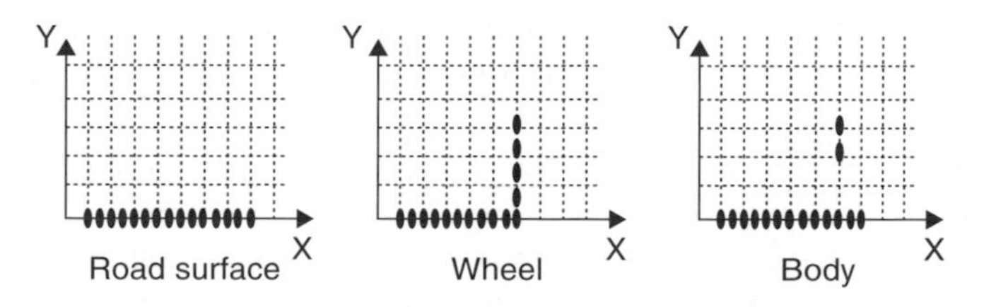

Fig. 4.31 Example of the imaging range.

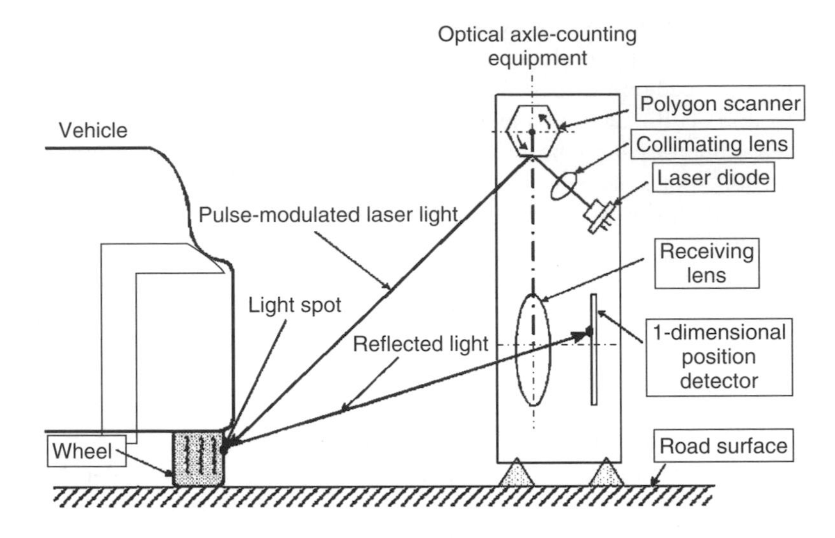

Fig. 4.32 Schematic diagram of the optical axle-counting sensor.

judgements in relation to the existence of contact between horizontal and vertical lines is regarded as an effective technique in making this distinction.

Using this range finder technique, this sensor detects range image data. Figure 4.32 shows the schematic diagram of the range finder. In order to get a fine spatial resolution, an optical triangulation range finder was chosen. In order to eliminate external factors such as sunlight, an infrared (830 nm) laser diode, which emits pulse-modulated light, was chosen as the light source. This laser is collimated and scanned across the lane by a polygon scanner. The scanning time is 6 msec. This scanner projects 150 laser illuminated points on the road surface which corresponds to maximum spacing of 25 mm. The output power of the beam is set below safety power output of 20 mW. The receiving lens focuses the reflected light from objects like the road surface, wheel and vehicle body onto an one-dimensional position detector. The output signal of the one-dimensional detector corresponds to the positions of light illuminated spots on the objects. The distance and direction of the illuminated spots on the object are based on the direction of the output emission from the scanner and receiver lens. The distance and position of the laser beam emitted on the object are based on the emission direction of the output beam as well as the centre of the receiver lens. Furthermore, the incoming direction of the reflected light is determined by the position of the light spot image on the position detecting diode. Henceforth, all the distances and directions are calculated by triangulation. The range image is produced by gathering the respective positional data obtained by one scan. The presence of an axle is determined by this range image.

Table 4.9 shows the specifications for the axle-counting equipment. Due to the narrow lane width and existence of the island as shown in Figure 4.12, the maximum vehicle speed is restricted to below 60 km/h. The detection area along the cross direction is determined to be 0.5–2.35 m and they are determined by the widths of both lane and vehicle.

Table 4.9 Optical axle-counting sensor specifications

Item	Specification
Function	Axle counting (axle detecting) Distinction of travelling direction
Vehicle speed	60 km/h (max.)
Processing time	6 ms
Detection area	0.5–2.35 m
External dimension	1200(H) × 300(W) × 450(D)

4.6 Conclusion

The rapid progress in electronics has been supporting the advancement of traffic systems.

In the case of vehicles, the advancement is actualized even in the basic automotive performances like running, turning and stopping through the combined use of sensors and actuators. Hereafter, it is expected that a more delicate and fine-tuned sensor technology will be required to answer the needs of further advancements.

Up to now, the scope of application and use of optical sensors has been limited in a traffic systems environment. However, as the demand for advanced sensing technology increases, their scope of application shall widen.

We have picked relatively known challenges for optical sensors in this chapter. We are of the opinion that the new and high-precision sensing technologies discussed in this chapter will be necessary for further progress in traffic systems from now on. We believe that the progress in ITS will realize a society in which the relations among people, roads and vehicles will be optimized.

References

Aizu, Y. and Asakura, T. (1987). Principles and development of spatial filtering velocimetry. *Applied Physics*, **B43**, 209–24.

Chpelle, W. and Gills, E. (1995). Sensing Automobile Occupant Position with Optical Triangulation. *Sensors*, December, pp. 18–22.

Garwood, R. (1995). Principles of photoelectric sensing: part I. *Sensors*, No. 7, p. 14.

Gustavson, R. and Gribben, T. (1996). Multi-lane Range-imaging Vehicle Sensor. *Proceedings of the 3rd World Congress on ITS*, Orlando, M18–3.

Hussain, T., Saadawi, T. and Ahmed, S. (1993). Overhead infrared vehicle sensor for traffic control. *ITE Journal*, **63**(9), 38.

Iida, A., Tanaka, H. and Iwata, T. (1995). Japanese electronic toll collection system research and development project. *Proceedings of the 2nd World Congress on ITS*, Yokohama, p. 1573.

Kreeger, K. and McConnell, R. (1996). Structural range image target matching for automated link travel time computation. *Proceedings 3rd World Congress on ITS*, Orlando, T29–1.

Mita, Y. and Imazu, K. (1995). Range-Measurement-Type Optical Vehicle Detector. *Proceedings of the 2nd World Congress on ITS*, Yokohama, pp. 146–51.

Morita, K. and Yoshida, F. (1996). Light wave attenuation characteristics in propagation through the atmosphere. *Kenkyu jitsuyouka houkoku*, **18**(5), 39 (in Japanese).

Naito, T., Nishida, H. and Ogata, S. (1996). Three-dimensional vehicle profile measurement with a pulsed laser scanning sensor. *Proceedings 3rd World Congress on ITS*, Orlando, M18–5.

Naito, M., Ohkami, Y. and Kobayashi, A. (1968). Non-contact speed measurement using spatial filter. *Journal of the Society of Instrumentation and Control Engineering*, **7**, 761–72 (in Japanese).

Rioux, M. (1984). Laser Range Finder Based on Synchronized scanners. *Applied Optics*, **23**(21), 3837–44.

Rudolf, H., Wanielik, G. and Sieber, A.J. (1997). Road condition recognition using microwaves. *Proceedings of the 1997 IEEE Conference on Intelligent Transportation Systems*, session 39–1.

Shinmoto, Y., Takagi, J., Egawa, K., Murata, Y. and Takeuchi, M. (1997). Road surface recognition sensor using an optical spatial filter. *Proceedings of the 1997 IEEE Conference on Intelligent Transportation Systems*, session 39–2.

Traffic Technology International, (1996). High-speed vehicle classification. *Traffic Technology International*, Aug/Sept, p. 107.

Tyburski, R. (1989). A review of road sensor technology for monitoring vehicle traffic. *ITE Journal*, **59**(8), 27.

Ueda, T., Ogata, S. and Yamashita, T. (1996). Application of a Laser Scanning Range Finder to the Extraction of Vehicle Characteristics. *Proceedings of the 3rd World Congress on ITS*, Orlando, M18–6.

Yasui, M., Noguchi, N. and Murakoshi, H. (1995). Electronic toll system on trial in Malaysia. *Proceedings of the 2nd World Congress on ITS*, Yokohama, p. 1470.

5

Towards intelligent automotive vision systems

Christoph Stiller
Robert Bosch GmbH, Germany

5.1 Introduction and motivation

One of the most fascinating capabilities that is common among intelligent beings is their seamless perception of the environment. Remarkably, the vast majority of higher animals as well as mankind possess similar senses – or, technically speaking, 'sensorial systems' – that analyse odouric, taste, haptic, acoustical and visual information. Clearly, biological sensing for the particular task of navigation is dominated by visual perception. Hence, it hardly comes as a surprise that intense research has been devoted to machine vision for guidance and control of vehicles over the past three decades.

We are currently witnessing market introduction of driver assistance functions into passenger cars. Most of these functions are based upon inertial sensors, i.e. sensors measuring the dynamics of the vehicle. Anti-lock braking systems (ABS) or the Electronic Stability Programme (ESP) are but few examples. Measurement and control of the vehicle dynamics by itself already yields remarkable improvements in vehicle stability. The capability to perceive and interact with the environment, however, opens up new perspectives towards somewhat 'intelligent' vehicle functions.

With the recent introduction of parking aids and adaptive cruise control (ACC), first remote sensors, namely ultra-sonar and radar, have been placed in the market. These form just the leading edge of an exciting evolution in vehicular remote sensing. Lidar and vision sensors offer their employment in the near future. A broad spectrum of novel vehicle functionalities can be foreseen for vehicles perceiving their environment. Substantial contributions to comfort, safety and traffic flow can be expected from vehicles with the ability of co-operative and traffic-adapted driving, e.g. Benz (1997).

The potential to increase safety through vehicle functions based on environmental sensing is illustrated in Figure 5.1 in the example of accident types within Germany. It is worth noting that a large proportion of accidents occasioning bodily harm can be associated with collisions involving other traffic participants such as cars, trucks, buses, pedestrians, bicycles and motorcycles. These accidents could be reduced substantially by timely automatic detection and appropriate action. Furthermore, about one third

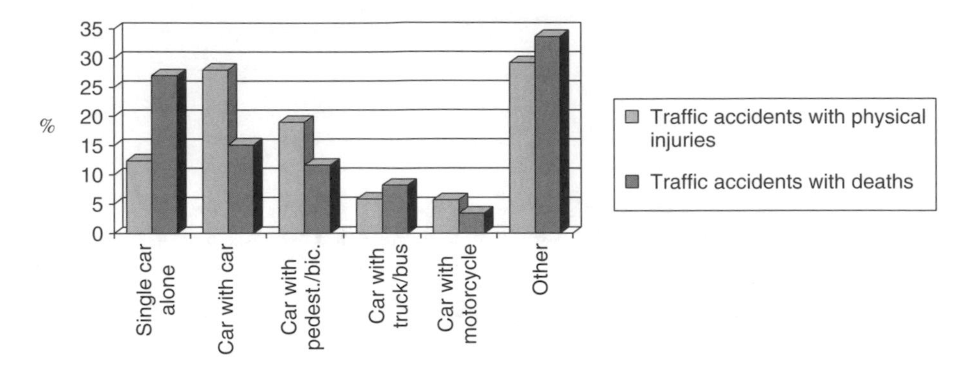

Fig. 5.1 Severe traffic accidents in Germany, 1998.
Source: StBA.

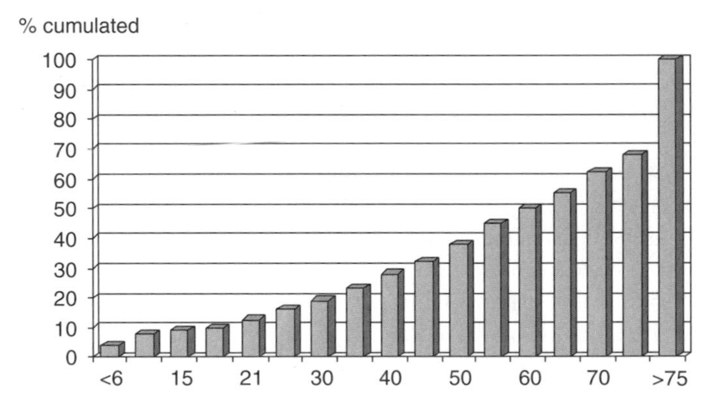

Fig. 5.2 Age of pedestrians killed in accidents with cars, Germany, 1990–1999.
Source: amtl. Stat., Germany.

of the single car accidents are associated to unintended lane departure and could be diminished by lane sensing and lane keeping support. The situation in other industrial countries shows hardly any qualitative difference, as pointed out by the German Statistisches Bundesamt (1998).

Although the total number of pedestrians killed has been reduced in the past few years, this kind of accident needs particular attention. Figure 5.2 depicts the age distribution of pedestrians killed by vehicle–pedestrian collisions. It can be seen that about 8 per cent are young children (up to 10 years) and 50 per cent of the killed pedestrians are at least 60 years old. At first glance, it may be astonishing to some readers that elderly people represent such a high proportion.

These statistics indicate the kind of information that is required from future automotive sensorial systems for the sake of safety enhancement. Since the appropriate action differs among the various collision partners, sensors should ideally not only detect other traffic participants and obstacles, but also provide information about their properties, such as type, speed and mass. These requirements are reflected in an overview on major properties for various principles for vehicle remote sensing in Table 5.1.

Table 5.1 Principles of automotive remote sensors

	Vision	Lidar	Radar	Ultra sonar
Wavelength (m)	$10^{-7}-10^{-6}$	10^{-6}	$10^{-3}-10^{-2}$	$10^{-3}-10^{-4}$
Weather dependency	Yes, visibility	Yes, visibility	Low	High, wind, rain
Detection of bad weather conditions	Yes	Yes	No	In some situations
Resolution:				
horizontal	High	High	Medium	None
vertical	High	Medium	None[1]	None[1]
Range × field of view product	High	High	Medium	Low
Instantaneous measurements:				
position	+	+	+	+
velocity	–	–	+[2]	+[2]
Functionality:				
obstacle detection	+	+	+	+
obstacle classification	+	+	+/–	+/–
pedestrian recognition	+	–	–	–
lane recognition	+	–	–	–
traffic sign recognition	+	–	–	–

[1] From a purely technical point of view, radar and sonar could provide high horizontal and vertical resolution, e.g. when employing a scanning antenna array. This may be realized in the medium or long term. However, the increase in hardware complexity is prohibitive for today's automotive applications.
[2] radial velocity.

As a fundamental property, vision and lidar sensors exhibit a degradation for bad visibility similar to the human visual system due to atmospheric attenuation. In contrast, radar performs almost constantly over all weather conditions. On the other hand, vision and lidar detect bad visibility conditions by themselves. This prevents driver assistance functions to operate in situations when the human is incapable of monitoring system behaviour. Sonar is highly dependent on weather conditions. Since air is employed as the transmission medium, its sensing capabilities may severely deteriorate in rain, wind and turbulence.

While sonar and radar only provide high spatial resolution at the cost of an array antenna or multiple sensors, video and lidar sensors offer high resolution in at least one spatial direction. Typically, vision sensors offer a 2D array of some 10^5-10^6 pixels. As a result, vision sensors offer a large range-field-of-view product $I \times \varphi$ of detection range I and angular field of view φ. In a first order approximation, these two factors can be traded against each other within a wide interval while maintaining the product constant by modification of the optics. Likewise lidar offers a large range-field-of-view product. Although radar offers a remarkably large viewing range that easily exceeds 120 m, the small field of view of today's radar sensors results in a medium range-field-of-view product. Likewise, the range of ultrasonic sensors restricts the range-field-of-view product for these sensors. Furthermore, the latter sensors hardly allow to trade range against field of view, because of the deterioration of the detrimental effects of the acoustical channel that have to be expected in automotive applications.

The major reason for the widespread employment of vision in both biology and technology, is the rich raw information represented by moving imagery. It not only allows for analysis of geometry, as is needed for object detection and tracking, but it also allows for classification and recognition of objects to a large degree. Hardly any other sensorial system is capable of recognizing some important information for automotive control such as traffic signs or lane boundaries.

The remainder of this chapter is organized as follows. Section 5.2 discusses applications of vision sensors for driver assistance systems. Section 5.3 provides an overview on the sensor hardware architecture and sketches the algorithmic operation of vision. Two practical implementations provide experimental insight in Section 5.4. First, a completely autonomous vehicle based on a multisensor concept is discussed. The second implementation is concerned with automatic and co-operative coupling of heavy-duty trucks. The chapter closes with a summary and concluding remarks.

5.2 Applications of vision in driver assistance systems

The demand for environmental sensing on one side and the perceptual capabilities of vision on the other side open up a broad spectrum of functions for vision sensors in the automotive area. The increasing diversity of potential functions may be subdivided into the following three classes:

- information and warning
- limited vehicle control
- autonomous functions.

An early function in the information and warning class is represented by a rearward parking aid. It encompasses a rear-view camera whose image is compensated for artifacts like lens distortion and enhanced by graphical overlays indicating distance, anticipated driving path, or the like. The signal will then be displayed on a monitor on the driver console. This information helps the driver to keep an overview on the rearward situation. Vehicle control is completely left with the driver. Other functions in this class include lane departure or collision warning, traffic sign assistance, rear view mirror substitution or warning in situations of driver drowsiness or impairment.

The second class comprises functions that directly influence vehicle dynamics but do not take over a complete vehicle control task, i.e. the driver keeps control of the vehicle and is able to intervene at any time. Clearly, the latter requirement necessitates a system behaviour that is transparent to the driver. As a concise example for such functions, adaptive light control adapts the headlight tilt and yaw angle keeping the beam stable on the road, and ascertaining proper illumination of curves, respectively. This function may later be extended to control of sophisticated headlights that allow an adapted light pattern distribution. An important function taking over longitudinal control to a limited extent is ACC (adaptive cruise control). It has recently been introduced into the market by three suppliers and car manufacturers as a comfort function. Accordingly, it passes control back to the driver in critical situations that necessitate a strong deceleration or when driving outside a particular speed interval. Future sensor enhancements will be employed to augment the operational range, in particular, towards operation in 'stop and go' traffic. Another interesting function in this class is lane-keeping support. This function controls lateral vehicle dynamics to a degree that is limited, e.g. by a maximum steering momentum or steering angle.

The third class of functions involves complete automatic control of the vehicle. At the extreme end of this class ranges autonomous driving which may be viewed as the most complete driver assistance system. Although several research groups have successfully developed experimental vehicles that demonstrate autonomous driving, e.g. Broggi *et al.* (1999), Dickmanns and Christians (1989), Franke *et al.* (1998), Maurer and Dickmanns (1997), Nagel *et al.* (1995), Shladover (1997), Stiller *et al.* (2000), Thorpe (1990), Weisser *et al.* (2000), these systems cannot cope with the arbitrary situations of today's traffic but are restricted to suitable environments and necessitate human supervision. The same holds true for collision avoidance systems that leave control to the driver but intervene just in time to avoid a threatening collision, e.g. Ameling and Kirchner (2000). Far beyond non-technical restrictions, like the current legal situation, the major source of this unavailability of autonomous functions stems from today's perceptual capabilities of the sensors, sensor data analysis and reliability issues.

Obviously, perceptual reliability forms a key requirement whose importance even increases from information and warning functions to autonomous functions. While information and warning functions tolerate a moderate frequency of sensing errors, limited vehicle control requires higher accuracy and reliability. Finally, responsible autonomous vehicle control necessitates sensing reliability that at least matches human capabilities. Hence, the information and warning driver assistance functions are technically feasible first. Those comfort functions with limited control will follow first that require only such information as allows machine perception with a tolerable reliability and accuracy. While these functions open the market for sensorial systems, additional functions follow with technological progress.

Figure 5.3 sketches a potential scenario of vision-based function realization. The broad spectrum of functions enabled by visual perception defines the information required from the sensor. At first a vehicle environment sensor will, in general, be uniquely associated to a particular function. In the sequel, the rich information of environmental sensors (vision and others) will be combined with information from other sources in a vehicle information platform that serves multiple functions in the field of information, warning, comfort and safety.

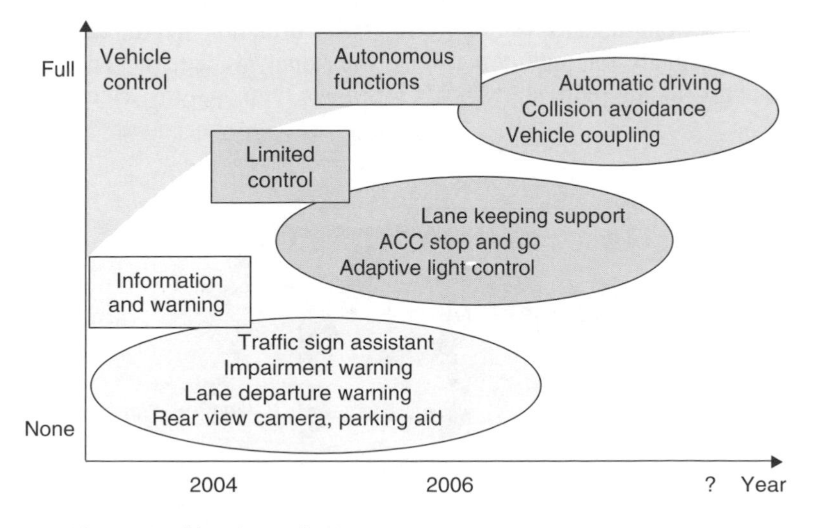

Fig. 5.3 Potential scenario of function realization.

5.3 Operating principles

5.3.1 Components of a vision sensor system

The major components of a vision sensor system are depicted in Figure 5.4. A suitable lens projects the incoming light pattern on to an imager that captures a sequence of digital images. As compared to imagers designed for the emerging market of digital consumer cameras, this device must meet some additional requirements. These basically result from the uncontrollable automotive environment. In particular, a wide luminance dynamic range, as provided by CMOS imagers with nonlinear luminance conversion, is expected to replace current CCD imagers. As a welcome by-product, image digitization (and possibly a controller) can be integrated on the same chip. Hence, a *system on chip* design that incorporates the major IC-components of the vision system as depicted in Figure 5.4 on a single chip comes within reach. In order to resolve fast movement, high temporal dynamics and a fast read-out are also required. Finally, the imagers must provide moderate noise even when operating at high temperatures up to 85°C.

The digital image sequence is passed to an evaluation unit that performs appropriate signal processing to extract the desired output. This evaluation unit may consist of components such as a processor, memory controller and interface, similar to existing high-end electronic controller units (ECUs) for automotive applications. It is, however, worth noting that the processed data flow as well as the computational load that has to be handled by a vision sensor ECU far exceeds the power of existing ECUs.

5.3.2 Sensor raw data analysis

The interrelation between sensor raw data and the information desired from a sensor system is governed by the physics of the raw data formation process. For most sensors a study of these physical principles naturally leads to algorithms for extraction of sensor output data from the sensor raw data. The major degree of freedom in design of these algorithms stems from model selection for raw data formation and the immanent noise. In contrast, algorithms for multidimensional data analysis are, in general, less well established. This finding particularly holds for image data analysis. Here, the situation

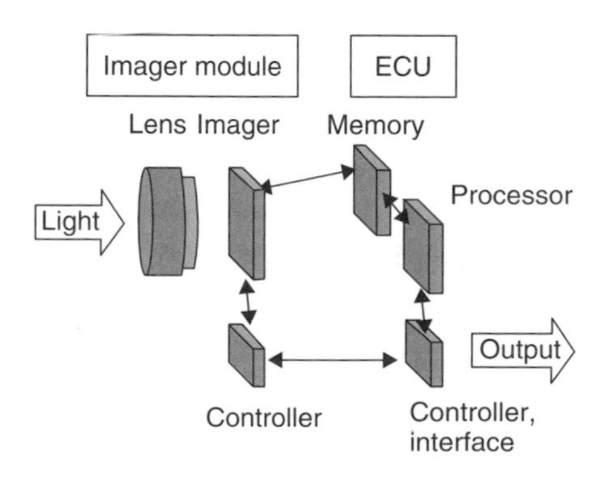

Fig. 5.4 Major components of a vision sensor system.

is additionally complicated by the image acquisition process that maps a complete ray of light from the 3D real world to a single pixel. Furthermore, the measured quantity, namely irradiance, is not directly related to the desired parameters, such as distance. Hence, it comes as no surprise that the inverse problem of reconstructing 3D parameters is often ill-posed in the absence of prior knowledge.[1] In conclusion, a key component in a vision system is its algorithm for image data analysis.

For the important task of 3D scene reconstruction, the algorithms that offer themselves may be categorized into two major classes. The first class is motivated by the finding that objects of interest in a traffic scene can often be associated with distinctive intensity patterns or features derived from those patterns in the 2D images. The algorithm design involves the definition of a set of objects of interest, such as road, vehicle or pedestrian, and the identification of patterns or features typical for each of these objects. For the example of the object 'vehicle', patterns and features chosen in the literature include bounding edges (Dickmanns and Christians, 1989), dark wheels and shadows below vehicles (Thomanek, 1996), as well as symmetry (Zielke *et al.*, 1992). With the above assumptions, object detection is reduced to detection of the pre-selected intensity patterns or features in the 2D images.

This scheme is illustrated by the example of identifying vehicles. Here, the lower edge of the shadow underneath is defined as a distinctive 2D feature. In order to avoid a number of undesired edges, the search may be restricted to a predetermined area of interest. It often exhibits triangular shape accounting for the expected width of vehicles that decreases with distance (Figure 5.5). An edge operator that amplifies horizontal edges from light to dark serves as a feature extraction filter. The lower edge of the vehicle is then readily detected by appropriate thresholding of the feature function. By applying additional features and plausibility checks and by stabilization through temporal filtering, good results have been reported for standard situations.

Obviously, problems remain in situations when the feature associated with an object is absent, e.g. when the vehicles lack a shadow underneath during low sun or night. Furthermore, the necessity of predefining the set of obstacles restricts these methods to some generic situations and kinds of objects. Since general visual features for arbitrary

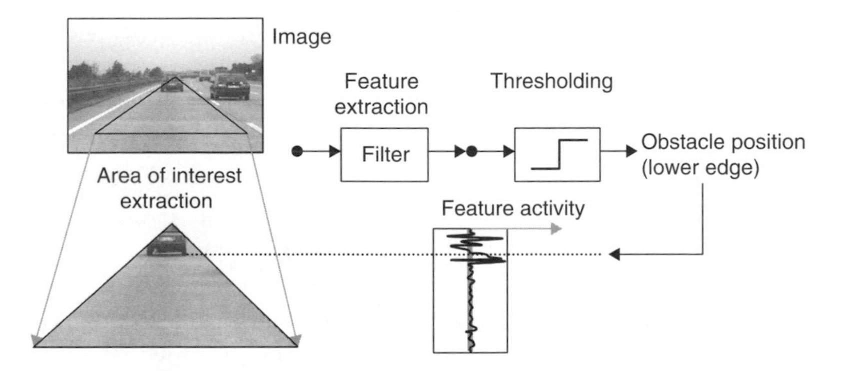

Fig. 5.5 Schematics of 2D feature extraction for obstacle detection.

[1] Following Hadamard's definition, a problem is well-posed if and only if its solution exists, is unique and varies continuously with the data.

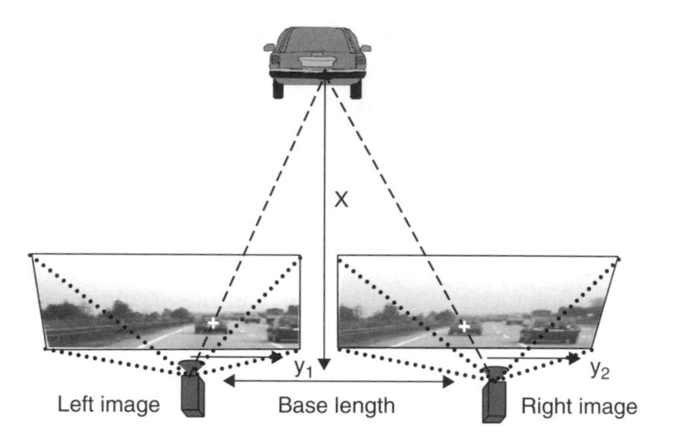

Fig. 5.6 Triangulation in stereo vision.

objects cannot be imposed, such approaches are only suited to detection of well-defined 2D objects, such as lane markers or traffic signs. Although perception of 3D objects may be improved and a valuable classification information may be added by 2D features, they cannot provide a fundamental basis for reliable automotive obstacle sensing.

Direct approaches towards 3D scene perception extract 3D features from disparity information obtained by a stereo camera. In a first step, a set of corresponding point pairs is identified, i.e. the positions in the left and right image that project identical points from the real world, see Figure 5.6. Most correspondence estimators are based on optimization of some similarity measure for small regions around the considered points. For a recent overview, the reader is referred to Stiller and Konrad (1999).

For each pair of corresponding points, the 3D real world position can be reconstructed by triangulation for known camera orientation, e.g. Faugeras (1995). Without loss of generality, we assume a calibrated and rectified camera pair in the sequel.[2] Then the epipolar constraint for corresponding points (x_1, y_1) and (x_2, y_2) in the left image g_1 and right image g_2, respectively, reads

$$y_1 = y_2, \tag{5.1}$$

i.e. corresponding points lie in the same row. A pair of corresponding points is thus defined through $\mathbf{k} = (x_1, y_1, x_2)^T$. The 3D world position $\mathbf{x} = (x, y, z)^T$ in a coordinate system attached to the left camera is given by

$$\begin{pmatrix} x \\ y \\ z \end{pmatrix} = \frac{b}{x_1 - x_2} \times \begin{pmatrix} x_1 \\ y_1 \\ z_1 \end{pmatrix}, \tag{5.2}$$

where b denotes the base length of the stereo camera. For small errors, the covariance of the point correspondence propagates to the world coordinates as

$$Cov(\hat{\mathbf{x}}) = J(\hat{\mathbf{k}}) \times Cov(\hat{\mathbf{k}}) \times J(\hat{\mathbf{k}})^T \tag{5.3}$$

[2] That is, two pinhole cameras with unity focal length and aligned retinal planes. The derivation can directly be transferred to any real stereo camera with known calibration.

with the Jacobian matrix

$$J(\mathbf{k}) = \frac{\partial \mathbf{x}}{\partial \mathbf{k}} = \begin{pmatrix} \frac{-zx_2}{x_1 - x_2} & 0 & \frac{x}{x_1 - x_2} \\ \frac{-y}{x_1 - x_2} & z & \frac{y}{x_1 - x_2} \\ \frac{-z}{x_1 - x_2} & 0 & \frac{z}{x_1 - x_2} \end{pmatrix} \tag{5.4}$$

In these equations, the superscript ^ denotes estimate of the respective quantity. They enable a rough estimate of the theoretical range of stereo vision. In our experiments a base length of 0.3 m and a focal length of 600 pixels was used. Assuming that at least displacements of 1 pixel can be detected, one can recognize obstacles up to a theoretical distance of 180 m when following Equation 5.2. Obviously, there are several artifacts not yet considered, such as non-ideal correction of lens distortion, which will deteriorate system performance. However, the above calculation shows that the order of magnitude of the range of stereo vision is reasonable for automotive applications.

Assuming rigid objects, a variant from stereo vision can be implemented with a single camera. When the camera moves in the observed scene, its consecutive images taken from different positions can be considered as images taken from various cameras at the respective positions (Nagel *et al.* 1995). This approach as illustrated in Figure 5.7 is therefore frequently referred to as *motion stereo*. However, as compared to real stereo vision, motion stereo experiences a number of additional difficulties. These mainly result from the particular nature of displacement between the camera location over time in automotive applications.

A fundamental difficulty stems from the fact that the area that is most interesting in automotive applications lies in driving direction. In this direction the triangle composed by optical centres of the camera at two different instances and a point on an object degenerates to a line and hence triangulation becomes singular, i.e. the solution becomes ambiguous on the complete ray. Triangulation becomes ill-conditioned for arrangements nearby this singularity.

A general framework for image data analysis is depicted in Figure 5.8. It is worth noting the structure of the closed control loop (Dickmanns and Christians, 1989). It aims to construct a model world that comprises all interesting information, such as lane geometry or the position of obstacles. The image data analysis process mimics the projective mapping of the imager module from the real world and computes the measurements, such as 2D correspondence pairs, that are expected from the model world. These predicted measurements are continuously compared with the measurements acquired from the observed image sequence. In the theoretical case of an ideal model world, the difference would vanish. Then, the model and its parameters perfectly 'explain' all variations in the imagery.

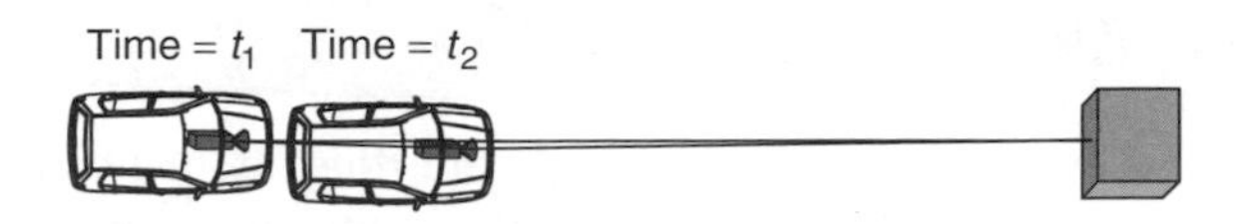

Fig. 5.7 Geometry in motion stereo. The two camera positions and a point in driving direction are close to collinearity.

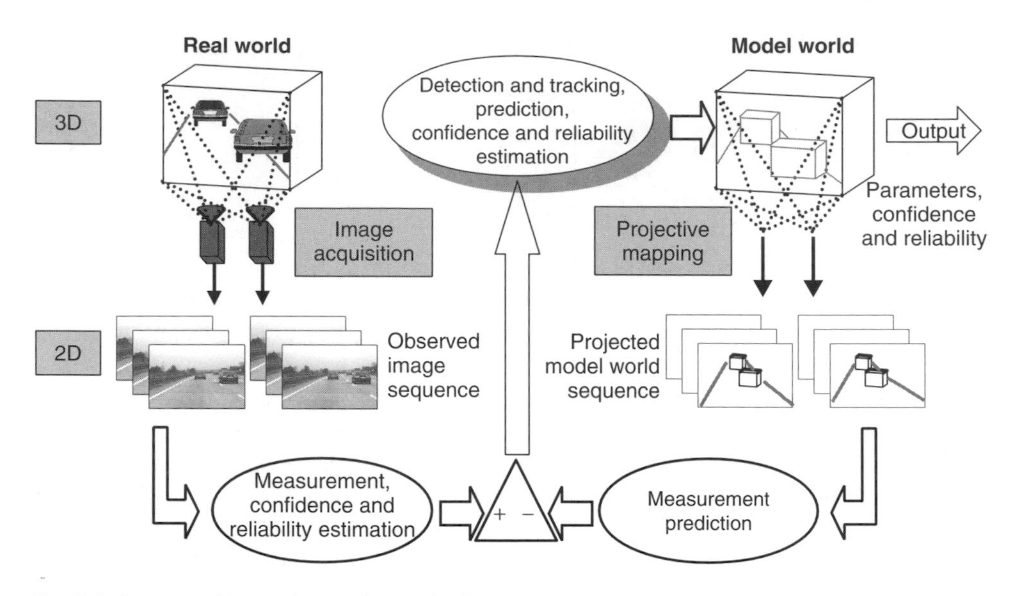

Fig. 5.8 Structure of image data analysis and self-assessment.

In practice the non-zero differences serve to update model parameters, and to assess the capability of the vision sensor in the actual situation. Moderate differences can be compensated for by parameter variation (object tracking) in the real world. Areas of large difference might be compensated for by variation of the number of objects (object detection). Beyond parameter update, the uncertainty of the estimated parameters can be accessed by covariance propagation from the measurements to the parameters analogously to Equation (5.3). Finally, the measurement residuum ϵ is related to its covariance. For the simple model of gaussian errors, the quantity

$$\chi^2 = \varepsilon^T \times Cov(\varepsilon)^{-1} \times \varepsilon \tag{5.5}$$

can be compared against the degree of freedom in the measurement process yielding the well known χ^2-test. Large measurement residuum indicate model failures or gross errors in model parameter estimation.

Several techniques from statistical signal processing and computer vision are applied to extract the desired parameters from the measurements. The large number of corresponding points that can be identified for most objects can be exploited by robust clustering and estimation (Huber, 1981; Zhang, 1997) to gain accuracy and reliability. Moreover, temporal prediction and tracking adds to the stability of the estimates over time (Bar Shalom and Li, 1995).

5.4 Applications and results

Two concise examples are discussed in the sequel to illustrate the capabilities of automotive vision systems. The first one elaborates the complete autonomous control of a standard passenger vehicle on extreme courses, while the second system is focused on tight coupling of heavy trucks.

5.4.1 Autonomous driving

At the extreme end of driver assistance functions ranges the complete automatic control of the vehicle. As mentioned before, several international groups have successfully accomplished this development task. They have equipped specialized autonomous vehicles that were able to drive under supervision in some environments, such as highways. Despite these efforts, autonomous control of series passenger cars is not yet in visible reach. This is not at least due to unsolved technical challenges.

The project consortium 'Autonomes Fahren' (Autonomous Driving) comprises international industry and SMEs as well as universities, all located in Lower Saxony, Germany. The major partners are Volkswagen AG, Robert Bosch GmbH, University of Braunschweig, University of Armed Forces Hamburg, Kasprich/IBEO, and Witt Sondermaschinen GmbH. This group has developed an autonomous driving system that approaches some challenging environments that occur in everyday traffic as well as on proving grounds. These environmental conditions include narrow curves with radii below 10 m, strong vertical curvature, and intentionally rough road surfaces that induce large perturbations to the autonomous system. As another significant extension to previous work, the system has been designed for fast installation into arbitrary series passenger cars to allow operation on proving grounds without the requirement of permanent human supervision. Several vehicles can be simultaneously monitored from a central monitoring station and vehicle operation is monitored on-board by a vehicle diagnosis unit (Michler *et al*, 2000).

The structure of the autonomous system is outlined in Figure 5.9. It involves a multisensor surround perception of the vehicle's environment. Various sensor principles are combined to cover a complete surround view, see Figure 5.10. It is worth noting that each important information is sensed by at least two different sensor principles (Weisser *et al.*, 2000; Stiller *et al.*, 2000).

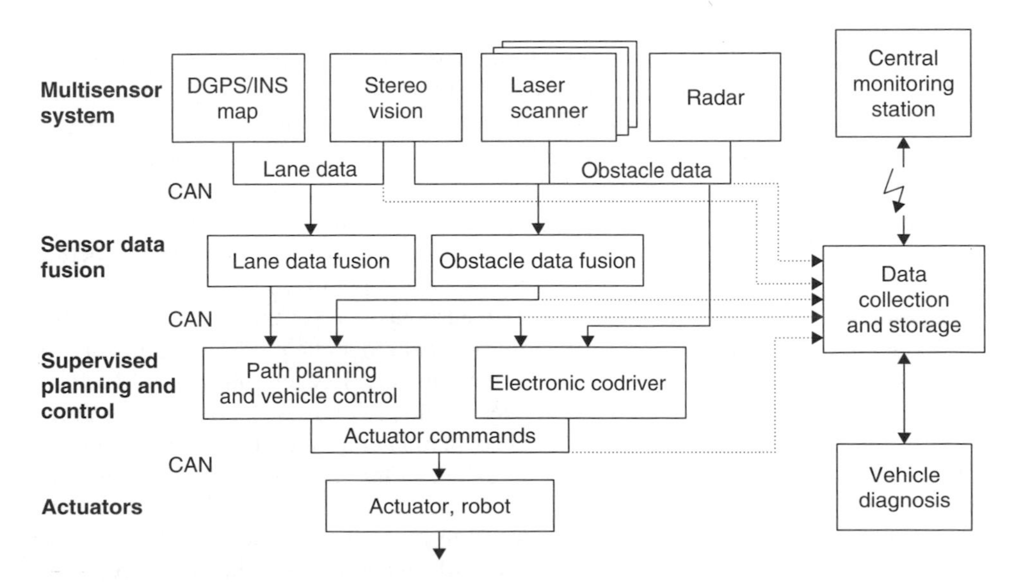

Fig. 5.9 Block diagram of autonomous system.

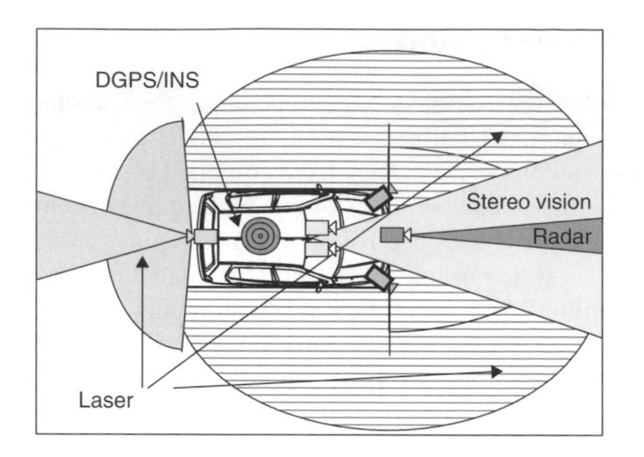

Fig. 5.10 Multisensor surround perception.

The lane geometry is documented in a digital map and the ego-position is acquired by a differential global positioning system (DGPS) using carrier phase recovery and inertial sensors. Independently, a stereo vision system senses lane geometry in parallel. The output from both sensors is evaluated for plausibility and fused in a sensor data fusion unit (Goldbeck *et al.*, 1998).

Beyond lane geometry, driving requires information about obstacles in order to avoid collision. For the sake of system reliability, obstacle detection and tracking is independently performed by various kinds of sensors, namely a stereo vision system, three laser scanners, and a radar sensor, whose field of view is depicted in Figure 5.10. The output of these sensors is also fed into a sensor data fusion unit.

The system structure not only provides an example for a multisensor vehicle environment sensing but also illustrates the multifunctional nature of the vision sensor system. It is the only sensor technology able to sense lane geometry and obstacles at the same time.

Following plausibility validation and information fusion, the lane and obstacle data is forwarded to a planning and control level. The path planning procedure first constructs a static corridor defined by the road boundaries. In a second step, a static path is derived in this corridor that is optimized with respect to comfort and safety criteria assuming the absence of obstacles. The third step is devoted to obstacle avoidance via dynamic path planning. The dynamic path may deviate from the static path locally only, wherever demanded for safe driving near obstacles. Whenever safe in the actual traffic situation, the dynamic path approaches the static path again.

The vehicle controller is designed as a H^2-controller with parameters adapted to the actual velocity. It is robust against load variations and moderate model deviations. This enables a change among a broad variety of test vehicles (Becker and Simon, 2000).

The reliable, redundant design that has been outlined for the sensorial system propagates to the planning and controller unit. It is permanently monitored by a co-driver unit that intervenes prior to an immanent collision. In this case it brings the vehicle to a controlled stand-still directly accessing the actuators (Ameling and Kirchner, 2000).

The actuators are located in a robot designed to allow a fast change of the operated vehicle. It moves steering wheel, pedal, gear shift and some peripheral devices such as ignition lock or blinker, similar to a human driver.

Fig. 5.11 Autonomous vehicle.

Figure 5.11 depicts the autonomous system in a vehicle during a public demonstration. The system successfully accomplishes the 'extreme course' on a proving ground, designed to put high stress on the vehicle chassis. In fact, the narrow curves and rough road surface also put extreme challenges on sensors, control and actuators. The system copes with narrow curves down to a radius of 10 m and successfully conducted a 24 hour drive.

5.4.2 Heavy truck coupling

The second application that illustrates the functionality of vision sensors addresses the growing demand for transportation capacity world-wide and its impact on our environment. Transportation load doubles about every ten years in the European Union with major amounts of this load laid on the road infrastructure. The CHAUFFEUR consortium comprising European industrial and scientific partners has approached this issue by creating technical means for coupling of heavy trucks. Mayor project partners include truck manufacturers, DaimlerChrysler, IVECO/Fiat, and Renault, automotive suppliers, Bosch, Wabco and ZFL, as well as other partners like CRL, CSST, TÜV, Benz Consult and Irion Management Consult. The CHAUFFEUR system is also denoted as *electronic tow bar* as it allows coupling of trucks with a headway of the following truck as short as 5–15 m. A towed truck is not linked to the leading truck by any mechanical means but it follows the leading truck by automatic lateral and longitudinal control based on sensor information (Schulze, 1997).

The main advantages of the system can be summarized as follows:

- benefit from lee driving effects, i.e. reduced fuel consumption and emission
- increased efficiency of road usage and thus improved traffic flow
- reduction of driver's work load yielding economical benefit for freight forwarders.

A sophisticated sensor system was needed in order to accomplish a short yet safe headway at any time, see Korries *et al.* (1988). In case of unpredictable driving

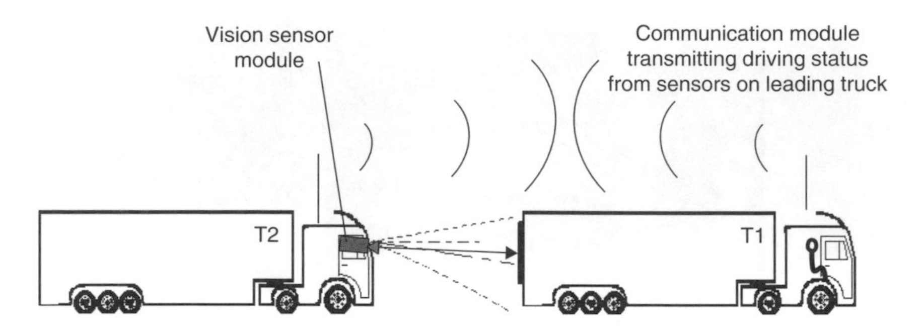

Fig. 5.12 Sensor concept for heavy truck coupling.

manoeuvres of the leading truck, i.e. abrupt braking, the sensor system must enable an instantaneous reaction of the towed truck. At the same time the system has to engage actuators smoothly for comfort and fuel efficiency. Beyond a high accuracy and high measurement rate, robustness must be accomplished against all kinds of conditions that occur in an uncontrolled automotive environment. For safety reasons, the sensors need to be self-assessing, i.e. in case of extreme conditions that the sensors cannot cope with, the sensors must note and signal their limited capabilities. This information is then employed by the vehicle controller for increasing the safety headway appropriately or, in the extreme, to invoking emergency manoeuvres (Lorei and Stiller, 1999).

The above requirements have been mapped to the sensorial system sketched in Figure 5.12. It is composed of two components that compliment each other. A microwave communication module working at 2.4 GHz signals the driving status of the leading truck virtually without delay. The transmitted driving status includes speed, longitudinal and lateral acceleration, and steering angle. Clearly, this information allows computation of the leading vehicle's trajectory via integration. Likewise, the towed vehicle can compute its own trajectory. Hence this information can be used in the feed forward control of the towed vehicle. It is worth noting that this system allows instantaneous reaction to abrupt driving manoeuvres. However, the transmitted data cannot provide any information that prevents inaccuracies from accumulating during integration.

This integration drift is compensated for by a video sensor module that directly measures the distance vector, $\vec{\mathbf{t}}$, and the orientation vector, $\vec{\omega}$, between the two trucks. These measurements are provided with a small delay of 0.04 s and at a high sampling rate of 25 Hz. For safety reasons, the leading truck is marked with a specific visual pattern which forms a base knowledge of the video sensor module. By analysis and verification of this pattern, a high degree of accuracy and reliability can be achieved. In order to allow for additional functionality, such as detection of intruders or recognition of lane markings, a camera that is sensitive in the visible domain is employed. Hence, the system can cope with a passive visual pattern as shown in Figure 5.13. It is worth noting, that the pattern shows similarity to a 2D bar code. This property allows fast and reliable detection in the image sequence.

It can be seen that detection of the leading truck is complicated due to uncontrollable real-world conditions. These and other situations frequently occur in automotive environments. They are partly handled via advanced imager technology, such as a

(a) Smear (CCD overflow) (b) Spatially varying illumination (c) low contrast

Fig. 5.13 Images of leading truck taken during test drives.

CMOS imager with a wide luminance dynamic range. The major element to cope with such situations, however, has been by thorough design of image analysis algorithms (Stiller and Lorei, 1999). The video sensor module assesses its own actual capability and signals a reliability measure along with its estimates. This self-assessment property is a very important feature for practical use.

In the controller the distance and orientation information from the vision sensor module is fused with the data transmitted by the communication module. It conducts completely automatic longitudinal and lateral control of the towed truck (Borodani *et al.*, 1997; Fritz, 1999). The communication module offers virtually instantaneous information about abrupt manoeuvres and allows a coarse estimate of headway and orientation by integration. The video sensor module complements the communication module data with direct estimates of headway and orientation, thus allowing accurate and smooth control. The immanent redundancy of the data provided to the controller can be exploited for a permanent cross check of the two modules thus yielding additional safety in the overall tow bar system.

The tow bar has been successfully integrated into two demonstrator truck pairs (Riva and Ulken, 1997). Extensive truck coupling on various roads including public tow bar demonstration events have proven the accuracy and reliability of the sensor system within the overall system. A multitude of different weather and illumination conditions have been covered by the driving experiments. These include rain, direct sunlight, fast motion, and vibrations. The benefits of the system have been analysed by simulations (Benz, 1997), and evaluated on the road over considerable mileage.

5.5 Conclusions

This chapter has elaborated on vision systems and their application to vehicle environment sensing. An analysis of accident statistics reveals importance of monitoring the vehicle's environment in the future. Reliable perception of the environment is thus viewed as the key technology that substantially contributes to traffic safety. While various principles for remote sensing are readily available, it is argued that vision systems offer a particularly broad functional spectrum and possess the highest potential to approach human capabilities in environment perception in the long term. Today, vision systems are already multifunctional. Tasks comprise 3D obstacle detection and tracking as well as recognition and classification of 2D structures like lane boundaries or traffic signs. Applications of vision sensors are expected to migrate from information and warning functions to comfort and control functions.

Automotive hardware of vision systems requires further development in imager modules and powerful ECUs. However, the key element that has to be accomplished for reliable machine vision systems lies in the image data analysis algorithms. These are expected to dominate system structure and reliability. Several principles for visual machine perception have been discussed. A stereo vision system appears suitable to 3D geometric analysis. This may be accompanied by 2D pattern recognition and classification tasks conducting obstacle or traffic sign classification or lane recognition.

Two concise examples have illustrated the capabilities of automotive vision systems. A multisensor system has been developed that allows complete autonomous control of a standard passenger vehicle on extreme courses. High reliability has been achieved through redundant system design. An attractive supplement to vision sensors has been identified as a vehicle-to-vehicle communication link exchanging information on the vehicle dynamics. This information has been employed to establish a virtual tow bar between two heavy trucks that perform tight autonomous following.

Despite a number of remaining challenges that have been identified for complete vehicle control, the above discussion clearly indicates that first vision-based functions are feasible in the near future. We are at the beginning of an exciting evolution of vehicle vision that is expected to form a basis for more intelligent vehicles.

Acknowledgements

The author gratefully acknowledges the support of the projects Autonomous Driving (Autonomes Fahren) and CHAUFFEUR from the Ministry of Economics, Technology, and Transport of the federal state of Lower Saxony, Germany and from the European Union, in the 4th framework Telematics Applications Programme, respectively, and thanks the members of the project consortiae for their fruitful collaboration.

References

Ameling, C. and Kirchner, A. (2000). The electronic copilot for an autonomous vehicle state of development. *IEEE Intelligent Vehicles Conference*, Dearborn, MI, USA, pp. 162–7, October.

Bar-Shalom, Y. and Li X.-R. (1995). Estimation and tracking: principles, techniques and software. *Artech House*.

Becker, J.C. and Simon, A. (2000). Sensor and Navigation Data Fusion for an Autonomous Vehicle. *IEEE Intelligent Vehicles Conference*, Dearborn, MI, USA, pp. 156–61, October.

Benz, T. (1997). Simulations within CHAUFFEUR. *4th ITS World Congress*, Berlin, Germany, CD ROM I, October.

Borodani, P., Carrea, P. and Gortan, L. (1997). Short headway control for trucks on motorways: the CHAUFFEUR project. *4th ITS World Congress*, Berlin, Germany, CD ROM I, October.

Broggi, A., Bertozzi, M., Fascioli, A. and Conte, G. (1999). *Automatic vehicle guidance: the experience of the ARGO autonomous vehicle*. World Scientific Co. Publisher, Singapore.

Dickmanns, E.D. and Christians, T. (1989). Relative 3-D state estimation for autonomous visual guidance of road vehicles. *2nd Conference on Intelligent Autonomous Systems*, Amsterdam, Netherlands, pp 683–93, December.

Faugeras, O. (1995). *Three dimensional computer vision: a geometric viewpoint*. MIT Press.

Franke, U., Gavrila, D., Görzig, S., Lindner, F., Paetzold, F. and Wöhler C. (1998). Autonomous driving goes downtown. *IEEE Intelligent Systems*, **13**, No. 6, 40–48, October.

Fritz, H. (1999). Longitudinal and lateral control of two electronically coupled heavy-duty trucks in the CHAUFFEUR project. *World Congress on Intelligent Transport Systems*, Toronto, Canada, CD ROM Nr. 2095, November.

Goldbeck, J., Draeger, G., Hürtgen, B., Ernst, S. and Wilms, F. (1998). Lane following combining vision and DGPS. *IEEE Intelligent Vehicles Conference*, Stuttgart, Germany, pp. 445–450.

Huber, P.J. (1981). *Robust Statistics*. John Wiley & Sons, Wiley Series in Probability and Mathematical Statistics.

Kories, R., Rehfeld, N. and Zimmermann, G. (1988). Towards autonomous convoy driving: Recognizing the starting vehicle in front. *IEEE International Conference on Pattern Recognition*, vol. **1**, pp. 531–5, November.

Lorei, M. and Stiller, C. (1999). Visual sensing in electronic truck coupling. *IEEE Internationa Conference Image Processing, ICIP'99*, Kobe, Japan, October.

Maurer, M. and Dickmanns, E.D. (1997). A System Architecture for Autonomous Visual Road Vehicle Guidance. *IEEE Conference on Intelligent Transportation Systems (ITSC'97)*, Boston, MA, USA, CD-Rom I, November.

Michler, T., Ehlers, T. and Varchmin, J. (2000). Vehicle diagnosis – an application for autonomous driving. *IEEE Conference on Intelligent Vehicles*, Dearborn, MI, USA, pp. 168–73, October.

Nagel, H.-H., Enkelmann, W. and Struck, G. (1995). FhG-Co-Driver: from map-guided automatic driving by machine vision to a co-operative driver support. *Journal of Mathematical and Computer Modelling*, **22**, 185–212.

Riva, P. and Ulken U. (1997). CHAUFFEUR tow-bar, The vehicle base system architecture. *4th ITS World Congress*, Berlin, Germany, CD ROM I, October.

Schulze, M. (1997). CHAUFFEUR – The European way towards an automated highway system. *4th ITS World Congress*, Berlin, Germany, CD ROM I, October.

Schmidt, R., Weisser, H., Schulenberg, P.J. and Göllinger, H. (2000). Autonomous driving on vehicle test tracks: overview, implementation and results. *IEEE Intelligent Vehicles Symposium*, Dearborn, MI, USA, vol. 1, pp. 152–5, October.

Shladover, S.E. (1997). Advanced vehicle control and safety systems. *IATSS Research*, **21**, No. 2, 40–48.

Statistisches Bundesamt, (1998). Unfallarten im Straßenverkehr, Wiesbaden.

Stiller, C., Hipp, J., Rössig, C. and Ewald, A. (2000). Multisensor obstacle detection and tracking. *Image and Vision Computing Journal*, pp. 389–396, April.

Stiller, C. and Konrad, K. (1999). Models, criteria, and search strategies for motion estimation in image sequences. *IEEE Signal Processing Magazine*, pp. 70–91 & pp. 116–117, July & September.

Stiller, C. and Lorei M. (1999). Sensor systems for truck coupling. *World Congress on Intelligent Transport Systems*, CD ROM Nr. 2095, November.

Thomanek, F. (1996). Visuelle Erkennung und Zustandsschätzung von mehreren Straßenfahrzeugen zur autonomen Fahrzeugführung. PhD thesis, University BW Munich, VDI Verlag, Fortschr.-Ber., ser. 12, no. 272.

Thorpe, C.E. (1990). *Vision and navigation – The Carnegie Mellon Navlab*. Kluwer Academic Publishers.

Zhang, Z. (1997). Parameter estimation techniques: a tutorial with application to conic fitting. *Image and Vision Computing*, **15**, 59–76.

Zielke, T., Brauckmann, M. and von Seelen, W. (1992). Intensity and edge-based symmetry defection with an application to car-following. *Computer Vision – Lecture Notes on Computer Science*, **588**, 865–73.

From door to door – principles and applications of computer vision for driver assistant systems

Uwe Franke, Dariu Gavrila, Axel Gern, Steffen Görzig, Reinhard Janssen, Frank Paetzold and Christian Wöhler,
DaimlerChrysler AG

6.1 Introduction

Modern cars will not only recover information about their internal driving state (e.g. speed or yaw rate) but will also extract information from their surroundings. Radar-based advanced cruise control was commercialized by DaimlerChrysler (DC) in 1999 in their premium class vehicles. A vision-based lane departure warning system for heavy trucks was introduced by DC in 2000.

This will only be the beginning for a variety of vision systems for driver information, warning and active assistance. We are convinced that future cars will have their own eyes, since no other sensor can deliver comparably rich information about the car's local environment. Rapidly falling costs for the sensors and processors combined with increasing image resolution provide the basis for a continuous growth of the vehicle's intelligence. At least two cameras will look in front of the car. Working in stereo, they will be able to recognize the current situation in 3D. They can be accompanied by other cameras looking backwards and to the side of the vehicle.

In this chapter, we describe the achievements in vision-based driver assistance at DaimlerChrysler. We present systems that have been developed for highways as well as for urban traffic and describe principles that have proven robustness and efficiency for image understanding in traffic scenes.

6.1.1 Vision in cars: why?

Three main reasons promote the development of computer vision systems for cars.

1. Safety

The constant improvement of vehicle safety led to a gradual decrease of injured traffic participants all over the world. A further considerable progress will be possible with sensor systems that perceive the environment around the car and are able to recognize dangerous situations. For example, they will alert the driver if he is leaving the lane, disregarding traffic signs and lights or overlooking a possible collision.

The important advantage of vision-based systems is their potential to understand the current traffic situation, a prerequisite for driver warning or interventions in complex situations, in particular to avoid false alarms. Today's radar-based systems, for example, suppress reflections of still objects since they cannot distinguish between a small pole and a standing car.

Stereo vision allows obstacle detection by three-dimensional scene analysis, whereas fast classification techniques are able to recognize the potential collision partner and to distinguish between cars, motorcycles and pedestrians. So, computer vision offers increased safety not only for the people inside the vehicle but also for those outside.

2. Convenience

Vision-based driver assistance systems allow an unprecedented increase in driving convenience. Speed limit signs can be recognized by the computer and taken into account in an adaptive cruise control (ACC) system. Tedious tasks like driving in stop-and-go traffic can be taken over by the system as well as distance or lateral control on highways.

3. Efficiency

It is obvious that less traffic accidents mean less traffic jams and less economical loss. In addition, computer vision can be used to automate traffic on special roads or to improve the efficiency of goods transport by coupling trucks by means of an electronic tow-bar system. The American Advanced Highway System (AHS) programme aimed at a throughput optimization on existing highways by reducing the vehicle spacing and lateral width of the lanes.

Another important aspect is that in the future drivers can do other jobs like administrative work, if the truck or the car is in autonomous mode. This is of interest to all drivers who use their car for business purposes.

6.1.2 One decade of research at DaimlerChrysler

The progress over the past ten years of vision research for vehicle applications is reflected in our demonstrator vehicles. The first experimental car, VITA I, was a 7.5 ton van built 1989 as a platform for experiments within the Prometheus project. It was equipped with a transputer system for lateral guidance and obstacle detection on highways and offered full access to the vehicle's actuators.

This vehicle was replaced by the well-known VITA II demonstrator for the final Prometheus demonstration in Paris 1994. VITA II was a Mercedes-Benz S-class and showed fully autonomous driving on public highways including lane changes (Ulmer

1994). It was equipped with 18 cameras looking in front, to the rear and to the sides in order to allow a 360° view. Built in cooperation with Ernst Dickmanns (University of the Armed Forces Munich) and Werner V. Seelen (University of Bochum), VITA II was able to recognize other traffic participants, the road course as well as the relevant traffic signs. In addition, it was provided with a behavioural module that was able to plan and perform overtaking manoeuvres in order to keep the desired speed. The side-looking and rear-looking cameras were used to ensure safety of these manoeuvres.

In parallel to the Prometheus demonstrators, the Mercedes-Benz T-model OSCAR was built to investigate vision algorithms and control schemes for robust and comfortable lateral guidance on highways at high speeds. The algorithms were based on the standard lane tracking approach developed by Dickmanns in a joint project. Based on the transputer technology of the early 1990s, OSCAR drove about 10000 km on public highways with maximum speeds of 180 km/h using conventional as well as neural controllers (Neußer *et al.*, 1993). OSCAR tracked not only the markings, but looked also for structures parallel to the lane. From the algorithms used in this car the lane departure warning system mentioned in the introduction has been derived (Ziegler *et al.*, 1995).

In 1995, Daimler-Benz finished the work on the OTTO-truck (Franke *et al.*, 1995). With the AHS-idea in mind, this truck was designed to follow a specific leader with minimum distance. To accomplish this task, OTTO measured the distance to the vehicle in front by looking at known markers. An infrared light-pattern was used as well as two checker-board markings. In order to reach minimum distance and to manage emergency braking of the leader, the acceleration of the leader was transferred to the follower by means of a communication link. Recently, OTTO has been replaced by a heavy duty truck (40 ton) within the European Chauffeur project. Investigations revealed an increased throughput on highways and a reduced fuel consumption of 10–20 per cent depending on the mass of the trucks.

The UTA project (Urban Traffic Assistant) aims at an intelligent stop-and-go system for inner-city traffic. At the Intelligent Vehicles Conference 1998, our UTA I (Mercedes-Benz S-class) demonstrator performed vision-based vehicle following through narrow roads in Stuttgart (Franke *et al.*, 1998). Recently, this car has been replaced by UTA II (Mercedes-Benz E-class), which uses standard Pentium III processors instead of PowerPCs and has increased image understanding capabilities. Details on this project and the developed techniques are given in Section 6.5.2.

6.1.3 A comprehensive driver assistance approach

A vision-based system should be able to assist the driver not only on the highway, but in all traffic situations. It is our goal to realize such a comprehensive system. Here is our vision:

Imagine you are driving to an unknown city to meet a business partner. From the beginning of your trip the vision system acts as an attentive co-driver. You will be warned of the bicyclist from the right, that you have failed to recognize. At the next intersection, it will save you from a possible rear-end collision if you are distracted.

On the highway, the car is able to take over control. Steering is based on the reliable recognition of lanes, longitudinal control exploits stereo vision to improve the radar system and takes into account the speed limit signs. If you prefer driving yourself, you still get this information as a reminder.

Near your destination you get stuck in a slowly moving tailback. The car offers you automated stop-and-go driving. This means that it is able to follow the vehicle in front of you longitudinally as well as laterally. This behaviour is not purely reactive. Traffic lights and signs are additionally taken into account by your intelligent stop-and-go system. Driving manually, the system is able to warn you if you have overlooked a red traffic light or a stop sign. Crosswalks are detected and pedestrians that intend to cross the road are recognized. Finally you reach your destination. The small parking lot is no problem, since you can leave your car and let it park itself.

6.1.4 Outline of the chapter

This chapter describes our work at DaimlerChrysler Research towards such an integrated vision system. It is outlined as follows: Section 6.2 describes the capabilities for the highway scenario developed within the early 1990s including lane and traffic sign recognition and presents improvements by sensor fusion. Section 6.3 concentrates on the understanding of the urban environment. Stereo vision as a key to three-dimensional vision is described. A generic framework for shape-based object recognition is presented. Section 6.4 regards object recognition as a classification problem. Various methods for the recognition of the infrastructure as well as the recognition of cars and pedestrians are presented. All modules for the urban scenario have been integrated in the UTA II demonstrator. A multi-agent software system controls the perception modules as described in Section 6.5.

6.2 Driver assistance on highways

In 1986, when Ernst Dickmanns demonstrated autonomous driving on a closed German highway with a maximum speed of 96 km/h, a revolution in real-time image sequence analysis took place. Whereas other researchers analysed single images and drove some metres blindly before stopping again for the next picture, he exploited the power of Kalman filtering to achieve a continuous processing. Only a small number of 8086 processors were sufficient to extract the information necessary to steer the car from the image sequence delivered by a standard camera. With this new idea adopted from radar object tracking, he influenced the field of image sequence analysis strongly. Today, Kalman filters are considered as a basic tool in image sequence analysis.

This first successfully demonstrated application of computer vision for vehicles was the starting shot for a quickly increasing number of research activities. During the following ten years, numerous vision systems for lateral and longitudinal vehicle guidance, lane-departure warning and collision avoidance have been developed all over the world.

This chapter presents systems for highways, which have already left their infancy and have been tested on many thousands of kilometres on European highways. It starts with classical lane recognition and explains the basic idea of Kalman filter based parameter estimation. Possible extensions are described that are under investigation for higher robustness. Important for future advanced cruise control and driver information systems is the knowledge of the current speed limit. A robust traffic sign recognition system is described in Section 6.2.2. It can easily be extended to other scenarios like urban traffic.

6.2.1 Lane recognition

Principles

The estimation of the road course and the position of the car within the lane is the basis for many applications, which range from a relatively simple lane departure warning system for drowsy drivers to a fully autonomously driving car. For such systems the relevant parameters are the same as for a human driver: the curvature of the road ahead and the position of the car within the lane, expressed by the lateral position and the yaw angle.

The idea of most realized vision-based lane recognition systems is to find road features such as lane markings or road surface textures that are matched against a specific geometrical model of the road (e.g. Kluge and Thorpe, 1992; Dickmanns, 1986). Using these, the parameters of the chosen model and the position of the car in the lane are determined, for example using least-square fitting. However, processing every single image independently is not very smart. A much better way is to take the history of the already driven road and the dynamic and kinematic restrictions of vehicles into account, especially when driving at higher speeds.

According to the recommendations for highway construction, highways are built under the constraint of slowly changing curvatures. Therefore, most lane recognition systems are based on a clothoidal lane model, that is given by the following equation:

$$c(L) = c_0 + c_1 \times L \tag{6.1}$$

$c(L)$ describes the curvature at the length L of the clothoid, c_0 is the initial curvature and c_1 the curvature-rate, which is called the clothoidal parameter. The curvature is defined as $c = 1/R$, where R denotes the radius of the curve. As already mentioned, the vehicle's position within the lane can be expressed by the lateral position x_{off} in the lane and the yaw angle $\Delta\psi$ relative to the lane axis.

Assuming the pinhole-camera model and knowing the camera parameter's focal length f, tilt angle α and height-over-ground H, the relation between a point on a marking and its image point $P(x, y)$ can be described by the following equations:

$$\begin{aligned} x_b &= \frac{f}{L}\left(a \times w - x_{\text{off}} - \Delta\psi \times L + \tfrac{1}{2} \times c_0 \times L^2 + \tfrac{1}{6} \times c_1 \times L^3\right) \\ L &= \frac{H}{\alpha + (y/f)} \end{aligned} \tag{6.2}$$

w is the lane width and $a = \pm 0.5$ is used for the left or the right marking, respectively. Hence, every measurement is projected onto a virtual measurement directly on the centre-line of the lane. In all equations, the trigonometrical functions are approximated by the argument ($\sin x = x$, $\tan x = x$), because we consider only small angles. These equations allow the relevant road course and vehicle position parameters to be determined.

Driving at higher speeds, dynamic and kinematic restrictions have to be taken into account. These constraints can be expressed by the following differential equations:

$$\begin{aligned} \dot{x}_{\text{off}} &= v \times \Delta\psi + v_x \\ \Delta\dot{\psi} &= \dot{\psi}_{\text{veh}} - c_0 \times v \\ \dot{c}_0 &= c_1 \times v \\ \dot{c}_1 &= 0 \end{aligned} \tag{6.3}$$

In these equations, v denotes the longitudinal speed of the vehicle, v_x the lateral speed caused by a possible side slip angle and $\dot{\psi}_{veh}$ the yaw rate. v_x and $\dot{\psi}_{veh}$ are measured by inertial sensors.

The integration of the above described models in the lane recognition system is done by means of a Kalman filter as first proposed by Dickmanns (1996). With this optimal linear estimation scheme, it is possible to estimate the state vector, i.e. the relevant model parameters. The Kalman filter is a recursive observer that uses the actual measurements to correct the predicted state (see e.g. Bar-Shalom and Fortmann, 1988).

Each cycle of the estimation process consists of two phases:

1. Prediction phase. Using the model of the system behaviour (in this case described by the differential equations (6.3), the state-vector estimated at time n is propagated to the next time step $n + 1$. With the above given measurement equations (6.2), one can estimate the positions of the markings in the next time step.
2. Update phase. Depending on the predicted state and the actual measurements a new state of the system is calculated such that the estimation error is minimized.

It is common to search for marking positions inside small parallelogram shaped regions only. They arc placed in the image according to the predicted positions. 1D-signals are obtained by integrating the intensity within these windows parallel to the predicted direction. The marking positions are found by analysis of these signals.

By calculating the expected measurement error variance, the size of the regions in which to search for markings can be minimized. The so-called 3σ-area describes where to find about 99 per cent of all measurements, if a gaussian noise process is assumed. As can be seen in Figure 6.1, the 3σ-area (the horizontal lines) significantly reduces the search range. This leads to a fast and robust lane recognition system because false-positive markings are not analysed.

The system described above is based on the assumption that the road in front of the car is flat. Finding markings on both sides of the car, it is possible to estimate the tilt

Fig. 6.1 The lane recognition system under rainy conditions, showing the tracked markings with found measurements, the predicted centreline of the lane and one tracked radar obstacle in front.

angle α and the lane width w in addition to the other parameters. Mysliwetz (1990) even estimates the vertical curvature assuming a constant lane width.

Sometimes problems occur because 'markings' are falsely found on cars cutting in or crash barriers within the 3σ-area, because they cannot be separated by using only a monocular camera. This violates the system, causing a wrong state estimation.

These problems can be solved using stereo information, delivering three-dimensional information for each point on the markings. It allows to estimate the vertical curvature c_v besides the already mentioned lane width w and the tilt angle α without further geometrical constraints. German highways are designed according to a parabola vertical curvature. The horizontal and vertical curvature models are separated. The parabola curvature is approximated using a clothoid as described in Mysliwetz (1990):

$$c_v(L) = c_{v,0} + c_{v,1} \times L \tag{6.4}$$

Besides a higher accuracy in all measurements, stereo vision allows the discard of all measurements of non-road objects, that lie above ground. This leads to a more reliable system.

Applied lane recognition

Lane keeping A couple of years ago, our test vchicle OSCAR required about ten transputers for monocular lane recognition at a cycle rate of 12.5 Hz. Today, the job is done with improved performance in less than 2 milliseconds on a 400 MHz Pentium II. An optimized nonlinear controller allows comfortable driving at high speeds. Since the car is often driven by non-experts, velocity is limited to 160 km/h at the moment.

Field tests revealed a high degree of acceptance for two reasons. First, the handling is very simple. When the car signals its readiness, you have just to push a button to give the control to the car. Whenever you like, you can gain the control back by pushing the button again or just steering. Second, autonomous lateral guidance is surprisingly precise. Figure 6.2 shows a comparison of manual and autonomous driving at the same speed on the same section of a winding highway. Although the human driver was told to stay in the middle of the lane as precisely as possible, he needed about 40 centimetres lateral space to both sides, which is typical. As can be seen, the autonomous vehicle performed significantly better. The stronger oscillations between 50 and 90 seconds

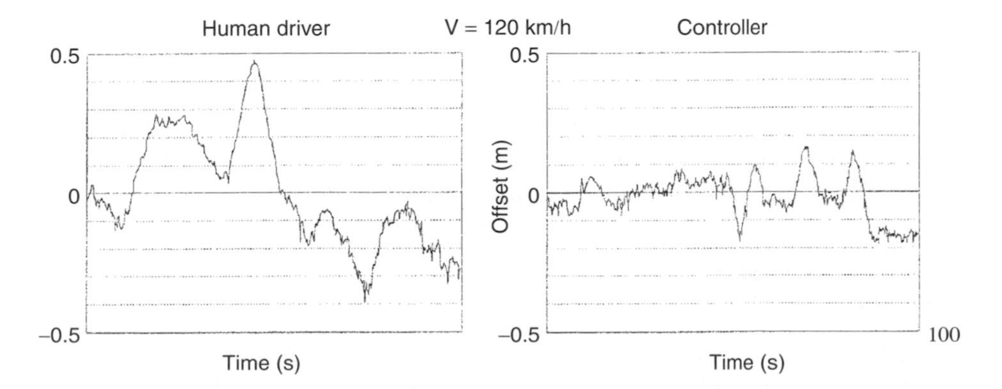

Fig. 6.2 Comparison of human lateral guidance and controller performance on a windy highway.

of driving stem from the tendency of the controller to cut the curves, which have a radius of about 500 metres. In accordance with human behaviour this deviation to the centreline is accepted.

Lane departure warning Many accidents occur due to inattentiveness or drowsiness of the driver, particularly at night. Two types of accidents dominate: rear-end collisions and leaving the lane.

A reliable lane departure warning would therefore lead to a significant increase in traffic safety and is of special interest for trucks and busses. Based on the above described methodology, we have realized an optical system that has been commercially available since May 2000. Leaving the lane without indicating the lane-change, it warns the driver acoustically by means of a rumble strip like sound from the left or right loudspeaker. Naturally, the direction depends on the marking the vehicle is crossing. Camera and processor are integrated in a small box that fits into a palm.

In order to achieve maximum availability and a minimum number of false alarms, the camera is mounted 2–3 metres above ground with a large tilt angle. This maximizes the robustness of the image processing since glare due to a low sun or reflections due to a wet road are avoided. Since the warning system has to be operational on all roads outside built-up areas and only the lateral position of the vehicle is of interest, the road is assumed to be straight. Thus only offset and yaw angle are determined in the estimation process. The error introduced by this assumption is negligible since the maximum look-ahead distance is smaller than 10 metres to guarantee optimal performance at night.

Advanced lane recognition

The traditional lane recognition system described above runs robustly and reliably under fair weather conditions. Problems occur when driving in adverse weather conditions such as rain or snow. Often, the contrast between the markings and the pavement is poor, sometimes the colours of the markings look negated. The range of sight is reduced enormously, causing a bad prediction of the lane parameters, especially the curvature.

A significant improvement of the reliability of the lane recognition system is possible by integration of other sensors that offer a better availability in darkness, rain and snow.

We are investigating two different systems using:

- radar information
- a GPS-based map information system.

a) Radar information DISTRONIC is a radar-based adaptive cruise control system which was introduced in the Mercedes-Benz S-class in May 1999. It measures the following three parameters for every radar obstacle i:

1. The distance d_{obj_i}.
2. The relative speed $v_{\text{rel, obj}_i}$.
3. The angle φ_{obj_i}.

Our approach for improved lane recognition using radar information, as first described in Zomotor and Franke (1997), is motivated by the human strategy when driving in bad weather conditions. Human drivers use the cars in front in order to estimate the road course, assuming that these cars stay in their lanes without significant lateral motion. Such a situation is shown in Figure 6.1.

In fact, every car in front contains strong information on the road curvature that can be used in the estimation process. If we track the vehicles, we can extract these parameters from the measured distance and angle over time. The basic assumption that the lateral motion of the leading vehicles relative to the lane is small can be expressed by:

$$\dot{x}_{\text{off,obj}_i} = 0 \tag{6.5}$$

Large lateral motion caused by lane changes of the tracked vehicles can be detected by appropriate tests.

The radar measurements are incorporated in the Kalman estimation by an additional measurement equation given by:

$$x_{\text{obj}_j} = x_{\text{off}} - x_{\text{off,obj}_i} - \Delta\psi \times d_{\text{obj}_i} c_0 \times d^2_{\text{obj}_i} + \tfrac{1}{6} \times c_1 d^3_{\text{obj}_i} \tag{6.6}$$

The lateral offset x_{obj_i} of each radar obstacle i is related to the measured angle φ_{obj_i} via

$$\varphi_{\text{obj}_i} \approx \sin\varphi_{\text{obj}_i} = \frac{x_{\text{obj}_i}}{d_{\text{obj}_i}}, \; \varphi_{\text{obj}_i} < 5^\circ$$

Getting raw data from the radar sensor, an adequate kinematic model of the obstacles is given by the differential equations:

$$\begin{aligned} \dot{\mathrm{d}}_{\text{obj}_i} &= v_{\text{rel,obj}_i} \\ \dot{v}_{\text{rel,obj}_i} &= 0 \end{aligned} \tag{6.7}$$

As can be seen in Figure 6.1, the range of sight can be enormously enlarged using radar obstacles. Particularly the curvature parameters c_0 and c_1 are improved significantly.

The improvements of the fusion approach can be seen best in simulations as shown in Figure 6.3. The graph shows the curvature c_0 of a simulated road as obtained from the lane recognition system under good weather conditions (range of sight about 50 m) as reference curve, the lane recognition system under bad visibility (range of sight about 10–14 m) and the sensor fusion system following one and two cars under bad visibility (range of sight again about 10–14 m). The distance to the cars in front is about 60–80 m.

The road consists of a straight segment, going into a right bend of radius 300 m and again a straight road. The curvature estimation is improved enormously taking other cars into account. Looking closer at the diagram, it can be seen that the radar-fusion estimates of the curvature run ahead of the lane recognition system. This effect can be explained by the other cars going earlier into the bend than one's own car. The same effect can be seen between 320 and 400 m, where the cars in front are already going out of the bend. The pure lane recognition system under bad visibility shows a delay in the curvature estimation due to the small range of sight. The detailed results are presented in Gern *et al.* (2000).

Assigning cars to specific lanes is an important task for ACC-systems. Since we observe the lateral position of the leading vehicles relative to our lane, this assignment is simultaneously improved by the described approach. This avoids unwanted accelerations and decelerations.

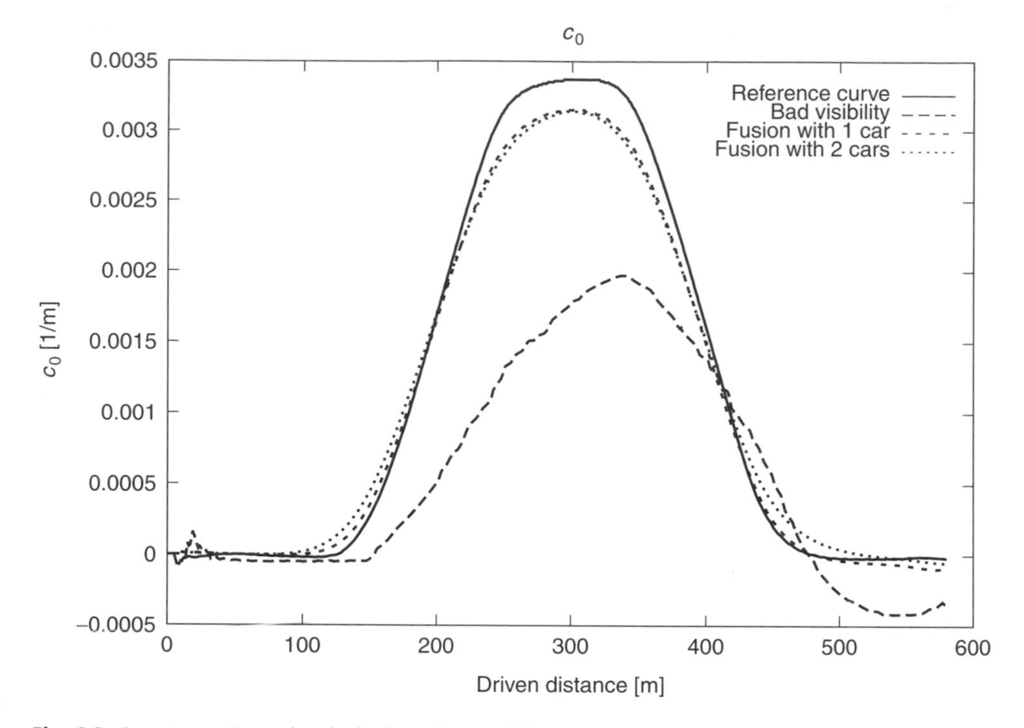

Fig. 6.3 Curvature estimated under bad weather conditions.

The fusion approach assumes that the radar measurements coincide with the centre axes of the cars. Unfortunately, the radar often detects the edges of an obstacle and not the middle axis. Sometimes it is sliding between the left and the right.

It is obvious that the run of the curve can't be estimated correctly if the radar delivers imprecise estimations of the positions of the vehicles. One consequence is that the measurement for every radar obstacle has a large variance. This weakens the strong information obtained by using radar information to increase the range of sight for the lane recognition system.

In order to solve this problem, we are developing a system that detects and tracks all radar obstacles in monocular images. A detailed description of this approach can be found in Gern *et al.*, 2000.

b) GPS-based map information system The second approach to enhance the robustness of the lane recognition system is to integrate GPS-based map information.

Since precise maps are not available at the moment, we generate our own data. During the generation phase, the road course is recorded. Using the optical lane recognition system and a highly accurate GPS-system, the centreline of the lane is determined to generate the necessary highly accurate map.

Later, when driving on this road, the lane recognition system can exploit the map information, especially in adverse weather conditions. The GPS-based map system provides the before measured centreline, describing the road course. This includes the curvature and the clothoidal parameter as well as a number of road points.

We are investigating two different fusion approaches, extending the already described Kalman filter:

1. Using the curvature c_0 and the clothoid parameter c_1 provided by map information system directly as measurements in the lane recognition system.
2. Using the world coordinates of the centreline given by the map information system directly as measurements using an adequate measurement equation.

Simulations and test drives show a higher accuracy of the curvature and yaw angle estimation for both approaches, even if the accuracy of the map information system is low. Results are comparable to those obtained by fusing radar obstacles.

6.2.2 Traffic sign recognition (TSR)

For the foreseeable future visible road signs will regulate the traffic. If vision systems are to assist the driver, they have to be able to perceive traffic signs and observe the implied rules.

Let us first characterize the nature of this object recognition task for a vision based system.

1. Traffic signs are man-made objects and standardized according to national law. Shape and colour are chosen such that they can be easily spotted by a human observer. Both facts alleviate the perception task also for a machine vision system.
2. Traffic signs mounted on sign posts or on bridges spanning the road may have high contrast against the sky. A vision sensor must cover a large dynamic range to make the sign clearly visible in the image. Poor visibility affects the system performance not less than that of a human driver.
3. A large family of traffic signs denote road attributes such as speed limits, which are valid inside a continuous interval, in a discontinuous way. One sign marks the beginning, and another sign the end of the interval. While we can apply an initialization and update process for tracking lanes or for following cars, here we have to search all the acquired images for new appearances of traffic signs in an exhaustive way. Since we do not know *a priori* where to expect traffic signs, this task will bind a considerable amount of computing effort if we do not want to miss a single sign, even driving at high speeds.

Of course there exist non-vision approaches to that problem. Road signs equipped with IR or microwave beacons signal the information regardless of visibility conditions but at high infra-structural costs. Digital maps attributed with traffic regulations supply the correct information only if they are up to date. Temporary changes due to road work, accidents, or switching electronic signs cannot be easily integrated into a map. Thus we are convinced that vision is an essential part of the solution.

The scope of traffic signs handled by the TSR module depends on the range of operation. For a highway scenario useful applications can start with a small set of signs, including speed limits, no overtaking, and danger signs. The urban scenario described in Section 6.3 requires the set to be extended by adding signs which regulate the right of way at intersections.

Detection

On highways, the TSR module is confronted with high resolution images acquired at high velocities of the vehicle. The key to a real-time traffic sign recognition system is a fast and robust detection algorithm. This algorithm detects regions which are likely to contain traffic signs. The traffic sign candidates will be tracked through the image sequence until a reliable result can be obtained. This allows the recognition process to focus on a limited number of small search areas, which speeds up the whole process significantly.

The detection stage can exploit colour and/or shape as the first cue for traffic sign hypotheses. The shape matching algorithm described in Section 6.3.2 is used for traffic sign detection when colour is not available from the camera.

For example, we will here elaborate on a detection algorithm which relies on the colour of traffic signs. The advantage of colour cues, in contrast to shape, is their invariance against scale and view and their highly discriminative power. Even partially occluded or deformed traffic signs can be detected using the colour information.

The algorithm consists of the following three principal steps (Janssen, 1993):

1. In the first step the significant traffic sign colours are filtered out from the acquired colour image. As a result of the colour segmentation the pixels of the image are labelled with the colours *red*, *blue*, *yellow*, *black*, *grey* and *white*.
2. In the second step the iconic information is transformed to a symbolic description of coloured connected components (CCC) by applying a fast connectivity analysis algorithm (Mandler and Oberländer, 1990).
3. Finally the CCC database is queried for objects with certain geometrical attributes and colour combinations. Ensembles of CCC objects extracted by those queries are called meta CCC objects. Meta CCC objects serve as hypotheses for the subsequent traffic sign recognition process.

Colour segmentation The task of colour segmentation is to mimic what the human observer does when he or she recognizes a specific *red* as 'traffic sign red'. The visual system seems to have a strong concept of colour constancy which enables recognition of this certain red although the colour description covers a wide range of different hues. These hues are influenced by the paint, the illumination conditions, and the viewing angle of the observer. Since the current knowledge is not adequate to model all facets of colour perception, a learning approach was chosen to generate colour descriptions suitable for the machine vision system.

The mapping from the colour feature vector to the colour label in the decision space is a typical classification problem which can be solved by means of statistical pattern recognition. The classification is performed by neural networks or polynomial classifiers as described in Section 6.4. The coefficients of the network have to be adapted during the learning phase with a representative set of labelled samples for every colour class. Manually labelled traffic scene images are used to adapt a polynomial classifier to the colours *red*, *blue*, *yellow*, *black*, *grey* and *white*. The traffic signs in the application scenarios can be described by these colour terms.

Colour connected components The colour labelled image is still an iconic representation of the traffic scene. Grouping all neighbouring pixels with a common colour into regions creates a symbolic representation of the image. The computation of the so-called colour connected components (CCC) is performed by a fast, one-pass, line-scan algorithm (Mandler and Oberländer, 1990). The algorithm produces for each CCC a list of all neighbouring components, thus providing full topological information. The connected component analysis does not induce any information loss since the labelled image can be completely reconstructed from the CCC database.

Now it is easy to retrieve candidate regions with a specific topological relationship and colour combination efficiently.

Meta CCC language Traffic sign candidates are extracted from the CCC database with queries searching for instance for regions with a certain colour, inclusions of a certain colour, and geometrical attributes inside specific intervals. The ensemble of CCC objects extracted by those queries is called a meta CCC object. The meta CCC query language efficiently parses the database at run-time. The language comprises pure topological functions (adjacency, inclusion, etc.) as well as functions exploiting the shape of colour components (size, aspect ratio, coverage, eccentricity, etc.). e.g. the filter

inside of(RED,WHITE) | aspect(0.8,1.2) | larger(12) | smaller(10 000) |group +
inside of(RED,GREY) | aspect(0.8,1.2) | larger(12) | smaller(10 000) | group

searches for red objects which have white or grey inclusions, a square bounding box, and a size reasonable for the traffic sign recognition procedure. Figure 6.4 shows that the use of adequate queries helps to detect traffic signs even under non-ideal conditions.

The bounding boxes of the meta CCC objects into which the objects are grouped with their inclusions form a region of interest (ROI). This ROI is the input to the verification and classification stage of traffic sign recognition.

Recognition

The task of the recognition stage is to map the pixel values inside the ROI either to a specific traffic sign class, or to reject the hypothesis. Employing statistical pattern recognition (see Section 6.4) ensures that this mapping is insensitive to all the variations of individual traffic signs, which are due to changes in illumination, weather conditions, and picture acquisition for instance. Figure 6.5 shows a collection of red danger signs varying in colour, spatial resolution, background, viewing aspect, and pictographic symbols.

Feature extraction A main problem to be solved in building statistical pattern recognition systems is the design of a significant feature vector.

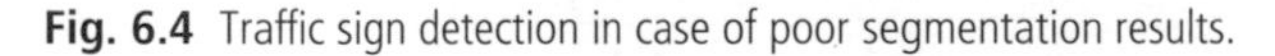

Fig. 6.4 Traffic sign detection in case of poor segmentation results.

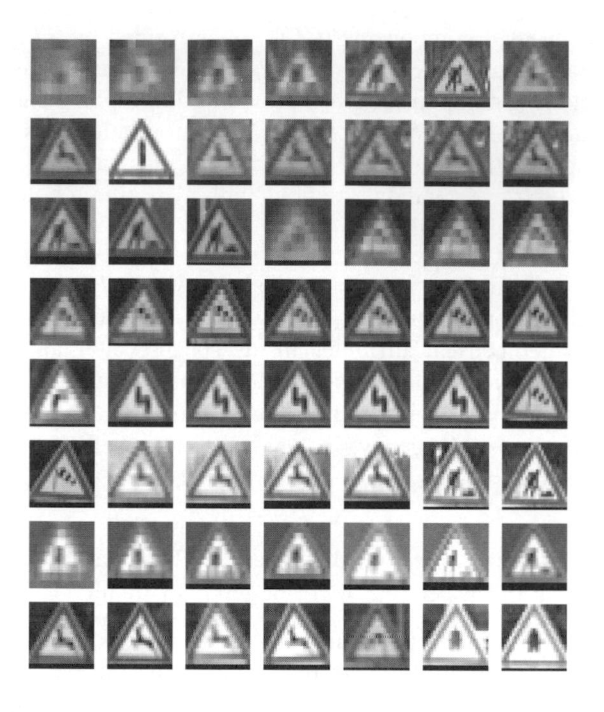

Fig. 6.5 Collection of red danger signs varying in colour, spatial resolution, background, viewing aspect, and pictographic symbols.

There is a trade-off between the expenditure on feature extraction and on classification. This means that clever normalization measures can simplify the classification process considerably. Hence, if we manage to design a well balanced system, we can do with a smaller learning set, will spend less computational effort, and increase the classification efficiency and performance.

In samples of traffic scenes we observe several kinds of variances:

1. in scale and resolution due to variable sizes of traffic signs and variable distances between camera and object,
2. in translation due to inaccurate segmentation,
3. in photometric parameters (brightness, contrast, colour) due to variable weather and lighting conditions,
4. in design of the traffic signs themselves due to different versions (fonts, layout).

A normalization of scale and photometric variances is feasible and part of our recognition system. The remaining variances can only be dealt with by learning from examples.

In fact we carry out three steps of normalization (see Figure 6.6): scale normalization, colour mapping and intensity normalization. The starting point is a region of interest delivered by our colour-based detection module.

Scale normalization is not dispensable, because the dimension of the input layer is fixed for most classifier networks. Each pixel is interpreted as one component of the feature vector.

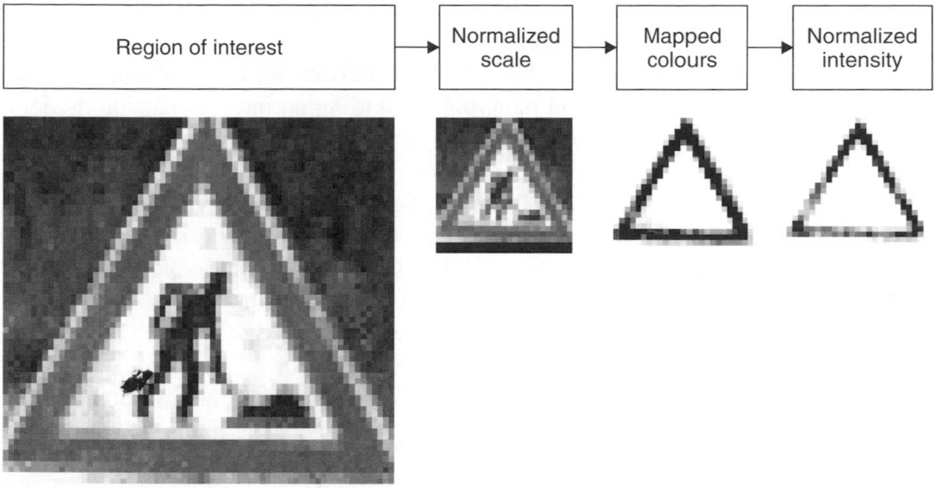

Region of interest → Normalized scale → Mapped colours → Normalized intensity

Fig. 6.6 Normalization steps.

In feature extraction for shape recognition a complementary colour mapping is applied. Chromatic pixels are mapped to high and achromatic pixels to low intensities. This mapping emphasizes shape discriminating regions and suppresses noise. Both background and pictograms are suppressed in the same manner, because these regions have low colour saturation.

There are three different mapping functions applicable according to the colour of the traffic sign, which is already known after traffic sign detection. In order to reduce the remaining variances of intensity (brightness, contrast), the mean and dispersion of the pixels' intensities are calculated and adjusted to standard values. If a pattern is entirely achromatic (e.g. end of restrictions and most pictograms) complementary colour mapping makes no sense. In this case we only normalize the intensity of the size-scaled ROI.

$$Y = R - \frac{G+B}{2} \quad \text{complementary colour mapping for red signs}$$

$$Y = B - \frac{R+G}{2} \quad \text{complementary colour mapping for blue signs}$$

$$Y = \frac{R+G+B}{3} \quad \text{colour mapping for achromatic signs} \tag{6.8}$$

We conclude that the goal of normalization is to decrease the intra-class distance while increasing the inter-class distance. Considering the feature space, we try to compress the distributions of each class while separating distributions of different classes, thus improving the ratio of spread and mean distance of the corresponding distributions. This effect supports the performance of the subsequent classification, no matter how it is implemented.

Model-based recognition A very important requirement for an autonomous TSR system is the capability of considering variable road traffic regulations due to differences in jurisdiction between countries and temporal changes of rules.

A hard coded structure of fixed classifiers with pre-programmed rules would mean a drawback in flexibility. For this reason we used a model-based approach to organize a number of shape and pictogram classifiers. In this approach we try to separate domain-dependent from domain-independent data, thus providing an interchangeable model of objects we intend to recognize and a more universal recognition machine.

In order efficiently to construct a traffic sign model we investigate significant features at first. Our system, motivated by human perception, is sketched in Figure 6.6. It starts with the dominant colour, e.g. red, white or blue, which has already been used for detection of the regions of interest. Second, the shape of the traffic sign candidate is checked before the pictogram is classified as described in Section 6.4.

For each image containing traffic sign candidates a so-called search tree is generated according to the structural plan of the model.

The development of each search tree node involves computational effort. For efficiency reasons the decision which node to develop next is taken in a best first manner. If there is an upper bound estimate of the future costs the even more efficient A* algorithm is applicable. Both methods guarantee to find optimal paths. If a terminal node is reached the search is finished.

Conclusions

The system described is capable of recognizing a subset of traffic signs in a robust manner under real-world conditions. The results obtained above, however, refer to single images only, i.e. relations between subsequent frames have been neglected. Tracking traffic signs over the time interval during which they are in the field of view adds to the stability of the recognition results. In test drives carried out by DaimlerChrysler on German and French motorways the recognition rate could be increased from 72 per cent in single images to 98 per cent in image sequences. Tracking is also a means of reducing computational effort in that the detection can be focused to smaller search regions.

We can even estimate the relative size and position of the sign if we evaluate all monocular measurements of the tracked object with regard to the ego-motion of the vehicle (depth from motion). This kind of information is used for further plausibility checks and interpretation tasks.

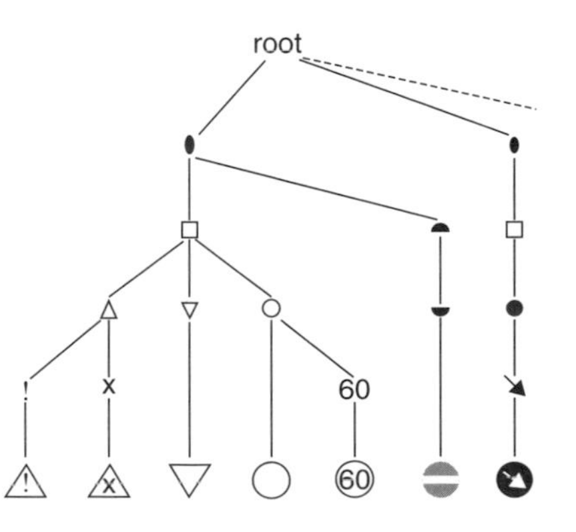

Fig. 6.7 Recognition tree.

Numerous context rules influence the validity of traffic signs, e.g. traffic signs may apply to specific lanes, at specific times of day, or not beyond the next road junction. Current vision modules cannot always gather this type of information with the required reliability. For a commercially feasible system the vision-based recognition of traffic signs and a digital map with traffic sign attributes must support each other. But an autonomous vision system will be part of the solution since even the best map is no exact mirror of the current traffic situation.

6.3 Driver assistance in urban traffic

As pointed out in the introduction, a vision-based driver assistance system would be even more attractive if it would be able to support the driver not only on the highway, but also in city traffic. Intelligent stop-and-go is our first approach to building such a system. It includes stereo vision for depth-based obstacle detection and tracking and a framework for monocular detection and recognition of relevant objects – without requiring a supercomputer in the trunk. Besides Intelligent Stop& Go, many other driver-assistance systems such as rear-end collision avoidance or red-traffic-light warning are also of interest for urban traffic. The most important perception tasks that must be performed to build such systems are to:

- detect the leading vehicle and estimate its distance, speed and acceleration;
- detect stationary obstacles that limit the available free space, such as parked cars;
- detect and classify different additional traffic participants, such as pedestrians;
- detect and recognize small traffic signs and traffic lights in a complex environment;
- extract the lane course, even if it lacks well painted markings and does not show clothoidal geometry.

This list shows that the ability to recognize objects is essential for Intelligent Stop-&-Go. Two classes of objects pertain to this application, namely infrastructure elements and traffic participants. How do we recognize those objects? Although devising a general framework is difficult, we often find ourselves applying three steps: detection, tracking and classification.

The purpose of the detection is efficiently to obtain a region of interest (ROI) – that is, a region in the image or parameter space that could be associated with a potential object. It is obvious that not all objects can be detected by means of a unique algorithm. We have developed methods that are based on depth from stereo, shape and colour. Detected objects are tracked from frame to frame to estimate their motion and increase the robustness of the system. Once an object of interest (depending on size or shape) has been detected, the system tries to recognize it. In the considered scenario, objects have a wide variety of appearances because of shape variability, different viewing angles and illumination changes. Since explicit models are seldom available, we derive models implicitly by learning from examples. This turns object recognition to a classification problem, which is described in detail in Section 6.4.

In this section, we present our detection schemes based on stereo and shape. Object recognition tasks exploiting colour have already been described in Section 6.2.2 (traffic sign recognition) and are sketched in Section 6.4 (traffic light recognition).

6.3.1 Stereo vision

Vision systems for driver assistance require an internal 3D map of the environment in front of the car, in order to safely navigate the vehicle and avoid collisions. This map must include position and motion estimates of relevant traffic participants and potential obstacles. In contrast to the highway scenario where you can concentrate on looking for rear ends of preceding vehicles, our system has to deal with a large number of different objects, some of them non-rigid like pedestrians, some of them unknown.

Several schemes for obstacle detection in traffic scenes have been investigated in the past. Besides the 2D model based techniques that search for rectangular, symmetric shapes, inverse perspective mapping based techniques (Broggi, 1997), optical flow based approaches (Enkelmann, 1997) and correlation-based stereo systems using specialized hardware (Saneyoshi, 1994) have been tested.

The most direct method to derive 3D-information is binocular stereo vision for which correspondence analysis poses the key problem. Given a camera pair with epipolar geometry, the distance L to a point is inversely proportional to the disparity d in both images according to:

$$L = \frac{f_x \times B}{d}, \tag{6.9}$$

where B denotes the base width and f_x the scaled focal length.

We have developed two different stereo approaches, one feature-based and one area-based. Both have in common that they do not require specialized hardware but are able to run in real-time on today's standard PC processors.

Real-time stereo analysis based on local features

A fast nonlinear classification scheme is used to generate local features that are used to find corresponding points. This scheme classifies each pixel according to the grey values of its four direct neighbours (Franke and Kutzbach, 1996). It is verified whether each neighbour is significantly brighter, significantly darker or has similar brightness compared to the considered central pixel. This leads to $3^4 = 81$ different classes encoding edges and corners at different orientations. The similarity is controlled by thresholding the absolute difference of pixel pairs.

Figure 6.8 shows the left image of a stereo image pair taken from our camera system with a base width of 30 cm. The result of the structure classification is shown in the right part. Different grey values represent different structures.

The correspondence analysis works on these feature images. The search for possibly corresponding pixels is reduced to a simple test whether two pixels belong to the same class. Since our cameras are mounted horizontally, only classes containing vertical details are considered. Thanks to the epipolar constraint and the fact that the cameras are mounted with parallel optical axis, pixels with identical classes must be searched on corresponding image rows only.

It is obvious that this classification scheme cannot guarantee uniqueness of the correspondences. In case of ambiguities, the solution giving the smallest disparity, i.e. the largest distance, is chosen to overcome this problem. This prevents wrong correspondences caused by for example periodic structures to generate phantom obstacles close to the camera. In addition, measurements that violate the ordering constraint are ignored (Faugeras, 1993).

Fig. 6.8 Left image of a stereo image pair and the features derived by the sketched nonlinear operation. Each colour denotes one of the 81 structural classes, pixels in homogeneous areas are assigned to the 'white' class.

Fig. 6.9 Results of the correspondence analysis. Image (a) shows the result of the feature based approach, image (b) shows the result of the correlation-based scheme. Distance is inversely proportional to the darkness.

The outcome of the correspondence analysis is a disparity image, which is the basis for all subsequent steps. Figure 6.9 visualizes such an image in the left half. Of course, the result looks noisy due to the extreme local operation.

On a 400 MHz Pentium II processor this analysis is performed within 23 milliseconds on images of size 384×256 pixel.

The advantage of this approach is its speed. Two facts might be a problem in some applications. First, the disparity image is computed with pixel accuracy only. This problem can simply be overcome by post-processing. Second, the described algorithm uses a threshold to measure similarity. Although the value of this threshold turns out to be uncritical, it is responsible for mismatches of structures of low contrast.

Real-time stereo analysis based on correlation

For applications that do not need a cycle rate of 40 milliseconds but require high precision 3D information, we developed an alternative area-based approach. The maximum processing time that we can tolerate is 100 ms.

In order to reach the goal, we must use the sum-of-squared or sum-of-absolute differences criterion instead of expensive cross-correlation to find the optimal fit. Since gain and shutter of our cameras are controlled by the stereo process, this is acceptable.

However, real time is still a hard problem. Full brute-force correlation of 9×9 pixel windows requires about 9 seconds for images of size 384×256, if the maximum disparity is set to 80 pixels. With an optimized recursive implementation we achieved typical values of 1.2 seconds.

To speed up the computation, we use a multi-resolution approach in combination with an interest operator (Franke and Joos, 2000). First, a gaussian pyramid is constructed for the left and right stereo images, based on a sampling factor of 2. Areas with sufficient contrast are extracted by means of a fast horizontal edge extraction scheme. Non-maximum suppression yields an interest image, from which a binary pyramid is constructed. A pixel (i, j) at level n is marked if one of its four corresponding pixels at level $n - 1$ is set. On an average, we find about 1100 attractive points at pyramid level zero (original image level), 700 at level one, 400 at level two and about 150 at level three. Only those correlation windows with the central pixel marked in these interest images are considered during the disparity estimation procedure.

Depending on the application, the correlation process starts at level two or three of the pyramid. If D is the maximum searched disparity at level zero, it reduces to $D/2^{**}n$ at level n. At level two this corresponds to a saving of computational burden of about 90 per cent compared to a direct computation at level zero. Furthermore, smaller correlation windows can be used at higher levels which again accelerates the computation.

The result of this correlation is then transferred to the next lower level. Here, only a fine adjustment has to be performed within a small horizontal search area of ± 1 pixel. This process is repeated until the final level is reached. At this level, subpixel accuracy is achieved by fitting a parabolic curve through the computed correlation coefficients.

The price we have to pay for this fast algorithm is that mismatches in the first computed level propagate down the pyramid and lead to serious errors. In order to avoid this problem, we compute the normalized cross-correlation coefficient for the best matches at the first correlation level and eliminate bad matches from further investigations.

If we start at level 2 (resolution 91×64 pixels), the total analysis including pyramid construction runs at about 90 milliseconds on a 400 MHz Pentium. If we abandon the multi-resolution approach, about 450 milliseconds are necessary to yield comparable results.

A disparity image derived by this scheme is shown in Figure 6.9 in comparison to the feature-based approach. Since a large neighbourhood is taken into account during processing, the result looks less noisy. In fact, only a few mismatches remain, typically in case of periodic structures. A common left-right test applied to the results of the highest evaluation level further reduces the error rate.

Obstacle detection and tracking

The result of both algorithms is a disparity or depth image. Therefore, the further processing is independent of the method used.

Driving on roads, we regard all objects above ground level as potential obstacles. If the cameras are mounted H metres above ground and looking downwards with a tilt

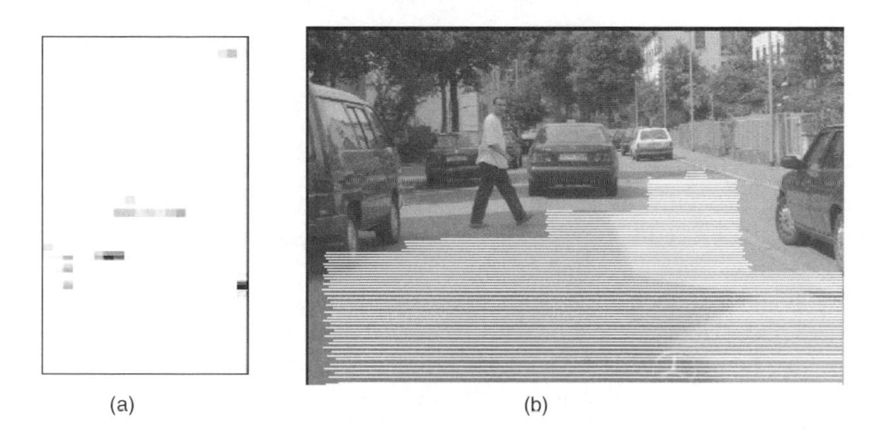
(a) (b)

Fig. 6.10 From the depth map (a) the free space in front of the car is derived (b).

angle α, all image points with a disparity given by

$$d = x_l - x_r = \frac{B}{H} f_x \times \left[\frac{y}{fy} \times \cos(\alpha) + \sin(\alpha) \right] \tag{6.10}$$

lie on the road.

The projection of all features above the road plane, i.e. those with disparities larger than given by equation 6.10, yields a two-dimensional depth map. In this histogram, obstacles show up as peaks.

The map shown in Figure 6.10 covers an area of 40 m in length and 6 m in width. The hits in the histogram are clearly caused by both cars parking left and right, the car in front, the pedestrian and the white car on the right side. Although the feature-based approach looks noisy, the depth maps of both approaches are comparable. This map is used to detect objects that are tracked subsequently. In each loop, already tracked objects are deleted in this depth map prior to the detection.

The detection step delivers a rough estimate of the object width. A rectangular box is fitted to the cluster of feature points that contributed to the extracted area in the depth map. This cluster is tracked from frame to frame. For the estimation of the obstacle distance, the disparities of the object's feature points are averaged.

In the current version, an arbitrary number of objects can be considered. Sometimes the right and left part of a vehicle are initially tracked as two distinct objects. These objects are merged on a higher 'object-level' if their relative position and motion fulfil reasonable conditions.

From the position of the objects relative to the camera system their motion states, i.e. speed and acceleration in longitudinal as well as lateral direction, are estimated by means of Kalman filters. For the longitudinal state estimation we assume that the jerk, i.e. the deviation of the acceleration, of the tracked objects is small. This is expressed in the following state model with distance d, speed v and acceleration a:

$$\begin{bmatrix} d \\ v_l \\ a_l \end{bmatrix}_{k+l} = \begin{bmatrix} 1 & T & T^2/2 \\ 0 & 1 & T \\ 0 & 0 & 1 \end{bmatrix} \times \begin{bmatrix} d \\ v_l \\ a_l \end{bmatrix}_k - T \times \begin{bmatrix} v_e \\ 0 \\ 0 \end{bmatrix}$$

The index l denotes the states of the lead vehicle, the index e denotes the ego vehicle. T is the cycle time. The longitudinal motion parameters are the inputs for a distance controller. An example of autonomous vehicle following is given in Section 6.5.

Further analysis of the depth information

The depth image contains more useful information. The fact that we can identify structures on the road plane improves the performance of lane and crosswalk recognition as described in Sections 6.2.1 and 6.3.3.

Camera height and pitch angle are not constant during driving. Fortunately, the relevant camera parameters can be efficiently estimated themselves using the extracted road surface points. Least squares techniques or Kalman filtering can be used to minimize the sum of squared residuals between expected and found disparities (Franke *et al.*, 1998).

Active collision avoidance is the ultimate goal of driver assistance. A careful evaluation of the depth map allows extraction of free space on the road that could be used for a jink. Figure 6.10 shows the depth map derived for the situation considered here and the determined free space. Alternatively, the driving corridor can be estimated from the depth map, if no other lane boundaries are present.

6.3.2 Shape-based analysis

Another important vision cue for object detection is shape. Compared with colour, shape information tends to remain more stable with respect to illumination conditions, because of the differential nature of the edge extraction process. We developed a shape-based object detection system general enough to deal with arbitrary shapes, whether parameterized (e.g. circles, triangles) or not (e.g. pedestrian outlines). The system does not require any explicit shape-models, and instead learns from examples. A template matching technique provides robustness to missing or erroneous data; it does so without the typical high cost of template matching by means of a hierarchical technique. The resulting system is called the 'Chamfer System' (Gavrila and Philomin, 1999); it provides (near) real-time object detection on a standard PC platform for many useful applications.

The Chamfer System

At the core of the proposed system lies shape matching using distance transforms (DT), e.g. Huttenlocher *et al.* (1993). Consider the problem of detecting pedestrians in an image (Figure 6.11(a)). Various object appearances are captured with templates

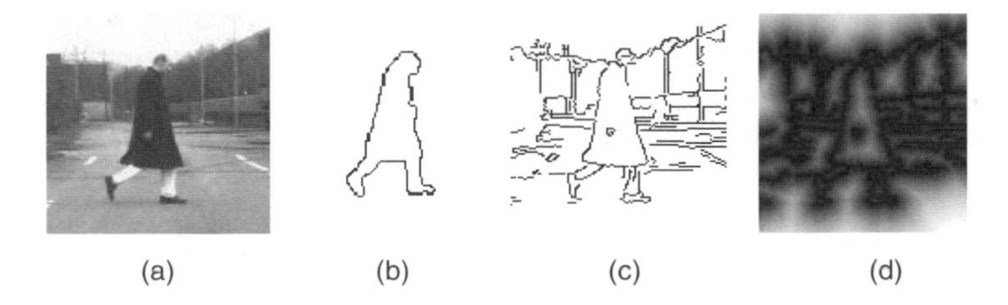

Fig. 6.11 (a) Original image (b) Template (c) Edge image (d) DT image.

such as in Figure 6.11(b). Matching template **T** and image **I** involves computing the feature image of **I**, (Figure 6.11(c)) and applying the distance transform to obtain a DT-image (Figure 6.11(d)). The template **T** is transformed (e.g. translated) and positioned over the resulting DT image of **I**; the matching measure D(**T**, **I**) is determined by the pixel values of the DT image which lie under the data pixels of the transformed template. These pixel values form a distribution of distances of the template features to the nearest features in the image. The lower these distances are, the better the match between image and template at this location. There are a number of matching measures that can be defined on the distance distribution. One possibility is to use the average distance to the nearest feature. This is the *chamfer* distance, hence the name of the system. Other more robust (and costly) measures further reduce the effect of missing features (i.e. due to occlusion or segmentation errors) by using the average truncated distance or the f-th quantile value (the *Hausdorff* distance), e.g. Huttenlocher *et al.* (1993). In applications, a template is considered matched at locations where the distance measure D(**T**,**I**) is below a user-supplied threshold.

The advantage of matching a template with the DT image rather than with the edge image is that the resulting similarity measure will be smoother as a function of the template transformation parameters. This enables the use of an efficient search algorithm to lock onto the correct solution, as will be described shortly. It also allows some degree of dissimilarity between a template and an object of interest in the image.

The main contribution of the Chamfer System is the use of a template hierarchy efficiently to match whole sets of templates. These templates can be geometrical transformations of a reference template, or, more generally, be examples capturing the set of appearances of an object of interest (e.g. pedestrian). The underlying idea is to derive a representation off-line which exploits any structure in this template distribution, so that, on-line, matching can proceed optimized. More specifically, the aim is to group similar templates together and represent them as two entities: a 'prototype' template and a distance parameter. The latter needs to capture the dissimilarity between the prototype template and the templates it represents. By matching the prototype with the images, rather than the individual templates, a typically significant speed-up can be achieved on-line. When applied recursively, this grouping leads to template hierarchy, see Figure 6.12.

The above ideas are put into practice as follows. Off-line, a template hierarchy is generated automatically from available example templates. The proposed algorithm uses a bottom-up approach and applies a K-means-like algorithm at each level of the hierarchy. The input to the algorithm is a set of templates $\mathbf{t_1}, \ldots, \mathbf{t_N}$ their dissimilarity matrix and the desired partition size K. The output is the K-partition and the prototype templates $\mathbf{p_1}, \ldots, \mathbf{p_K}$ for each of the K groups $S_1, \ldots, S_K$. The K-way clustering is achieved by iterative optimization. Starting with an initial (random) partition, templates are moved back and forth between groups while the following objective function E is minimized:

$$E = \sum_{k=1}^{K} \max_{t_i \in S_k} \mathrm{D}(\mathbf{t_i}, \mathbf{p_K^*}) \tag{6.11}$$

Here, $\mathrm{D}(\mathbf{t_i}, \mathbf{p_K^*})$ denotes the distance measure between the ith element of group k, $\mathbf{t}_i$, and the prototype for that group at the current iteration, $\mathbf{p_K^*}$. The distance measure is the same as the one used for matching (e.g. chamfer or Hausdorff distance). One way of

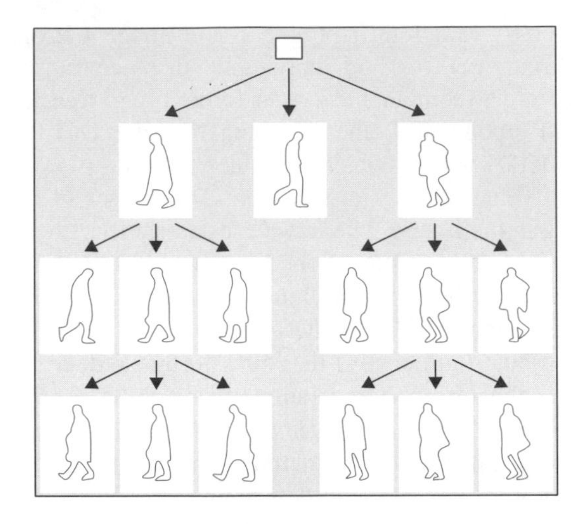

Fig. 6.12 A hierarchy for pedestrian shapes (partial view).

choosing the prototype $\mathbf{p}_{\mathbf{K}}^{*}$ is to select the template with the smallest maximum distance to the other templates. $D(i, j)$ then represents the i-jth entry of the dissimilarity matrix, which can be computed fully before grouping or only on demand.

Note that a low E-value is desirable since it implies a tight grouping; this lowers the distance threshold that will be required during matching which in turn likely decreases the number of locations which one needs to consider during matching. *Simulated annealing* (Kirckpatrick *et al.*, 1993) is used to perform the minimization of the objective function E.

Online, matching can be seen as traversing the tree structure of templates. Each node corresponds to matching a (prototype) template with the image at some particular locations. For the locations where the distance measure between template and image is below a user-supplied threshold, one computes new interest locations for the children nodes (generated by sampling the local neighbourhood with a finer grid) and adds the children nodes to the list of nodes to be processed. For locations where the distance measure is above the threshold, the search does not propagate to the sub-tree; it is this pruning capability that brings large efficiency gains. Initially, the matching process starts at the root and the interest locations lie on a uniform grid over relevant regions in the image. The tree can be traversed in breadth-first or depth-first fashion. In the experiments, we use depth-first traversal, which has the advantage that one needs to maintain only $L - 1$ sets of interest locations, with L the number of levels of the tree. It is possible to derive an upper-bound on the distance threshold at each node of the hierarchy, such that one has the desirable property that using untruncated distance measures such as the chamfer distance, one can assure that the combined coarse-to-fine approach using the template hierarchy and image grid will not miss a solution (Gavrila and Philormin, 1999). In practice however, one can get away with using more restrictive thresholds to speed up detection.

A number of implementation choices further improved the performance and robustness of the Chamfer System, e.g. the use of oriented edge features, template subsampling, multi-stage edge segmentation thresholds and the incorporation of regions of interest (e.g. ground-plane). Applying SIMD processing (MMX) to the main bottlenecks of the system, distance transform computation and correlation, resulted in a speed-up of a factor of 3–4.

Application: traffic sign detection

We tested the Chamfer System on the problem of traffic sign detection as an alternative to the colour system described above (Gavrila, 1999b). Specifically, we aimed to detect circular, triangular (up/down) and diamond-shaped traffic signs, as seen in urban traffic and on secondary roads. We used templates for circles and triangles with radii in the range of 7–18 pixels (the images are of size 360 by 288 pixels). This led to a total of 48 templates, placed at the leaf level of a three-level hierarchy. In order to optimize for speed, we chose to scale the templates (off-line), rather than scale the image (on-line).

We did extensive tests on the traffic sign detection application. Off-line, we used a database of several hundred traffic sign images, taken during both day- (sunny, rainy) and night-time conditions. Under good visibility conditions, we obtained high single-image detection rates, typically of over 95 per cent, when allowing solutions to deviate by 2 pixels and by radius 1 from the values obtained by a human. At this rate, there were a handful of detections per image which were not traffic signs, on average. These false positives were overwhelmingly rejected in a subsequent verification phase, where a RBF network (see Section 6.4) was used as pictograph classifier (the latter could distinguish about 10 pictographs); see Figure 6.13.

The traffic signs that were not detected, were either tilted or otherwise, reflected difficult visibility conditions (e.g. rain drops, partial occlusion by window wiper, direct sunlight into camera). Under the latter conditions, detection rates could decrease by as much as 30 to 65 per cent. We spent many hours testing our system on the road. The traffic sign detection (and recognition) system currently runs at 10–15 Hz on a 600 MHz Pentium processor with MMX. It is integrated in our Urban Traffic Assistant (UTA II) demonstration vehicle.

Application: pedestrian detection

The second application of the Chamfer System was the detection of pedestrians. Not surprisingly, it is more challenging than traffic sign detection since it involves a much larger variety of shapes that need to be accounted for. Furthermore, pedestrian contours are less pronounced in the images and more difficult to segment. Note that with a few exceptions much of the previous work on 'looking at people' has involved a static camera, see Gavrila (1999a) for an overview; initial segmentation was possible by background subtraction. That 'luxury' is not given to us here, because of a moving vehicle.

Fig. 6.13 Shape-based traffic sign detection (and recognition) with the Chamfer System.

We compiled a database of about 1250 distinct pedestrian shapes at a given scale; this number doubled when mirroring the templates across the y-axis. On this set of templates, an initial four-level pedestrian hierarchy was built, following the method described in Section 6.3.2. In order to obtain a more compact representation of the shape distribution and provide some means for generalization, the leaf level was discarded, resulting in the three-level hierarchy used for matching (e.g. Figure 6.12) with about 900 templates at the new leaf level, per scale. Five scales were used, covering a size variation of 50 per cent. Our preliminary experiments on a database of a few hundred pedestrian images (distinct from the sequences used for training) resulted in a detection rate of about 75–85 per cent per image, with a handful false-positives per image. These numbers are for un-occluded pedestrians. See Figure 6.14 for a few detection results.

Figure 6.15 shows some potential false positives; typically they are found on trees or windows. Using the flat-world assumption and knowledge about camera geometry, we

Fig. 6.14 Shape-based pedestrian detection with the Chamfer System.

Fig. 6.15 Potential false positives.

have set region of interests for the template in the hierarchy, so that many erroneous template locations can *a priori* be excluded, speeding up matching greatly. The current pedestrian system runs at 2–5 Hz on a 600 MHz Pentium processor with MMX. It is part of our pedestrian detection and recognition system that is described in Section 6.5.

6.3.3 Road recognition

The standard applications of road recognition are lane keeping and lane departure warning. In the context of a stop-and-go system, road recognition has additional relevance:

- If following a vehicle, lateral guidance can be accomplished by driving along the trajectory of the leading vehicle. If its distance and the angle between the leading vehicle and the heading direction is measured, a lateral tow bar controller can be used approximately to follow the trajectory. This controller tends to cut corners. Its performance can be improved if the position of the ego-vehicle relative to the lane is known (Gehrig and Stein, 1999).
- When the leading vehicle changes lane, a simple-minded following leads to hazardous manoeuvres. As long as the camera exclusively observes the field in front of the vehicle, collision avoidance is no longer guaranteed if the autonomous vehicle departs the lane. Thus, an intelligent stop-and-go system should be able to register lane changes of the leading vehicle and return the control to the driver.
- If the leading vehicle gets lost, but the vehicle's position in the lane is known, a lateral controller should guide the vehicle at least for a while, so that the driver has enough time to regain control.
- The response of the autonomous car to a detected object depends on the object's position in the road topology. A pedestrian on a crosswalk is clearly more critical than a pedestrian on a sidewalk.

In Section 6.2 a lane recognition system for highways is presented. The vision component of that system has a top-down architecture. A highly constrained model is matched against lane markings in image sequences. As pointed out, such model-based tracking is not only efficient since the considered image regions are small but also very reliable since the vehicle kinematics and the road geometry and its continuity are integrated in the model.

In the urban environment a global model for the broad range of possible road topographies is not available. The roads are characterized by:

- lane boundaries of which the shape and appearance are often discontinuously changing
- an arbitrary road shape
- a lack of a unique feature such as markings.

In this scenario, a pure tracking approach that simply extrapolates previously acquired information is not sufficient. The detection capability of data driven algorithms is required. Unfortunately, such bottom-up control strategies are computationally expensive since they have to process large image portions. Our urban lane recognition system combines a data driven approach with the efficiency of model-based tracking (Paetzold and Franke, 1998).

The data-driven global detection generates road structure hypotheses at a cycle rate just high enough to keep up with the dynamic environment. The model-based tracking estimates a dense description of the road structure in video real time, a requirement for comfortable lateral guidance. As common in multiple target tracking, the tracking results are referred to as tracks (Bar-Shalom and Fortmann, 1988). The detected hypothesis and the already existing tracks comprise a pool of tracks.

Both parts are fused such that the required computational power is minimized. The goal of global detection is the separation of road structures such as markings, curbs and crosswalks from heavy background clutter. Unlike highway lane boundaries, their local intensity distribution is not very distinctive. Rather, global geometric properties as length, orientation and shape are utilized. The detection schemes rely on the assumptions that:

- lane boundaries are long
- lane boundaries and crosswalks are parallel, stop lines are orthogonal to the vehicle's trajectory
- road structures have linear shape or can partially be approximated by lines
- road structures are bands of constant brightness
- road structures lie in the 3D-road surface (which is the ground-plane by default).

These global characteristics can be derived from a polygonal edge image which is well suited to describe the linear shapes of road structures, see bottom left image of Figure 6.16. For related work, see Enkelmann *et al.* (1995).

The detection of road structures is facilitated through an inverse-perspective mapping of the edge image into bird's eye view. By removing the perspective distortion, the geometrical properties of the road structures in 3D-world are restored, illustrated in top right image of Figure 6.16. In that representation, the data is scanned for subsets of

Fig. 6.16 Detection of road structures. Clockwise from left top: detected crosswalk in an urban scenario, bird's eye view on road, typical pattern of vertically projected edges in presence of a crosswalk, edge image.

features that are consistent with a simple feature model of the road structure, embodying the above listed characteristics.

To speed up this procedure, the polygon data is perceptually organized. The interesting intrinsic elements are length, orientation and position; the interesting clustering elements are parallelism and collinearity. The polygon data can be filtered for instances of these regularities in an arbitrary, iterative order to minimize the computational load. This is particularly crucial when we are interested in more than one object type that all have the same basic regularities in common.

Stereopsis helps to discard non-road measurements. Image regions where obstacles are present, detected as described in Section 6.3.1, are ruled out. Furthermore, measurements that do not lie in the 3D-road surface are suppressed. Stereopsis also enables the separation of vertical from horizontal shape and motion (pitching). Contrary to monocular vision where the pitch angle and the camera height are determined by assuming lanes of constant width, a model of the vertical geometry is recursively fitted through the stereo observations. Within the presented lane recognition system, the vertical model is linear and estimated by a first order Kalman filter.

The model underlying the tracking process must account for the arbitrary shape of urban lane boundaries. No appropriate geometric parameter model of low order exists that approximates the global road shape which can have sharp corners, discontinuities and curvature peaks. Therefore, local geometric properties such as lane tangent orientation and curvature are estimated by means of Kalman filtering.

This model definition is equivalent to a local circular fit to the lane boundary. Since a circle approximates arbitrary curves only for short ranges with negligible error, these properties are estimated not at the vehicle's position as done on highways but at a distance z_0 ahead. The system dynamics is given by

$$\begin{aligned} \dot{x}_0 &= -v\Delta\psi - \dot{\psi}_{\text{sensor}}\, z_0, \\ \Delta\dot{\psi} &= -vc_0 + \dot{\psi}_{\text{sensor}},\; c < \frac{1}{80\,\text{m}} \\ \dot{c}_0 &= 0 \end{aligned} \tag{6.12}$$

The measurement equation is approximately given by

$$x_{\text{measured}}\,(z) = x_0 + (z - z_0)\Delta\psi + 1/2c_0(z - z_0)^2 \tag{6.13}$$

where z is the considered distance. Both equations turn into the standard highway equations for $z_0 = 0$ and $c_1 = 0$.

For each track in the pool the state vector is updated recursively. At each tracking cycle the track pool is subject to initiation, assessment, merging, verification, classification and deletion. The track assessment is the central operation. Its objective is to indicate whether a track is assigned to a lane boundary or to background clutter. An appropriate track quality is determined by lowpass-filtering the ratio of successful measurements and attempted measurements. This measure is used to trigger initiation and deletion.

Verification and classification are rule-based components. All tracks that are markings, parallel to already verified tracks or parallel to the vehicle's trajectory for a certain travelled distance, are labelled as verified. Verified tracks are classified into left/right

Fig. 6.17 (a) The vehicle ahead is driving in the adjacent lane. In turns, estimated lanes give valuable information about the topological traffic situation. (b) The lane is defined by tracked broken lane markings and a kerb. The leading vehicle drives in that lane.

lane boundary, other lane boundaries and clutter by evaluating the relative position of the tracks to the vehicle.

In Figure 6.17(a), the broken lane markings are tracked. The possible driving corridor takes a right turn, bypassing the light truck that is driving in another lane. In Figure 6.17(b), a kerb and broken lane markings are tracked, which define the lane of the autonomous car as well as the leading vehicle. The lane is defined by broken lane markings on the left side and a kerb on the right side. The vehicle ahead is centred in the lane.

The detection of crosswalks also draws from the principles of a data-driven strategy. It consists of an early detector relying on spectrum analysis of the intensity image in horizontal direction and an edge-based verification stage.

The early detector observes the intensity function in multiple image lines evenly distributed over the interesting range of sight. Inspecting these intensity functions shows that a zebra crossing gives rise to a periodic black and white pattern of a known frequency. The power spectrum of that signal is calculated. When the spectral power at the expected frequency exceeds a threshold, the scene is subject to further investigation. By projecting the edges parallel and orthogonal to their predominant direction, marginal distributions are produced. Those marginal distributions exhibit distinctive patterns in presence of crosswalks, see Figure 6.16 where the parallel projection is displayed. These patterns are analysed by a rule-based system.

Arrow recognition

Besides lane course, stop lines and crosswalks, arrows painted on the road are of interest. Our recognition of those arrows follows the two-step procedure, detection and classification, mentioned earlier. In contrast to the lane recognition, shape and brightness cues are used in a region-based approach. The detection steps consist of grey-value segmentation and filtering.

An adaptive grey-scale segmentation reduces the number of colours in the original image to a handful. In this application, we base this reduction on the minima and plateaux of the grey-scale histogram. Following this grey-scale segmentation the colour connected components analysis described in Section 6.2.2 is applied to the segmented image. The algorithm produces a database containing information about all regions

Fig. 6.18 (a) Street scene displaying a direction arrow. (b) The segmented and classified arrow.

in the segmented image. Arrow candidate regions are selected from the database by appropriate queries and merged to objects of interest.

The resulting set is normalized for size and given as input to a radial-basis-function classifier (see Section 6.4). It has been trained to about a thousand different arrow images taken under different viewing angles and lighting conditions. Figure 6.18 shows the original and the obtained result.

6.4 Object recognition as a classification problem

6.4.1 General aspects

Principles, pros and cons

In Section 6.3 we described methods for object detection. They yield information about the position and motion behaviour of an object, but we still have to find out about what type of object is concerned. The latter task is what we call *object recognition*.

After detecting an object that fulfils certain simple criteria concerning e.g. size and width-to-height ratio and eventually motion, the image region of interest (ROI) delivered by the detection stage is first cropped and scaled to a uniform size; eventually, a contrast normalization is performed afterwards. Our general approach is then *training instead of programming*; we thus regard the object recognition problem as a classification problem to be solved by classification techniques that all require a training procedure based on a large number of examples. The advantage of this approach is that no explicit models of the objects to be recognized have to be constructed, which would be a rather difficult, if not impossible task especially for largely variable objects, e.g. pedestrians. Robust recognition systems are obtained by performing bootstrapping procedures, i.e. beginning with a certain set of training samples, then testing the resulting system in the real-world environment and re-training the recognition errors in order to generate a new version of the system that then has to undergo the same procedure, and so on. The disadvantage of the model-free approach is of course that huge amounts of training and test data have to be acquired and labelled in order to be able to achieve a high generalization performance.

Our object recognition algorithms are based on the classification of single images when regarding rigid objects like traffic signs, traffic lights or rear views of cars; this approach is as well used for a fast preselection of ROIs delivered by the detection stage possibly containing pedestrians, to be processed by more complicated methods afterwards. To achieve a more reliable recognition of pedestrians, image sequences displaying the characteristic motion of the legs, i.e. the pedestrian's *gait pattern*, are classified as a whole by employing the novel *adaptable time delay neural network* (ATDNN) concept based on spatio-temporal receptive fields. This neural network is used both as a 'standalone' classification module and for computationally efficient dimension reduction of high-dimensional input feature vectors. The latter technique makes it possible to employ standard classification techniques like *polynomial classifiers* (Schürmann, 1996) and *support vector machines* (Schoelkopf, 1999) as well as a *radial basis function network* algorithm specially designed for the recognition of traffic signs (Kreßel *et al.*, 1999).

Description of the classification techniques

The **polynomial classifier** as described in Schürmann (1996) constructs a so-called *polynomial structure list* of multiplicative features from the original input features in the first layer as shown in Figure 6.19. These are referred to as *enhanced features*. The second layer is a linear combination of the enhanced features defined by the coefficient matrix W. While the polynomial structure list must be chosen by the designer of the classifier (e.g. complete quadratic as in Figure 6.19), the adaptation of the weight matrix W is performed using the training set. An important advantage of the polynomial classifier is that there exists an analytical solution for adapting the weight matrix W, if the *empirical risk*, i.e. the average over the training set of the sum of square differences between the actual classifier outputs and the corresponding desired values is minimized.

The **support vector machine (SVM)** (see e.g. Scholkopf, 1999) in its elementary form is a classification concept for two-class problems that constructs a hyperplane in feature space that separates the samples of the two different classes in a manner that is optimal in the sense that the euclidean distance between the samples and the separating plane is as large as possible. The underlying concept is that of *structural risk minimization*. In the context of perceptron learning, the perceptron whose weight vector defines this optimal hyperplane is called the *perceptron of maximum stability*; it is obtained by a special training procedure called the *AdaTron algorithm* (Anlauf and Biehl, 1990). The hyperplane is defined in terms of the training samples situated nearest to it only; these training samples are therefore called *support vectors*. As most

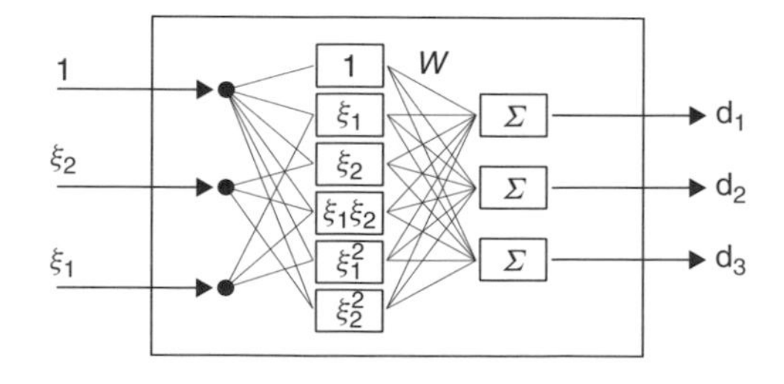

Fig. 6.19 Complete quadratic polynomial classifier for two features and three classes.

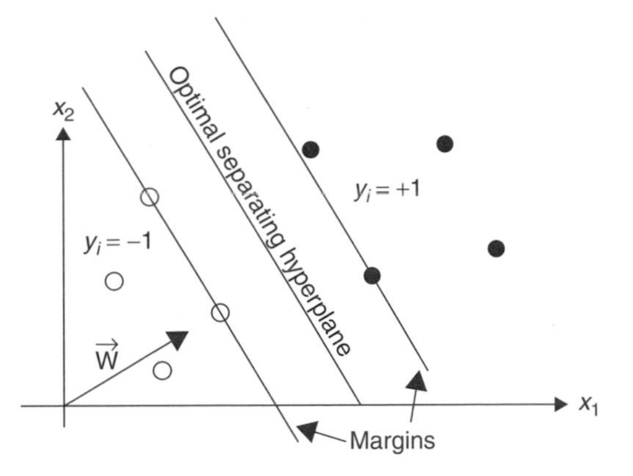

Fig. 6.20 Optimal separating hyperplane for a two-class problem.

realistic problems are not linearly separable in input space, i.e. it is impossible to find a hyperplane that perfectly separates the two classes in input space, a feature space of a usually much higher dimension is generated from the input space by a nonlinear transformation. The separating plane is constructed in this feature space; special procedures exist for handling distributions of training patterns that are still not linearly separable after transformation into feature space. Throughout this section, we will make use of the so-called *polynomial SVM* of a given order d, the feature space of which is spanned by polynomial combinations of the input features of up to dth order. Compared to other classification algorithms, the SVM concept yields a rather high generalization performance especially in the case of problems involving a relatively low number of training samples.

For the application of traffic sign recognition we developed a special **radial basis function (RBF)** classifier introduced in Kreßel *et al.* (1999). It consists of N reference vectors in feature space to each of which an object class c_i and two parameters a_i and b_i with $a_i < b_i$ are assigned. The number of object classes is denoted by K. For an unknown sample fed into the classifier, all N euclidean distances d_i in feature space between the sample and the reference vectors are calculated. The decision to which class the sample belongs is based on the value $R(d_i)$ which we define as:

$$R(d_i) = \begin{cases} 1 & \text{if} \quad d_i \leq a_i \\ \dfrac{b_i - d_i}{b_i - a_i} & \text{if} \quad a_i < d_i < b_i \\ 0 & \text{if} \quad d_i \geq b_i \end{cases} \tag{6.14}$$

The ramp function $R(d_i)$ is a radial basis function as it only depends on the euclidean distance d_i. In classical RBF classifiers (e.g. Poggio and Girosi, 1990) gaussians are used as radial basis function but the described ramp functions are more suitable for real-time applications due to their high computational efficiency. To be able to decide to which class the input sample has to be assigned we set:

$$\tilde{P}_k = \sum_{i=1, c_i=k}^{N} R(d_i), \quad S_{\tilde{P}} = \sum_{k=1}^{K} \tilde{P}_k \tag{6.15}$$

As a measure for the probability that the sample belongs to class k we then define:

$$P_k = \begin{cases} \tilde{P}_k/S_{\tilde{P}} & \text{if} \quad S_{\tilde{P}} > 1 \\ \tilde{P}_k & \text{if} \quad S_{\tilde{P}} \leq 1 \end{cases} \tag{6.16}$$

and as a measure for the probability that the sample belongs to none of the object classes (*reject probability*):

$$P_{\text{reject}} = \begin{cases} 1 - S_{\tilde{P}} & \text{if } S_{\tilde{P}} \leq 1 \\ 0 & \text{if } S_{\tilde{P}} > 1 \end{cases} \tag{6.17}$$

The input sample is assigned to the class with the highest P_k value; if P_{reject} is larger than all P_k values, the sample is assigned to an additional reject class. The sample is as well rejected if the highest P_k value is lower than a given threshold t with $0 < t < 1$. Varying t and measuring the rate of rejected test samples yields the *receiver operating characteristics* (ROC) curve of the classifier.

The reference vectors of the RBF classifier are the centres of clusters which are derived from the training examples divided into the K training classes. They are determined by an agglomerative clustering algorithm. The ramp parameters a_i and b_i of the ith radial basis function are defined in terms of the distance to the nearest cluster centre of the same class, the distance to the nearest cluster centre of one of the other classes, the average mutual distance of all clusters within each class k and the corresponding average over all K classes. Details are given in Kreßel *et al.* (1999).

For classification of image sequences we developed the **adaptable time delay neural network (ATDNN)** algorithm presented in detail in (Wöhler and Anlauf, 1999a, b). It is based on a time delay neural network with spatio-temporal receptive fields and adaptable time delay parameters. The general time delay neural network (TDNN) concept is well known from applications in the field of speech recognition (Waibel *et al.*, 1989). An important training algorithm for the TDNN, named *temporal backpropagation*, is presented in (Wan, 1990). A training algorithm for adaptable time delay parameters is developed in (Day and Davenport, 1998) for continuous time signals (*Continuous-time temporal backpropagation*). The related *Tempo 2 algorithm* is described in (Bodenhausen and Waibel, 1991) for input data defined at discrete time steps, as is the case for image sequences; the input window, however, is restricted to gaussian shape.

The architecture of the ATDNN is shown in Figure 6.21. The three-dimensional input layer receives a sequence of images acquired at a constant rate, as is the case especially for video image sequences. The activation of the input neuron at spatio-temporal position (x, y, t) corresponds to the pixel intensity at position (x, y) on the tth image of the input sequence. In the ATDNN architecture, a neuron of a higher layer does not receive input signals from all neurons of the underlying layer, as is the case, e.g. for multi-layer perceptrons (MLPs), but only from a limited region of it, called the *spatio-temporal receptive field* of the corresponding neuron. Such a spatio-temporal receptive field covers a region of $R_x \times R_y \times T_{\text{eff}}^{(1)}$ pixels in the input sequence with $T_{\text{eff}}^{(1)} = 1 + (R_t - 1)\beta$ and R_t as the number of weight sets that belong to the same time slot, respectively. We call β the *time delay parameter*. The distance of the centres of two neighbouring spatio-temporal receptive fields is given by D_x, D_y, and D_t, where we constantly take $D_t = 1$. For each neuron of layer 2 and higher,

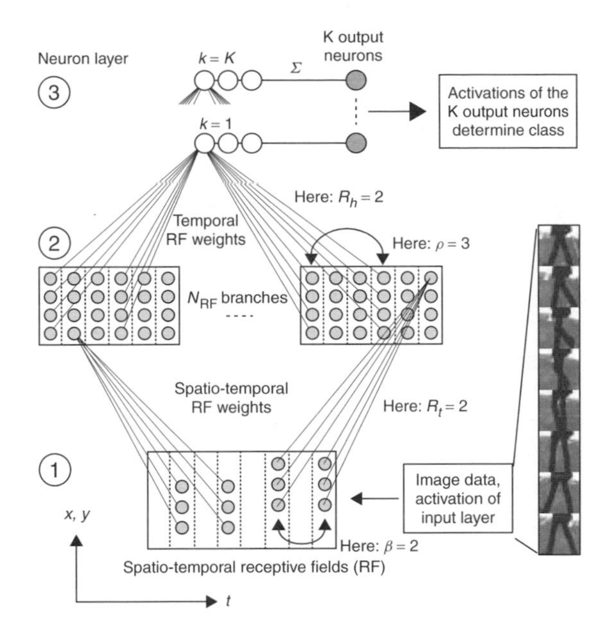

Fig. 6.21 Architecture of the adaptable time delay neural network.

we set $g(x) = \tanh(x)$ as a sigmoid activation function. The ATDNN is composed of NRF different branches, each of which consists of a three-dimensional layer of neurons (layer 2 in Figure 6.21). As we follow the *shared weights* principle inside each branch, the same set of weight factors is assigned to each layer 2 neuron of a certain branch. Effectively, this configuration of spatio-temporal receptive fields produces activation patterns in neuron layer 2 representing one spatio-temporally filtered version of the original input image sequence per network branch. Neuron layer 2 and 3 are fully connected in the spatial directions; in the temporal direction, however, we implemented a structure of temporal receptive fields and shared weights with time delay parameter ρ. The extension of the temporal receptive fields between neuron layer 2 and 3 amounts to $T_{\text{eff}}^{(2)} = 1 + (R_h - 1)\rho$, R_h standing for the number of weight sets belonging to the same time slot, respectively. To each branch s and each output class k one shared set of temporal receptive field weights is assigned. The resulting activations of neuron layer 3 are then averaged classwise to obtain the output activations ω_k for the K output classes to which the ATDNN is trained.

The ATDNN as shown in Figure 6.21 is only defined for integer-valued time delay parameters β and ρ. The output of the ATDNN for real-valued time delay parameters is obtained by bilinear interpolation between the output values resulting from the neighbouring integer-valued time delay parameters. We use a backpropagation-like on-line gradient descent rule to train the ATDNN weights and time delay parameters, referring to a standard quadratic error measure. Details can be found in Wöhler and Anlauf (1999a, b).

Techniques for dimensionality reduction

The comparably high dimensionality of images or image sequences to be classified poses difficulties for many standard pattern recognition techniques, a problem sometimes known as the 'curse of dimensionality'. It is therefore often necessary to reduce the dimensionality of the patterns, preferably by techniques that conserve class-specific properties of the patterns while discarding only the information that is not relevant for the classification problem.

A well-known standard method for dimensionality reduction is the principal component analysis (PCA) algorithm (for a thorough introduction, see e.g. (Diamantaras and Kung (1996), Schürmann (1996)). In a first step, the covariance matrix C of the distribution of training patterns is computed. The size of C is $N \times N$, where N denotes the dimension of the feature space in which the original patterns are defined. Then the N eigenvalues and corresponding eigenvectors of C are calculated and ordered according to the size of the eigenvalues; as C is necessarily positive semidefinite, all eigenvalues of C are non-negative. For further processing, the training patterns are then expanded with respect to the M most significant eigenvectors ('principal components') of C, i.e. the eigenvectors belonging to the M largest eigenvalues of C, neglecting the remaining $N - M$ eigenvectors. The obtained M expansion coefficients ($M < N$) are used as new features for classification. The basic property of the PCA algorithm is that it minimizes the Euclidean distance in the original N-dimensional feature space between an original pattern and its reconstructed version obtained by expansion with respect to the M principal components ('reconstruction error'). This does not necessarily mean that the information needed for classification is as well preserved in an optimal manner; in many practical applications, however, the PCA algorithm turns out to yield a very reasonable performance. Difficulties may again arise in the case of very high-dimensional patterns, i.e. large values of N, as this leads to problems concerning numerical stability when trying to diagonalize the covariance matrix C by standard numerical methods such as the Jacobi algorithm.

A further technique for dimensionality reduction that is specially adapted to process image or image sequence data consists of an extension of the previously described ATDNN algorithm. Apart from using the ATDNN 'standalone' as a classification module after training, we can as well regard the activation values of the neurons in the second layer as feature vectors to be processed, e.g. by the first three described classification techniques. For this purpose we employ a slightly simplified version of the ATDNN with integer-valued temporal extensions of the receptive fields (see Wöhler and Anlauf, 1999a). The ATDNN then serves as a preprocessing module that reduces the dimension of the input patterns (Wöhler *et al.*, 1999a). Especially, we combine the ATDNN in this manner with the RBF classifier for traffic sign recognition and with support vector machines for pedestrian recognition.

6.4.2 Traffic lights and signs

Traffic signs

We have developed traffic sign recognition systems for highways as well as urban traffic. As an example, the recognition of circular traffic signs, i.e. speed limits, passing restrictions, and related signs for ending restrictions (see Figure 6.22) is presented here. Grey-scale images of size 360×288 pixels are the basis for the investigation. The extracted regions of interest are scaled to 16×16 pixels and normalized in contrast. The general traffic sign class can be split up into 5 or 12 subclasses as shown in Figure 6.23.

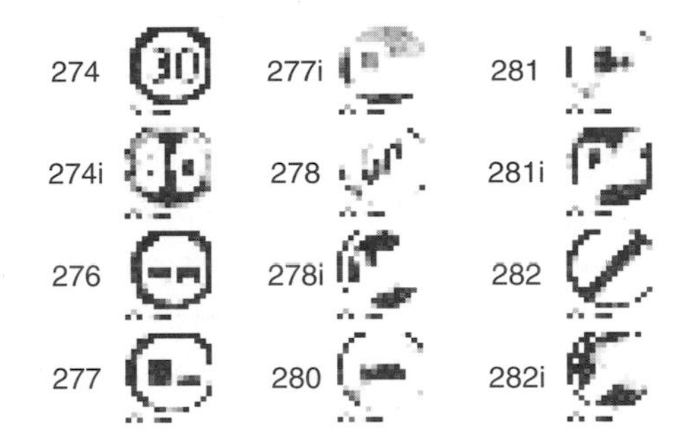

Fig. 6.22 The examined traffic sign classes with their labels according to German law.

label	#elements	13 classes	6 classes	2 classes
274	3787	class 01	class 1	class 1
274i	384	class 02		
276	1235	class 03	class 2	
277	178	class 04		
277i	34	class 05		
278	378	class 06	class 3	
278i	3	class 07		
280	118	class 08	class 4	
281	40	class 09		
281i	14	class 10		
282	119	class 11	class 5	
282i	5	class 12		
garb	19473	class 13	class 6	class 2
sum	25768			

Fig. 6.23 Composition of the training set.

The RBF classifier is trained to the five traffic sign subclasses and one garbage class only. For RBF networks the number of subclasses does not influence the recognition performance or the computational complexity of a classification cycle. A two-dimensional version of the ATDNN for processing single images is used to reduce the dimension of the original input patterns from 256 to a value of 64.

A combination of principal component analysis (PCA) and polynomial classifier is used as a reference technique that is well known from applications in the field of text analysis, especially handwritten digit recognition (see Franke (1997)). The polynomial classifier is applied to the 2, 6, and 13 class split up; here it turned out that the recognition performance as well as the computational complexity is rising with the number of subclasses. The dimension of the original input patterns is reduced by PCA to values of 40 and 50, respectively.

Concerning the recognition performance, the most interesting point is the trade-off between the false positive rate versus the rate of traffic signs that are rejected or explicitly classified as garbage. The corresponding results obtained from about 7000 separate test samples are shown in Figure 6.24. The errors among the various traffic sign subclasses are always smaller than 0.1 per cent. As a general result it comes out that the polynomial classifier yields a slightly higher overall recognition performance

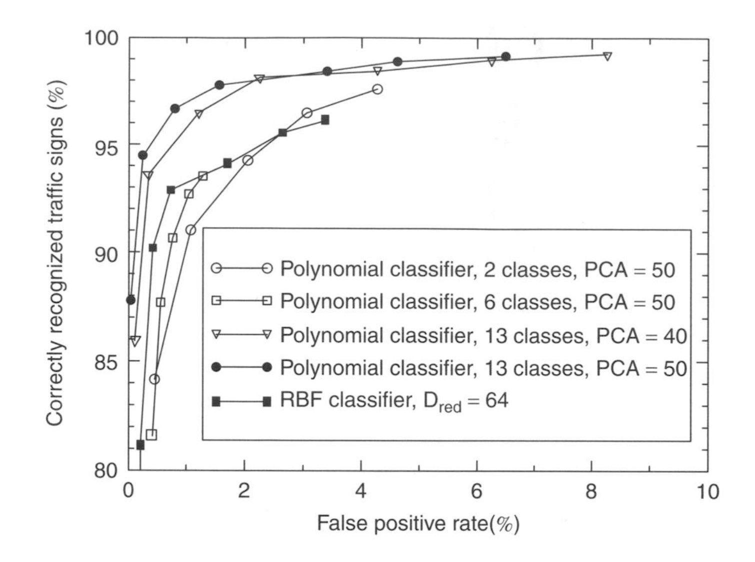

Fig. 6.24 ROC curves on the test set for several classifier settings.

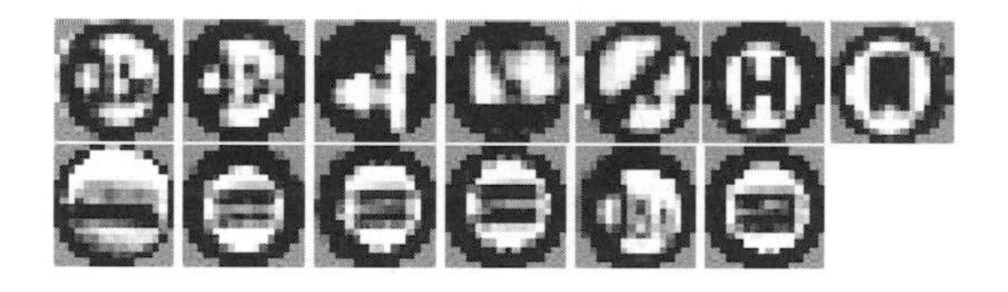

Fig. 6.25 The false positives yielded by a second degree polynomial classifier applied to 50 features obtained by PCA.

than the RBF classifier. For our real-time application, however, we choose so far the RBF network, since it is almost completely implemented by fast table look-ups such that one classification process needs only about 3 milliseconds of CPU time on the Pentium II system of our test vehicle. Moreover, it is simpler with the local RBF network method to take into account special garbage patterns which are very close to the speed limit signs in feature space. An example is the bus stop ('H') sign appearing in Figure 6.25. Adding a number of such examples to the training set creates a rather limited 'island' in feature space for this pattern with speed limit sign clusters around it.

The recognition of speed limits has to be further extended as generally one wants to know not only that there is a speed limit but also what is the maximum allowed speed. We thus added a second classification stage to read the inlays, i.e. the numbers on the traffic signs, which is activated when the first classification stage described above has recognized a speed limit sign. The dimension of the input patterns is again reduced by the two-dimensional ATDNN version for single images. One classification process needs only about 0.5 milliseconds of CPU time on the Pentium II system of our test vehicle. For a rate of 0.4 per cent of incorrectly classified inlays, 92 per cent of the inlays are correctly classified, with the remaining samples being rejected.

Traffic lights

Traffic lights are detected using the colour segmentation techniques described in Section 6.5. The detection stage is extracting blobs in the traffic light colours red, yellow and green; around each blob a region of interest (ROI) is cropped that contains not only the blob itself but also, if a traffic light has been detected, the dark box around it. The size of the ROI is thus related to the size of the detected blob. As the colour of the traffic light candidate is already known by the segmentation procedure, classification is performed based on the grey-scale version of the ROI only.

The ROI is first scaled to a size of 32 × 32 pixels. In order to enhance the contrast between the box of the traffic light and the background, which is often very weak, we perform a local contrast normalization by means of a simulated Mahowald retina (see Mead (1985)). We have three training and test sets, one for red, one for yellow, and one for red-yellow traffic lights, as shown in Table 6.1.

As our colour camera displays most green traffic lights as white blobs we could not generate a large set of well-segmented green traffic lights; on the detection of a green blob, we thus flip the corresponding ROI at its horizontal axis and classify this flipped ROI with the module designed for red traffic lights. This workaround will of course become obsolete with a high dynamic range colour camera of a sufficiently high resolution.

The recognition performance of the three classification modules for the different traffic light types is very similar, such that in this summary we only present the ROC curve of the classification module for red traffic lights. We compare the performance of the two-dimensional version of the ATDNN with spatial receptive fields of size $R_x = R_y = 13$ pixels, applied at an offset of $D_x = D_y = 6$ pixels, to the performance of a first and second order polynomial SVM and a linear polynomial classifier that have all been applied directly to the preprocessed ROIs of size 32 × 32 pixels (Figure 6.26). It becomes very obvious that with respect to the recognition performance, the 'local' ATDNN concept of spatial receptive fields is largely superior to 'global' approaches not explicitly taking into account the neighbourhood of the respective pixel. Both the first and the second order polynomial SVM separate the training set with only one error; we obviously observe overfitting effects due to systematic differences between training and test set which illustrate a very low generalization capability of the global approaches in this special application.

6.4.3 Pedestrian recognition

Single images

Possible pedestrians in the scene are detected by applying the stereo vision algorithm described in Section 3.1. Around each detected object an image area (ROI) of 1 m

Table 6.1

	Traffic lights	Training set garbage	Sum	Traffic lights	Test set garbage	Sum
Red	540	1925	2465	172	400	572
Yellow	292	1292	1584	80	400	480
Red-yellow	395	1292	1687	80	400	480

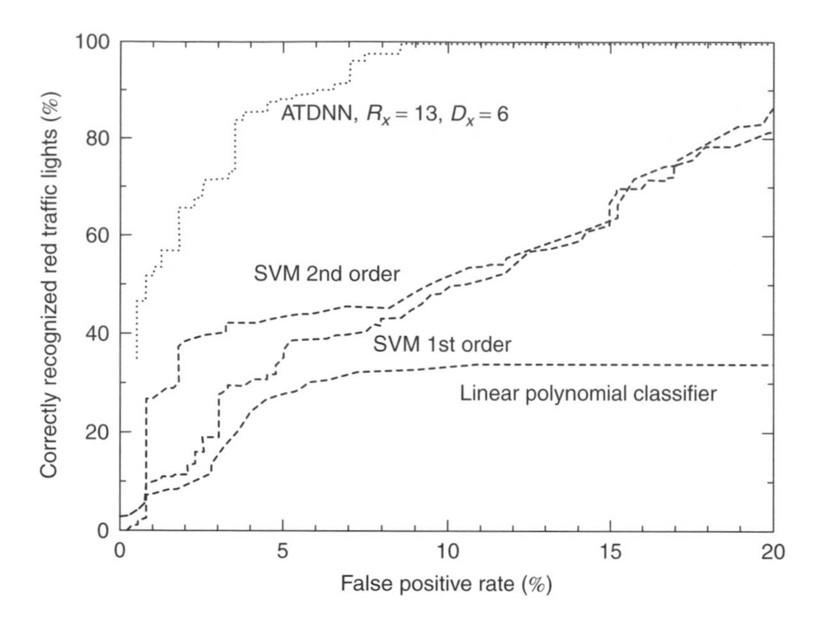

Fig. 6.26 Recognition performance of several classification modules for red traffic lights on the test set.

Fig. 6.27 Observing a traffic light at a crossing.

width and 2 m height, taking into account the object distance, is cropped, such that if the object is a pedestrian the resulting bounding box circumscribes it. These ROIs are scaled to a size of 24×48 pixels but not further preprocessed. Typical training samples are shown in Figure 6.28. The training set consists of 1942 pedestrian and 2084 garbage patterns, the test set of 600 pedestrian and 907 garbage patterns. The two-dimensional version of the ATDNN is again used for classification. Combining the

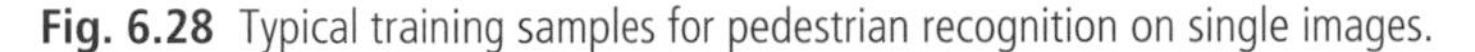

Fig. 6.28 Typical training samples for pedestrian recognition on single images.

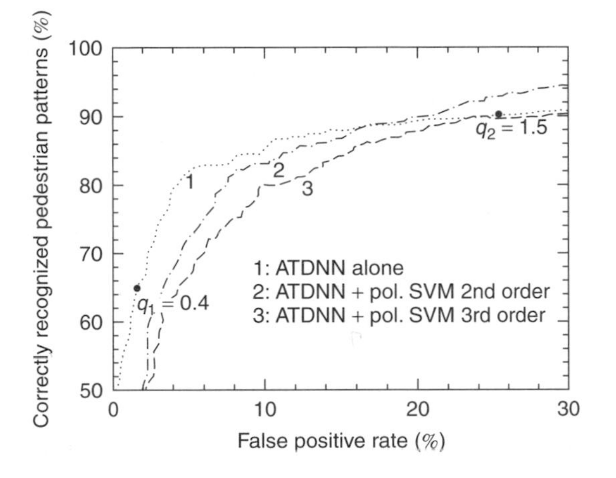

Fig. 6.29 ROC curves of the single image ATDNN on the test set.

ATDNN with a polynomial SVM of order 2 and 3 does not increase the recognition performance (see Figure 6.29).

The single image ATDNN tends to incorrectly classify vertically elongated shapes such as pedestrians to the extent that it should be combined with more complex approaches. The recognition result is thus only accepted if the network is rather 'sure' to have made a correct decision (see Figure 6.30), i.e. if the network output is lying well inside the pedestrian or the garbage region in decision space. Intermediate network outputs are regarded as a 'don't know' state. If the stereo vision algorithm detects no lateral motion of the object, the chamfer matching algorithm is activated, otherwise the full image sequence version of the ATDNN is used to more reliably classify the object based on combined shape and motion features. The single image ATDNN acts as a very fast preselection stage (the CPU time per classification process is about 1 ms) that eventually triggers computationally more complex classification modules. This decision structure is illustrated in Figure 6.30.

Image sequences

After detection of an object spatially emerging from the background by the stereo vision algorithm, we crop the lower half of the ROI delivered by the stereo vision

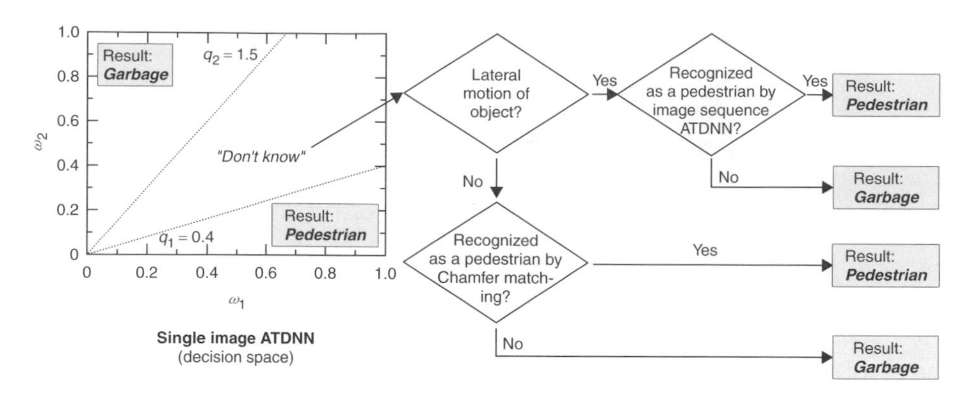

Fig. 6.30 Decision structure of the system for pedestrian recognition.

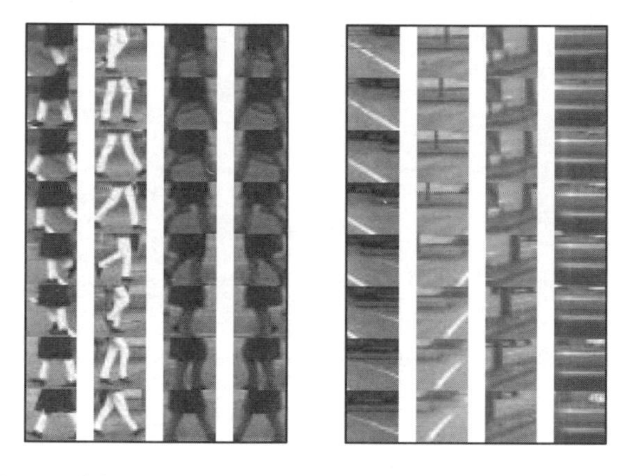

Fig. 6.31 Typical pedestrian (left) and garbage patterns (right).

algorithm, which will contain the pedestrian's legs, and normalize it to a size of 24×24 pixels. The time step between two subsequent frames of the sequence is 80 ms. Eight subsequent normalized ROIs are ordered into an image sequence covering a temporal range thus approximately corresponding to one walking step. After each stereo detection procedure, the batch of images is shifted backward by one image, discarding the 'oldest' image while placing the new image at the first position, resulting in an overlap of seven images between two sequences acquired at subsequent time steps. By a tracking algorithm based on a Kalman filter framework which is combined with the stereo vision algorithm it is guaranteed that each image of an input sequence displays the same object (see also Wöhler *et al.*, 1998). Our aim is again to distinguish between pedestrian and garbage patterns, resulting in two training classes typical representatives of which are shown in Figure 6.31. Our training set consists of 3926 pedestrian and 4426 garbage patterns, the test set of 1000 pedestrian and 1200 garbage patterns. It turned out that the performance on the test set is best for $N_{RF} = 2$ network branches and spatio-temporal receptive fields of a spatial size of $R_x = R_y = 9$ pixels, applied at an offset of $D_x = D_y = 5$ pixels. We performed seven training runs with different initial configurations of the time delay parameters, resulting in four 'optimal' ATDNN

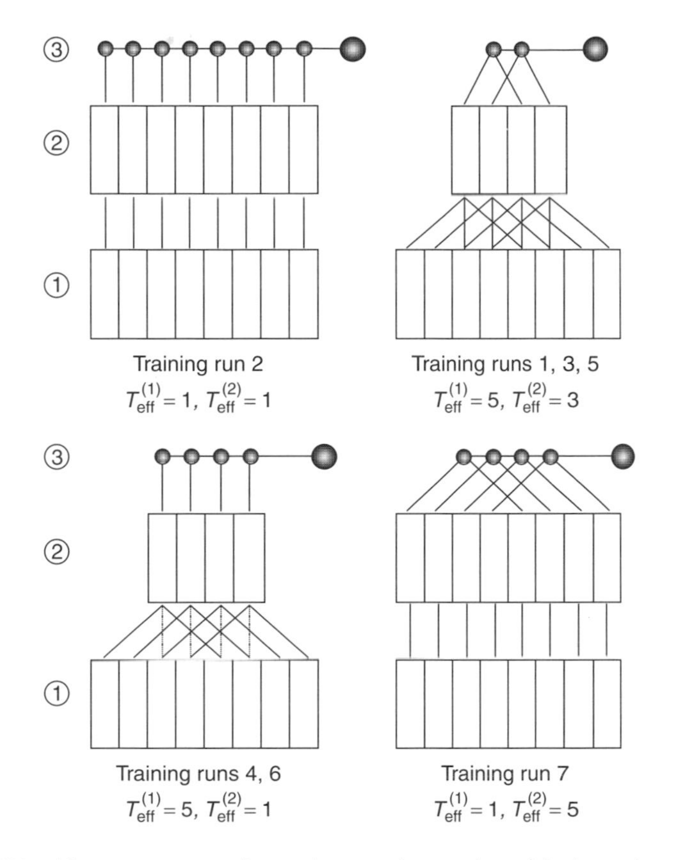

Fig. 6.32 ATDNN architectures corresponding to the approximate values of the learned temporal extensions of the respective fields, resulting from the seven performed training runs. The connections drawn as dashed lines only exist for $R_t = 3$.

architectures as depicted in Figure 6.32, the ROC curves of which are shown in Figure 6.33. Obviously, for the configuration $T_{\mathrm{eff}}^{(1)} = T_{\mathrm{eff}}^{(2)} = 1$ (training run 2) the lack of a temporal receptive field structure significantly reduces the recognition performance.

According to Wöhler and Anlauf (1999a), we trained a TDNN with fixed time delay parameters derived from the configuration on the upper right in Figure 6.32, the ROC curve of which is also shown in Figure 6.33. The performance could be slightly enhanced by combining this TDNN with a second and third order polynomial SVM (see Figure 6.35); the length of the feature vector processed by the SVM is reduced to $D_{\mathrm{red}} = 128$.

The recognition rates as given by Figures 6.33 and 6.35 refer to single input patterns such that by integration of the results over time our system recognizes pedestrians in a very stable and robust manner. In Figure 6.34, typical scenes are shown. The black bounding boxes have been determined by the stereo vision algorithm for the left stereo image, respectively. In the upper part of the image, the input of the ATDNN, i.e. the scaled ROI on the current and on the seven preceding images, is shown. On the second sequence, a pedestrian and a garbage pattern are detected simultaneously.

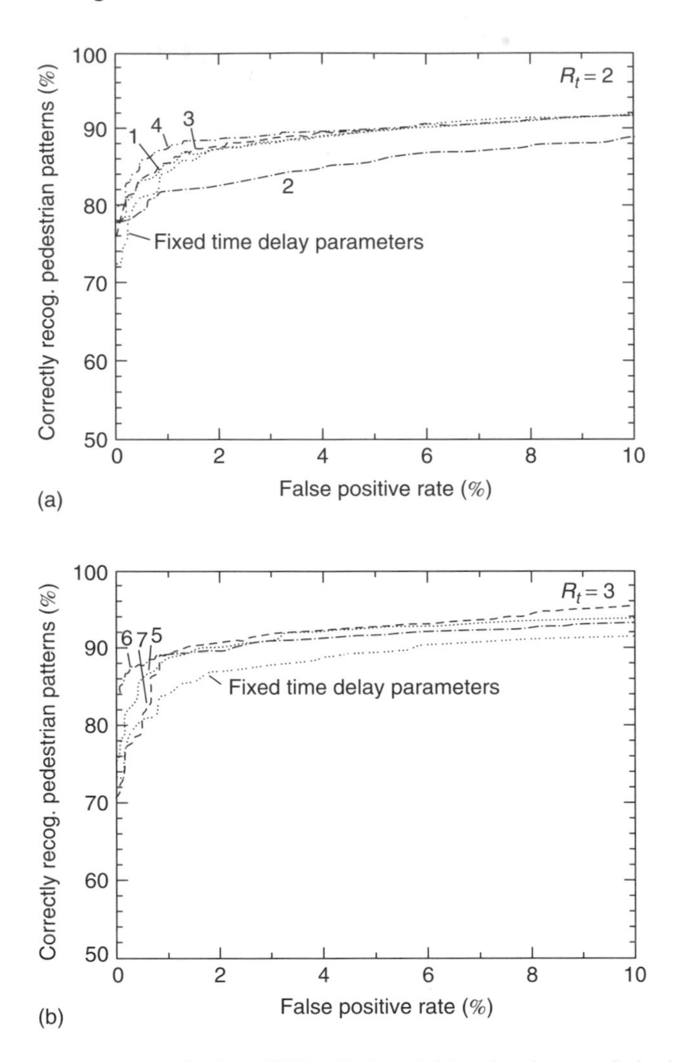

Fig. 6.33 ROC curves on the test set for the ATDNN, with $R_t = 2$ (above) and $R_t = 3$ (below). The ROC curve for the TDNN with $R_t = 5$ and $R_h = 3$ and fixed time delay parameters $\beta = \rho = 1$ is shown for comparison.

Global approaches versus local spatio-temporal processing

We will now compare the ATDNN approach based on local spatio-temporal feature extraction by receptive fields and its combination with polynomial SVMs to standard 'global' classification approaches, i.e. polynomial SVMs applied directly to the image sequences and after dimension reduction by principal component analysis (PCA).

For direct classification by a polynomial SVM the image sequence is regarded as a pixel vector of length $24 \times 24 \times 8 = 4608$. This approach is related to the one described in Papageorgiou and Poggio (1999), where SVM classifiers are applied to vectors consisting of temporal sequences of two-dimensional spatial Haar wavelet features. We adapted a second and third order polynomial SVM to the training set which could be perfectly separated in both cases. To reduce the very high dimension of the image sequences by PCA we took into account only the subspace of the

Fig. 6.34 Recognition of pedestrians in an urban traffic environment.

$D_{\text{red}} = 128$ most significant eigenvectors, which we obtained by using the well-known unsupervised perceptron learning technique known as 'Oja's rule' (Diamantaras and Kung, 1996). We then adapted a second and third order polynomial SVM to the correspondingly transformed training set, which again led to perfect separation in both cases. The local ATDNN-based classification approaches are somewhat superior to the global approaches with respect to their recognition performance on the test set (Figure 6.35). Regarding computational complexity and memory demand, values which are of high interest when it is intended to integrate the classification modules into real-time vision systems running on hardware with limited resources, the local approaches are largely more efficient, as becomes obvious in Figure 6.36.

6.4.4 Further examples

Recognition of rear views of cars

The two-dimensional version of the ATDNN used for object recognition on single images as described in Sections 6.4.2 and 6.4.3 about traffic sign and traffic light recognition and pedestrian recognition is furthermore used for the recognition of the rear views of cars. This is an important application for intelligent stop-and-go, as before focusing on an object to follow that has been detected by the stereo vision algorithm, the system should be able to verify that it is indeed looking at the rear of a car. At a false positive rate of 3% (6%), a fraction of 80% (90%) of rear views of cars are correctly classified. The recognition errors occur in a temporally uncorrelated manner.

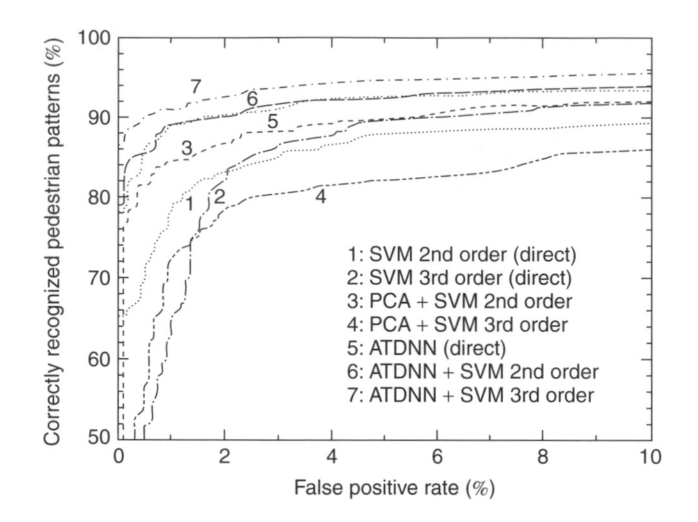

Fig. 6.35 ROC curves of the described classification modules for pedestrian recognition.

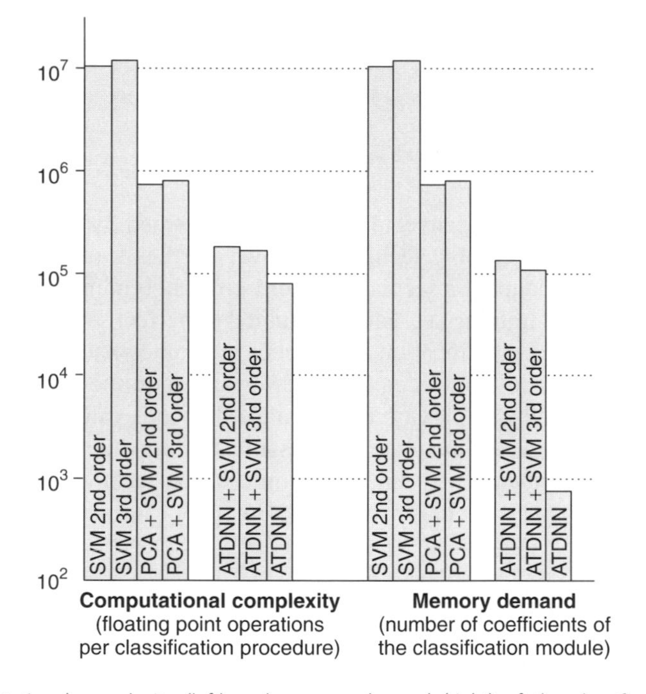

Fig. 6.36 Computational complexity (left) and memory demand (right) of the classification modules for pedestrian recognition. Note that the *y* axis is logarithmic.

As during autonomous car following the object to be followed is detected and tracked for a rather long time, integration of the single image recognition results over time yields a very stable behaviour of the system, recognizing all well-segmented cars and producing a hardly noticeable rate of false alarms. The module is again very fast as it requires less than 1 ms of CPU time per classification process.

Segmentation-free detection of overtaking vehicles

On motorways, our autonomous driving algorithm allows following the leading vehicle not only in stop-and-go traffic but also at speeds of up to about 130 km/h. To enable the system to select a new leading vehicle driving at a higher speed than the current one, it is necessary to have a system that permanently observes the left lane, assuming that the ego-vehicle is driving on the right lane. A single wide-angle camera is sufficient to perform this task; it is not necessary to employ stereo vision.

The input image sequences for the ATDNN are obtained by cropping a region of interest (ROI) sized 350 × 175 pixels at a fixed position out of the left half of each half frame, i.e. *no segmentation stage is involved.* This ROI is then downsampled to a size of 32 × 32 pixels. As an overtaking process takes about one second and the grabber hardware is able to grab, crop, and scale four ROIs per second, four subsequent ROIs are ordered into an image sequence, respectively, forming now an input pattern to the ATDNN. An example of such an overtaking process as well as the ROC curve of our system is given in Figure 6.37. Here, the rate of correctly detected vehicles does not refer to single patterns but to complete overtaking processes; the false positive rate denotes the fraction between the time during which an overtaking vehicle is erroneously detected and the time during which in fact no overtaking vehicle is present. Our test set corresponds to a 22 minutes drive on the motorway in dense traffic, containing 150 overtaking processes. The false positive rate of the system can be reduced by an order of magnitude by further analysing over time either the trajectory of the two ATDNN output values in decision space or the temporal behaviour of an appropriately averaged single output value by means of a simple second classification stage. This procedure is described in detail in Wöhler *et al.* (1999b).

6.5 Building intelligent systems

Driver assistance systems are challenging not only from the algorithmic but also from the software architecture point of view. The architectures of most driver assistant systems are usually tailored to a specific application, e.g. lane keeping on highways. The functionality is achieved by a few computer vision and vehicle control modules, which are connected in a hard-wired fashion. Although this kind of architecture is suitable for a lot of applications, it also has some disadvantages:

- The architecture is not scalable for a larger number of modules.
- There is no uniform concept for the cooperation of modules (e.g. for sensor fusion).
- New applications usually require extensive re-implementations.
- Reuse of old modules can be difficult due to missing interfaces.
- Hard-wired modules cause great efforts in maintenance since the dependencies of the modules are high. This is especially true for large systems.

The growing complexity of autonomous systems and our aim to realize a comprehensive assistance system hence reinforces the development of software architectures, which can deal with the following requirements:

- integration and cooperation of various computer vision algorithms
- different abstraction levels of perception and action

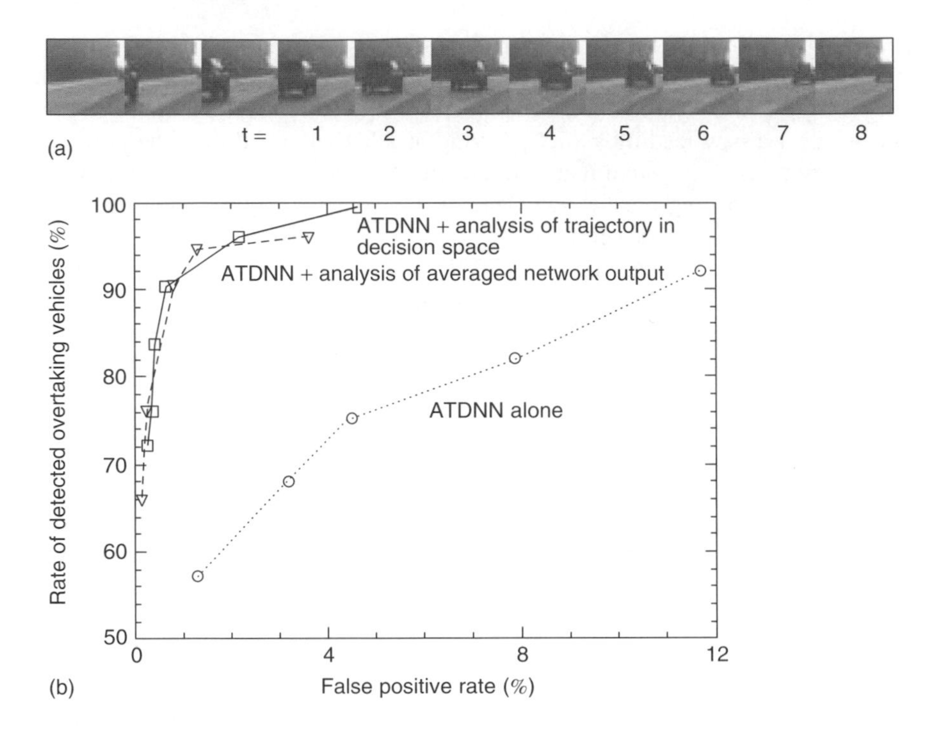

Fig. 6.37 (a) Typical example of an overtaking process. (b) Detection performance of the ATDNN with and without a second classification stage.

- sensor fusion
- economical use of resources
- integration of new algorithms without a complete redesign of the system
- simple enhancement to new computer vision applications
- distributed computing.

To meet these requirements, a multi-agent system was developed. In our demonstrator UTA II, the 'Agent NeTwork System' (ANTS) administrates computer vision, vehicle control and driver interface processes (Görzig and Franke, 1998). It selects and controls these algorithms and focuses the computational resources on relevant tasks for specific situations. For example, there is no need to look for new traffic signs or continuously determine the lane position, while the car is slowing down in front of a red traffic light.

6.5.1 ANTS: a multi-agent system

Agent software is a rapidly developing area of research. Since heterogeneous research is summarized under this term there is no consensus definition for 'agent' or 'multi-agent' system. A working definition of a multi-agent system (MAS) can be defined as 'a loosely-coupled network of problem solvers that work together to solve problems that are beyond their individual capabilities' (O'Hare and Jennings, 1996). The smallest entity of a MAS is an agent. An agent can be described as a computational entity, which provides services and has certain degrees of autonomy, cooperation and communication.

With these definitions it is not difficult to see how a MAS can be applied for autonomous vehicle guidance. Each computer vision or vehicle control module contains some functionality which can be useful for driver assistant systems. The combination of these modules allows more complex applications like autonomous stop-and-go driving in urban environments. The idea is to add the missing autonomy, cooperation and communication to the modules to create a MAS.

The main components of ANTS are a distributed data base, the administrators and the modules. Figure 6.38 visualizes the architecture.

The modules are the computational entities of the system, whereas the administrators contain the autonomy and the cooperation of the software agents. Combined with the communication ability of the distributed data base they build the MAS as described above. This distinction between a computational component and the 'intelligent' component of an agent has several advantages:

- The components can be executed in parallel.
- Existing software modules can be reused.
- The modification of one component does not necessarily require a modification of the other components.

Although ANTS is a generic MAS for various applications, we will focus in the following on driver assistance systems on our UTA II demonstrator and explain the main components in more detail.

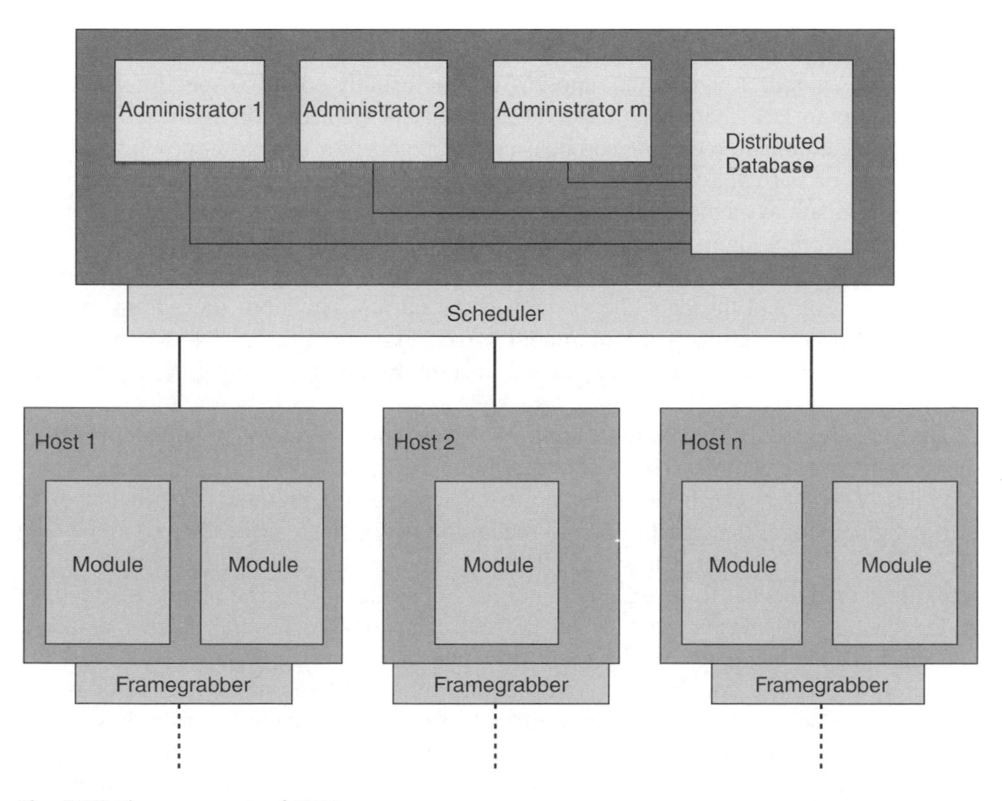

Fig. 6.38 The components of ANTS.

Distributed database

The central component of ANTS is the distributed database. It contains all incoming and outgoing data of the modules and distributes it to the computational nodes. This database typically contains information on the symbolic level and is used, e.g. for driver information or for the cooperation of modules.

The database has several access methods for its data such as exclusive write and concurrent read. For example, the stereo obstacle module can access its symbolically represented results in the exclusive write mode to track old and detect new obstacles, whereas the visualization module can access the results of all computer vision modules concurrently to submit important information to the driver at the same time.

Modules

Each module (like the vehicle control module or the stereo obstacle detection) represents a computational entity. They can be distributed transparently on the available computational nodes. The interface to a module encapsulates the in/out data, the configuration parameters and the module function calls. This is useful in many ways:

- Reuse of existing algorithms.
- Independence for the algorithm developers. They don't have to care about ANTS components like administrators or the distributed database.
- Application developers can reuse the existing modules to perform new applications without having detailed knowledge about the algorithms.

Administrator

Autonomous and driving assistant applications are usually bound to specific situations, e.g. autonomous lane keeping or a speed limit assistant are useful on highways, whereas autonomous stop-and-go driving can be used if you get stuck in a slow-moving tailback. Some parts of the applications are common (e.g. the obstacle detection is useful in several applications) and some are very specific for the current situation (e.g. there is a lane detection algorithm for highways and another one for the city). So if you want to implement more than one application you need a component to determine the current situation and to adapt the system to the current situation: the administrators. An administrator controls a set of modules (see Figure 6.39). He has to choose the modules, that have to be executed in the current situation, and to give them to the scheduler.

The actual module selection is done by filters and a decision component within the administrator control. This control component is the core 'intelligence' of the modules. The filters can be used for a pre-selection of the modules. For instance the traffic light recognition and the arrow recognition modules can be filtered out while the vehicle is driving on a highway. The decision component decides which of the remaining modules has to be executed and parameterizes them. The results of the filter and decision components depends on the current situation, i.e. the current database entries. For example there can be more than one module for a certain task: a fast but less precise obstacle tracking and a slower but more precise one. When a pedestrian enters the scene, the fast variant is used to focus the computational resources on the pedestrian detection. Otherwise, the slower one is more likely to be called.

The chosen modules are submitted to the scheduler. The scheduler communicates with the modules and executes the module tasks on the specified computational nodes.

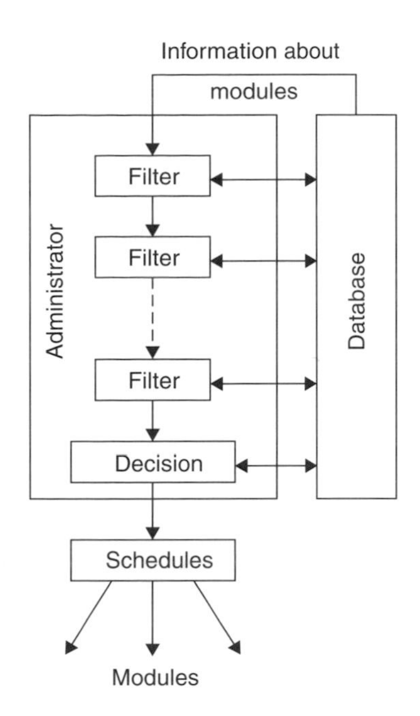

Fig. 6.39 Administrator.

Due to the real-time constraint, all processes have already started waiting for messages from the administrator. The initial static distribution of these processes is done in the start-up phase of ANTS using script files.

ANTS can handle several administrators. Modules that do not belong together are assigned to different administrators. This is useful to avoid complex decision components. In UTA II there is (among others) one administrator for the control of the computer vision modules, one to observe the system status, and another one for the driver interface modules.

An administrator can also handle exceptions in modules. A critical error within a module is submitted to the administrator. The administrator can now decide to stop the entire system, or just disable the affected module, in the case that it is not needed for safe vehicle guidance or if multiple modules are available for the same task.

Configuration and cooperation

ANTS can be configured statically and dynamically. For the static configuration a script file is parsed. The script causes the creation of the database including objects and the administrators. The modules are distributed on the available nodes, parameterized and started. The dynamic configuration is done by the administrators, as described above. They allow to switch between modules during runtime. On highways, as mentioned, the computer vision administrator can for example switch off the traffic light recognition. If you want your vehicle to park, ANTS can activate totally different modules.

ANTS allows several kinds of cooperation. A module can depend on results from other modules as well as results from several modules can be used to achieve a higher accuracy. Figure 6.40 shows some cooperation examples of administrators and modules we are using in UTA II to perform autonomous stop-and-go driving.

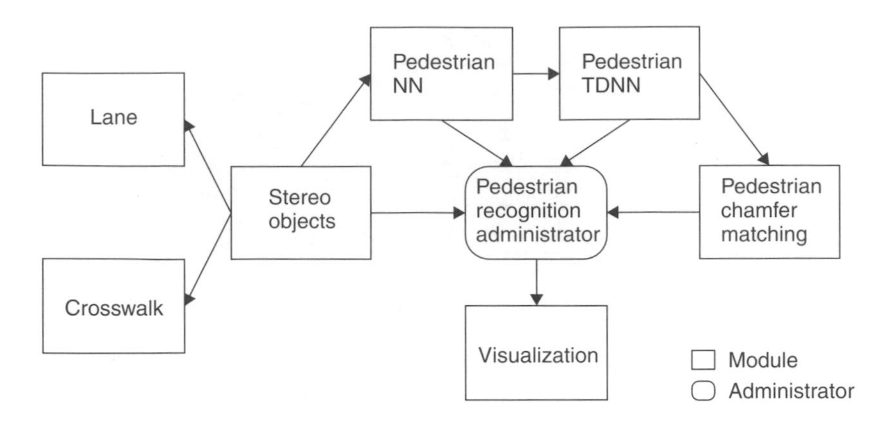

Fig. 6.40 Cooperation of modules and administrators.

The distributed database is used for data transfer (arrows). The stereo objects are transferred to the lane and to the crosswalk recognition. They mask the image to avoid wrong detections, e.g. car tail-lights as lanes. The pedestrian recognition uses more complex cooperation. Several pedestrian classifiers work together to improve the reliability of the results. The pedestrian recognition administrator receives results from the stereo module and organizes the recognition as described in Section 6.4.3.

6.5.2 UTA II on the road

The DaimlerChrysler demonstrator UTA II (Urban Traffic Assistant) was designed with special attention for information, warning and intervention systems in an inner-city environment (see Figure 6.41). UTA II is an E-class Mercedes containing sensors

Fig. 6.41 The demonstrator UTA II.

for longitudinal speed, longitudinal and lateral acceleration, yaw and pitch rate and the steering wheel angle. It is equipped with a stereo black/white camera-system as well as a colour camera. UTA II has full access to throttle, brake and steering.

The computer systems in UTA II are three 700 MHz Linux/Pentium III (SMP) PCs for the perception of the environment and one Lynx/604e PowerPC to control the sensors and actuators. So far, five administrators for computer vision, pedestrian recognition, driver interface (visualization), driving phase determination and a system status watchdog have been integrated.

Most of the integrated computer vision modules have been described above. Currently, the following modules can be activated:

- stereo-based object detection and tracking
- pedestrian recognition based on:
  - neural network (for standing pedestrians)
  - time delay neural network
  - chamfer matching
- lane detection and tracking:
  - on the highway
  - in the city
- traffic signs based on:
  - colour images (see Ritter *et al.*, 1995)
  - black/white images
- traffic light recognition
- recognition of road markings
- crosswalk recognition
- vehicle classification
- vehicle control (lateral/longitudinal)
- driver interface (2D/3D visualization).

Figure 6.46 shows a view out of UTA II. You can see the stereo camera system mounted behind the windscreen. The visualization on the monitor is enlarged in Figure 6.43.

Fig. 6.42 View out of UTA II.

Fig. 6.43 Animated scene showing the recognized objects and their positions in the world.

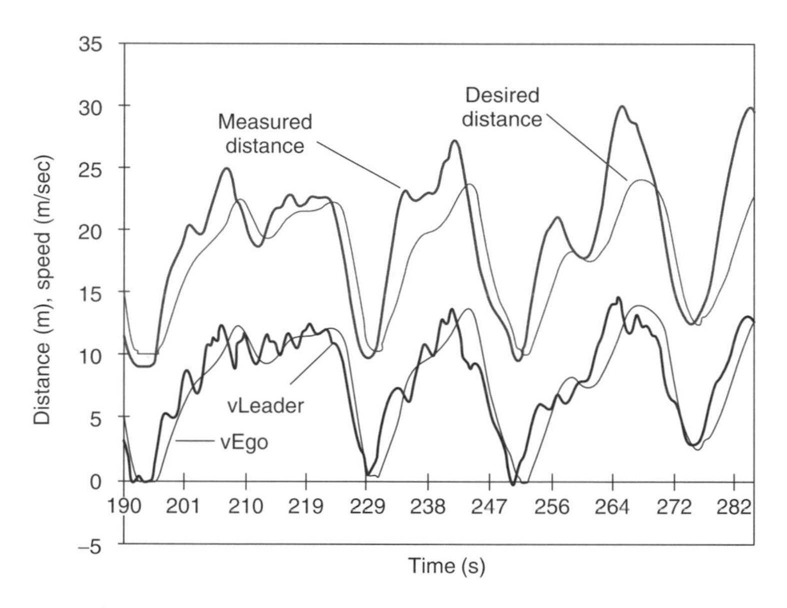

Fig. 6.44 Diagram of autonomous stop-and-go driving. Notice the small distance error when the leader stops.

It shows the objects recognized by UTA II: the detected lane, the obstacle in front classified as a car, obstacles classified as pedestrians, a traffic light and a traffic sign. Thanks to stereo vision, the scene reconstruction is geometrically correct.

The main application of UTA II is autonomous stop-and-go driving in an inner-city environment. Once the car in the visualization turns red, the driver can switch the system on. From now on, the own car follows the car in front laterally and longitudinally. Figure 6.43 shows the results of a test drive in the city of Esslingen,

Germany. The graph shows the measured and desired distance to the car in front as well as our own speed and the estimated speed of the lead vehicle. The latter distance is composed of a safety distance of 10 metres and a time headway of one second. The speed profile shows three stop-and-go cycles.

6.6 Summary

Over the past ten years, computer vision on board vehicles has evolved from rudimentary lane keeping on well structured highways to scene understanding in complex urban environments. In this chapter, we described our effort to increase the robustness on highways and to develop a next-generation cruise control called Intelligent Stop-&-Go, which takes into account relevant elements of the traffic infrastructure and other traffic participants while allowing autonomous vehicle control.

What did we learn during this time? At least three guiding principles have emerged for robust vision-based driver assistant systems:

First, vision in cars is vision over time. Kinematic and dynamic constraints applying to vehicles can be taken into account by means of Kalman filters. Obstacle candidates can be tracked over time. The repeated recognition stabilizes the decisions and allows the estimation of their motion state. A high imaging rate simplifies the establishment of object correspondences.

Second, stereo vision providing 3D information became a central component of robust vision systems. It allows the detection of arbitrary obstacles and the determination of their size and position. Monocular model based approaches as investigated in the early 1990s turned out to be less robust and reliable in the traffic scenario.

Third, object recognition can be considered as a classification problem. Powerful tools are at hand for the adaptation of generic classification schemes to a specific task, as described in Section 6.4. Developers are no longer forced to formulate heuristics but the relevant aspects of the considered objects are learned from representative examples.

In spite of the achieved success many problems related to reliability remain to be solved. Besides continuous improvement of the robustness of the image analysis modules, sensor problems have to be overcome. Standard CCD cameras lack the dynamic range that is necessary to operate in traffic under adverse lighting conditions (e.g. allowing the camera to capture structure in shadowed areas when exposed to bright light). CMOS camera technology can be of help.

As other information sources like radar, digital maps and communication become available in modern cars, their utilization will help to raise the performance of vision based environment perception. It will be a challenge to combine the power of each source in order to obtain a most reliable and complete interpretation of the current traffic situation.

Nevertheless, we are convinced that vision will be the key component of intelligent vehicles.

First vision products on board vehicles are already on the market: witness the Lane Departure Warning System available in Mercedes and Freightliner's trucks. Many more will undoubtedly follow.

Thanks to the foreseeable performance improvement, the future will see systems that assist the driver during his whole trip, from door to door.

References

Anlauf, J.K. and Biehl, M. (1990). The AdaTron: An Adaptive Perceptron Algorithm. *Europhysics Letters*, **10**, 687.

Bar-Shalom, Y. and Fortmann, T.E. (1988). *Tracking and Data Association*. Academic Press.

Bodenhausen, U. and Waibel, A. (1991). The Tempo 2 Algorithm: Adjusting Time Delays by Supervised Learning. In *Advances in Neural Processing Systems 3*, (R.P. Lippman, J.E. Moody and D.S. Touretzky, eds), pp. 155–61. Morgan Kaufmann, San Mateo, CA.

Broggi, A. (1997). Obstacle and Lane Detection on the ARGO. *IEEE Conference on Intelligent Transportation Systems ITSC'97*, Boston, 9–12, November.

Day, S.P. and Davenport, M.R. (1993). Continuous-Time Temporal Back-Propagation with Adaptable Time Delays. *IEEE Transactions on Neural Networks*, **4**, No. 2, 348–54.

Diamantaros, K. and Kung, S.Y. (1996). Principal Component Neural Networks, Wiley Interscience, New York.

Dickmanns, E.D. and Zapp, A. (1986). A curvature-based scheme for improving road vehicle guidance by computer vision. *Proceedings SPIE Conference on Mobile Robots*, Vol. 727, pp. 161–16.

Enkelmann, W. (1997). Robust Obstacle Detection and Tracking by Motion Analysis. *IEEE Conference on Intelligent Transportation Systems ITSC'97*, Boston, 9–12, November.

Enkelmann, W., Struck, G. and Geisler, J. (1995). ROMA – a system for model-based analysis of road markings. *IEEE Intelligent Vehicles Symposium*, September, pp. 356–60.

Faugeras, O. (1993). *Three-Dimensional Computer Vision*. MIT Press.

Franke, J. (1997). Isolated Handprinted Digit Recognition. In *Handbook of Character Recognition and Document Image Analysis*, (H. Bunke and P.S.P. Wang, eds), pp. 103–22, World Scientific, Singapore.

Franke, U., Böttiger, F., Zomotor, Z. and Seeberger, D. (1995). Truck platooning in mixed traffic, *Intelligent Vehicles '95*, Detroit, 25–26, September, 1995, pp. 1–6.

Franke, U. and Kutzbach, I. (2000). Fast Stereo based Object Detection for Stop & Go Traffic. *Intelligent Vehicles '96*, Tokyo, pp. 339–44.

Franke, U., Gavrila, D., Görzig, S., Lindner, F., Paetzold, F. and Wöhler, C. (1998). Autonomous Driving Goes Downtown. *IEEE Intelligent Systems*, Vol. 13, No. 6, September/October, pp. 40–48.

Franke, U. and Joos, A. (2000). Real-time Stereo Vision for Urban Traffic Scene Understanding. *IEEE Conference on Intelligent Vehicles 2000*, 3/4 October, Detroit.

Gavrila, D. (1999a). The visual analysis of human movement – a survey. *Computer Vision and Image Understanding*, Vol. 73, No. 1, 82–98.

Gavrila, D. (1999). Traffic sign recognition revisited. *Mustererkennung 1999*, W. Förstner, *et al.*, eds). Springer Verlag.

Gavrila, D. and Philomin, V. (1999). Real-time object detection for 'smart' vehicles. *Proceedings of IEEE International Conference on Computer Vision*, pp. 87–93, Kerkyra, Greece.

Gehrig, S. and Stein, F. (1999). Cartography and dead reckoning using stereo vision for an autonomous vehicle. In *SCA Eighth International Conference on Intelligent Systems*, Denver, Colorado.

Gern, A., Franke, U. and Levi, P. (2000). Advanced Lane Recognition – Fusing Vision and Radar. *Proceedings of IEEE Conference on Intelligent Vehicles*.

Görzig, S. and Franke, U. (1998). ANTS – Intelligent Vision in Urban Traffic. *Proceedings of IEEE Conference on Intelligent Transportation Systems*, October, Stuttgart, pp. 545–549.

Huttenlocher, D., Klanderman, G. and Rucklidge, W. (1993). Comparing images using the Hausdorff distance. *IEEE Transactions on Pattern Analysis and Machine Intelligence*, **15**, No. 9, 850–63.

Janssen, R., Ritter, W., Stein, F. and Ott, S. (1993). Hybrid Approach for Traffic Sign Recognition. *Proceedings of Intelligent Vehicles Conference.*
Kirckpatrick, S., Gelatt, C. and Vecchi, M. (1993). Optimization by Simulated Annealing. *Science*, No. 220, pp. 671–80.
Kluge, K. and Thorpe, C. (1992). Representation and recovery of road geometry in YARF. *Proceedings IEEE Conference on Intelligent Vehicles.*
Kreßel, U., Lindner, F., Wöhler, C. and Linz, A. (1999). Hypothesis Verification Based on Classification at Unequal Error Rates. *International Conference on Artificial Neural Networks,* pp. 874–79, Edinburgh.
Mandler, E. and Oberländer, M. (1990). One-pass encoding of connected components in multivalued images. *Proceedings 10th International Conference on Pattern Recognition*, Atlantic City.
Mysliwetz, B. (1990). Parallelrechner-basierte Bildfolgen-Interpretation zur autonomen Fahrzeugsteuerung. Dissertation, *Universität der Bundeswehr München*, Fakultät für Luft- und Raumfahrttechnik.
Neußer, R., Nijhuis, J., Spaanenburg, L., Höfflinger, B., Franke, U. and Fritz, H. (1993). Neurocontrol for lateral vehicle guidance. *IEEE MICRO*, February, pp. 57–66.
O'Hare and Jennings, N. (1996). *Foundations of Distributed Artificial Intelligence*. John Wiley & Sons.
Paetzold, F. and Franke, U. (1998). Road Recognition in Urban Environment. *Proceedings of IEEE Conference on Intelligent Transportation Systems*, October, Stuttgart.
Papgeorgiou, C. and Poggio T. (1999). A Pattern Classification Approach to Dynamical Object Detection. *International Conference on Computer Vision*, pp. 1223–28, Kerkyra, Greece.
Poggio, T. and Girosi, F. (1990). Networks for Approximation and Learning. *Proceedings of the IEEE,* **78**, No. 9.
Ritter, W., Stein, F. and Janssen, R. (1995). Traffic Sign Recognition Using Colour Information. in *Mathematic Computation and Modelling*, **22**, Nos. 4–7, pp. 49–161.
Saneyoshi, K. (1994). 3-D image recognition system by means of stereoscopy combined with ordinary image processing. *Intelligent Vehicles'94*, 24–26, October, Paris, pp. 13–18.
Schölkopf, B., Burges, A. and Smola, A. (1999). *Advances in Kernel Methods: Support Vector Machines.* MIT Press, Cambridge, MA.
Schürmann, J. (1996). *Pattern Classification.* Wiley-Interscience, New York.
Ulmer, B. (1994). VITA II – Active collision avoidance in real traffic. *Intelligent Vehicles'94,* 24–26, October, Paris, pp. 1–6.
Waibel, A., Hanazawa, T., Hinton, G. and Lang, K.J. (1989). Phoneme Recognition Using Time-Delay Neural Networks. *IEEE Trans. Acoust., Speech, Signal Processing*, **37**, 328–339.
Wan, E.A. (1990). Temporal Backpropagation for FIR Neural Networks. *IEEE Internation Joint Conference on Neural Networks,* pp. 575–80, San Diego, CA.
Wöhler, C. and Anlauf, J.K. (1999). A Time Delay Neural Network Algorithm for Estimating Image-pattern Shape and Motion. *Image and Vision Computing Journal*, **17**, 281–94.
Wöhler, C. and Anlauf, J.K. (1999). An Adaptable Time Delay Neural Network Algorithm for Image Sequence Analysis. *IEEE Transactions on Neural Networks,* **10**, No. 6, 1531–6.
Wöhler, C., Anlauf, J.K., Pörtner, T. and Franke, U. (1998). A Time Delay Neural Network Algorithm for Real-Time Pedestrian Recognition. *IEEE International Conference on Intelligent Vehicles*, pp. 247–52, Stuttgart, Germany.
Wöhler, C., Kreßel, U., Schürmann, J. and Anlauf, J.K. (1999). Dimensionality Reduction by Local Processing. *European Symposium on Artificial Neural Networks*, pp. 237–44, Bruges, Belgium.

Wöhler, C., Schürmann, J. and Anlauf, J.K. (1999). Segmentation-Free Detection of Overtaking Vehicles with a Two-Stage Time Delay Neural Network Classifier. *European Symposium on Artificial Neural Networks*, pp. 301–6, Bruges, Belgium.

Ziegler, W., Franke, U., Renner, R. and Kühnle A. (1995). Computer Vision on the Road: A Lane Departure and Drowsy Driver Warning System. *4th Mobility Technology Conference*, SAE Brasil '95, Sao Paulo, Brasil, 2–4, October.

Zomotor, Z. and Franke, U. (1997). Sensor fusion for improved vision based lane recognition and object tracking with range-finders. *Proceedings of IEEE Conference on Intelligent Vehicles.*

Radio communication technologies for vehicle information systems

Shingo Ohmori
Communications Research Laboratory, Ministry of Posts and Telecommunications, Japan

Tetsuo Horimatsu
Fujitsu Limited, Japan

Masayuki Fujise
Communications Research Laboratory, Ministry of Posts and Telecommunications, Japan

and

Kiyohito Tokuda
Oki Electric Industry Co., Ltd, Japan

7.1 Introduction

7.1.1 Overview

In the advanced information and communications society, radio communication technologies in the fields of transportation such as roads, traffic and vehicles are considered to play a very important role. The intelligent transport systems (ITS) are totally integrated systems for transportation based on radio communication technologies. In this chapter, ITS are introduced from a standpoint of radio communication technologies for information and communications for transportation systems.

In Section 7.1, ITS will be overviewed and activities of R&D and standardization in the world will be introduced. Section 7.2 describes the present status and introduces the future prospects for intelligent transport communication systems. Present systems such as a Taxi AVM and VICS will be introduced, and dedicated short-range communications

(DSRC) and ETC systems on an ITS platform will also be introduced. In Section 7.3, road–vehicle and vehicle–vehicle communication systems, which will become key technologies in the second phase of the ITS, will be described. Key technologies such as millimetre-wave and optical devices will also be introduced. Section 7.4 introduces the present research and development of the optical and millimetre-wave devices, which makes it possible to downsize the components for road–vehicle and vehicle–vehicle Communications systems and the automotive radar systems.

Contents of this section make reference mainly to the technical report on ITS of the Telecommunications Technology Council of the Ministry of Posts and Telecommunications and the ITS web site of the Ministry of Construction of Japan (http://www.moc.go.jp/road/ITS/j-html/index.html).

7.1.2 Vision for ITS communications

The ITS is a new concept of information and communication systems for transport systems, which provide advanced information and telecommunications networks for drivers, passengers, roads and vehicles as shown in Figure 7.1.

The future vision of ITS will not be as a dedicated and closed information system only for transportation systems. The ITS will be a totally integrated information communication system that will be integrated with existing information networks as shown in Figure 7.2.

The ITS is greatly expected to contribute much to solving problems such as traffic accidents and congestion, which will result in improving efficiency of transportation,

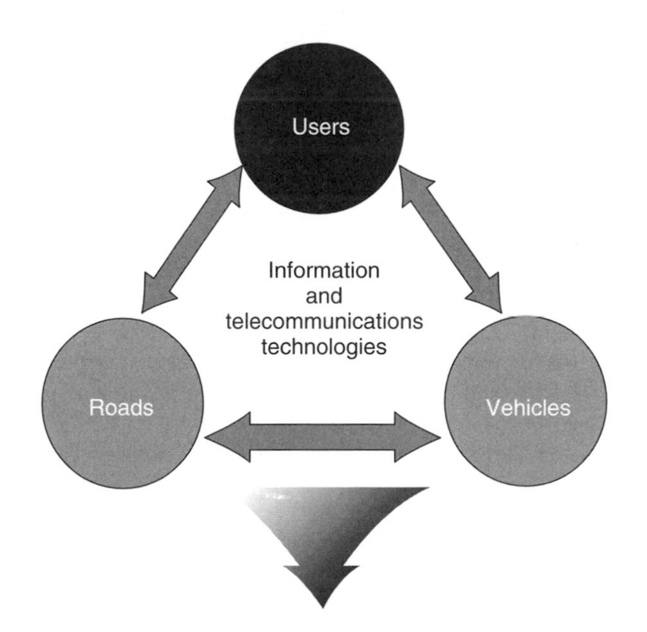

Fig. 7.1 The ITS will connect users, roads and vehicles by information and telecommunications technologies. Source: http://www.moc.go.jp/road/ITS/index.html.

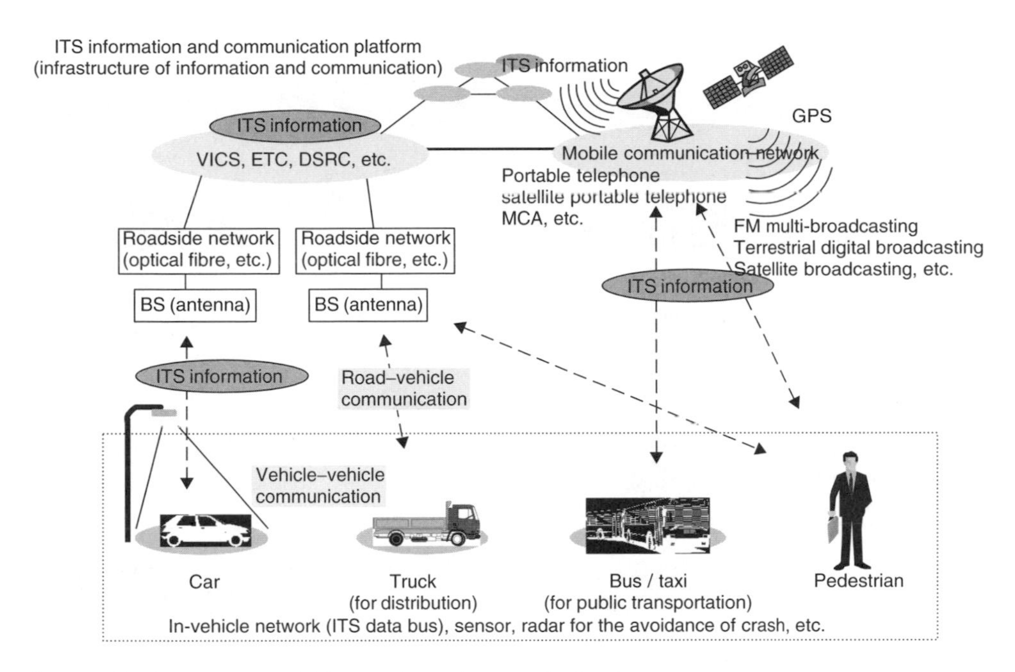

Fig. 7.2 The future image of ITS information and communication System.

improving environmental pollution, and creating new industry and business. Development in ITS is divided into nine areas, and the schedule of R&D and operational deployment of each area is shown in Figure 7.3

The ITS will become an advanced information network system for transportation infrastructures, which are essential to our social activities. As mentioned before, the introduction of ITS will greatly change the environments not only of transportation systems, but also of social activities. Thus, in accordance with its deployment, quality of our social life will be greatly improved in the field of transport. In Japan and other countries, the ITS will be advanced step by step and our social lifestyle in the twenty-first century is assumed as described below.

First phase (around 2000)
Beginning of the ITS: the initial stage of ITS with VICS and ETC

Since April 1996, in Japan, traffic information has been distributed via a Vehicle Information Communication System (VICS), which can provide navigation and traffic information services for drivers. Information on road traffic conditions processed and edited by the VICS centre is sent out from beacons set up on roads, using infrared rays on main trunk roads and radio waves (quasi-microwaves) on expressways. The providing of the road traffic information needed by drivers becomes possible through the use of these beacons. Also, information on road traffic conditions covering wide areas is provided by FM multiplex broadcasts via FM radio waves. As well as traffic congestion information, value-added information such as optimum routes, nice restaurants and sightseeing spots will be displayed on the in-vehicle terminal so that the driver can achieve pleasant travel with saving of travel time.

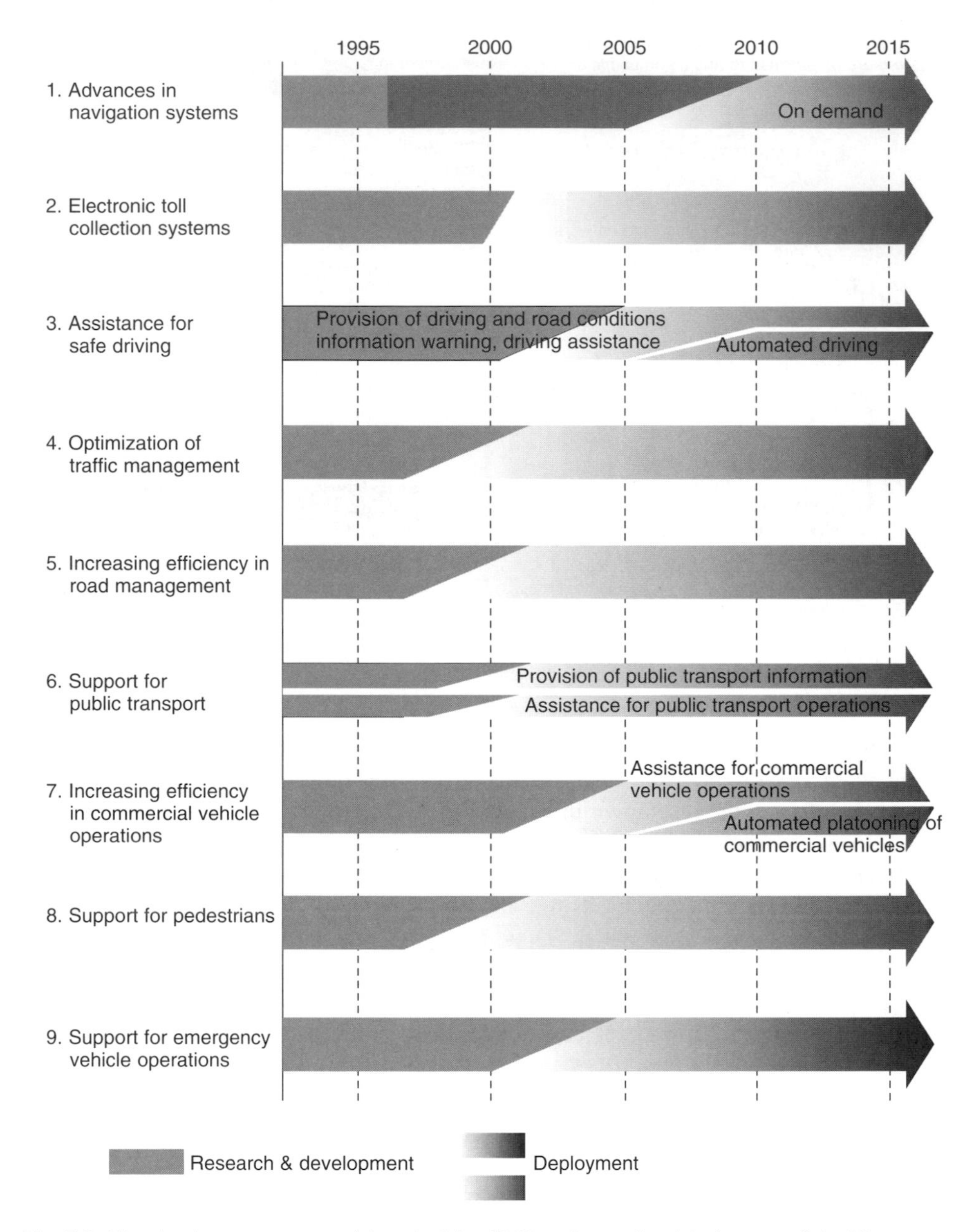

Fig. 7.3 Nine development areas, and the schedule of R&D and operational deployment of the ITS.

In the last half of the first phase, an electronic toll collection (ETC) system operated in a 5.8 GHz frequency band, shown in Figure 7.4, will become a major system in the ITS, so that traffic congestion at tollgates will begin to be eliminated. The capacity of the present manned tollgates is about 230 vehicles per hour, and this will be improved to be about 1000 vehicles per hour by introducing ETC system. The ETC system has already been introduced in 17 countries in the world such as the USA, Canada, China, Singapore, Sweden, France and Spain.

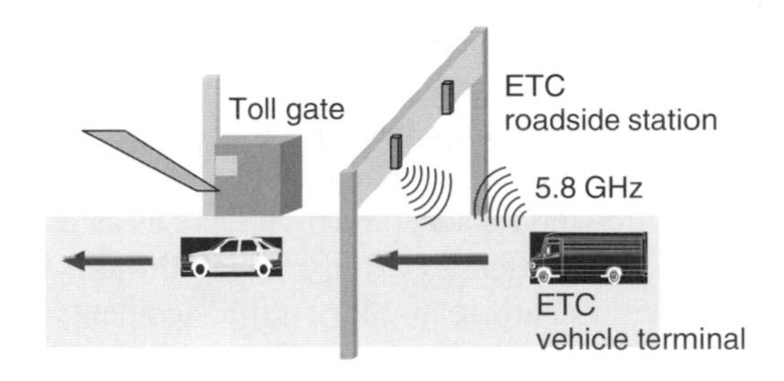

Fig. 7.4 Electronic toll collection (ETC) system using a 5.8 GHz frequency band.

Second phase (around 2005)
Start of user services/traffic system revolution

In this period of the 21st century, a revolution of traffic systems will start by introducing new ITS services for users. In this stage, dedicated short range communications (DRSC) will play a major role in the ITS system. An ETC system is, off course, one of the DSRC systems. The ETC is a specified system to collect a charge at a tollgate. However, the DRSC systems will be used more widely at limited areas such as parking lots, petrol stations, and dispatch centres. Information distributed to users by the DRSC system of ITS will include not only public transport information but also value-added information provided by private sectors. For example, when a trip is being planned, any information that meets the user's needs and requests can be obtained through a DRSC system. Examples of information are optimum route, nice restaurants, accommodation, and sightseeing spots up to the destination by considering travel time and other important factors.

The number of traffic accidents on the expressways and ordinary roads will be reduced by supporting the driver's safe driving and improving the pedestrian's safety. If a traffic accident occurs, quick notification and proper traffic restrictions will prevent further damage. Quick response of emergency and rescue activities will save the life of a person who would have not survived in the traditional situation.

Evolutionary improvements of traffic information services and time saving and punctuality of dispatch management will greatly enhance the convenience of public transportation. These effects will greatly reduce the cost of operation in transportation.

Third phase (around 2010)
Advances in ITS and enhanced social system – automated highway systems – realization of a dream

In the third phase, the ITS will be advanced to a higher level. With penetration of the ITS into social life, the ITS will be firmly recognized as an essential social infrastructure. With a realization of advanced functions, automated driving will start in full-scale service; and the inside of the vehicle will become a safer and more pleasant place.

Fourth phase (after 2010)
Maturity of the ITS – innovation of social systems

In the fourth phase, which is the final stage of this project, all systems of ITS will have been fully deployed. A full-scale advanced information and telecommunications society will be established with the nationwide optical fibre networks and innovative social systems.

In this period, the number of automated driving users will have considerably increased so that automated driving will be established as a general system. The ITS will come to a stage of maturity and be accepted by people as an essential infrastructure system pertinent to road, and other means of transport. With full-scale ITS deployment, it is expected that the number of deaths caused by traffic accidents will greatly decrease from that of the present in spite of increased traffic volumes and density. All roads will have less traffic congestion, enabling pleasant and smooth travel. In addition, a reduction in business traffic will enable harmonization with the roadside environment and the global environment.

7.1.3 International activities for standardization

The international standardization of ITS has been under examination since 1992 when the International Organization for Standardization (ISO) established TC204 (Technical Committee 204/Transport Information and Control Systems). In 1991, in order to promote standardization, the Committee European de Normalization (CEN) set up TC278, which is similar to ISO/TC204. Both of them maintain a close relationship to advance world standardization of the ITS. The international standardization of the ITS has been conducted mainly by the ISO/TC204. Depending on the themes to be examined, some tasks are carried out by a joint ISO/IEC technical committee (JTC) and coordinated with the International Telecommunication Union (ITU). The ITU is in charge of coordinating subjects such as requirements of ITS wireless communications, function and technical requirements, frequency matter, communication capacity and frequency allocation. Figure 7.5 shows a framework of an international standardization of the ITS.

The International Electrotechnical Commission (IEC) is an organization closely related to the ISO, conducting standardization of electrical and electronic engineering.

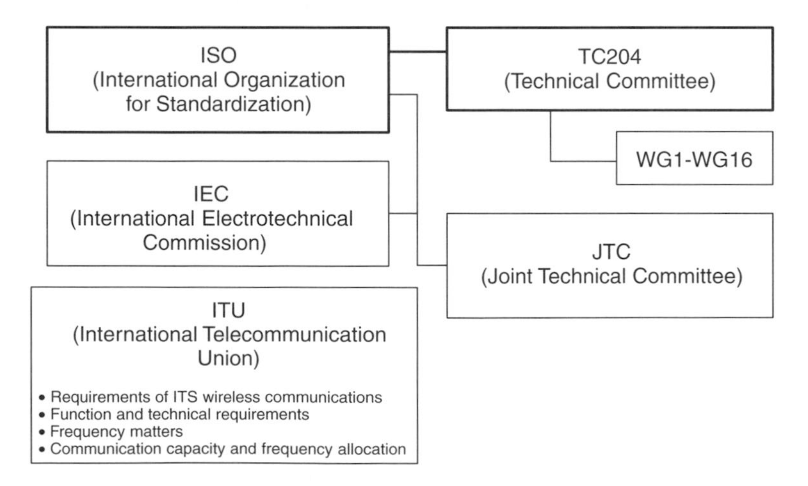

Fig. 7.5 Framework of international standardization of the ITS.

The IEC and the ISO share roles based on a treaty signed in 1976. The ISO/TC204 has 16 working groups from WG1 to WG16 for each standardization field. Based on the cooperative relationship between the ISO and the CEN, the ISO/TC204 and the CEN/TC278 have agreed to decide which organization takes the lead in standardization procedures in order to prevent redundancy as shown in Table 7.1.

Table 7.1 Equivalence between ISO/TC204WG and CEN/TC278 WG. Cited from http://www.hido.or.jp/ITSHP_e/TS/cont_TS.htm

ISO/TC204			CEN/TC278		Leading	
WG	Work programme	Secretary Member Country	Equivalent WG	Work programme	ISO	CEN
WG1	Architecture	Great Britain	WG13	System architecture and terminology	O	
WG2	Quality and reliability requirements	USA	–	–		–
WG3	Database	Japan	WG7 WG8	Geographic road data base Road traffic data	O	
WG4	Automatic vehicle identification	Norway	WG12	Automatic vehicle and equipment identification		O
WG5	Automatic fee collection	Netherlands	WG1	Automatic Fee collection and access control		O
WG6	Freight operation	USA	WG2	Freight and fleet management	O	
WG7	Vehicle operation	Canada			O	
WG8	Public transport	USA	WG3	Public transport	O	
WG9	Traffic control	Australia	WG5	TC (traffic control)	O	
WG10	Traveller information	Great Britain	WG4	TTI (Traffic and travel information)		O
WG11	Route guidance and navigation	Germany	–	–		–
(WG12)	Parking management	Absence	WG6	Parking Management	–	
(WG13)	Man–machine Interface	Absence	WG10	Man–machine interface	–	
WG14	Vehicle control	Japan	–	–		–
WG15	DSRC	Germany	WG9	Dedicated short range communication		O
WG16	Wide range communication	USA	WG11	Subsystem-intersystem interfaces	O	

– = no equivalent WG.

7.2 ITS communication systems

7.2.1 Overview

Intelligent Transport Systems (ITS) conduct studies for the development of safe, effective and comfortable driving by utilizing advanced information and communication technologies. An intelligent social infrastructure and a smart automobile accelerate the widespread implementation of ITS. For the expansion of ITS, 'ITS communication systems' called 'Smart Gateways' are expected to be important subsystems to support the various ITS as shown in Figure 7.6. When these systems become widespread, many benefits for users will be given. For example, comfortable driving with requested information such as traffic conditions, a weather forecast, and multimedia information will be universally available when driving.

In the 'Smart Gateway,' road–vehicle and inter-vehicle communication are the primary means for supplying various information to drivers. In those systems, both dedicated short-range communications (DSRC) on an ITS platform, and conventional network such as a public network will play an important role.

In this section, the present status and future prospects of ITS communication systems will be discussed. This section includes Section 7.2.2 – Multimedia communication in a car, Section 7.2.3 – Current ITS communication systems and services and Section 7.2.4 – Prospects of growing technology.

7.2.2 Multimedia communication in a car

Taxi AVM

Taxi companies have introduced AVM (automatic vehicle monitoring) systems to capture the condition of each vehicle and to dispatch the closest taxi to a customer

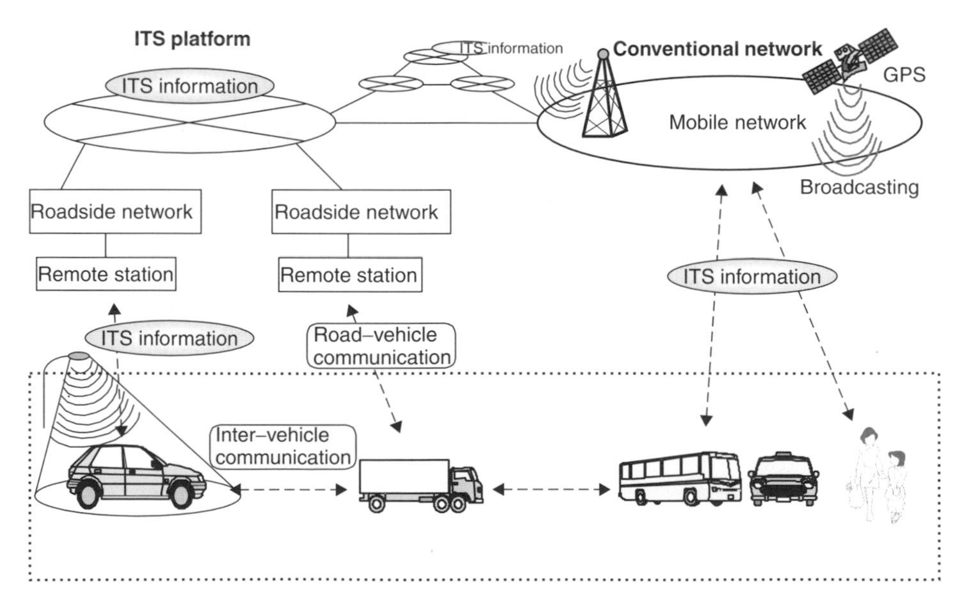

Fig. 7.6 'Smart Gateway'.

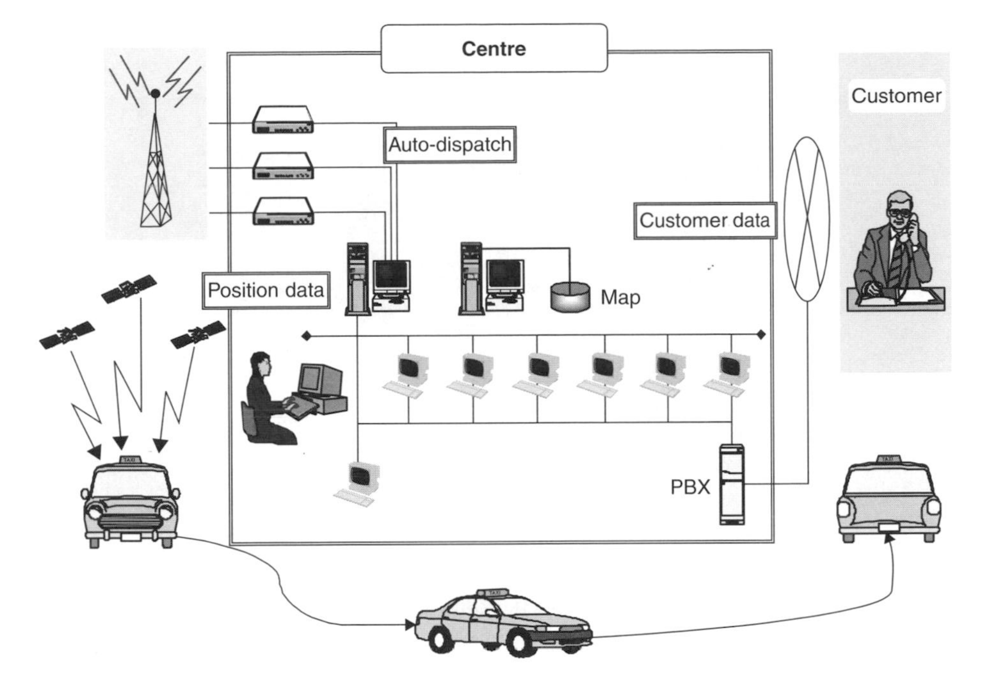

Fig. 7.7 System outline of the taxi AVM.

in the least time for increasing customers' satisfaction, reducing operation cost, and burden of a taxi driver. Recent development pushes the GPS (global positioning system) into public use as a location sensor on 'navigation systems'. In addition to using GPS for positioning, the use of a database of residential maps will present an automated response without an operator at a centre, and quick dispatch of a taxi. The system outline is shown in Figure 7.7 (Iwai *et al.*, 1997). The AVM centre delivers the taxi requested to the customer in the least time using the position data of each taxi, master map and a customer data. As an air interface, the 144 and 430 MHz communication bands are used in Japan.

Bus location system

Similar to the taxi AVM, a bus location system, designed for giving customers more satisfaction, attracts attention as a total bus operation managing system. The bus location systems are constructed using various methods of detecting each bus location, communication among the buses, the bus stops, and the dispatching centre. A bus stop using the system is shown in Figure 7.8 (Chujo *et al.*, 1995). This system is expected to raise customers' satisfaction, to appeal their advanced transportation system and to realize efficient bus scheduling. To differentiate it from taxi AVM, as the driving routes of buses are fixed in advance, the position of the bus can be allocated by counting pulses coming from the rotating tyres, without using any GPS systems. As an air interface, 150 MHz band is used in Japan.

Taxi Channel

'Taxi Channel' will deliver universal services to all taxi passengers and will be a superior multimedia service (see http://nttdocomo.co.jp/). The system delivers social

Fig. 7.8 The bus stop using the bus location system.

events, political news, exchange rate information, sports news, a weather forecast, and so on. The contents, displayed on a liquid-crystal panel in a taxi, are as up-to-date as those of a radio or TV. The system uses a pager system as a communication media. This makes the system low-cost. By using this service, the taxi itself will be a new medium which serves passengers various information through the Taxi Channel. The taxi companies are trying to create new services for obtaining more passengers. Unexpected service-down in mobile conditions seldom occurs when using this media if the taxi is driving within a service area. The Kanto Kotsu taxi company has introduced this system into 101 vehicles. The company estimate indicates that more than 40 000 people per month will receive the service, and this system will be a new advertising media too. The system will make a safe and comfortable mobile world.

On-board car multimedia

Here, a car-navigation system based on an on-board computer is introduced. A lot of people are paying attention to the development of on-board computers because of the expectation for a large market size and its usefulness in ITS applications.

There are two ways. One is to use as a client connected with a transaction processing system in a network, and the other is to use a server/controller to various on-board car equipment. For businesses, usefulness can be evaluated with the effect on cost reduction and increased sales. For instance, usefulness can be proven if the transaction processing system which uses the on-board computer increases the vehicle operating rate and load factor of the truck, and can expand business opportunities.

One example of introducing an on-board computer into a commercial vehicle is shown. Requirements of on-board computers for commercial vehicles are as follows.

1. **Upgrade and efficiency improvement of transportation services.** For example, data of the vehicle positioning, the amount of collecting cargo, and the work status should be collected and properly processed with a central computer. And, for the request from the customer, the freight delivery situation should be collected from the vehicles in real time.

2. **Safety improvement and accident avoidance.** Risky driving, such as rapid acceleration and heavy breaking, should be detected and the driver warned.
3. **Cost reduction of overall system.** Customizing and supplementing of a new function should be easy.

One example developed is shown in Figure 7.9. This is composed of main unit, TFT display, controller, CD-ROM player, printer, and other I/O devices. This system has been used to record the data of vehicle operation, such as positioning, progress status of work of a commercial vehicle. And the vehicle operation data is used to make the daily report, which was done manually before introducing this system. In this application, total cost reduction of logistics is a key, but conventional car navigation systems are also introduced for reducing work load of the driver and for improving traffic conditions.

Another example is the commercial vehicle dispatching system. The system overview is shown in Figure 7.10.

Fig. 7.9 Example of on-board car computer.

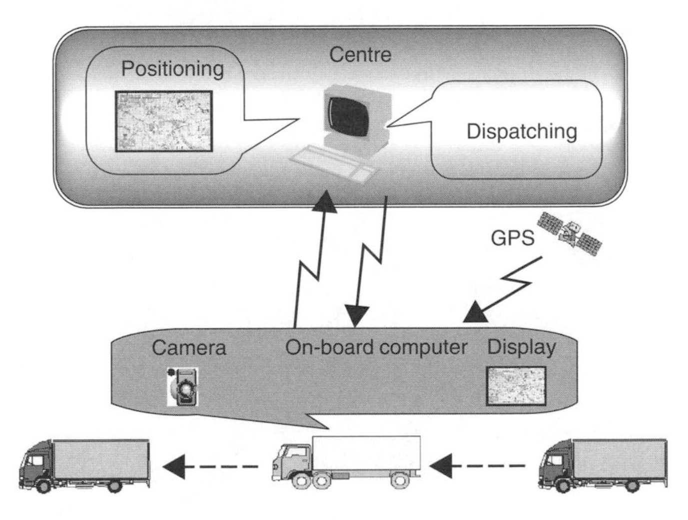

Fig. 7.10 Commercial vehicle dispatching system.

The purpose of this application is as follows:

1. Improve quick response for spot transport request.
2. Reduce work load of dispatching.
3. Assist drivers with automatic guidance and easy reporting of work status.
4. Assist drivers in sales work.

IDB (ITS data bus)

The IDB defines the standard network for connecting consumer electronics products such as navigation and anti-theft systems, personal digital assistants, pagers, and wireless communication devices, to each other, and to other vehicles.

The first specification of IDB was based on a 115.2 kbps UART/RS485 physical layer, and was called IDB-T. With input from the AMI-C, the specification was changed to use the CAN 2.0B. The CAN-based version is called IDB-C to differentiate it from IDB-T. IDB-C will be the first version to be deployed in vehicles, expected to start appearing on model year 2002 cars. A higher speed version of the IDB, IDB-M for multimedia, is being developed at 100 Mbps on plastic optical fibre. The Technical Committee of the IDB Forum™ will coordinate all of these developments with other organizations, including JSK/JAMA's Task Force, AMI-C, Ertico's CMOBA Committee, the SAE IDB Committee, and the TSC (see http://www.idbforum.org/).

Requirements for IDB are as follows:

1. peer to peer, i.e. no host or bus master
2. 'hot' plug and play
3. self-configuration
4. short boot/discovery time
5. automotive physical and electrical specs.

Application examples are shown in Figure 7.11. This shows in-car computing, Internet access, remote diagnostics, remote monitoring, and activation, etc.

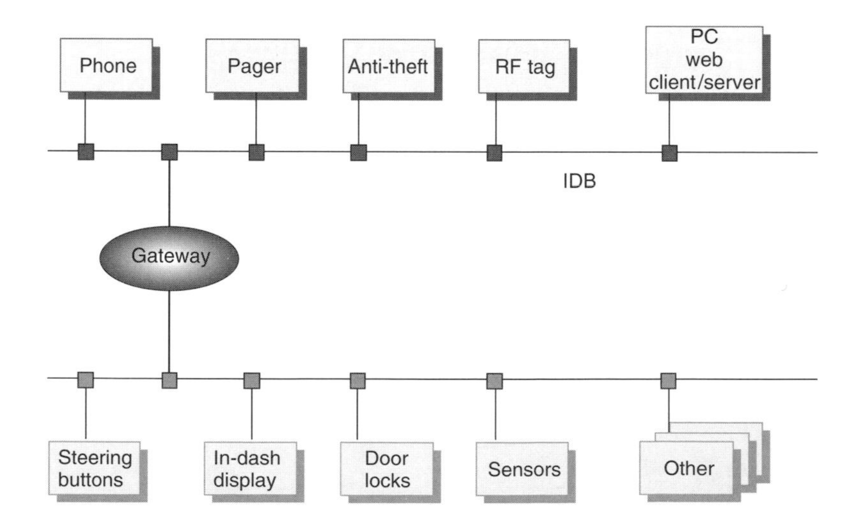

Fig. 7.11 Application examples of IDB.

Next generation IDB is called IDB-M for multimedia. Some organizations are studying specifications for IDB-M. For examples, SAE has designed IDB for low-cost and low-risk systems aimed at telematics (data/control) applications. The IDB-M will include MOST, MML, and IEEE1394 for convenience and the multiple-frame protocol, together with the multiple-star architecture, based on optical POFs (plastic optical fibres) would be a good data bus solution for a mobile-media system.

With IDB, the vehicle also becomes a platform for explosive growth in in-vehicle electronics.

7.2.3 Current ITS communication systems and services

VICS

VICS (**v**ehicle **i**nformation and **c**ommunication **s**ystem) is based on the need to drive by selecting a road without traffic congestion and makes overall traffic efficiency as high as possible. On the other hand, the VICS relieves traffic jams and optimizes the traffic flow only by informing the driver of road conditions. The system has been designed from three viewpoints as follows:

1. The system can be realized with low-cost technology.
2. The systems will penetrate the social infrastructure, and will be a fundamental of the society, used by many people.
3. The system will be used for a long period.

It is from these viewpoints that the system has been designed. First, the traffic information such as the flow condition should be collected and can be delivered by one specific organization instead establishing a new organization with high cost. Second, some new technology is not required for delivering information. Third, the need for a specific in-car equipment such as a car navigation system is decreasing.

The VICS information is delivered through three media. One is a wireless beacon, which transmits 64 kbps of data within about 70 m area, as shown in Figure 7.12. The other is an optical beacon which is installed adjacent to the road and transmits 1 Mbps of data around 3.5 m, as shown in Figure 7.13.

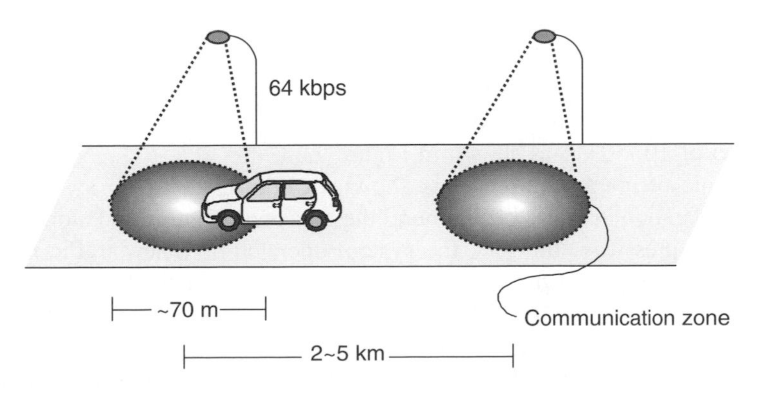

Fig. 7.12 Wireless beacon.

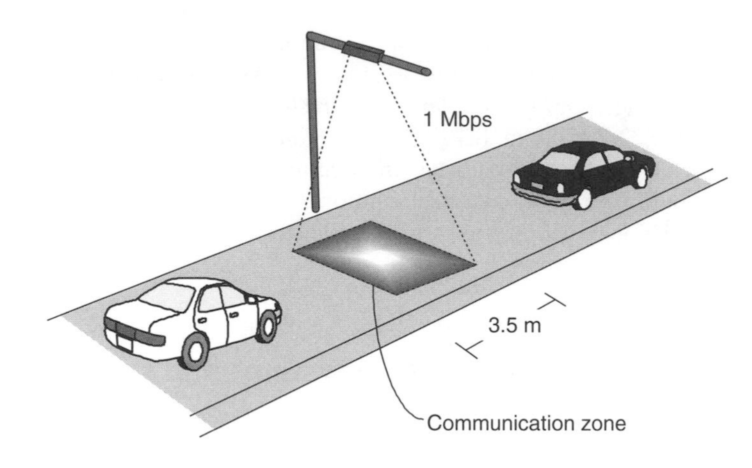

Fig. 7.13 Optical beacon.

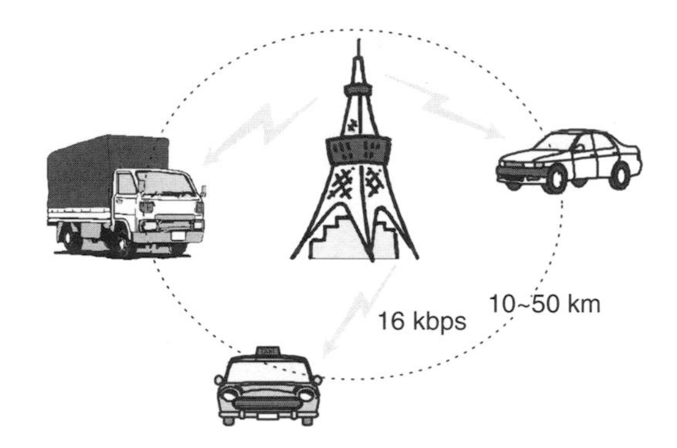

Fig. 7.14 Broadcasting by FM radio.

Both wireless and optical beacons are effective road–vehicle communication media to deliver respective information to individual cars in a restricted area. The cars can receive requested information on beacon site and when needed, the car can also transmit data to the beacon. Using this system, the driver can receive desired information in real time at a desired site requested. The beacon, however, restricts a service area within several tens of metres; the driver can't receive information everywhere. To cover these characteristics, another media as broadcasting of 16 kbps data with FM radio is served within a area of 10–50 km as shown in Figure 7.14.

One example of in-car equipment is shown in Figure 7.15. The system started its commercial operation in April 1996 around the metropolitan area and along the Toumei and Meishin Expressway. In 1998, the system operated as a new traffic management system during the period of the 1998 Olympic Winter Games in Nagano. It is expected that for 90 per cent of users, the VICS system will provide excellent information system when they drive not only in a city but in a rural area. According to an expansion of the service area, the number of on-board receivers associated with the car navigation equipment has exceeded one million by March 1999.

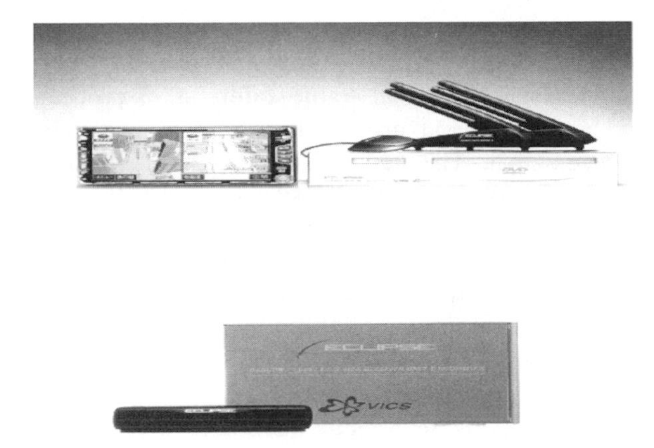

Fig. 7.15 Example of in-car equipment with VICS receiver.

DSRC

Road–vehicle communication (RVC) is communication between the road infrastructure and mobile equipment equipped in vehicles. RVC is classified as a spot-communication type and a continuous communication. The former performs communication using a dedicated short zone, the latter performs communication using a continuously structured communication zone.

The research and development on RVC has a long history, with experiments since the 1960s. In Japan, the first application of RVC is probably the Comprehensive Automobile Traffic Control System (CACS), in which, in 1977/8, the Ministry of International Trade and Industry played a major role in the pilot system. In 1984, the Road/Automobile Communication System (RACS) was proposed. In the fall of 1989, large-scale bi-directional RVC experiments were conducted. This resulted in the commercialization of the Vehicle Information and Communication System (VICS) in April 1996. The Electric Toll Collection System (ETC), the focus of current attention, also must apply as being conducted mostly by the Ministry of Construction.

In spot-communication type RVC, small radio communication zones are intermittently allocated along the road, and communication is performed the moment a vehicle passes through the zone. This type again is classified into two types, one is a one-way type and the other is bi-directional. A typical example of the former is the VICS, the latter is the communication of ETC.

The typical specification of RVC is shown in Table 7.2 (Fukui, 1997). The communication uses radio beacons, which transmit such information as positional information,

Table 7.2 Typical specification of road–vehicle communication

	VICS	ETC
Frequency board	2.5 GHz	5.8 GHz
Antenna power	<10 MW	<300 MW <10 MW
Modulation	GMSK, AM	ASK
Speed	64 kpbs	1024 kpbs
Zone	~70 m	~30 m

road guidance, and dynamic information which spontaneously changes time by time, and some other information as travelling guidance. The bi-directional RVC performs sophisticated functions such as a road traffic information probe and its provision of route guidance on request, road management support, and logistic support. The enhanced ETC is one example. The standard specification defined by the Association of Radio Industries and Business of the ETC system is also shown in Table 7.2. The ETC system is designed to prevent traffic jams around tollgates, to realize toll collection, and to decrease environmental pollution by exhaust gas and noise, by realizing automatic toll collection on highways and toll roads without requiring vehicles to stop. The Ministry of Posts and Telecommunications in Japan has selected two pairs of 5.8 GHz band waves, a transmission rate of 1024 kbps, and ASK modulation scheme. The spot type RVC is not appropriate for danger warnings to support collision avoidance and for cooperative driving such as an automated driving, because it does not provide continuous services. To cover this, necessity of continuous RVC occurs. There are several types to realize continuous RVC: use of a wide area communication zone, use of a series of small beacon zones, and use of a leaky cable. A research group led by the Ministry of Construction used the leaky coaxial cable and realized an experimental continuous RVC system. In this experiment, 500 m communication units form continuous RVC whose respective units are managed by a processor for communication control. This communication allows travelling vehicles to receive warning information on danger ahead in real time. It also becomes possible to receive information on the distance of other vehicles to prevent collision, and to coordinate driving, and an automated cruise in future.

The another example we noted is road–human communication. This is one example of a spot communication. Some Japanese cities are trying to help handicapped people to walk across intersections using a road–human communication system by informing the traffic signal condition and by presenting warning information to avoid traffic accidents.

Advanced safety vehicle (ASV) project

The reduction of traffic accidents is one of the vital targets of ITS. When applying newly developed advanced technologies, it becomes important to what extent they could save human lives. To realize ASV, some communication technologies should be used. One is inter-vehicle communication and the other is road–vehicle communication.

For the Advanced Safety Vehicle (ASV) project proposed by the Japanese Ministry of Transport, some themes were chosen according to the analysis of traffic accidents. One is an accident prediction and preventing technology with an intelligent navigation system, the other is collision avoidance by radar technology to compensate for driver errors in perception and judgement, and pedestrian protection safety technology to prevent accidents and reduce injury (Kamiya, 1997).

7.2.4 The prospect of growing technology

As mentioned above, various communication systems are now commercially available and a new communication system such as a DSRC will be widely in service in a few years. In these circumstances, some trials are now under way to increase traffic efficiency, to reduce traffic accidents, and to reduce air pollution.

Among the trials, inter-vehicle communications and automotive radar systems are supposed to be indispensable, therefore, a novel solution for making cars intelligent by using these technologies will be a main theme. However, some important problems still remain to be solved before such automotive systems will be practical and widely used.

One example we should note is the Chauffeur Project. The concept developed in the project has been intended as a promising solution to improve the traffic capacity. The basic idea is coupling two commercial vehicles by means of an electronic virtual coupler as an inter-vehicle communication and automotive radar system. This is called an ACC/Advanced Cruise Assist system. For the automotive radar, millimetre-wave band has been used, because its propagation loss does not degrade in such bad weather conditions as fog and snow. This leads to stable operation of detecting vehicles and obstacles ahead, and makes safe and comfortable cruising by maintaining adequate vehicle–vehicle distance and managing relative speed.

Another example of growing technology is the collision avoidance system. General Motors and DOT in the USA will together pay for 100 Michigan drivers to test vehicles equipped with crash-avoidance radar systems. Beginning in 2002, ten test vehicles will be real-world tested. Initially, the work will focus on rear-end collision (because more than one-quarter of all injuries in motor vehicle crashes occur in rear-end collisions).

It is expected that for an extra half-second or second of warning, this will cut rear-end collisions by 50 per cent. Test vehicles will be equipped with a new adaptive cruise control, and its forward-looking collision-warning system, plus an advanced head-up display. System components include: radar, miniature TV cameras, GPS, digital maps, on-board sensors, and on-board signal processing. Researchers at UMTRI (University of Michigan) will collect data transmitted from on-road vehicles, and also manage the experiments (IEEE, 1999). Detailed descriptions of radar systems are left to another chapter.

One other point on ITS communication we note is the evolution of the system for safe driving. Efforts to provide intelligence in vehicles have been made by various

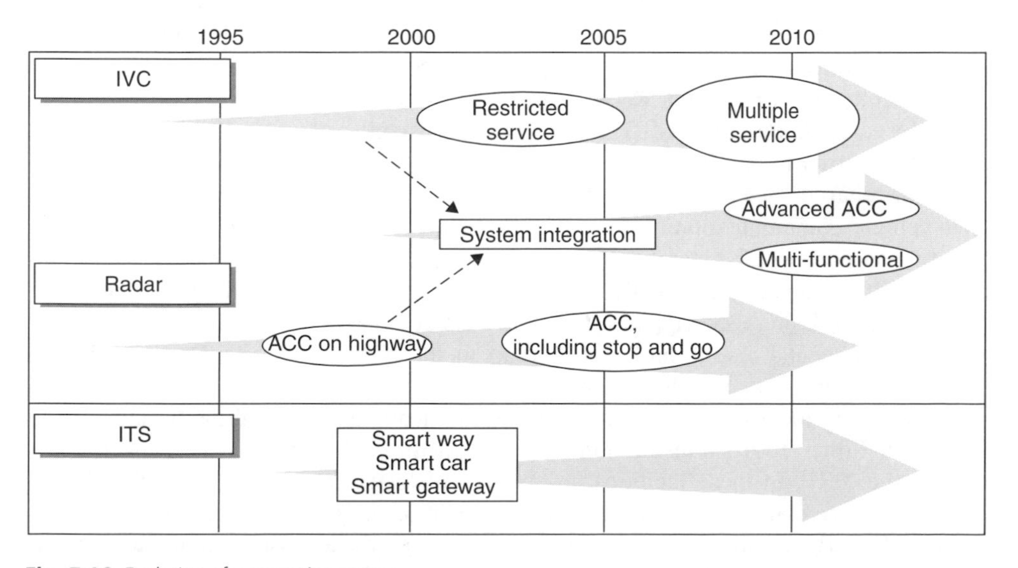

Fig. 7.16 Evolution of automotive system.

organizations and a private company. Through the efforts, single automotive radar systems are now installed in passenger cars, and provide comfortable driving. And the technical base for the inter-vehicle communication will be achieved in a few years. Under these circumstances, another solution to integrate superior functions of each inter-vehicle communication and radar into one system can be a new automotive system as shown in Figure 7.16 (Horimatsu, 1999).

This system will provide new services such as a multifunctional ACC which enables automated branch-in and branch-out at crossroads smoothly, and multimedia information services in addition to real-time information for the vehicle's control. When implementing this system, a safe and comfortable automated highway cruising without any vehicle control by the driver will be in practical use.

7.3 Vehicle–vehicle and road-vehicle communication systems

7.3.1 Overview

The development of transportation in recent years has also held negative sides such as traffic accidents and environmental pollution. It is expected that with ITS many problems of the present transit system will be overcome. ETC (electronic toll collection) and VICS (vehicle information and communication systems) are already in practical use in Japan, and the research is advanced in order to carry out the more advanced communication. ITS is a fusion technology between the vehicles and communications, and provides drivers and passengers with a comfortable and safe travelling environment. In ITS, inter-vehicle communications (IVC, communications among vehicles, not depending on infrastructure of road side) and road–vehicle communications (RVC) are expected to play an important role for assisting safe driving, and supporting automatic driving such as automated highway systems (AHS). The quality of such systems is a matter of life or death for many users of transportation systems. Therefore, real-time and robust communication must be secured for ITS.

CRL Yokosuka (Yokosuka Radio Communications Research Center, Communications Research Laboratory, MPT Japan) has intensively set up millimetre-wave test facilities in order to accelerate research activities on the ITS wireless communications.

In this chapter, we first introduce the millimetre-wave test facilities for the ITS inter-vehicle communication. For the IVC experiments, we have prepared two vehicles on which experimental apparatus for the evaluations of propagation characteristics and transmission characteristics in the millimetre-wave frequency band of 60 GHz have been mounted. Using these apparatus, we have executed experiments and have obtained useful experimental results on a public road in the YRP (Yokosuka Research Park). Some results are shown in this chapter.

Then, we introduce the road–vehicle communications test facilities based on the Radio On Fibre (ROF) transmission system and micro-cell network system along a road in the YRP. In these facilities, millimetre-wave frequency bands of 36–37 GHz as the experimental band are used. A control station is located on the third floor of a research building and 12 antenna poles for the roadside base stations are put up in equal intervals of 20 metres along a straight road about 200 metres long. Optical fibre cables are installed in the state of a star connection between the control station (CS) and each roadside local base station (LBS). Propagation characteristics between the roadside antenna and a vehicle are presented and overall transmission systems including optical fibre cable section and air section are also mentioned in this chapter.

7.3.2 Road-vehicle communication system (Fujise and Harada, 1998; Fujise *et al.*, 1998)

Configuration of the proposed system

Nowadays the amount of communications equipment on the car, especially antenna, has been increased, because many services such as vehicle information and communication systems (VICS), TV and mobile communications are available on different frequency bands. As a result, the car looks like a hedgehog. However, by using the common frequency band, the number of air interfaces between the car and the wireless service network is drastically decreased. This is an important factor from the viewpoints of not only car design but also efficient frequency use.

Figure 7.17 illustrates the concept of the ITS multiple service network based on the Common Frequency Band Radio On Fibre (CFB-ROF) transmission.

In this technique, first of all, we convert the radio frequencies of various wireless services into the common frequency band. The users of the ITS can use this common specified frequency band for the ITS multiple service communications. For the down-link of this system, the combined electrical radio signal, which is converted to the common frequency band, drives the electric absorption modulator (EAM) and the modulated optical signal is delivered to the local base station (LBS). Then, by using a photodiode, the optical signal is converted to the radio signal and is transmitted

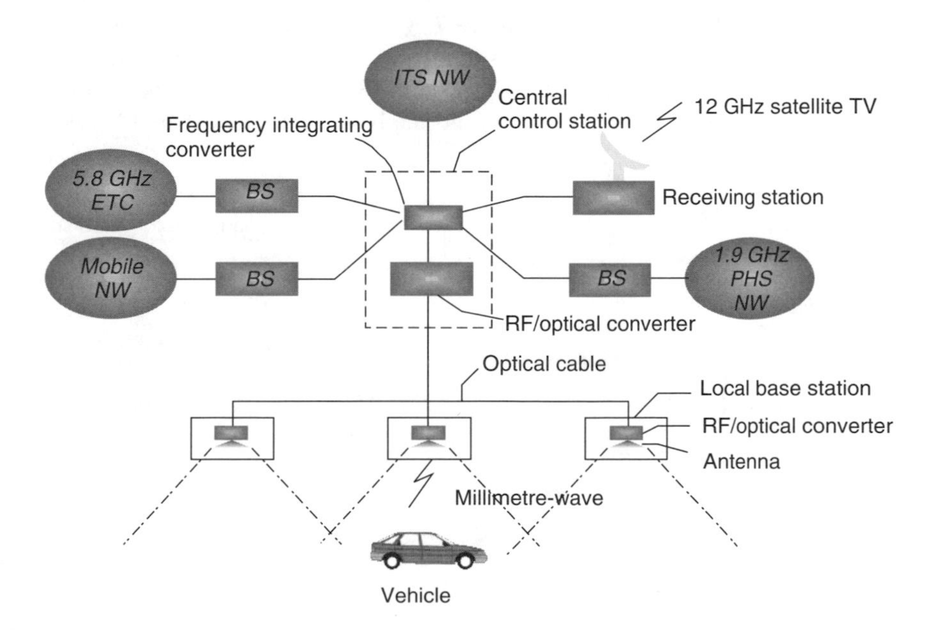

Fig. 7.17 Concept of the ITS multiple-service network based on CFB-ROF system.

to the vehicle from the roadside antenna. The vehicle has only to have the antenna, which matches with the common frequency band and receives the radio signal from the LBS. In the vehicle, the radio signal is converted and divided into the original band of each service. Finally, the signal is carried to each terminal on the original band by the distributor. The distributor may be equipped with several connectors for the distribution to each terminal. We can connect the distributor and each off-the-shelf terminal with a cable. If we use a multi-mode terminal, it is not necessary to distribute the received signals to each terminal. The multi-mode terminal is expected to be realized by adopting the software radio technology. Furthermore, the multi-mode terminal will contribute to efficient space use in the vehicle. For the up-link, the procedure is the reverse of the down-link.

Experiments Figure 7.18 shows the experimental set-up for the optical transmission of three kinds of mobile communication services such as IS-95, PHS (personal handy phone system) and PDC (personal digital cellular) in Japan. In this experiment, the 5.8 GHz band is used as the common frequency band and the interval of each carrier frequency is set at 10 MHz. The wavelength of the laser diode is 1552 nm and its output power is 0 dBm. The insertion loss of electroabsorption external modulator is about 9 dB. We use four kinds of optical fibre length at almost 1 m, 5 km, 10 km and 20 km. At the receiving side, each channel of three services is filtered out after detection by the PD which has a frequency response up to about 60 GHz. After demodulation of each channel signal, the transmission quality was measured by a modulation analyser. Figure 7.19 shows the frequency response of the ROF link of the experimental set-up. Due to the chromatic dispersion of the single mode fibre, the received power decreases at every constant frequency interval depending on the fibre length. In the case of fibre length of 10 km, the first power decreasing frequency is about 15 GHz. Figure 7.20 shows a measured spectrum of three different kinds of mobile communication channels for IS-95, PHS and PDC. As shown in Figure 7.20, high dynamic ranges were obtained. Table 7.3 shows the results of the measurements of transmission qualities of these channels. The error vector magnitude (EVM) for PHS and PDC and the ρ for IS-95, which are standard evaluation parameters, normally must be less than 12.5 per cent

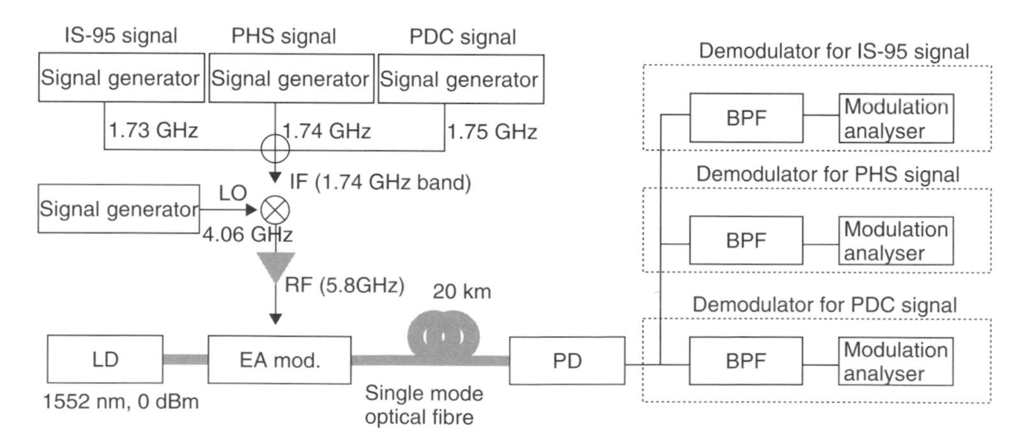

Fig. 7.18 Configuration of the experimental setup.

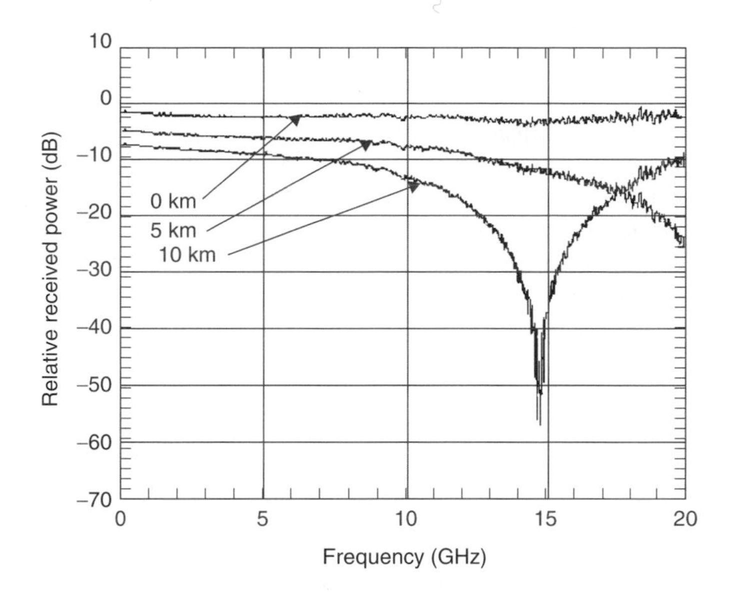

Fig. 7.19 Frequency response of the ROF link of the experimental setup.

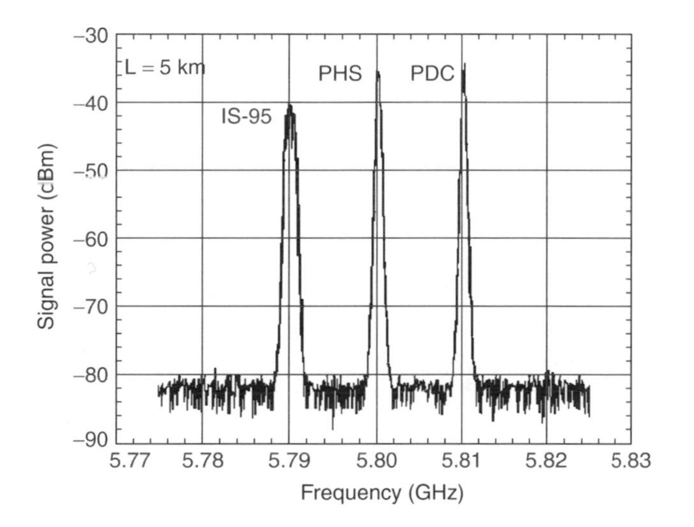

Fig. 7.20 Measured spectrum of triple service radio transmission by CFB-ROF.

Table 7.3 Measurement results of modulation qualities

Fibre length (km)	IS-95 (ρ)	PHS (EVM: %)	PDC (EVM: %)
5	0.99764	2.30	2.28
20	0.99750	3.07	2.29

and more than 0.99 respectively. The measured values were sufficiently good relative to these normal values.

Development of prototype system

We have successfully developed a prototype system for dual wireless service, e.g. ETC and PHS, utilizing CFB-ROF techniques in the frequency band of 5.8 GHz region. Figure 7.21 illustrates the configuration of the developed system. This system consists of a roadside network and a mobile terminal in the vehicle. This prototype can support two-way communications.

The original frequency band of PHS is in the 1.9 GHz region and it is converted into the 5.8 GHz region. Figure 7.22 illustrates the frequency allocation for PHS and ETC in this system. By using this prototype with PHS handsets and a set of ETC terminal and server, we can get an announcement of the tollgate charge and can make a phone call simultaneously. Figure 7.23 shows an overview of the prototype system for PHS and ETC and Table 7.4 shows its specifications.

Experimental facilities for RVC in the 37 GHz band

Figure 7.24 shows the configuration of the experimental facilities for the ROF transmission system of three kinds of mobile services such as electric toll collection (ETC), personal handy phone system (PHS) and TV broadcasting in Japan. In these experimental facilities, the 37 GHz band is used as the common frequency band and the

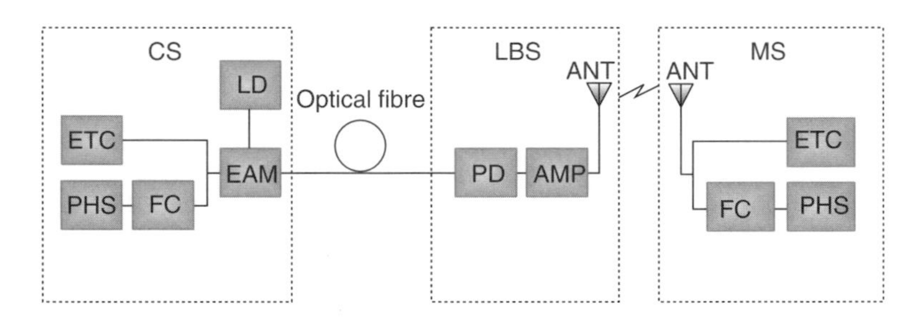

Fig. 7.21 Configuration of developed prototype system. LD, laser diode; EAM, electroabsorption modulator; PD, photo diode; FC, frequency converter.

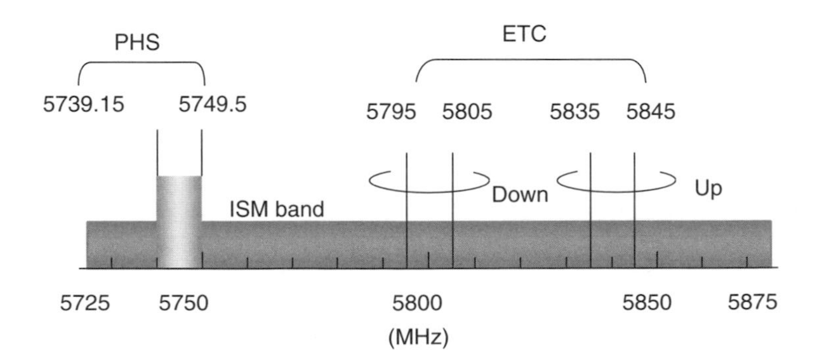

Fig. 7.22 Frequency allocation of PHS and ETC in the prototype system.

Fig. 7.23 Overview of the prototype for ITS dual service radio transmission system in 5.8 GHz band.

Table 7.4 Specification of the developed system

Modulation	PHS	$\pi/4$ DQPSK (384 kbps)/TDMA-TDD
	ETC	ASK (1.024 Mbps)/Slotted ALOHA
Frequency	PHS	5739.15 ~ 5749.95 MHz
	ETC	5795 5805(down), 5835 5845(up) MHz
Antenna gain	Road	18 dBi
	Vehicle	5 dBi
Output RF power	Road	3 dBm
	Vehicle	10 dBm
Optical fibre length	1 km	

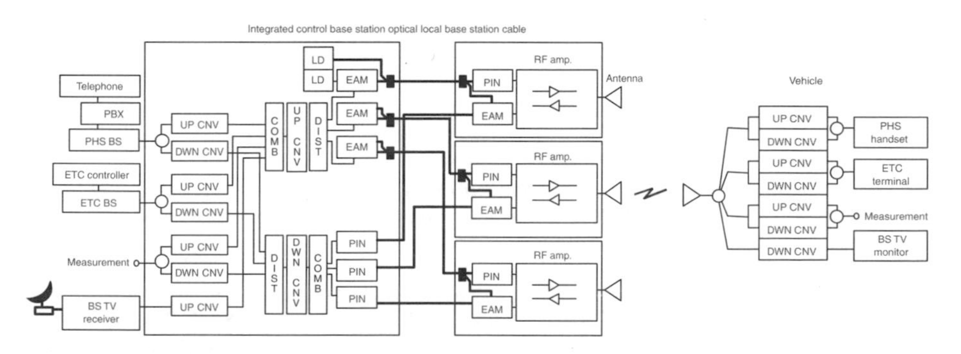

Fig. 7.24 Configuration of the experimental facilities for the multiple service CFB-ROF transmission system.

frequency for each service is allocated as shown in Figure 7.25. The wavelength and the output power of the laser diode for this scheme is in 1.5 μm region and about 0 dBm, respectively. The length of the optical fibre cable section between the control station in the research building and the roadside LBS is about 700 meters. The frequency bands for the down-link and up-link are 36.00–36.50 GHz and 36.75–37.25 GHz,

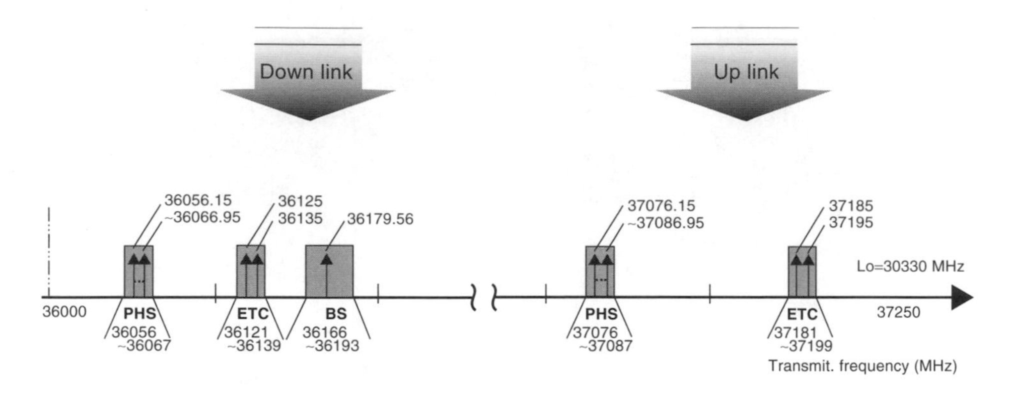

Fig. 7.25 Frequency allocation of multiple service ROF transmission system in 37 GHz band.

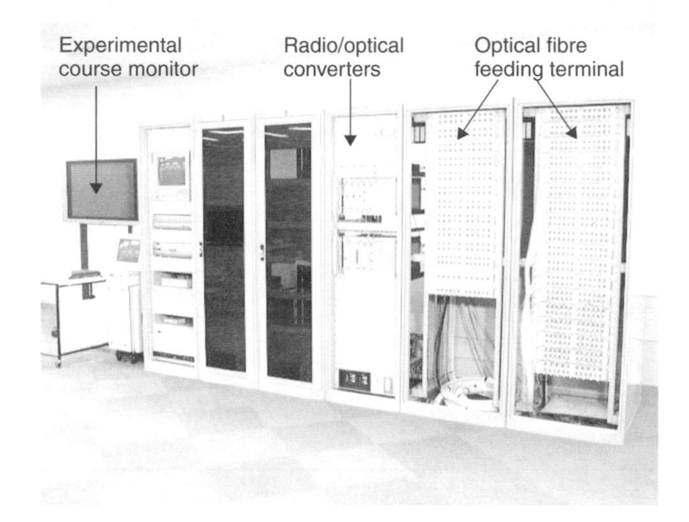

Fig. 7.26 Facilities for ROF control station.

respectively. The frequency bands of the RF amplifiers and antennas for the LBS and the vehicle match with these frequency bands.

We have prepared two kinds of roadside antenna. One is the reflector antenna and the other is the patch antenna with 20 element antennas and both of them have the cosec-squared beam pattern on the vertical plane. The interval between roadside antennas is 20 metres.

The antenna, the frequency converter and the mobile terminals are mounted in the vehicle. The original frequency bands of PHS and ETC are in the 1.9 GHz and 5.8 GHz region, respectively. Therefore, the received RF signals are divided and delivered into the each mobile terminal after frequency down conversion in the vehicle. The control station and the roadside antennas are shown in Figures 7.26 and 7.27, respectively.

Estimation of received power for ROF road-vehicle communication

Next, we estimate the received power at the mobile station (MS). In this system, three LBSs connected to one CS transmit the same frequency radio wave to the vehicle. It

Fig. 7.27 Scenery of test course.

is predicted that the strong interference can be observed at the boundary area between two cells covered by different LBSs. In this simulation, each LBS located at 5 m on the poles installed along the road, the poles spaced at 20 m intervals. The height of vehicle antennas is 2.1 m. So the height difference between LBS and MS antennas is 2.9 m. In this estimation, the transmitted power is 10 dBm and the frequency is 36.06155 GHz. The transmitting antenna has a cosec-squared beam pattern on the vertical plane. This antenna enables us to get almost the same received power in the coverage area. The receiving antenna on the vehicle has a beam pattern with 3 dBi gain. We did not consider the reflection from road or other objects.

Figure 7.28(a) shows the contour map of calculated received power of 5.0 m × 40.0 m area on the road. Antenna poles stand in 20 m intervals along the roadside. Figure 7.28(b) shows the received power at 2.1 m heights from road surface and on the centre of lane, i.e. 2.5 m from edge of road. The variation of the received power as a function of position is caused by the interference of the radio waves from several LBSs. The interference between LBSs causes very complicated fluctuations of received power. This result shows that we need to develop some new technologies, for example, some kind of diversity with very high-speed signal-selection.

Conclusions

In this section, we have proposed an integration method of wireless multiple services in ITS, which is based on Common Frequency Band Radio On Fibre (CFB-ROF)

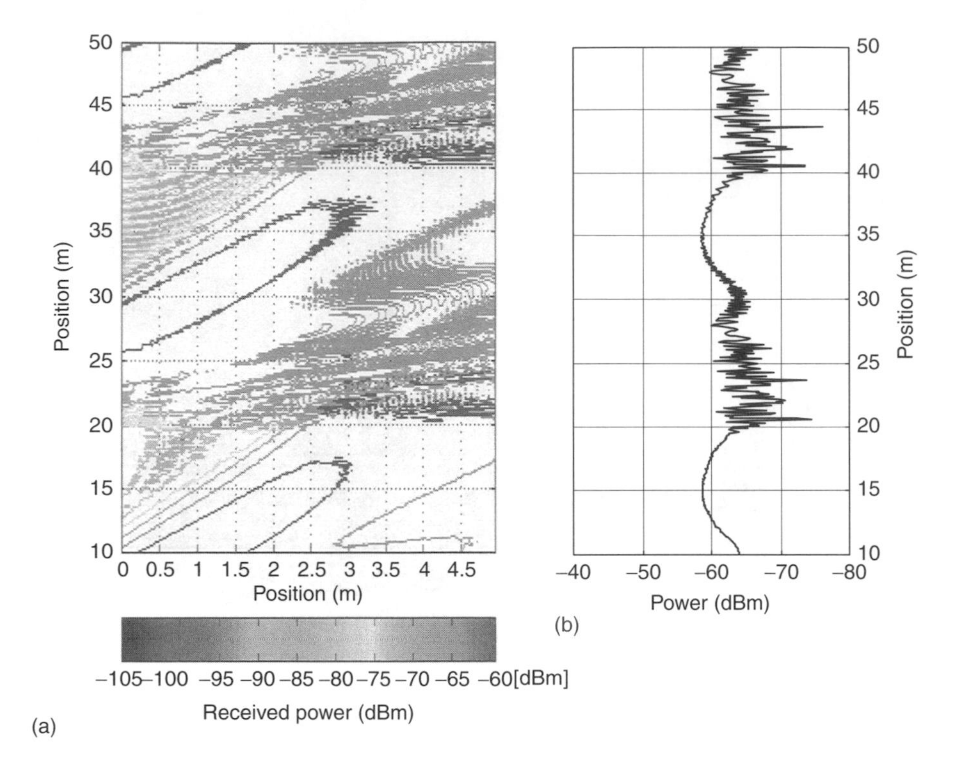

Fig. 7.28 Contour map of calculated received power (a) and received power variation on the centre of the lane (b).

technique. Moreover, we have confirmed the feasibility of our proposed system. The CFB-ROF will become a key technique for mobile multimedia communications in ITS. As a further study, we here open a new concept for ITS services, which we have named MLS (multimedia lane and station). Figure 7.29 shows the concept of MLS. MLS consists of multimedia lanes and stations which provide multimedia communication services to cars moving on a road and to cars stopping at a place such as a parking lot, respectively.

7.3.3 Inter-vehicle communication system (Kato *et al.*, 1998)

Experimental facility for inter-vehicle communication

In millimetre-wave propagation between vehicles, the propagation condition is affected by various types of environmental change as encountered by travelling vehicles. To investigate the behaviour of propagation characteristics comprehensively, we prepared an experimental facility for IVC systems using millimetre waves. Figure 7.30 shows the block diagram of our experimental system. There are two vehicles for the IVC measurement. The precedent car has one RF section (A), and the following car equipped two RF sections (B and C) for space diversity. Each RF section has the transmitter and receiver. The propagation experiments are executed by use of the RF section (A) as the transmitter and B and C as the receivers. The two-way data transmission is

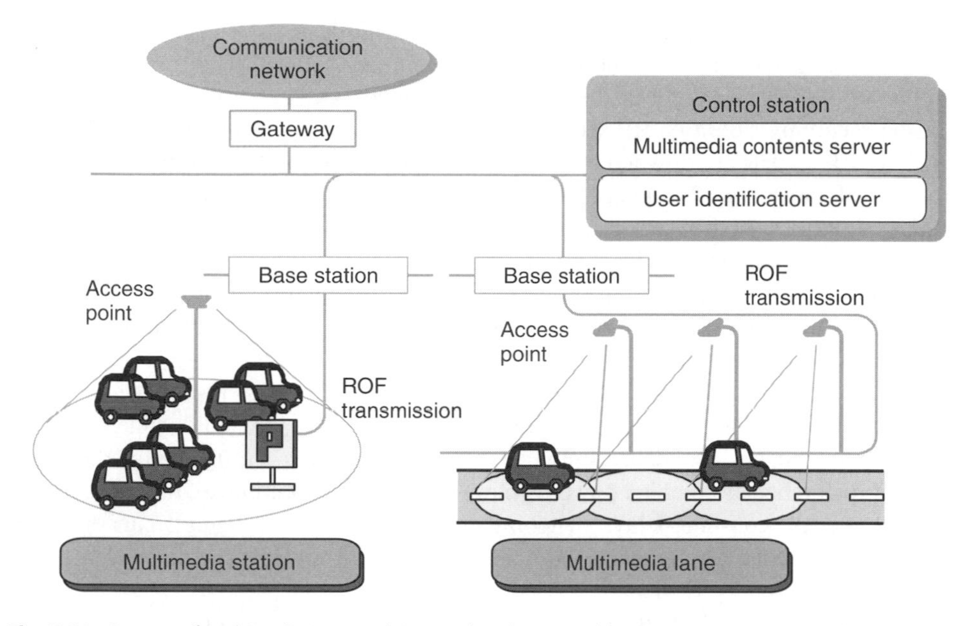

Fig. 7.29 Concept of Multimedia Lane and Station (MLS) proposed by Communication Research Laboratory.

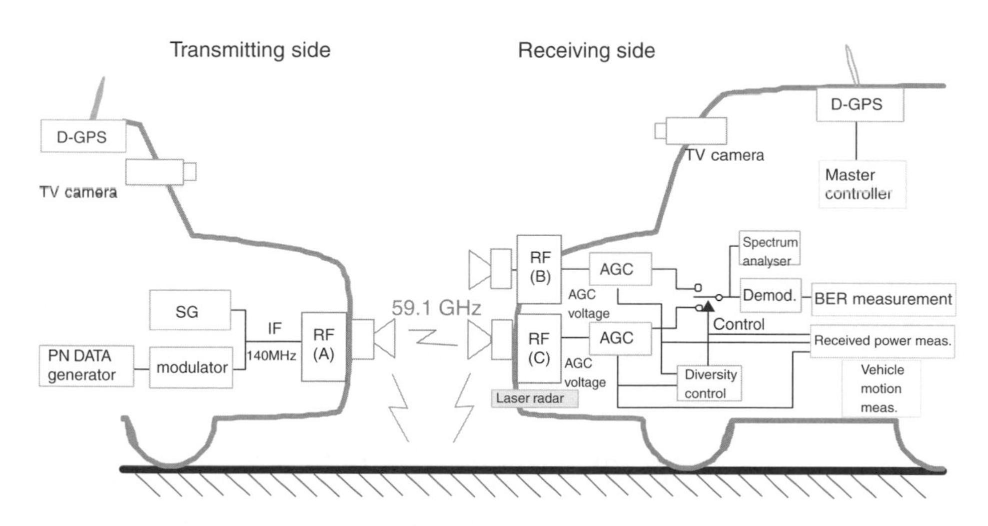

Fig. 7.30 Block diagram of experimental systems.

also available for the demonstration of general data transmission as 10 Mbps Ethernet. Frequency domain duplex is used for the two-way data link. The centre frequencies of RF are 59.1 GHz (for A) and 59.6 GHz (for B, C).

On the transmitting side, the signal generator makes the carrier frequency of 140 MHz. This signal generator also makes various modulation signals as ASK, xFSK, xPSK, and xQAM for modulation analysis. In the data transmission experiment, the PN code at 1,5 or 10 Mbps is made by the data generator, and the IF carrier at 140 MHz is modulated by Manchester-DFSK in the modulator. The IF signal is upconverted to an RF signal at 59.1 GHz by RF section using MMIC devices. The RF section is in a waterproofing radome with the constant temperature control. This radome can be installed at a constant height in the rear of the vehicle.

In the receiving side, two RF sections are located at the constant heights in front of a vehicle. RF signals are downconverted to IF frequency at 140 MHz in the RF section. The gain of these two IF signals are controlled by AGC section. These two signals of AGC voltage, which is corresponding, to the received power, are storage at DAT storage at a sampling rate of 150 kHz or less. These two IF signals are selected to one signal alternatively in the diversity section. Selected IF signal is demodulated, and bit error rate is measured. This IF signal is inputted to the real-time spectrum analyser in the modulation analysis.

Diversity switching is triggered by comparison between each instantaneous AGC voltage with adjustable threshold value of each AGC voltage, threshold value of difference of AGC voltages, and delay timing. If the received power is less than the threshold level and difference of received power is more than the threshold level, the switching is executed after the constant delay. Switching method is 'switch-and-stay'. This diversity switching is not considered the synchronization for the bit timing. The diversity-control signal is also stored by DAT.

In the IVC by the millimetre wave, the condition of the propagation channel should be affected by a change in environmental conditions such as buildings or fences and vibration of vehicles. Thus, various environmental conditions around the vehicles are also observed. The CCD camera is installed at the front of the vehicle and the digital video system, which is synchronized with the other measurement system, records the visual information. An optical gyroscope is also equipped for the measurement of instantaneous motion of each vehicle separated into three axes of gyration. The laser radar with the resolution of 2.5 cm is installed at the front of the following vehicle. This radar measures the instantaneous distance between the vehicles. In the measurement, these data of environmental conditions on both sides of the vehicles and the propagation data are synchronized by D-GPS signal with each other. The off-line data-playback system is installed in a room. This system can play obtained data visually and synchronously, and it analyses the propagation parameter such as distribution of cumulative probability of received power.

Experiments The 1 Mbps wireless digital data transmission with a carrier frequency of 59.1 GHz was examined between a transmitter (Tx.A) on a fixed precedent car and two receivers (Rx.B, C) on a following car. Figure 7.27 also shows the experimental scenery. The test course is straight two-lane pavement and almost 200 m long. There is one building and several prefabricated houses, and several banks around the course. It seems that there were few objects that cause the reflection, and there was no obstacle between the Tx and Rx. The precedent car was parked at the edge of the road, and the following car moved at a constant speed of 2.5 m/s from the other edge of the road to the precedent car.

Table 7.5 shows the experimental set-up for the measurement. The transmitted power is −4 dBm. Each antenna at Tx and Rxs is a standard horn antenna with the gain of 24 dBi, and these were placed at a height of 46 cm (Tx.A), 85 cm (Rx.B), and 38 cm (Rx.C) respectively.

Table 7.5 Experimental set-up

Centre frequency	59.1 GHz
Transmitted power	−4 dBm
Data rate	1 Mbps
Modulation	DFSK (Manchester code)
Detection	Differential
Antenna	Standard horn
Antenna gain	24 dBi
Polarization	Vertical or horizontal
Diversity threshold (level)	−70 dBm
Diversity threshold (def.)	10 dB
Diversity timing delay	10 micro seconds

The bit error rates (BERs) were measured each second and the received powers were also measured simultaneously at the rate of 18750 points per second.

A diversity threshold value of absolute level is set at −70 dBm and that of difference level is set at 10 dB and timing delay is set at 10 micro seconds.

Two-ray model for millimetre-wave propagation The two-ray propagation model between the direct wave and the reflected wave from the pavement was applied for estimation of propagation characteristics of the millimetre wave. Figure 7.31 is a schematic view of the two-ray propagation model. In this model, the received power P_r is expressed approximately as:

$$P_r = \frac{P_t G_t G_r}{L(r)} \left(\frac{\lambda}{4\pi r}\right)^2 \sin\left(\frac{2\pi h_t h_r}{\lambda d}\right)$$

where P_t is the transmitted power, G_t and G_r are the antenna gains at the transmitter and the receiver, $L(r)$ is the absorption factor by oxygen, λ is the wavelength, r is the distance between the antennas, d is the horizontal distance between the antennas, h_t and h_r are heights of the transmitter and the receiver, respectively. In this model, the reflection coefficient of the pavement is assumed as −1 and the directivity of antenna is ignored. Absorption of oxygen is assumed as 16 dB/km.

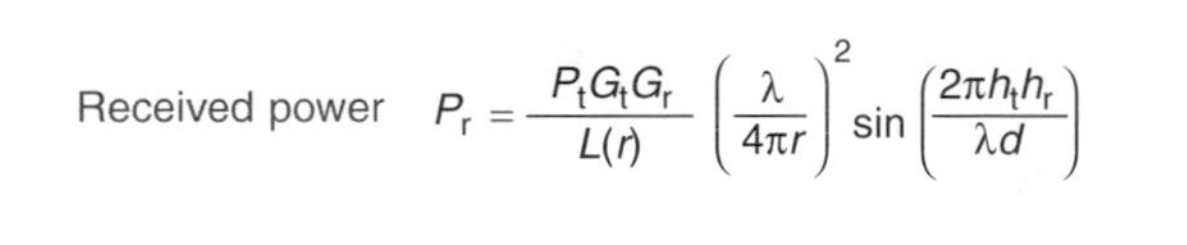

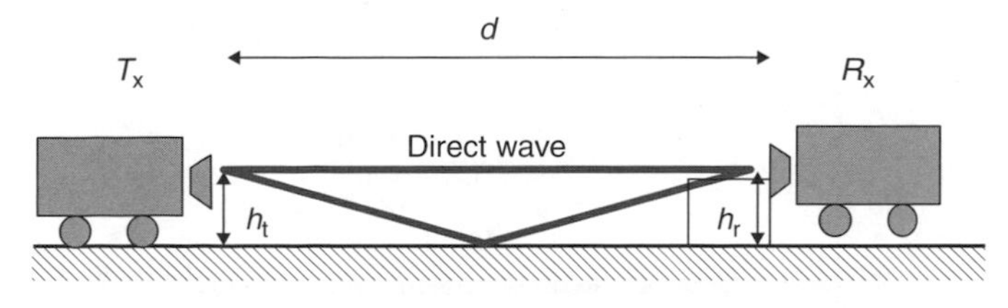

Fig. 7.31 Two-ray propagation model.

Results Figure 7.32 shows the measurement results of the relationship between the received power and BER, and horizontal distance between the vehicles at each Rx position with each receiving antenna height (Rxh). The estimated received power using the two-ray propagation model is also indicated by the dashed line in Figure 7.32. Bit error rates are also shown in Figure 7.32 as circular markers, where the 10^{-10} shows the error free. The results of measured receiving power give fairly good agreement with those obtained by the two-ray propagation model. In this graph, it is found that the bit error rates are degraded when the received power is not sufficient. Figure 7.33 shows the measurement result when the vertical space-diversity was applied. The received power and BER are not so much degraded as those when the vertical space-diversity is not applied. This result shows that the vertical space-diversity is effective for IVC systems using millimetre-waves for data transmission experimentally.

Figure 7.34 shows the measurement result of cumulative distribution of BER travelling on the expressway. Although the shadowing by other vehicles occurred many

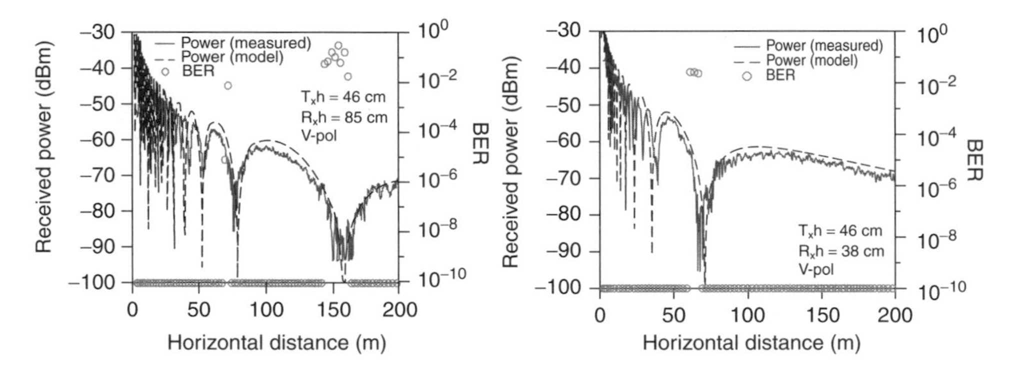

Fig. 7.32 Measurement results of relationship between received power/BER and horizontal distance.

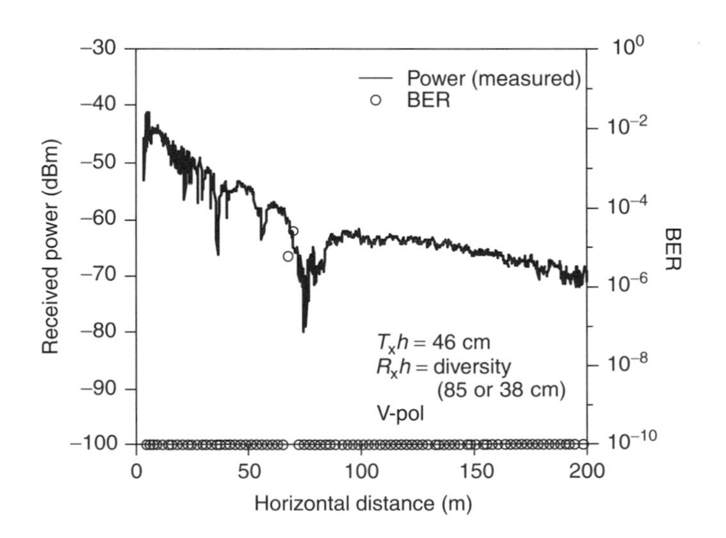

Fig. 7.33 Measurement results when vertical space-diversity is applied.

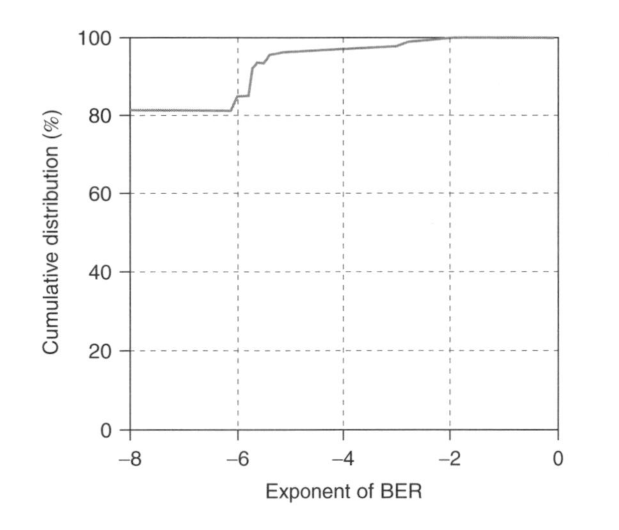

Fig. 7.34 Results of cumulative distribution of BER travelling on the expressway.

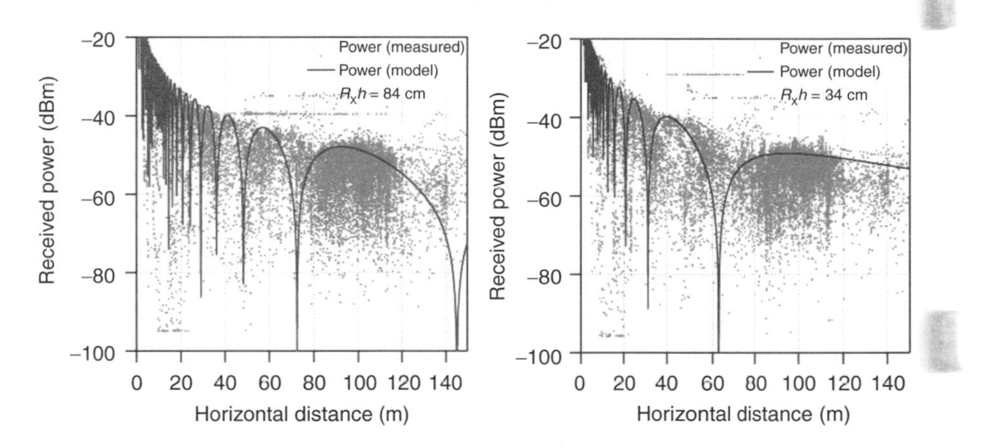

Fig. 7.35 Measurement results of relationship between received power and horizontal distance travelling on the expressway.

times, the error-free transmission was realized for 81 per cent of the travel time on the expressway. Figure 7.35 shows the measurement results of the relationship between the received power and horizontal distance between the vehicles on the expressway. The characteristics of received power are different from those from the two-ray model. This will be caused by the fluctuation of the vehicles.

7.4 Device technologies

7.4.1 Overview

Recently, RVC (road–vehicle communication) systems and IVC (inter-vehicle communication) systems are worthy of remark in the area of ITS (Highway Industry

Development Association, 1996). Various services of the ITS – driving information, driver's assistant information and vehicle control data – can be provided for vehicles by using both the RVC and the IVC systems. Moreover, automotive radar systems have already been practically used for collision avoidance. This section introduces the present research and development of the optical and millimetre-wave devices, which makes it possible to downsize the RVC system and the automotive radar system.

7.4.2 Optical devices

ROF system

The optical communications systems are suitable for transmitting broadband signals because of both the wide-band and the low transmission characteristics of the optical fibre. In the conventional optical communications systems, an IM-DD (intensity modulation direct detection) is adopted for modulation method of the optical fibre transmission. For the purpose of seamless communication between wired and wireless communication network systems, ROF (Radio on Fibre) System is newly proposed for the RVC system (Fujise *et al.*, 1999). Figure 7.36 shows a basic configuration of the ROF system. Multimedia signals are transmitted from a central base station (BS)

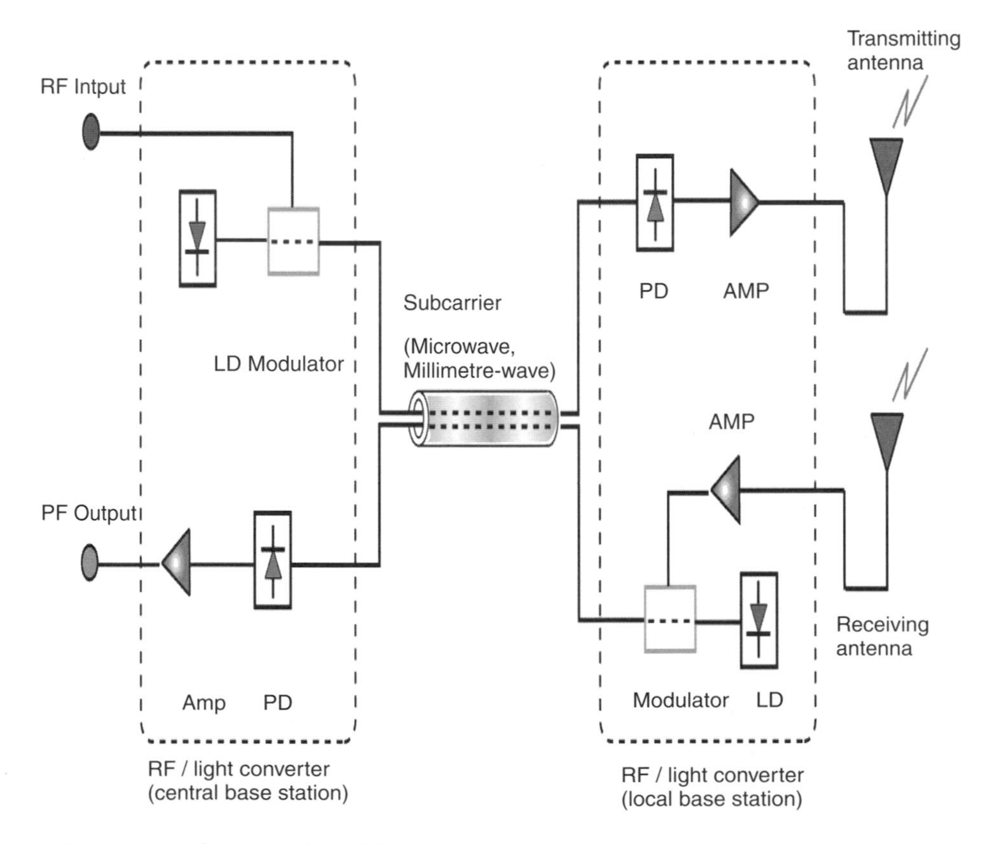

Fig. 7.36 Basic configuration of the ROF system.

to some local BSs allocated along the road. The radio frequency band is considered from microwave band to millimetre-wave band for the RVC system. On the forward link of the ROF system, an optical signal provided by LD (laser diode) is modulated with the RF (radio frequency) input signals defined by the various existing wireless communications systems in the central BS. The optical modulated signal is transmitted to the local BS using the optical fibre cables. At the local BS, the optical modulated signals are detected by PD (photo diode) and become the original RF signals directly. The RF signals are transmitted from an antenna after amplifying to a proper power level. The transmitted radio wave is received at the terminal equipment mounted on the vehicles and demodulated to the BB signals. On the reverse link of the ROF system, optical signals are modulated by amplified received RF signals at the local BS and transmitted in optical fibre cables from the local BS to the central BS. The detected optical modulated signals become the original RF output signals at the central BS.

The ROF systems are intensively studied for the RVC system because of the simple configuration. The key optical devices for the ROF systems are EO (electrical to optical) devices, i.e. optical modulators and OE (optical to electrical) devices, i.e., photo-detectors. Other optical devices will be also needed depending on the system design (Akiyama, 1999).

Optical modulators

In the present optical communication systems, the data transmission rate has already reached up to 10 Gbs per channel. The direct current modulation of the LD is used for low-speed transmission systems. For high-speed transmission systems, MZ (Mach-Zehnder) interferometer type modulators, EA (electro-absorption) type modulators or EA-LD (EA modulators with monolithically integrated laser diode) type modulators are commonly used because of their low chirping characteristics that reduce the penalty due to the dispersion of the optical fibres.

The above mentioned optical modulator devices can be applicable for the RVC systems based on the IM-DD modulation method. However, the ROF systems could also be constructed with these devices only in cases where the RF carrier frequency was not so high up to the microwave regions. The high-frequency characteristics of these modulators are much improved these days, but it is still difficult to operate them in the millimetre-wave region. Recently, some $LiNbO_3$ MZ modulators can operate up to millimetre-wave region with low insertion losses. The MZ modulators present the following technical issues:

- Long-term drift of the characteristics occurs.
- A fairly complicated control unit is necessary to compensate the drift.
- PDL (polarization dependent loss) makes it difficult to use them as in-line devices.

Especially for the PDL problem, it is very important to do the practical use in the ROF system because the LD can be put in the central BS, not each local BS when it is possible to use the modulators in-line. One way to solve this problem is that we have to use drift controlled MZ modulators putting the LDs near-by and connecting them with polarization maintained optical fibres. The performance will be satisfied not only the microwave ROF systems but also the millimetre-wave ROF systems. On the other hand, another candidate is to use the EA modulators. The EA modulators have a disadvantage of fairly large insertion loss (8 dB or more). However, the module size becomes small and the PDL is smaller than 0.5 dB by using bulk or stress controlled MQW (multiple quantum wells) absorption layer (Yamada *et al.*, 1995; Oshiba *et al.*, 1998). This low PDL characteristic makes it possible to use them as in-line devices. The LD can be put separately from the modulators. This increases the flexibility for constructing the systems.

To operate the EA modulators in the millimetre-wave region, a matching circuit is adopted, which is commonly done for high-frequency circuits (Mineo *et al.*, 1999). The used modulator chip is a wave guide type with a stress controlled MQW absorption layer and passive wave guides for both sides. The length of the absorption layer is shortened to 75 μm to reduce the capacitance. By these methods, the characteristics are much improved. Figure 7.37 shows the schematic structure of the modulator chip and Figure 7.38 shows the schematic construction and the photograph of the module. Because the modulator characteristics are somewhat changed by temperatures, the chip is commonly put on a Peltier device to control the temperature.

Figure 7.39 shows the input return loss (S11). The bandwidth where VSWR is less than 1.5 is about 3 GHz at 60 GHz band. Figure 7.40 shows the modulated optical spectrum with an input optical power of 9 dBm and a RF modulation power of 6 dBm

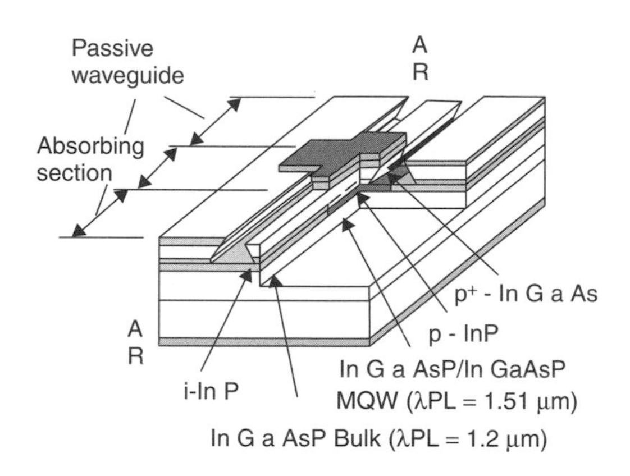

Fig. 7.37 Structure of the EA modulator chip.

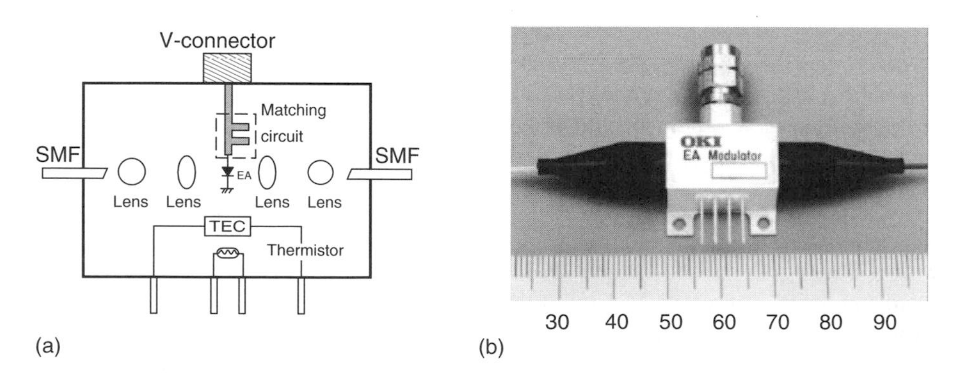

Fig. 7.38 Schematic construction (a), and the photograph (b) of the EA modulator module with a matching circuit.

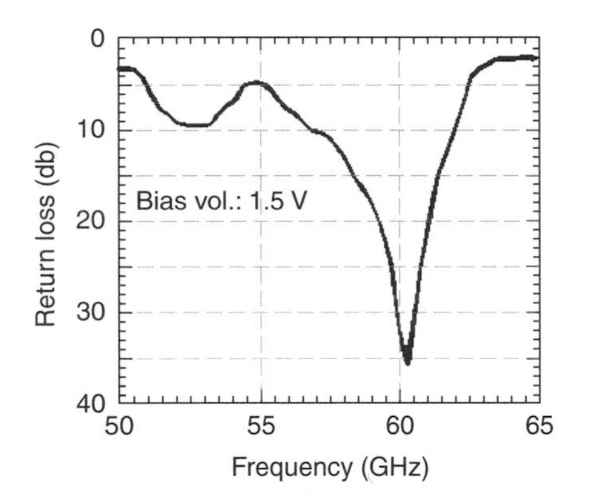

Fig. 7.39 Return loss of the modulator module at 60 GHz band.

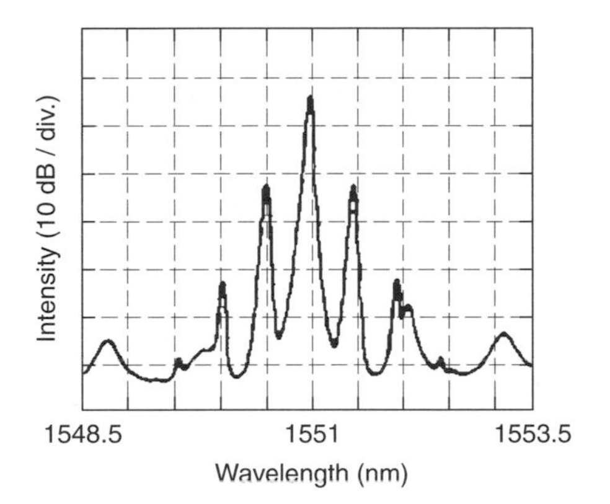

Fig. 7.40 Optical spectrum modulated at 59.6 GHz.

at 59.6 GHz. The wide-range linearity is not commonly expected for any optical modulator because of their nonlinear properties between applied voltage and the modulated optical power. It is, therefore, difficult to obtain a deep modulation. In the above mentioned EA modulator case, the modulation index of about 35 per cent is the limit to avoid the nonlinearity effect. Using this modulator, two channels of 156 Mb/s signals are transmitted with no error (Kuri *et al.*, 1999).

Photo-detectors

For detecting the high-speed optical signals, PIN-PDs (PIN photodiodes) are commonly used. To increase the operating frequency, it is necessary to reduce the capacitance. The wave-guide type PIN-PD is effective for reducing the capacitance and for realizing sufficient absorbing length for photons simultaneously. Recently, the PIN-PDs of this type operating up to the millimetre-wave region are reported (Kato *et al.*, 1992). These PIN-PDs are indispensable devices for the millimetre wave ROF systems.

Other optical devices

Other optical devices such as directional couplers, filters, isolators and so on are also necessary for the ROF systems. EDF (erbium doped fibre) optical amplifiers will be needed for transmitting the signals to many local BSs. In these optical devices, filters will play an important role. In a very simple system, it will be possible to construct it without filters. However, when we want to use multi-wavelengths, filters are necessary to separate them. Using multi-wavelengths, we can increase the transmitting data without increasing the bit rate. Furthermore, we can send the CW optical signals for the reverse link or (if necessary) we can send a local signal using one of the wavelengths from the central BS. In these cases, the LDs or local oscillators are not necessary at each local BS. When CW optical signal is transmitted from the central BS, the signal can be modulated using an in-line modulator such as a low PDL EA modulator. Another effective application of filters is to generate single side band signals after modulation. The dispersion effect of the optical fibre is a severe problem for the ROF systems, especially when the conventional single-mode fibre is used and the RF frequency is high enough or long distance transmission is necessary. The relative phase relation among both side bands and the carrier spectra is changed by the effect (Gliese *et al.*, 1999). To solve the problem, single side band transmission is one of the effective methods (Kuri *et al.*, 1999). For this purpose, sharp filters are usable. When filters are not used, some devices to compensate the effect, such as dispersion compensating fibres, are necessary. The use of dispersion compensating fibres is effective when the RF carrier frequency is not so high and the sprit of the side band signal is too narrow to separate them by a filter. When dispersion shift fibres are used and the wavelengths are selected whose dispersions are zero or very low, this problem does not arise.

7.4.3 Millimetre-wave devices

Radar system

The development of the Advanced cruise-assist Highway Systems (AHS) and the Advanced Safety Vehicle (ASV) are being carried out to aid drivers and increase vehicle safety (Highway Industry Development Association, 1996). Radio sensors are an essential technology in making it possible. Automotive radar is one of the radio sensors and is used for distance detection between vehicles. Spread Spectrum (SS) radar system is investigated for interference robustness and its possible operation for the IVC systems (Mizui *et al.*, 1995; Akiyama and Tokuda, 1999).

The conventional automotive radar system at 76 GHz millimetre-wave band is recently developed for collision avoidance in practical use. The detection of distance between vehicles is based on the frequency modulated continuous wave (FMCW) method. In the FMCW method, the vehicle distance is calculated by the frequency difference between the echo radio wave from a forward target vehicle and transmitted radio wave of a measuring host vehicle. When another radio interference source with the same frequency band exists in the neighborhood of the host vehicle, it is not possible to distinguish the echo wave from the interference wave by using the received wave at the host vehicle. The FMCW method is never able to compensate for the performance degradation caused by the radio interference of the oncoming vehicles. Therefore, the performance of the FMCW radar system depends on the beam width and the polarization of the antenna in terms of the interference. On the contrary, the SS radar system has a preferable performance advantage relating to the interference, inherently (Shiraki *et al.*, 1999). For all cases where the transmitted waves of the

Table 7.6 The SS radar system specification

Item	Specification
Frequency band	76.5 GHz
Band width	30 MHz
Tx power output	−3 dBm
Modulation	PSK
Number of antenna	Transmitter: 1, Receiver: 2
Polarization of antenna	Linear: 45°
Gain of antenna	31 dBi
Beam width of Antenna	Horizontal: 2.6° Vertical: 4.0°
Spreading code	M sequence (code period: 127 chips)
Sampling rate	60 MHz
Chip rate	15 Mcps
Distance resolution	2.5 m

host vehicle and the oncoming vehicles are generated by SS modulation with each orthogonal spreading codes respectively, the desired echo radio wave can be detected by correlation detection and the vehicle distance is calculated by the peak position of the correlation values (Tokuda *et al.*, 1999). The specification of the automotive SS radar system is shown in Table 7.6.

RF/IF module for SS radar system

Figure 7.41 shows the RF module of the SS radar system. The radar wave at 76.5 GHz is up-converted to an IF signal using the local frequency at 70.7 GHz. The local frequency is output from a × 4 multiplier. As shown in Table 7.6, the maximum output power and the bandwidth of the SS radar system are −3 dBm and 30 MHz, respectively. The antenna gain is 31 dBi, and horizontal and vertical beam width of each antenna

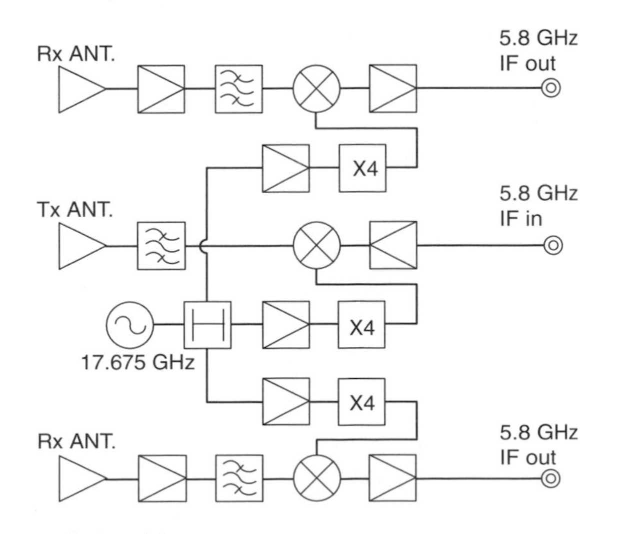

Fig. 7.41 Block diagram of RF module.

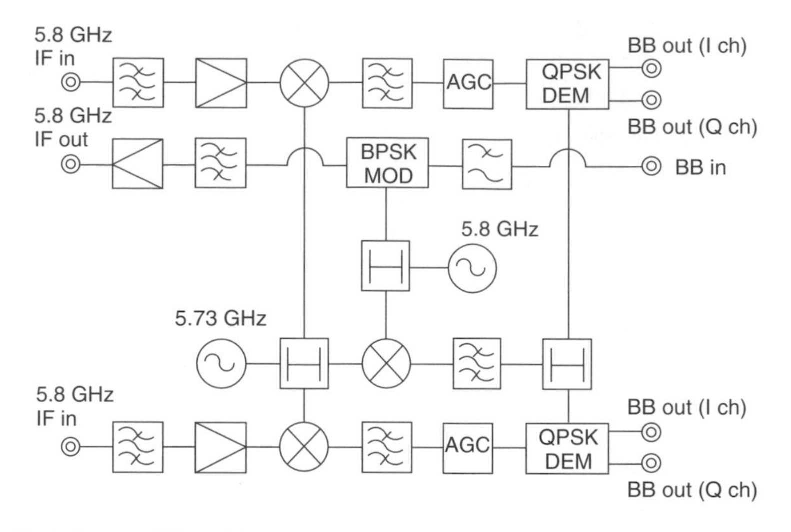

Fig. 7.42 Block diagram of IF module.

are 2.6° and 4.0°, respectively. The block diagram of the IF module is shown in Figure 7.42. The input baseband signal (15 Mcps) is BPSK modulated at 5.8 GHz. The received IF signal is down-converted to a 70 MHz signal and its power is controlled in this band. This signal at 70 MHz is QPSK demodulated to obtain the baseband signal. The wireless communication has a long history. The devices for this SS radar system have been constructed with MICs (microwave integrated circuits) for relatively low frequencies or rectangular wave guide circuits for high frequencies.

Monolithic microwave integrated circuits (MMIC)

Recently, practical MMICs (monolithic microwave integrated circuits) have been developed and used for some practical FMCW radar systems. Because the wiring and propagation losses become large at high frequencies, MMIC techniques will be necessary not only to keep the systems compact but also to obtain high performances. By the same region, MMICs are necessary to construct the compact ITS systems. For fabricating MMICs, we can use FETs (including HEMTs) whose f_t is over 100 GHz and f_{max} is over 200 GHz, now. With these devices, most of the active circuits can be fabricated. As an example of MMICs, our millimetre-wave amplifier chip is shown in Figure 7.43. This chip is a 76 GHz band two-stage amplifier using InGaAs P-HEMTs (pseudomorphic HEMTs) with a gate length of 0.1 μm and co-planer circuits. The amplifier shows a gain of over 12 dB at the 76 GHz band, as shown in Figure 7.44.

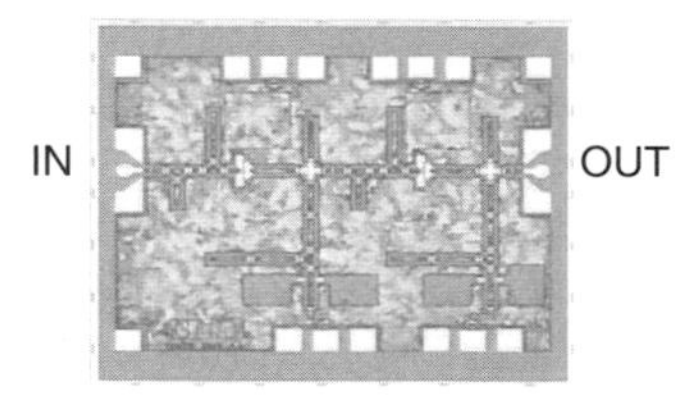

Fig. 7.43 Photograph of 76 GHz two-stage amplifier chip.

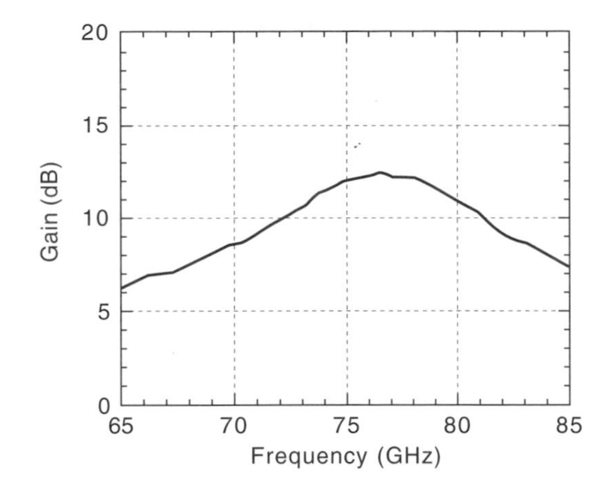

Fig. 7.44 Gain characteristics of the 76 GHz two-stage MMIC amplifier.

To construct the systems, passive circuits are also necessary. Most of the passive circuits can be integrated in MMICs. However, it is difficult to obtain some sharp filters in MMICs. These filters must be added outside. Such filters have been commonly realized with rectangular or circular wave-guide circuits. Recently, very small band-pass filters have been reported using a patterned dielectric sheet in the millimetre-wave range (Ishikawa *et al.*, 1995). These filters will be useful for keeping the systems compact.

It is also difficult to fabricate irreversible circuits such as isolators and circulators in MMICs. These devices are also important and convenient for fabricating the systems. However, we can construct most of the systems without these irreversible circuits by the proper design.

Antenna

Other important devices are antennas. As frequency becomes higher, the antenna size becomes smaller to obtain the same gain. Because the propagation loss in free space increases as frequency becomes high, the gain of the antenna is desired to be high in the millimetre-wave region. Flat antennas with a high gain of more than 30 dBi have already been developed. In order to minimize the effect of the reflection from the surroundings, antennas with circular polarization have been also developed. However, as the gain becomes high, the beam shape becomes sharp. When such high gain antennas are used, it will be somewhat troublesome to align them. To avoid the trouble-some, adequate gains must be chosen for the systems. In the future, adaptive antennas will become necessary.

References

Akiyama, M. (1999). Trends of Millimetre-Wave and Optical Devices. *MWE'99, Microwave Workshop Digest*, October

Akiyama, M. and Tokuda, K. (1999). Inter-Vehicle Communications Technology for Group Cooperative Moving. *VTC'99*.

Chujo, Y. *et al.* (1997). Wireless Bus Location System. *Fujitsu Ten Technical Report*, **13**(2), 34–42, in Japanese.

Fukui, R. (1997). Technical Report of DSRC for ITS-RVC & IVC. *MWE'97 Microwave Workshop Digest*, 206–211.

Fujise, M. and Harada, H. (1998). Multimode DSRC by Radio On Fibre. *Proceedings of the 1998 Communication Society Conference of IEICE, SAD-2-8*, 32–33, September.

Fujise, M., Sato, K. and Harada, H. (1998). New Road-Vehicle Communication Systems Based on Radio on Fibre Technologies for Future Intelligent Transport Systems (ITS). *Proceedings of the First International Symposium on Wireless Personal Multimedia Communications (WPMC'98)*, 139–144, November.

Fujise, M., Sato, K., Harada, H. and Kojima, F. (1999). ITS Multi-Service Road-Vehicle Communications Based on Radio on Fibre Systems. *Proceedings of the Second International Symposium on Wireless Personal Multimedia Communications (WPMC'99)*, September.

Gliese, U., Oslashrskov, S.N and Nielsen, T.N. (1996). Chromic Dispersion in Fibre-Optic Microwave and Millimetre-Wave Links. *IEEE Trans. Microwave Theory and Techniques*, **44**, 1716.

Highway Industry Development Organization, (1996). ITS Handbook in Japan. Supervised by Ministry of Construction.

Horimatsu, T. (1999). Trends of Inter-Vehicle Communication and Automotive Radar Systems. *MWE'99 Microwave Workshop Digest*, 397–402.

Ishikawa, Y., Hiratsuka, T., Yamashita, S. and Iio, K. (1995). Planer type dielectric resonator filter at millimetre-wave frequency. *MWE'95 Microwave Workshop Digest*.

Iwai, A. *et al.* (1997). Automatic Dispatching System with GPS. *Fujitsu Ten Technical Report*, **15**(1), 30–39, in Japanese.

IEEE, *Vehicular Technology Society News*, 1999.

Kato, K., Hata, S., Kawano, K., Yoshida, J. and Kozen, A., (1992). A High-Efficiency 50 GHz InGaAs Multimode Waveguide Photodetector. *IEEE Journal of Quantum Electronics*, **28**, 2728.

Kamiya, H. *et al.* (1997). Intelligent Technologies of Honda ASV. *ITS World Congress Berlin*.

Kato, A., Sato, K. and Fujise, M. (1998). Experiments of 156 Mbps Wireless Transmission using 60 GHz band on a pavement. *Proceedings of the First International Symposium on Wireless Personal Multimedia Communications (WPMC'98)*, pp. 389–392, November.

Kuri, T., Kitayama, K., Stöhl, A. and Ogawa, Y., (1999). Fibre-Optic Millimetre-Wave Downlink System Using 60 GHz-Band External Modulation. *Journal of Lightwave Technology*, **17**, 799.

Mizui, K., Uchida, M. and Nakagawa, M., (1995). Vehicle-to-Vehicle Communication and Ranging System Using Spread Spectrum Technique. *IEICE*, **J78-B-II**(5), 342–9.

Mineo, N., Yamada, K., Nakamura, K., Shibuya, Y., Sakai, S. and Nagai, K., (1999). 60 GHz Band Electroabsorption Modulator Module with Built-in Machine Circuit. *1999 IEMT/IMC Proceedings*.

Oshiba, S., Nakamura, K. and Horikawa, H. (1998). Low-Drive-Voltage MQW Electroabsorption Modulator for Optical Short-Pulse Generation. *IEEE Journal of Quantum Electronics*, **43**, 277.

Shiraki, Y., Hoshina, S., Tokuda, K., Nakagawa, M. and Mizui, K. (1999). Evaluation of Interference Reduction Effect of SS Radar. Proceedings of the 1999 IEICE General Conference, A-17-22, 421.

Tokuda, K., Shiraki, Y., Ohyama, T. and Kawakami, E. (1999). Evaluation of Fundamental Performance for 76 GHz Millimetre Wave Band Automotive SS Radar System. *Proceedings of the Second International Symposium on Wireless Personal Multimedia Communication (WPMC'99)*, 112–118.

Yamada, K., Murai, H., Nakamura K. and Ogawa, Y. (1995). Low polarization dependence (<0.3 dB) in an EA modulator using a polyimide-buried high-mesa ridge structure with an InGaAsP bulk absorption layer. *Electronics Letters*, **31**, 237.

Global positioning technology in the intelligent transportation space

Patrick Herron, Chuck Powers and Michael Solomon
Motorola Inc., USA

There is no doubt that one of the most important enabling technologies in the intelligent vehicle space is the global positioning system (GPS). Without the ability to accurately determine a vehicle's position on demand, there would be no way to cost-effectively implement autonomous or server-based vehicle navigation, nor would the ability to deliver customized, location-based services to the vehicle be possible.

This chapter will provide a brief overview of the GPS, and how it can be leveraged in intelligent vehicle applications. This chapter begins with a section describing the history of space-based positioning projects that have led to the current GPS, followed by a detailed description of the system as it exists and operates today. This is followed by a discussion of the science behind the GPS, and the techniques and components required to accurately and cost-effectively determine a user's position. The chapter concludes with some example applications where GPS is being used in the intelligent vehicle and related spaces, as well as future services that will be made possible because of GPS-based positioning capabilities.

8.1 History of GPS

Long before the development of the GPS in use today, the concept of time transfer and positioning via signals from space was being researched around the world. These costly research projects were mainly sponsored by government agencies, to address their long-standing need to improve techniques for quickly and accurately positioning military vehicles and personnel on or above the battlefield. Troops and vehicles of centuries past relied on maps, charts, the stars and various electronic devices to find their location; however, with each improved method of determining position came inherent limitations. Boundaries and landmarks change with the passage of time, making mapping a continual, time-consuming task. Positioning via the stars has long been a necessity for mariners, but accurate time keeping and clear skies are at times elusive. Until the deployment of today's GPS, the ultimate solution did not exist – an 'always on, always available' system for determining an exact position anywhere on the globe.

The constellation of satellites being used for global positioning today has it roots in the satellite positioning and time transfer systems of the early 1960s. Like many successful endeavours, the GPS was conceived from building blocks of other programmes such as the Navy Navigation Satellite System (NNSS, or Transit), Timation and Project 621B. It is worthwhile to have a brief understanding of these predecessors of GPS in order fully to understand and appreciate the complexity of space-based radionavigation.

Transit was conceived to provide positioning capabilities for the US submarine fleet, and originally deployed in 1964. While Transit proved to be a tremendous success in demonstrating the concept of radio-navigation from space, the system was inherently inaccurate and required long periods of satellite observation in order to provide a user with enough information to calculate a position. Periods of observation in excess of 90 minutes were not uncommon, which limited the system's effectiveness for positioning a submarine at sea, since extended surface time could leave the vessel vulnerable. In its simplest form, Transit consisted of a small constellation of satellites broadcasting signals at 150 MHz and 400 MHz. The Doppler shift of these signals as measured by observers at sea, coupled with the known positions of the satellites in space, was sufficient to provide range measurements to the satellites, enabling the user to compute their position in two dimensions. Since all Transit satellites broadcast their signals at the same frequencies, the potential for interference allowed for only a small number of satellites. It was this limited number of satellites that necessitated the long periods of data collection, reducing the overall effectiveness of the system. This system was finally decommissioned in 1996.

In order to overcome the limitations of signal interference inherent in Transit and thereby increase the availability and effectiveness of satellite observation, an alternate technique for signal broadcast was necessary. US Air Force Project 621B, also begun in the late 1960s, demonstrated the use of pseudorandom noise (PRN) to encode a useful satellite ranging signal. PRN code sequences are relatively easy to generate, and by carefully choosing PRN codes which are nearly orthogonal to one another, multiple satellites can broadcast ranging signals on the same frequency simultaneously without interfering with one another. This simple concept forms the fundamental basis for GPS satellite ranging, and for the future implementation of the Wide Area Augmentation System (WAAS), which will be discussed later in this chapter.

The US Navy's Timation satellite system, initially launched in 1967, was also in full swing by the early 1970s. Timation satellites carried payloads with atomic time standards used for time keeping and time transfer applications. This enabled a receiver to use the signal broadcast by each Timation satellite to measure the distance to that satellite by measuring the time it took the signal to reach the receiver. Timation provided a key proof of concept and a foundation building block for the GPS, because without accurate time standards, the current GPS would not be possible.

In 1973, building on the success and knowledge gained from Transit, Timation and Project 621B, and with inputs and support from multiple branches of the military, the US Department of Defense (DoD) launched the Joint Program for GPS. Thus, the NAVSTAR GPS project was born.

8.2 The NAVSTAR GPS system

The NAVSTAR project was conceived as an excellent way to provide satellite navigation capabilities for a wide variety of military and civilian applications, and it has been doing so quite effectively since full operational capability (FOC) was declared in 1995. Building on previous satellite technology, the initial GPS satellites were launched between 1978 and 1985. These so-called Block I satellites were used to demonstrate the feasibility of the GPS concepts. Subsequent production models included Block II, Block IIA and Block IIR, each designed with improved capabilities, longer service life and at a lower cost. The next-generation models, known as Block IIF, are now being designed for launch in 2002.

This system, which currently consists of 28 fully operational satellites, cost an estimated $10 billion to deploy. The constellation is maintained and managed by the US Air Force Space Command from five monitoring sites around the world, at an annual cost of between $300 million and $500 million.

8.2.1 GPS system characteristics

The 28 satellites in the GPS are deployed in six orbital planes, each spaced 60° apart and inclined 55° relative to the equatorial plane. The orbit of each satellite (space vehicle, or SV) has an approximate radius of 20 200 km, resulting in an orbital period of slightly less than 12 hours. The system design ensures users worldwide should be able to observe a minimum of five satellites, and more likely six to eight satellites, at any given time, provided they have an unobstructed view of the sky. This is important because users with no knowledge of their position or accurate time require a minimum of four satellites to determine what is commonly known as a position, velocity and time solution, or PVT. The PVT data consists of latitude, longitude, altitude, velocity, and corrections to the GPS receiver clock.

The GPS satellites continuously broadcast information on two frequencies, referred to as L_1 and L_2, at 1575.42 MHz and 1227.6 MHz, respectively. The L_1 frequency is used to broadcast the navigation signal for non-military applications, called the Standard Positioning Service (SPS). Because the original design called for the SPS signal to be a lower resolution signal, it is modulated with a pseudorandom noise (PRN) code referred to as the Coarse Acquisition (C/A) code. For the purposes of reserving the highest accuracy potential for military users, the DoD may also impose intentional satellite clock and orbital errors to degrade achievable civilian positioning capabilities. This intentional performance degradation is commonly known as Selective Availability (S/A). For US military and other DoD-approved applications, a more accurate navigation signal known as the Precise Positioning Service (PPS) is broadcast on both the L_1 and L_2 frequencies. The PPS, in addition to the C/A code available on L_1, includes

Table 8.1 Original navigation signal accuracy targets for SPS and PPS

	Horizontal accuracy	Vertical accuracy	Timing accuracy
Standard positioning service	100 metres	156 metres	340 ns
Precise positioning service	22 metres	27.7 metres	200 ns

Note: By design, all accuracies are statistically achievable 95% of the time

a more accurate signal modulated with a code known as the Precise code (P-code) if unencrypted, and as the P(Y)-code if encrypted. Authorized users who have access to the PPS can derive more accurate positioning information from the L_1 and L_2 signals. Refer to Table 8.1 for a list of the original positioning and timing accuracy goals of the SPS and PPS services.

On 1 May, 2000, US President Bill Clinton announced the cessation of the S/A, which immediately resulted in greatly increased positioning accuracy for non-military GPS applications. The cessation of S/A should allow users of the SPS a level of accuracy similar to those using the PPS. Within the first week of the discontinuation of S/A, positioning accuracies within 10 metres were already being reported, without any upgrade to the GPS receivers being used.

8.2.2 The navigation message

The navigation message broadcast by every GPS satellite contains a variety of information used by each GPS receiver to calculate a PVT solution. The information in this message includes time of signal transmission, clock correction and ephemeris data for the specific SV, and an extensive amount of almanac and additional status and health information on all of the satellites in the GPS.

Each SV repeatedly broadcasts a navigation message that is 12.5 minutes in length, and consists of 25 1500-bit data frames transmitted at 50 bits per second. A single data frame is composed of five 300-bit subframes, each containing different status or data information for the receiver, preceded by two 30-bit words with SV-specific telemetry and handover information. The first three subframes, containing clock correction and ephemeris data relevant to the specific SV, are refreshed as necessary for each data frame transmitted during the navigation message broadcast. The almanac and other data transmitted in the final two subframes are longer data segments, relevant to the entire GPS, requiring the full 25 data frames to be broadcast completely. Below is a brief description of the contents of each subframe. For an illustration of the complete Navigation Message, refer to Figure 8.1.

Clock correction subframe

The clock correction subframe, the first subframe transmitted in the navigation message data frame, contains the GPS week number, accuracy and health information specific to the transmitting SV, and various clock correction parameters relating to overall system time, such as clock offset and drift.

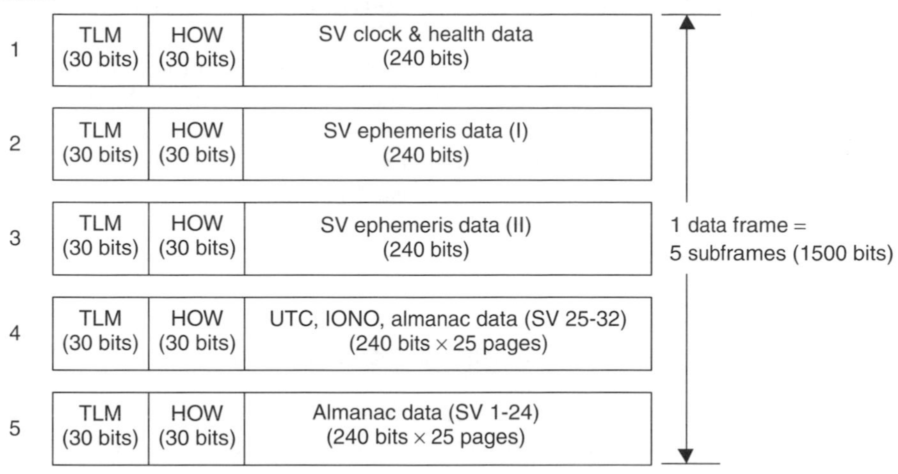

Fig. 8.1 Navigation message. TLM, telemetry word; HOW, handover word.

SV ephemeris subframes

The second and third subframes of the navigation message contain ephemeris data. This data provides the GPS receiver with precise orbit information and correction parameters about the transmitting SV that the receiver uses accurately to calculate the satellite's current position in space. This information, in turn, is used with the clock information to calculate the range to the SV. Included in the ephemeris subframes are telemetry parameters specific to the transmitting SV, such as correction factors to the radius of orbit, angle of inclination, and argument of latitude, as well as the square root of the semi-major axis of rotation, the eccentricity of the orbit of the SV, and the reference time that the ephemeris data was uploaded to the SV.

Almanac and support data subframes

Subframes four and five of the navigation message data frame contain comprehensive almanac data for the entire GPS constellation, along with delay parameters that the receivers use for approximating phase delay of the transmitted signal through the ionosphere, and correction factors to correlate GPS and Universal Time Coordinated (UTC).

The almanac data contains orbit and health information on all of the satellites in the GPS constellation. GPS receivers use this information to speed up the acquisition of SV signal transmissions. The almanac data in subframe four contains health and status information on the operational satellites numbered 25 through 32, along with ionospheric and UTC data. The almanac data in subframe five contains health and status information on the operational satellites in the GPS numbered 1 through 24.

For a more detailed description of the information contained in the Navigation Message, refer to the ICD-GPS-200c specification, which is available from the US Coast Guard Navigation Center.

8.3 Fundamentals of satellite-based positioning

To understand the true value and cost of the positioning capabilities of the GPS, it is important for the user to have a basic understanding of the science behind positioning, and the types of components and techniques that may be used to calculate accurate positions. This section divides this discussion into three main areas: the basic science behind GPS; the different unassisted and assisted position calculation techniques that may be used, depending upon the needs of the specific application; and the hardware and software components necessary for calculating a position.

8.3.1 The basic science of global positioning

The design of the GPS makes it an all-weather system whereby users are not limited by cloud cover or inclement weather. Broadcasting on two frequencies, the GPS provides sufficient information for users to determine their position, velocity and time with a high degree of accuracy and reliability. As mentioned previously, frequency L_1 is generally regarded as the civilian frequency while frequency L_2 is primarily used for military applications. Applications and positioning techniques in this chapter will focus on GPS receiver technology capable of tracking L_1 only, as cost and security issues typically preclude most users from taking full advantage of both GPS frequencies. Without a complete knowledge of the encrypted L_2 frequency, only mathematical exercises enable high accuracy applications of GPS such as surveying to take advantage of any information provided by L_2.

Position calculation

The fundamental technique for determining position with the GPS is based on a basic range measurement made between the user and each GPS satellite observed. These ranges are actually measured as the GPS signal time of travel from the satellite to the observer's position. These time measurements may be converted to ranges simply by multiplying each measurement by the speed of light; however, since most GPS receiver internal clocks are incapable of keeping time with sufficient accuracy to allow accurate ranging, the mathematical PVT solution must solve for errors in the receiver clock at the time each observation of a satellite is made. Satellite ranges are commonly called pseudoranges to include this receiver clock error and a varicty of other errors inherent in using GPS. These receiver clock errors are included as one component in a least squares calculation, which is used to solve for position using a technique called trilateration.

To calculate the values for PVT, the concept of triangulation in two-dimensions as is commonly practised in determining the location of an earthquake epicentre is extended into three-dimensions, with the ranges from the satellites prescribing the radius of a sphere (see Figure 8.2). This technique is known as trilateration, since it uses ranges to calculate position, whereas triangulation uses angular measurements. If a sphere centred on the satellites' position in space is hypothetically created with the range from the user to each satellite as its radius, the intersection of three of these spheres may be used to determine a user's two-dimensional position. While it may seem counter-intuitive that ranges to three satellites will allow for only a two-dimensional position,

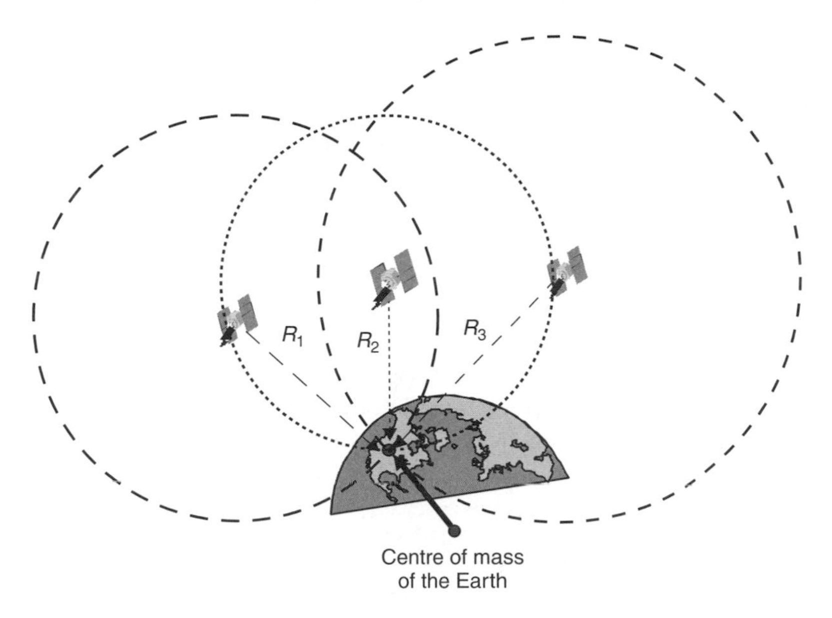

Fig. 8.2 3-Dimensional trilateration of GPS satellites.

in fact one observation is needed to solve for each of latitude, longitude and receiver clock error. Thus, to determine a user's position in three-dimensions a minimum of four satellites is required, in order to solve for altitude, as well as latitude, longitude and clock error.

Once pseudoranges have been determined to three or more SVs, the user's PVT can be calculated by solving N simultaneous equations as a classic least squares problem, where N is the number of satellite pseudoranges measured. The relationship between the receiver and each satellite's position can best be written by extending the Pythagorean Theorem as illustrated in equation 8.1, where i is the number of each satellite detected (3 to N), $\{x_i, y_i, z_i\}$ is the known position of each satellite i, R_i is the pseudorange measurement for each satellite i, and b is the receiver clock error:

$$R_i = \sqrt{(x_i - x)^2 + (y_i - y)^2 + (z_i - z)^2} - b \tag{8.1}$$

While a three-dimensional PVT calculation may be made using pseudoranges from four satellites, improved accuracy can be achieved if five or more are used, as the redundancy can help reduce the effects of position and receiver clock errors in the calculation.

Coordinate systems

The coordinate frame used by the GPS to map a satellite's position, and thus a receiver's position, is based on the World Geodetic System 1984 (WGS 84). This coordinate reference frame is an Earth-centred, Earth-fixed (ECEF) Cartesian coordinate system, for which curvilinear coordinates (latitude, longitude, height above a reference surface) have also been defined, based on a reference ellipsoid, to allow easier plotting of a user's position on a traditional map. This coordinate frame, or datum, is the standard reference used for calculating position with the GPS. However, many regional and local maps based on datums developed from different ground-based surveys are also in use today, whose coordinates may differ substantially from WGS 84. Simple mathematical transformations can be used to convert calculated positions between WGS 84 and these regional datums, provided they meet certain minimum criteria for the mapping of their longitude, latitude and local horizontal and vertical references. At last count, more than 100 regional or local geodetic datums were in use for positioning applications in addition to WGS 84.

8.3.2 Positioning techniques

Several different techniques have been developed for using the GPS to pinpoint a user's position, and to refine that positioning information though a combination of GPS-derived data and additional signals from a variety of sources. Some of the more popular techniques, such as autonomous positioning, differential positioning and server-assisted positioning, are briefly described below.

Autonomous GPS positioning

Autonomous positioning, also known as single-point positioning, is the most popular positioning technique used today. It is the technique that is commonly thought of when a reference to using the GPS to determine the location of a person, object or address is made. In basic terms, autonomous positioning is the practice of using a single GPS receiver to acquire and track all visible GPS satellites, and calculate a PVT solution. Depending upon the capabilities of the system being used and the number of satellites in view, a user's latitude, longitude, altitude and velocity may be determined. As mentioned earlier, until May of 2000 this technique was limited in its accuracy for commercial GPS receivers. However, with the discontinuation of S/A this technique may now be used to determine a user's location with a degree of accuracy and precision that was previously available only to privileged users.

Differential GPS positioning

The use of differential GPS (DGPS) has become popular among GPS users requiring accuracies not previously achievable with single-point positioning. DGPS effectively eliminated the intentional errors of S/A, as well as errors introduced as the satellite broadcasts pass through the ionosphere and troposphere.

Unlike autonomous positioning, DGPS uses two GPS receivers to calculate PVT, one placed at a fixed point with known coordinates (known as the master site), and a second (referred to here as the mobile unit) which can be located anywhere in the vicinity of the master site where an accurate position is desired. For example, the master site could be located on a hill or along the coastline, and the mobile unit could be a GPS receiver mounted in a moving vehicle. This would allow the master site to have a clear view of the maximum number of satellites possible, ensuring that pseudorange corrections for satellites being tracked by the mobile unit in the vicinity would be available.

The master site tracks as many visible satellites as possible, and processes that data to derive the difference between the position calculated based on the SV broadcasts and the known position of the master site. This error between the known position and the calculated position is translated into errors in the pseudorange for each tracked satellite, from which corrections to the measured distance to each satellite are derived. These pseudorange corrections may then be applied to the pseudoranges measured by the mobile unit, effectively eliminating the affects of SA and other timing errors in the received signals (see Figure 8.3).

Corrections to measured pseudoranges at the master site are considered equally applicable to both receivers with minimum error as long as the mobile unit is less than 100 km from the master site. This assumption is valid because the distance at which the GPS satellites are orbiting the earth is so much greater than the distance between the master site and the mobile unit that both receivers can effectively be considered to be at the same location relative to their distance from each SV. Therefore, the errors in the pseudorange calculated for a particular satellite by the mobile unit are effectively the same as errors in the same pseudorange at the master site (i.e. the tangent of the angle between the master site and second receiver is negligible (see Figure 8.4)).

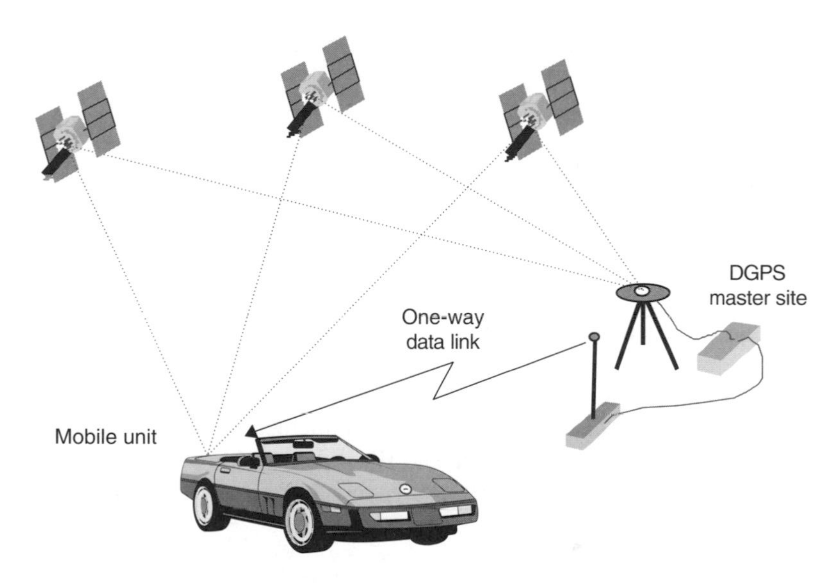

Fig. 8.3 Differential GPS positioning.

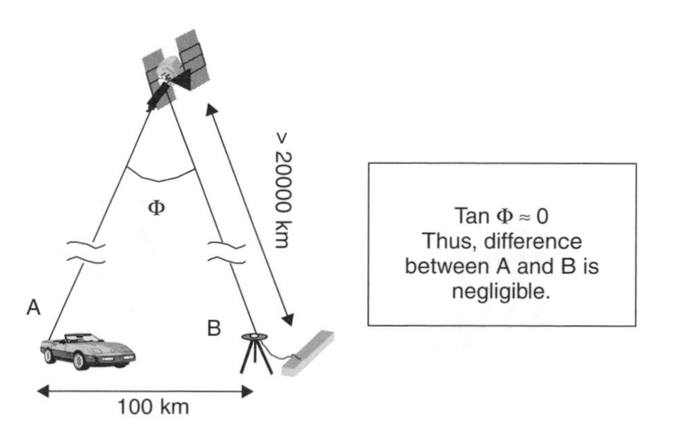

Fig. 8.4 Pseudorange correction in DGPS (not to scale).

Of course, to calculate a position using DGPS, a mobile unit must establish communication with a master site broadcasting DGPS correction information. One source is the US Coast Guard, which operates a series of DGPS master sites that broadcast DGPS corrections across approximately 70 per cent of the continental US, including all coastal areas. Alternatively, a GPS receiver that has wireless communication capabilities, such as one that is integrated into an intelligent vehicle, may be able to access DGPS correction data on the Internet, or have it delivered on a subscription basis from a private differential correction service provider.

With the discontinuation of S/A, using the DGPS positioning technique will still provide enhanced positioning accuracy, since other timing errors are inherent in the SV broadcasts that DGPS may help correct. However, these much smaller improvements in accuracy may no longer offset the additional cost of receiving and processing the DGPS correction information for many applications.

Inverse differential GPS positioning

Inverse differential GPS (IDGPS) is a variant of DGPS in which a central location collects the standard GPS positioning information from one or more mobile units, and then refines that positioning data locally using DGPS techniques. With IDGPS, a central computing centre applies DGPS correction factors to the positions transmitted from each receiver, tracking to a high degree of accuracy the location of each mobile unit, even though each mobile unit only has access to positioning data from a standard GPS receiver (see Figure 8.5).

This technique can be more cost-effective in some ways than standard DGPS, since there is no requirement that each mobile unit be DGPS-enabled, and only the central site must have access to the DGPS correction data. However, there is an additional cost for each mobile device, since each unit must have a means of communicating position data back to the central computer for refinement. For applications such as delivery fleet management or mass transit, IDGPS may be an ideal technique for maintaining highly accurate position data for each vehicle at a central dispatch facility,

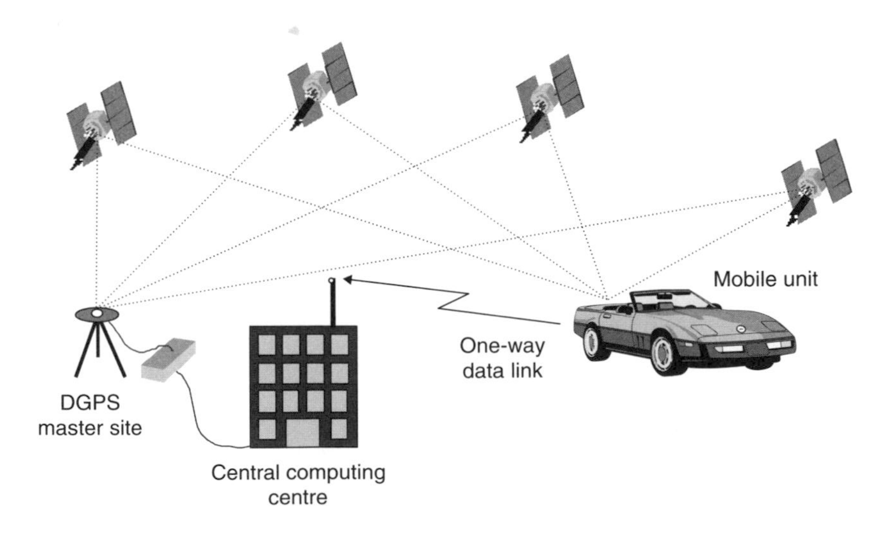

Fig. 8.5 Inverse differential GPS positioning.

since the communication channel is already available, and the relative cost of refining the positioning information for each mobile unit at the central location is minimal. Of course, with the discontinuation of S/A, DGPS refinement may no longer be necessary for many of these applications.

Server-assisted GPS positioning

Server-assisted GPS is a positioning technique that can be used to achieve highly accurate positioning in obstructed environments. This technique requires a special infrastructure that includes a location server, a reference receiver in the mobile unit, and a two-way communication link between the two, and is best suited for applications where location information needs to be available on demand, or only on an infrequent basis, and the processing power available in the mobile unit for calculating position is minimal.

In a server-assisted GPS system, the location server transmits satellite information to the mobile unit, providing the reference receiver with a list of satellites that are currently in view. The mobile unit uses this satellite view information to collect a snapshot of transmitted data from the relevant satellites, and from this calculates the pseudorange information. This effectively eliminates the time and processing power required for satellite discovery and acquisition. Also, because the reference receiver is provided with the satellite view, the sensitivity of the mobile unit can be greatly improved, enabling operation inside buildings or in other places where an obstructed view will reduce the capabilities of an autonomous GPS receiver.

Once the reference receiver has calculated the pseudoranges for the list of satellites provided by the location server, the mobile unit transmits this information back to the location server, where the final PVT solution is calculated. The location server then transmits this final position information back to the mobile device as needed. Because the final position data is calculated at the location server, some of the key benefits of DGPS can also be leveraged to improve the accuracy of the position calculation. An illustration of the relationship between the reference receiver and the location server in a server-assisted GPS system can be seen in Figure 8.6.

Enhanced client-assisted GPS positioning

The enhanced client-assisted GPS positioning technique is a hybrid between autonomous GPS and server-assisted GPS. This type of solution is similar to the server-assisted GPS, with the location server providing the mobile unit with a list of visible satellites on demand. However, in an enhanced client-assisted system, the mobile unit does the complete PVT calculation rather than sending pseudorange information back to the location server.

This technique essentially requires the same processing power and capabilities as an autonomous GPS solution, in addition to a communication link between the mobile unit and the location server. However, the amount of time required to complete the PVT calculation is much less than with an autonomous GPS solution, because of the satellite view information provided by the location server, and fewer exchanges with the location server are required than with a server-assisted solution.

Dead reckoning

Dead reckoning (DR) is a technique used in conjunction with other GPS-based positioning solutions to maintain an estimate of position during periods when there is

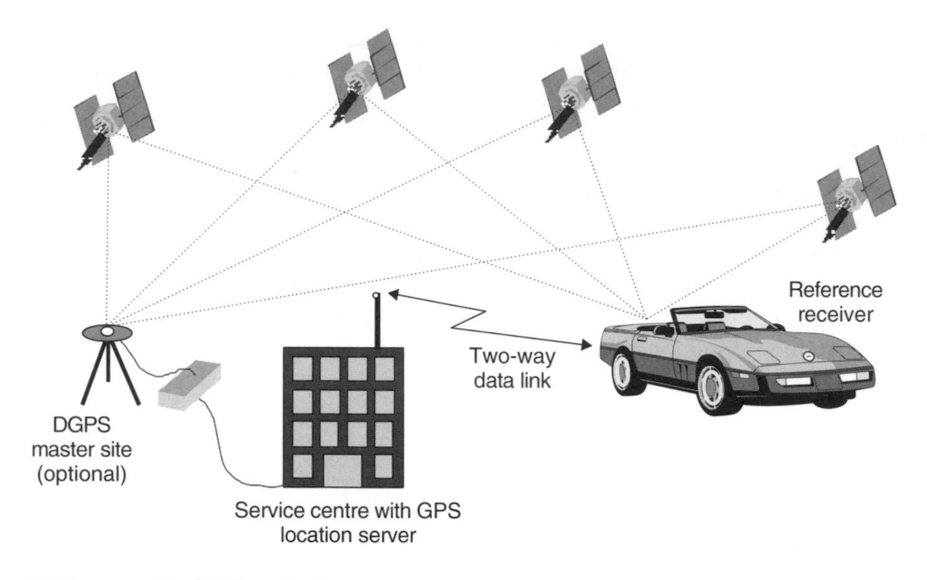

Fig. 8.6 Server-assisted GPS positioning.

poor or no access to the GPS satellite broadcasts. DR is used primarily to enhance navigation applications, since maintaining an accurate position in real time is crucial to the performance of a navigation system, and there may be times during a trip when the GPS-derived position may be intermittent, or not available at all. These GPS outages can be caused by a variety of environmental and terrain features. Examples of areas where GPS coverage could be interrupted include:

- **tunnels** through mountains or in urban areas, which prevent signal reception
- **urban canyons**, such as downtown areas populated by tall buildings, which can result in either blocked signals, or multipath errors caused by signal reflection
- **heavy foliage**, where overhanging trees or bushes block reception of the signal broadcasts
- **interference/jamming**, which can be caused by either harmonics of commercial radio transmissions, or by transmissions specifically designed to interfere with the reception of the satellite broadcasts for security reasons
- **system malfunction**, where the GPS receiver itself is functioning intermittently.

When a positioning data outage of this sort is encountered, a system that is DR-enabled will monitor inputs from one or more additional sensors in order to continue to track the direction, distance and speed the unit is moving. The system will process that data starting from the last known position fix, which will enable it to keep a running estimate of its position. The system will continue to monitor these sensor inputs and update its estimated position until the GPS receiver can again obtain an accurate position fix. At this point, the system updates its position with the satellite-based data.

For example, in an intelligent vehicle with an autonomous navigation system, the GPS receiver normally calculates the position data used by the navigation algorithm to determine the progress of the vehicle along the desired path. However, when driving in some environments, the GPS receiver may have trouble maintaining a continuous satellite lock, resulting in intermittent periods where the vehicle's position cannot be determined based on valid satellite data. In situations like this, DR is used to 'fill in the gaps', providing a method for estimating the current position based on the vehicle's movements since the last known positioning fix.

A variety of input sensors can be used to provide DR capability. In the intelligent vehicle example, several different sensor inputs can be made available to the navigation system to assist in DR calculation. The types of sensors that could be used to enable DR in a vehicle system include:

- **magnetic compass**, which can provide a continuous, coarse-grained indication of the direction in which the vehicle is moving
- **gyroscope**, which can be used to detect the angular movement of the vehicle
- **speedometer**, which can provide the current speed of the vehicle
- **odometer**, which can provide continuous data on the elapsed distance
- **wheel speed sensors**, such as Hall-effect or variable reluctance sensors (VRS), which can provide fine-grained vehicle speed information
- **accelerometers**, which can detect changes in the velocity of the vehicle.

Many of these sensors are already widely used in vehicles for other applications. Accelerometers are being used today in impact detection (airbag) systems; wheel speed sensors are being used in traction-control and anti-lock braking systems; and of course the trip meters available today in many cars use inputs from the speedometer, odometer and compass to calculate distance travelled, distance remaining and fuel economy.

Systems that leverage inputs from remote vehicle sensors to enable DR can certainly provide more consistent positioning information under some circumstances than may be possible with a single-point GPS receiver. However, depending upon the mix of sensor inputs used, the accuracy of the resulting position data may vary. Some of these sensors are more accurate than others, and most are subject to a variety of environmental, alignment and computational errors that can result in faulty readings. Some vendors of DR-enabled positioning systems have been exploring methods of reducing the effects of these errors. The development of self-correcting algorithms and self-diagnosing sensors may help reduce the impact that sensor errors can have on these systems in the future.

Additional GPS augmentation techniques

Additional techniques are being developed for increasing the accuracy of the positioning information derived from the GPS for certain applications. One technique, which has been developed by the US Federal Aviation Administration (FAA), uses transmissions from communication satellites to improve the positioning accuracy of GPS receivers in aircraft. This technique, known as the wide area augmentation system (WAAS), uses a network of wide area ground reference stations (WRS) and two wide area master stations (WMS) to calculate pseudorange correction factors for each SV, as well as to monitor the operational health of each SV. This information is uplinked to communication satellites in geostationary earth orbit (GEO), which transmit the information on the L_1 frequency, along with additional ranging signals. This system has improved the positioning accuracy of GPS receivers on board aircraft to within 7 metres horizontally and vertically, allowing the system to be used by aircraft for Category I precision approaches. A Category I system is intended to provide an aircraft operating in poor weather conditions with safe vertical guidance to a height of not less than 200 feet with runway visibility of at least 1800 feet.

Another method for improving positioning accuracy is known as carrier-phase GPS. This is a technique where the number of cycles of the carrier frequency between the SV and the receiver is measured, in order to calculate a highly accurate pseudorange. Because of the much shorter wavelength of the carrier signal relative to the code signal, positioning accuracies of a few millimetres are possible using carrier-phase GPS techniques. In order to make a carrier-phase measurement, standard code-phase GPS techniques must first be used to calculate a pseudorange to within a few metres, since it would not be possible to derive a pseudorange using only the fixed carrier frequency. Once an initial pseudorange is calculated, a carrier-phase measurement can then be used to improve its accuracy by determining which carrier frequency cycle marks the beginning of each timing pulse. Of course, receivers that can perform carrier-phase measurements will bear additional hardware and software costs to achieve these improved accuracies.

8.4 GPS receiver technology

In order to design and build a GPS receiver, the developer must understand the basic functional blocks that comprise the device, and the underlying hardware and software necessary to implement the desired capabilities. The sections below describe the main functional blocks of a GPS receiver, and the types of solutions that are either available today or in development to provide that functionality.

8.4.1 GPS receiver components

GPS receivers are composed of three primary components: the antenna, which receives the radio frequency (RF) broadcasts from the satellites; the downconverter, which converts the RF signal into an intermediate frequency (IF) signal; and the baseband processor or correlator, which uses the IF signal to acquire, track, and receive the navigation message broadcast from each SV in view of the receiver. In most systems, the output of the correlator is then processed by a microprocessor (MPU) or microcontroller (MCU), which converts the raw data output from the correlator into the positioning information which can be understood by a user or another application.

The sections below provide an overview of the three key components of a GPS receiver, describing in generic terms the functionality and capabilities typically found in these systems. As the capabilities of the MPU or MCU needed to process the correlator output is largely dependent on the needs of the applications and the particular GPS chip set being considered, MPU/MCU requirements and capabilities are not discussed here.

Antennas

As with most RF applications, important performance characteristics to be considered when selecting the antenna for a GPS receiver include impedance, bandwidth, axial ratio, standing wave ratio, gain pattern, ground plane, and tolerance to moisture and temperature. In addition, the relatively weak signal transmitted by GPS satellites is right-hand circularly polarized (RHCP). Therefore, to achieve the maximum signal strength the polarization of the receiving antenna must match the polarization of the transmitted satellite signal. This restriction limits the types of antennas that can be used. Some of the more common antennas used for GPS applications include:

- **microstrip**, or **patch**, antennas are the most popular antenna because of their simple, rugged construction and low profile, but the antenna gain tends to roll-off near the horizon. This makes it more difficult to acquire SVs near the horizon, but it also makes the antenna less sensitive to multipath signals. This type of antenna can be used in single or dual frequency receivers.
- **helix-style** antennas have a relatively high profile compared to the other antennas, maintaining good gain near to the horizon. This can provide easier acquisition of SVs lower on the horizon, but also makes it more sensitive to multipath signals that can contribute to receiver error. The spiral helix antenna is used in dual-frequency receivers, while the quadrifilar helix antenna is used in single frequency systems.
- **monopole** and **dipole** antennas are low cost, single frequency antennas with simple construction and relatively small elements.

Systems with an antenna that is separate from the receiver unit, such as a GPS receiver installed in a vehicle with a trunk-mounted antenna, often use an active antenna which includes a low noise pre-amplifier integrated into the antenna housing. These amplifiers, which boost the very weak received signal, typically have gains ranging from 20 dB to 36 dB. Active antennas are connected to the receiver via a coax cable, using a variety of connectors, including MMCX, MCX, BNC, Type N, SMA, SMB, and TNC. Systems that have the antenna integrated directly into the receiver unit (such as a handheld GPS device) use passive antennas, which do not include the integrated pre-amplifier.

The demand for the integration of positioning technology into smaller devices is challenging antenna development. The industry is already pushing for smaller antennas for applications such as a wristwatch with integrated GPS, which is smaller than most patch antennas available today. Another demand is for dual-purpose antennas that do double duty in wireless communication devices, such as in a mobile telephone with an integrated GPS receiver. Inevitably, the future will bring smaller and more flexible antennas for GPS applications.

Downconverter

The function of the downconverter is to step down each GPS satellite signal from its broadcast RF frequency to an IF signal that can be output to the base-band processor. The signal from each SV in view of the antenna (active or passive) is filtered and amplified by a low noise pre-amplifier, which sets the overall noise of the system, and rejects out of band interference. The output of this pre-amplifier is input into the downconverter, where the conversion to the IF signal is typically made in two stages. The two-stage mixer is clocked by a fixed-frequency phase-locked loop controlled by an external reference oscillator that provides frequency and time references for the downconverter and base-band processor.

The mixer outputs, which are composed of in-phase (I) and quadraphase (Q) signals, are amplified again and latched as the IF input to the base-band processor to be used for satellite acquisition and tracking. To enable the baseband processor to account for frequency variation over temperature, an integrated temperature sensor is often included in the downconverter circuit.

The downconverter in a GPS receiver is often susceptible to performance degradation from external RF interference from both narrowband and wideband sources. Common sources of narrowband interference include transmitter harmonics from Citizens Band (CB) radios and AM and FM transmitters. Sources of wideband interference can include broadcast frequency harmonics from microwave and television transmitters. In mobile GPS applications such as in intelligent vehicle systems, the GPS receiver will often encounter this type of interference, and must rely on the antenna and downconverter design to attenuate the effects.

Correlator/data processor

The correlator component in a GPS receiver performs the high-speed digital signal processing functions on the IF signal necessary to acquire and track each SV in view of the antenna. The IF signal received by the correlator from the downconverter is first integrated to enhance the signal, then the correlator performs further demodulation and despreading to extract each individual SV signal being received. Each signal is then multiplied by a stored replica of the C/A signal from the satellite being received, known as the Gold code for that satellite. The timing of this replica signal is adjusted relative to the received signal until the exact time delay is determined. This adjustment period to calculate the time delay between the local clock and the SV signal is defined as the acquisition mode. Once this time delay is determined, that SV signal is then considered acquired, or locked.

After acquisition is achieved, the receiver transitions into tracking mode, where the PRN is removed. Thereafter, only small adjustments must be made to the local reference clock to maintain correlation of the signal. At this point, the extraction of the satellite timing and ephemeris data from the navigation message is done. This raw data and the known pseudoranges are then used to calculate the location of the GPS receiver. This information is then displayed for the user, or otherwise made available to other applications, either through an external port (for remote applications) or through a software API (for integrated applications).

In the past, GPS correlators were designed with a single channel, which was multiplexed between each SV signal being received. This resulted in a very slow process for calculating a position solution. Today, systems come with up to 12 channels, allowing the correlator to process multiple SV signals in parallel, achieving a position solution in a fraction of the time. Also, while the correlator functionality is sometimes performed in software using a high-performance digital signal processor (DSP), the real-time processing requirements and repetitive high rate signals involved make a hardware correlator solution ideal, from both a cost and throughput standpoint.

8.4.2 GPS receiver solutions

When access to the GPS first became available for military and commercial use, only a few companies had the technology and expertise to develop reliable, accurate GPS receivers. Application developers who needed GPS services would simply purchase a board level solution from a GPS supplier, and integrate it into their design.

More recently, the demand for putting GPS capabilities into customized packaging has grown dramatically. To meet that demand a variety of solutions are now available, ranging from traditional board-level solutions that connect to an application via a serial interface, to integrated circuit (IC) chip sets, which application developers can embed directly into their designs. The sections below will give a brief overview of the types of solutions available on the market today.

System level solutions

The first commercially available GPS receivers were designed as either standalone units with connectors for power, an antenna, and a serial interface to a computer or other device, or as more basic board-level solutions, which could be integrated into an application enclosure, but which still required an external antenna connection and serial network interface. These units were entirely self-contained, with the RF interface, downconverter and baseband processing done entirely independent of the application. With this type of solution, the PVT information was transmitted out of the serial port, to be displayed or used as appropriate depending upon the application. In some cases, the user could provide some configuration data to the system, such as the choice of a local datum, and in that way 'customize' the resulting positioning information for their needs. This type of solution is still widely available, and for many applications provides a cost-effective way of adding GPS positioning or timing services to an existing design.

One variation of the board-level solution that is becoming more popular today is to supply the RF section of the GPS receiver, including the discrete RF interface and downconverter, as a self-contained module, along with a standalone correlator ASIC or an MCU with an integrated correlator and software to perform the baseband processing of the IF signal. Typically, the RF section of a GPS receiver is the most challenging portion of the design because of the sensitivity to component layout and extraneous signals, and many of the RF circuits that exist today were designed with a combination of technical know-how and trial-and-error experience that few application developers can afford. By comparison, designing the hardware layout for the baseband processor and interface to the RF module is a relatively minor task, which is what has made this an attractive solution for application developers who want to integrate GPS into their designs, but cannot afford the cost or space necessary for a board-level solution.

IC chip set solutions

For those developers that have the skill (or want the challenge) of designing the entire GPS receiver circuit into their application, several semiconductor manufacturers now offer GPS chip set solutions. These chip sets, offered with either complete or partial reference designs and control software, enable the designer to integrate GPS into an application at the lowest possible cost, while also conserving power, board space and system resources. However, this high level of integration is achieved at the expense of doing the RF and IF circuit layout and software integration in-house, which can take significant resources and effort.

The custom chip sets used for the original GPS receivers often had up to seven ICs, including the external memory chips, amplifiers, downconverter, correlator ASIC and system processor, in addition to a variety of discrete components. Continuous advances in the performance and integration level of MCUs have greatly increased the performance of the newer GPS chip sets while reducing the power consumption and physical size of the complete system. System-on-a-Chip (SoC) technology has resulted in the integration of the GPS correlator directly onto the MCU, along with embedded RAM, ROM and FLASH memory. In some cases, this increased level of integration has reduced the device count down to a mere two ICs and a handful of discrete components, further decreasing the cost and development effort required.

Even more recently, high-performance RISC MCUs have begun showing up in low-cost GPS chip set solutions. These powerful processors have many more MIPS available for GPS computations, which in turn increases the overall performance and reliability of the GPS solution. This level of computational power is making it possible to execute dead reckoning or WAAS algorithms on the same processor as the GPS algorithms, further improving the accuracy of the positioning solution at little or no increase in chip set cost.

A block diagram illustrating the primary components of a GPS receiver as described in the previous sections is pictured in Figure 8.7. This diagram illustrates all of the functional blocks required by a basic GPS system, including an active antenna, a downconverter with an integrated temperature sensor, and a correlator integrated onto a basic microcontroller, along with the additional MCU peripherals required to perform a basic tracking loop routine and calculate a PVT solution.

Development tools

The development tools available for GPS application design vary depending on the complexity of the target system and the GPS solution being used. Most GPS solution vendors offer software tool suites that allow a developer to communicate with the GPS receiver through the serial port of a personal computer. These software tools typically use messages compatible with the standard NMEA (National Marine Electronic Association) format, but many vendors also offer their own customized sets of messages and message formats.

The more advanced development tools, available for some GPS chip sets, are intended to help the application developer integrate their software with the GPS tracking software running on the same MCU. Because of the hard real-time constraints typical of GPS software implementations, the most efficient way to enable the smooth integration of the GPS tracking loop with the application software is through a clearly

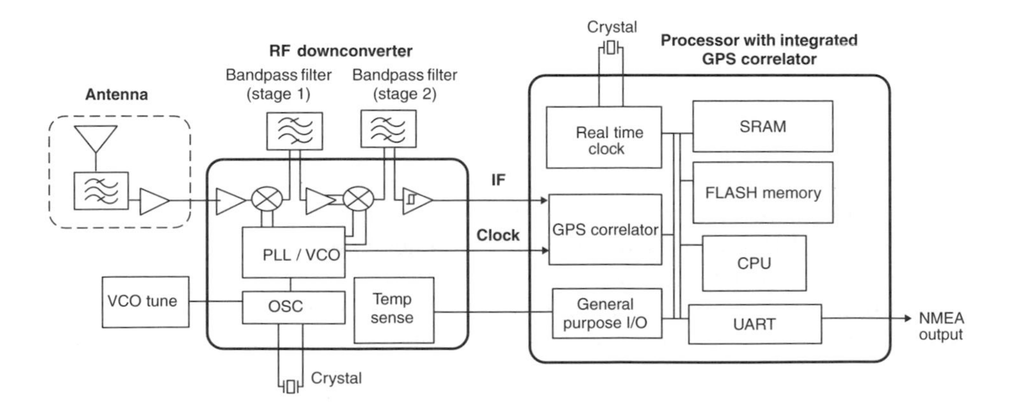

Fig. 8.7 Functional block diagram of GPS receiver.

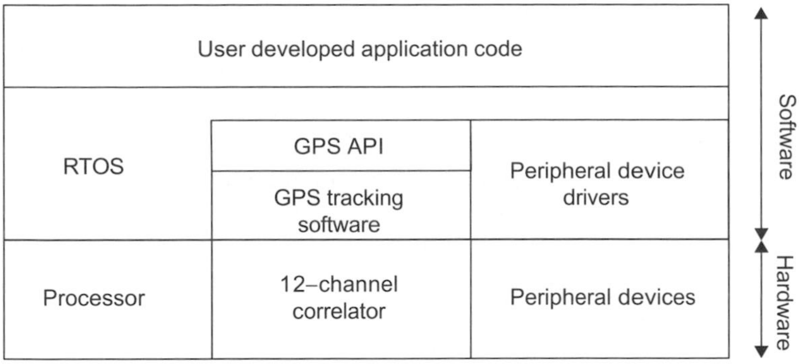

User developed application code

RTOS | GPS API | GPS tracking software | Peripheral device drivers

Processor | 12–channel correlator | Peripheral devices

Software

Hardware

Fig. 8.8 GPS-enabled application software architecture.

defined software API. With a standard interface to the GPS software and the necessary development/debugger tools to support it, an application developer can easily configure the GPS receiver software, enabling access to the appropriate PVT information by the application as needed. For an illustration of the basic software architecture of a GPS-enabled application running on a single MCU that is supported by this type of tool suite, refer to Figure 8.8.

8.4.3 Performance considerations

There are many parameters used by the industry to assess the performance of a GPS receiver, and to evaluate the relative performance of comparable receivers. The most common parameters being used to evaluate GPS receiver performance include positioning and timing accuracy, time-to-first-fix, reacquisition time, and receiver sensitivity.

Positioning and timing accuracy

The most obvious of these parameters is positioning accuracy – how accurate are the positions calculated by an autonomous GPS receiver, based on the number of satellites that can be seen by that receiver? This is typically measured by performing a map-matching test, where positions calculated by a receiver for landmarks on a map are compared to their known positions. This is a standard test that is often used to compare the accuracy of multiple GPS receivers simultaneously. When the S/A feature was still enabled, the accuracy of the SPS signal served as the baseline for positioning accuracy for commercial GPS receivers. With the discontinuation of S/A, the accuracy of autonomous GPS receivers has increased significantly, but as of this writing there is no new accepted baseline for the measurement of post-S/A receivers, except perhaps for the PPS signal accuracy.

A performance parameter closely related to positioning accuracy is timing accuracy – how close to UTC is the time calculated by the GPS receiver. This parameter essentially measures the deviation of the calculated time from UTC as maintained by the US Naval Observatory. However, since the accuracy of the time component of the SPS signal with the S/A feature enabled was within 340 ns, this is obviously a test requiring sophisticated time measurement equipment to perform. The timing accuracy of most GPS receivers is more than adequate for commercial applications such as intelligent vehicle systems.

Time-to-first-fix

Time-to-first-fix (TTFF) is the measure of the time required for a receiver to acquire satellite signals and calculate a position. The three variants of a TTFF measurement, which depend upon the condition of the GPS receiver when the TTFF is measured, are referred to as hot start, warm start, and cold start. These TTFF measurements include the amount of time it takes the GPS receiver to acquire and lock each satellite signal, calculate the pseudorange for each satellite, and calculate a position fix.

Hot start occurs when a GPS receiver has recent versions of almanac and satellite ephemeris data, and current time, date and position information. This condition might occur when a receiver has gone into a power-conserving stand-by mode due to application requirements. In this situation, most receivers should be able to acquire a position fix within 15 seconds.

Warm start occurs when a GPS receiver is powered on after having been off or out of signal range for several hours to several weeks. In this condition, the receiver has an estimate of time and date information, and a recent copy of satellite almanac data, but no valid satellite ephemeris data. In this state, a receiver can begin tracking satellites immediately, but must still receive updated ephemeris data from each satellite, which is only valid for approximately four hours. Under these conditions, most receivers should be able to acquire a position fix within 45 seconds.

Cold start occurs when a GPS receiver has inaccurate date and time information, no satellite ephemeris data, and no almanac data, or data which is significantly out of date. In this state, the receiver must perform a search for the available satellite signals, and can take 90 seconds or more to acquire a position fix. This condition is encountered when the GPS receiver is powered up for the first time after leaving the factory, or in other situations where the device has not been powered up or used for long periods of time.

Many GPS receivers allow the user to enter time, date and even current position information, which can reduce the TTFF in a cold start situation down close to that of a warm start.

Reacquisition time

Reacquisition time is the amount of time required by a GPS receiver to regain a position fix when the satellite signal is temporarily disrupted due to a loss of visibility of one or more satellites. This condition can occur when the receiver is operating in areas of dense foliage, or in urban canyons, or anywhere that the satellite views may be intermittently blocked. Most GPS receivers should have reacquisition times of five seconds or less. This is an important parameter for assessing the capability of GPS receivers for intelligent vehicle applications, since navigation systems must routinely operate in locations, such as downtown areas, where reception can be intermittent.

Receiver sensitivity

The sensitivity to satellite transmissions is another measure of performance of a GPS receiver. This is basically an assessment of how many satellites a receiver can detect under varying conditions. Because operating conditions for GPS receivers in intelligent vehicle applications can range from high elevations with an unobstructed view of the sky to locations inside or between buildings where reception can be more difficult, it is important for the application developer to understand under what conditions the GPS receiver can detect the SV signals, and under what conditions alternate methods of positioning must be relied upon.

8.5 Applications for GPS technology

There are a variety of uses for GPS technology today, from basic positioning applications which might provide a traveller with their current location, speed, and direction to their destination, to highly complex applications where the user's position information is feed into a system that provides location-specific features and services tailored for that user. What follows are some examples of how GPS technology is being used to enhance the capabilities of intelligent vehicle platforms. The initial examples illustrate some of the more traditional positioning applications, such as basic location and autonomous navigation systems, which are already seeing widespread use today. This is followed by examples describing how GPS-derived positioning information is being used to provide location-based services in vehicles today, and how the richness and complexity of those services will increase in the near future.

8.5.1 Basic positioning applications

The most basic applications for the positioning capabilities provided by the GPS are those providing the user/operator with information regarding their current location, where they have been, and more recently, where they are going. These systems include today's wide assortment of handheld devices, which provide the user with their current location, speed and elevation, as well as track their most recent movements. More complex devices integrate navigation capabilities, which map the user's position information onto a map database, and use that information to provide the user with text-based or graphical directions to their destination. Each type of system is described in more detail in the following sections.

Handheld GPS system

A handheld GPS device, while not strictly an intelligent vehicle system, provides a good example of an application utilizing the most basic positioning services provided by GPS technology. These units range from less than one hundred dollars to a few thousand dollars, depending upon their features and capabilities. These devices are battery operated, have a small LCD display and basic, menu-driven user interface. Most of the newer models can detect at least 6–8 satellites concurrently, with many now offering the ability to track up to 12 satellites at once. The more complex units also provide a variety of configuration options, such as the choice of localized datums or different information display formats. Because these devices are small enough to be easily carried in a briefcase or purse, users can take them as a positioning aid when

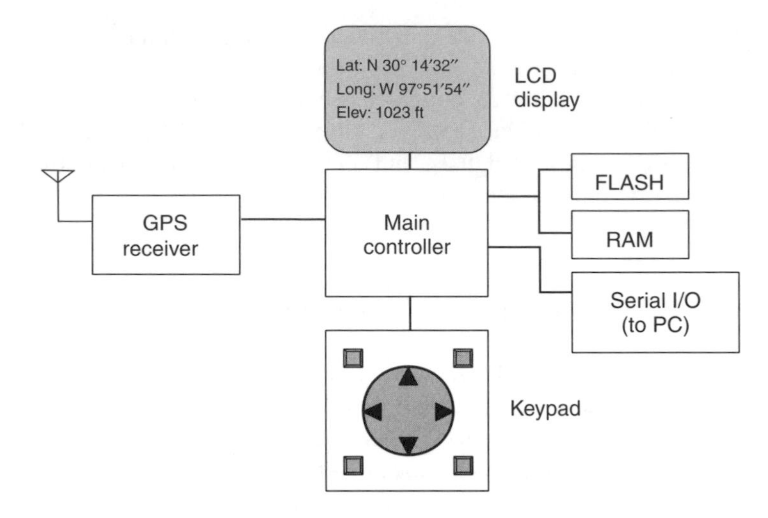

Fig. 8.9 Block diagram of basic handheld GPS device.

travelling, whether on foot, or by private or public transport. The basic functional blocks of a handheld GPS receiver are illustrated in Figure 8.9.

All of these devices provide the user with their current location information, usually in the form of a latitude and longitude reading. Most also provide the user with additional positioning information, including the current local time, elevation above sea level, and velocity, provided enough satellites can be detected by the device. Many of today's handheld devices also provide some tracking services, enabling the user to store the location of points they have previously reached, allowing them to return easily to their starting location. This can be a very useful feature to those travelling in unfamiliar areas, whether in a wilderness area or in an unfamiliar town or city.

Some of today's newer handheld systems now have larger displays and removable memory devices that enable the unit to plot graphically the user's current location onto a map of the local area. These devices may also provide the ability to enter in destination information, so the user can more easily understand where they are in relation to where they want to be. However, most of these devices fall short of being true navigation systems, since they do not provide any assistance to the user in reaching their destination. Instead, they simply give the user a more complete picture of their current location.

Autonomous navigation systems

True navigation, which provides the user with detailed instructions on how to reach a specific destination, is one of the fastest-growing areas in intelligent vehicle technology. Navigation devices utilize map-matching and best-path algorithms, along with user-defined filtering, to allow the user to choose between the fastest or most direct route to a desired destination. Some systems even allow the user to indicate specific routes to be avoided. The map databases used all provide basic mapping information (streets, major landmarks, etc.), but can also include points of interest and/or helpful location information (restaurants, etc.), depending upon their level of detail and how often they are revised.

Autonomous navigation devices range from in-dash units that are small enough to fit into a 1-DIN slot, to multi-component systems with CD-ROM changers and large multi-plane colour displays. The price of these systems can vary from a few hundred dollars to several thousand dollars, depending upon the complexity and capabilities. These systems utilize position information from an integrated GPS receiver, along with map database information provided from a CD-ROM or memory cartridge, to determine the user's current geographical location on the map. The smaller, in-dash units typically have a limited ability to display the user's position graphically, instead indicating the current location using a text description, such as the current address or location relative to a near-by landmark. Systems supporting a larger display can graphically indicate the user's current position superimposed on a map of the surrounding area. Also, because these systems are typically mounted in the dashboard, displacing the existing vehicle entertainment system, many of them include entertainment functionality such as an AM and FM stereo tuner or audio CD player. The more advanced systems with direct interconnections into the vehicle may also include HVAC system controls or other vehicle-specific comfort and convenience controls, although this is usually limited to systems installed by the vehicle manufacturer or dealer. An illustration of the functional blocks of an autonomous navigation system with an integrated GPS receiver can be seen in Figure 8.10.

To determine the appropriate travelling instructions with one of these devices, the user enters the desired destination using a menu-driven system via a hardware or software keypad, depending upon the system. Some systems also support voice-based destination entry using basic voice-recognition technology. While these voice-driven systems are becoming more sophisticated, much progress is still necessary to improve

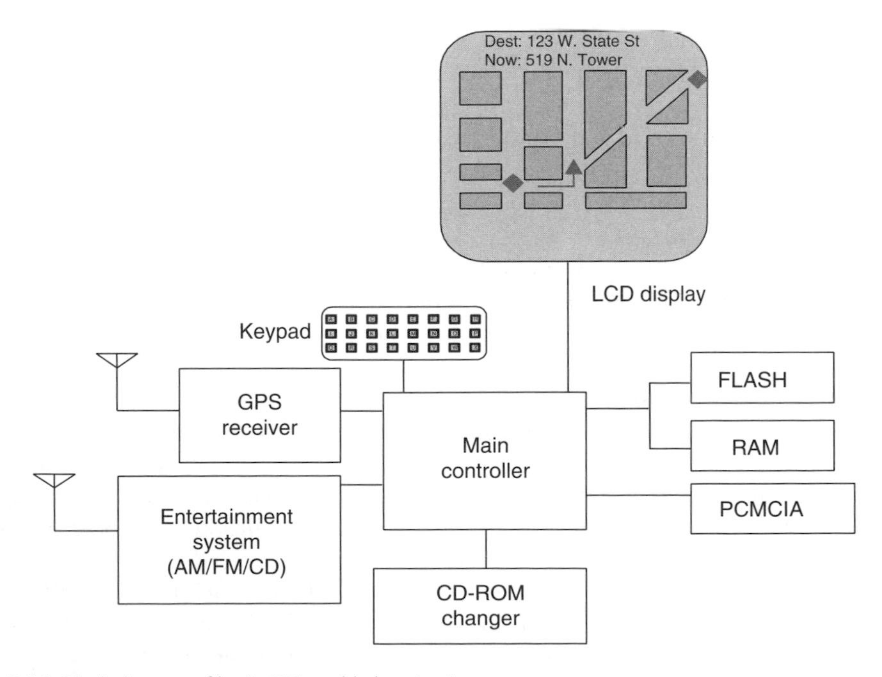

Fig. 8.10 Block diagram of basic GPS-enabled navigation system.

them to the point where non-technical users are satisfied with their accuracy and reliability. The methods in which the directions are communicated to the user also depend upon the complexity of the system. Navigation systems with limited displays may use simple graphics combined with text to indicate the directions in a turn-by-turn manner. Some systems may combine these graphical turn-by-turn instructions with spoken instructions, using text-to-speech technology. Systems with larger displays can indicate the current position and immediate directions on a map of the immediate area, as well as the desired destination, once it is within the boundaries of the current map being displayed.

The value of GPS technology to these systems is obvious. GPS provides the essential positioning elements of location and speed necessary to make dynamic navigation possible. However, the occasional difficulties in maintaining a GPS position lock, particularly in 'urban canyon' areas such as in the downtown districts of big cities, often require the use of additional techniques to maintain the accuracy of the user's location and movements between position locks. These dead-reckoning techniques, described earlier in this chapter, include the use of internal gyroscopes or accelerometers to track the movement of the vehicle between the times that a solid position fix can be obtained by the GPS receiver. Vehicle speed and direction information, often obtained directly from the vehicle's internal communication network, can also be used to enhanced dead-reckoning capabilities, although the capabilities for the input and processing of this type of data are typically only found in systems installed by the vehicle manufacturers and dealers.

8.5.2 Location-based services

The delivery of location-based services is really the next phase in the evolution of intelligent vehicle systems. These services, which use GPS technology to pinpoint the user's current position, can then use that information to provide location-specific services to the user, such as relevant points of interest, or the nearest locations where a desired service or product may be available.

Current location-based services

The most common types of location-based services available in intelligent vehicles today are emergency and concierge/assist services. These services are accessed using a system combining GPS and wireless communication technology with a very basic user interface. This provides the vehicle operator with an on-demand wireless voice link to a call centre staffed 24 hours a day. At the time the wireless connection is initiated, the coordinates of the vehicle are transmitted to the call centre, indicating the exact position of the vehicle. This allows the call centre to provide timely and appropriate services relevant to the location of the customer. These services are available today in multiple vehicle models from several manufacturers, and will likely become standard features in the near future on many vehicle lines.

The most common emergency services being offered today include the notification of emergency response personnel in the case of an accident, and the notification of automotive service personnel in the case of a vehicle malfunction. When one of these events occurs, the appropriate local authorities are vectored to the exact position of the vehicle by the call centre, using the uploaded GPS positioning data to pinpoint the location of the vehicle. In some systems, the contact with the call centre can be made automatically if the system detects that an incident has occurred, upon the deployment of an airbag, for example. Other systems rely on a vehicle occupant to initiate the contact, even in the case of an accident. In the case of a vehicle malfunction, most systems today require the vehicle operator to initiate the call to the service centre.

The other class of location-based services currently being offered which rely on this combination of GPS and communication technologies are concierge/assist services. Examples of the services available include: getting directions to a desired destination ('Help, I'm lost, I need to get to ...'), getting recommendations on a local point of interest ('We are hungry and don't know the area, can you tell us where a nearby restaurant is?'), and remote vehicle services, such as the remote unlocking of the car doors, or recovering a vehicle which has been stolen. Some providers even offer such highly personalized services as helping their customers purchase tickets for local events like plays or concerts.

All of the above services are available today in one form or another. In the future, the providers of location-based services will take advantage of the data capabilities of newer communication technologies to greatly expand their services. This will result in more advanced vehicle and user services, examples of which are illustrated below.

Future location-based services

In the near future, location-based services available in the vehicle will begin expanding beyond operator-assisted services to include wireless data-oriented services, as well. The wireless communication technology to support these services is already available today due to the accelerating roll-out of digital cellular, which is already in use in Europe and Japan, and is growing rapidly in the US. The current digital standards are still somewhat limited in their ability to support data services, but the roll-out of the next generation of digital wireless communication technology will make data services much more widely available, and will provide significantly improved bandwidth for the delivery of digital content to the user. Digital communication, along with more advanced positioning technology, will enable a wide variety of advanced location-based services for deployment in intelligent vehicles. Examples of these more advanced services include:

- **Server-based navigation systems**, which allow the user to dynamically download the most up-to-date, and therefore accurate, map information of the area in which they are travelling. This will help to ensure that the user gets the most current directions and points of interest possible without having to maintain a subscription with a map database provider. It will also reduce the cost of the navigation system itself, since a costly memory storage subsystem would not be required to access the map database.
- **Dynamic traffic routing and management services**, which will enable navigation systems to take into account current traffic conditions when calculating directions. This will allow travellers to easily route around congested areas, getting them to their destination faster and helping to prevent additional congestion. One approach to this is for the travellers' systems to periodically provide their position information to a central server, which uses the positioning deltas to map traffic flow. This data can then be fed back to each system to provide real-time traffic movement updates.

- **Location-based marketing services**, which can be used by local providers of goods and services to target advertising to travellers who are entering their local service area. For example, if a traveller is looking for a restaurant near their current location, a request including their position information and food types of interest could be submitted, which would return directions to nearby restaurants of the desired type. A restaurant could even include with the directions a coupon for a meal specially to encourage the traveller to visit that establishment.

These are just a few examples of the types of location-based services that will become available in intelligent vehicles as these technologies mature. Of course, the development of the technology to support these advanced location-based services also raises a variety of privacy issues. While the positioning information of individuals can be very valuable to merchants with goods or services to sell, many individuals may consider this information to be very personal and private, and wish to limit its distribution. Therefore the protection and methods of distribution of this information will very likely be the subject of intense debate between merchants and privacy advocates, and may ultimately result in legislation regarding how and to whom that information is disseminated.

8.6 Conclusion

The global positioning system was hailed as a technological success soon after it became fully operational in 1995. With the continual improvements the system is undergoing, and in particular with the discontinuation of the selective availability feature in May 2000, many more commercial, military and space applications will be able to derive benefit from this system's services and capabilities in the future. Intelligent vehicle applications will be one of the biggest beneficiaries of these improvements in service.

This chapter has presented an overview of the GPS, including the history of satellite-based positioning, the basic system architecture, the science and mathematics used for determining location, an overview of the components and solutions which are available for use in GPS-enabled applications, and some examples of current and future applications which will utilize the positioning services made possible by the GPS. It is hoped that the reader has gained a basic understanding of the system architecture and requirements, and the impact on applications that require the use of GPS services.

For more detailed information about the science and technology behind the global positioning system, please refer to the publications listed under Further reading.

Further reading

Kaplan, E. (ed.) (1996). *Understanding GPS: Principles and Applications*. Norwood: MA: Artech House.

Parkinson, B. and Spilker, J. (eds) (1996). Global Positioning System: Theory and Applications Volume I. *Progress in Astronautics and Aeronautics*, Vol. 163. Washington DC: American Institute of Astronautics and Aeronautics, Inc.

Farrell, J. and Barth, M. (1999). *The Global Positioning System & Inertial Navigation*. New York, NY, McGraw-Hill.
Enge, P. and Misra, P. (eds) (1999). Special Issue on Global Positioning. *Proceedings of the IEEE*, **87**, No. 1.
Anonymous (1995). *Global Positioning System Standard Positioning Service Signal Specification*. US Department of Defense, 2nd edition.

Part Three Intelligent vehicle decision and control technologies

9

Adaptive control system techniques

Muhidin Lelic
Corning Incorporated, Science and Technology, Manufacturing Research, Corning, USA

and

Zoran Gajic
Department of Electrical and Computer Engineering, Rutgers University, USA

In this chapter we review some of the most interesting and challenging control theory and its applications/results obtained within the framework of automated highway traffic and moving vehicle control problems that arose in the last ten years. In that direction, we review the results concerned with intelligent cruise control systems, vehicle conservation flow modes, inter-vehicle distance warning systems, computer controlled brake systems, constant spacing strategies for platooning, robust lateral control of highway vehicles, vehicle path following and vehicle collision avoidance. We also present an overview of the main adaptive control techniques that can be either used or can be potentially used for solving control problems of automated highways and moving vehicles. Adaptive control techniques are mostly used for adaptive control of a vehicle's longitudinal motion, adaptive cruise control, and real-time control of a convoy of vehicles.

9.1 Automatic control of highway traffic and moving vehicles

Automatic control of highways dates back to the 1960s and 1970s (see for example, Levine and Athans, 1966; Fenton and Bender, 1969; Chu, 1974; Peppard, 1974). During the last ten years automatic control of highway traffic and automatic control of moving vehicles have become very popular, interesting and challenging research areas. The world wide research in that direction has been especially active in USA, Germany, Japan and Sweden. Inter-vehicle distance warning systems for trucks have been operational in Japan since the end of the 1980s and intelligent cruise control systems for passenger cars have been commercially available in Japan since 1995 (Tsugawa *et al.*,

1997). In Sweden, 'drive by wire' controllers have been commercially available in passenger cars for some time. A nice survey of control problem that control engineers are faced with in dealing with automated highways is given in Varaiya (1993). Here, we review some of the interesting control theory and application results accomplished within the general problems of automated highway control during the 1990s.

The first step in the design of automatic controllers for moving vehicles is to develop the appropriate mathematical models for vehicles' longitudinal and lateral dynamics. The empirical vehicle longitudinal dynamics model, obtained using a system identification technique, is developed in Takasaki and Fenton (1977). For such an obtained model, a digital observer/controller is designed in Hauksdottir and Fenton (1985). Longitudinal dynamics of a platoon of nonidentical vehicles is considered in Sheikholeslam and Desoer (1992). A control law is designed in the same paper that shows that it is possible to keep vehicles within the platoon approximately equally spaced even at high speeds. In Shladover (1991), technical issues needed to be resolved in the design of automated highway controllers, such as process and measurement noise, sampling and quantization, acceleration and jerk limits, have been outlined, clarified, and studied.

In Chien *et al.* (1997) a discrete-time model of a traffic flow of a computer driven vehicles is developed and a roadway controller that eliminates traffic congestion is proposed. The controller for the obtained nonlinear model is realized using the integrator back-stepping technique such that the actual traffic density converges exponentially to the desired one. The similar problem is solved in Alvarez *et al.* (1999) using vehicle conservation flow models and Lyapunov stability theory. For the desired vehicular density and velocity profiles stabilizing velocity controllers are obtained. Another version of constant spacing strategies for platooning in automated highway systems has been presented in Swaroop and Hedrick (1999). The paper establishes conditions for stability of individual vehicles and a string of vehicles. Several constant spacing vehicle following algorithms are presented in that paper.

A computer controlled brake system has been studied in Raza *et al.* (1997). Unknown parameters of the obtained first order nonlinear system are first identified and then the nonlinear model is feedback linearized. A PI controller is used in the feedback loop such that the zero-steady state error and no overshoot are achieved for the step input. The efficiency of the controller is demonstrated on the brake system of a Lincoln town car.

The kinematic model of a moving vehicle is derived in Murray and Sastry (1993) as follows

$$\begin{aligned}
\frac{\mathrm{d}x(t)}{\mathrm{d}t} &= u_1(t)\cos(\theta(t)) \\
\frac{\mathrm{d}y(t)}{\mathrm{d}t} &= u_1(t)\sin(\theta(t)) \\
\frac{\mathrm{d}\theta(t)}{\mathrm{d}t} &= \frac{1}{d}u_1(t)\tan(\phi(t)) \\
\frac{\mathrm{d}\phi(t)}{\mathrm{d}t} &= u_2(t)
\end{aligned} \tag{9.1}$$

where $x(t)$ and $y(t)$ are position coordinates, $u_1(t)$ is the velocity of the rear wheels, $u_2(t)$ is the velocity of the steering wheel, and d is the distance between the front and rear wheels, $\theta(t)$ is the vehicle's angle with respect to $x(t)$ coordinates, and $\phi(t)$ is the angle between the front wheels and the car's direction. In some applications $u_2(t)$ is constant so that the last equation can be eliminated. Using the above kinematic model (with $u_2(t)$ constant) in Sugisaka *et al.* (1998) a fuzzy logic controller is developed such that a moving vehicle is able to search for an object in space and recognize a stimulant traffic signal.

The vehicle following control law that includes actuator delays has been proposed in Huang and Ren (1997). The paper derives the upper bound on the time delay that guarantees stability of individual vehicle. The vehicle position $x_i(t)$ and velocity $v_i(t)$ are modelled by:

$$\begin{aligned} \frac{dx_i(t)}{dt} &= v_i(t) \\ \frac{dv_i(t)}{dt} &= k_i[T_{ti}(t-\tau_i) - T_{Li}] \end{aligned} \tag{9.2}$$

where $T_{ti}(t)$ is the throttle input, $T_{Li}(t)$ is the load torque, τ_i is the actuator's time delay, and k_i is a constant that depends on gear ratio, effective tyre radius, and effective rotational inertia of the engine. The dynamics of the spacing error $\delta_i(t)$ is modelled by:

$$\frac{d\delta_i(t)}{dt} = x_{i-1}(t) - x_i(t) - H_i \tag{9.3}$$

where H_i is the safety spacing for the i-th vehicle in the platoon. The proposed control law of Huang and Ren (1997) also produces zero steady state spacing error for the platoon of vehicles.

In Zhang *et al.* (1999) a mathematical model of an intelligent cruise control that mimics human driving behaviour is developed. An intelligent cruise controller is developed that uses information about distances from both the vehicle in front and the vehicle behind given vehicle. In Ioannou and Chien (1993) such a controller is developed using only information about the distance from the vehicle in front. The controller of Zhang *et al.* (1999) guarantees both stability of the individual vehicle and the stability of the platoon of vehicles under the constant spacing policy.

Robust *lateral* control of highway vehicles has been considered in Byrne *et al.* (1998). The proposed controller guarantees stability over a broad range of parameter changes. The paper proposes the following linear model for the car's lateral error dynamics:

$$E(s) = \frac{114.2552(s^2 + 13.4391s + 31.439179)}{s^2(s^2 + 24.3156s + 151.9179)}\Delta(s) \tag{9.4}$$

where $E(s)$ is the lateral error and $\Delta(s)$ is the front steering angle.

Statistical learning automata theory has been used for vehicle path following in an automated highway system in Unsal *et al.* (1999). An intelligent neural network based driver warning system for vehicle collision avoidance is presented in An and Harris (1996). A neural speed controller that uses throttle and brake inputs has been proposed in Fritz (1996). Road tests on an experimental Daimler Benz vehicle shows that such a controller performs well for both low and high speeds. An automatic road following fuzzy controller of a vehicle's lateral motion has been considered in Blochl and Tsinas (1994). A computer controlled camera provides data about the vehicle's position.

9.2 Adaptive control of highway traffic and moving vehicles

Adaptive control is a promising technique that can be used to solve some of the problems that appear in automatic control of highways and moving vehicles. Adaptive control has been already used in several papers dealing with control of moving vehicles (Ishida *et al.*, 1992; Raza and Ioannou, 1996; Shoureshi and Knurek, 1996; Holzmann *et al.*, 1997; Pfefferl and Farber, 1998; Bakhtiari-Nejad and Karami-Mohammadi, 1998). It has been indicated in Ackermann (1997) that the control problem of a moving vehicle has two major parts: path following and disturbance attenuation. The path following problem can be superbly solved by a human controller, but the disturbance attenuation problem requires an automatic controller since it takes about 0.5 seconds before a human driver can react to a disturbance. Hence, such an automatic controller for disturbance attenuation can prevent accidents. It can be concluded that the adaptive control techniques that require excessive time either for identification and parameter estimation or controller parameters tuning are not very well suited for solving the disturbance attenuation problem of moving vehicles.

An *adaptive* cruise control system is developed in Holzmann *et al.* (1997) that adjusts the vehicle's velocity depending on the distance from adjacent vehicles. That paper also develops an adaptive controller for the vehicle's longitudinal motion. In the adaptation level, using recursive parameter estimation, all changes in vehicle parameters are obtained on-line. In the automation level, in the lower layer feedforward and feedback linearization together with a PI controller are used. The upper automation layer is based on a fuzzy logic controller. The same paper also describes a technique for supervision of lateral vehicle dynamics. Load-adaptive real time algorithms based on imprecise computations are used for identification of a mathematical model of a convoy of vehicles that is described a system of fourteen nonlinear differential equations (Pfefferl and Farber, 1998). A corresponding linear discrete-time controller is also proposed.

In Ishida *et al.* (1992) a *self-tuning* based automatic cruise controller with a time delay is proposed. A *model reference* adaptive control technique is used in (Bakhtiari-Nejad and Karami-Mohammadi, 1998) for vibration control of vehicles with elastic bodies. A nonlinear PID controller with gain *scheduling* has been used in (Raza and Ioannou, 1996) for the engine's throttle control. This controller was simultaneously used with the brake controller, designed through a feedback linearization technique, for automated longitudinal vehicle control. The corresponding feedback block diagram and the system dynamics equations are given in the paper. Active noise and vibrations in moving vehicles have been considerably reduced via the use of a *generalized predictive controller* in (Shoureshi and Knurek, 1996). This is particularly important for lighter vehicles that are susceptible to noise and vibrations. In Section 9.3 we will present the basics of PID controllers and in Section 9.4 we will present fundamentals of main adaptive control techniques used these days in automated highway problems.

9.3 Conventional control schemes

Conventional control and signal processing techniques assume that processes and systems have fixed parameters. One of the most popular traditional control techniques is PID (proportional-plus-integral-plus-derivative) control. PID control proves to be remarkably efficient in regulating a wide range of industrial processes. Most of PID controllers do not require exact knowledge of the process model, which makes this control law feasible even for processes whose models are hardly known. PID-PD-PI controllers have been used widely in industry for almost sixty years. Surprisingly, despite a huge number of already classic optimal control theory algorithms developed during the 1960s, 1970s and 1980s, and a large number of very sophisticated optimal control algorithms developed in the 1990s, the popularity of the old fashioned PID-PI-PD controllers is growing every day. In a recent paper (Lelic and Gajic, 2000), a survey of more than three hundred papers published between 1990 and 1999 in the leading control theory and application journals on PID control is presented. Despite of a huge number of theoretical and application papers on tuning techniques for PID controllers, this area still remains open for further research. The centennial work of Ziegler and Nichols (1942) is still widely used in industrial applications either as an independent tuning technique or a benchmark model for newly developed techniques.

The theoretical (classroom) form of the PID controller in the frequency domain is given by

$$G_{PID}(s) = K_{\mathrm{P}}\left(1 + \frac{1}{T_{\mathrm{i}}s} + T_{\mathrm{d}}s\right) \tag{9.5}$$

There are three parameters in this transfer function, K_{P}, T_{i}, T_{d}, that have to be tuned such that the controlled system has required closed-loop characteristics. It is more appropriate in practice to use the weighted setpoint form of the PID control law with filtered derivative term:

$$u_{\mathrm{c}} = k_{\mathrm{p}}\left[(\beta y_{\mathrm{r}} - y) + \frac{1}{T_{\mathrm{i}}}\int_0^t e\,d\tau - T_{\mathrm{d}}\frac{\mathrm{d}y_{\mathrm{f}}}{\mathrm{d}t}\right]$$

$$y_{\mathrm{f}} = \frac{1}{1 + sT_{\mathrm{d}}/N}y \tag{9.6}$$

In the above formulas the parameters are chosen as $0 < \beta \leq 1$ and $N \approx 10$. Other forms of PID controllers and their practical importance were discussed in Astrom and Hagglund (1995) and Lelic (1998).

Despite of many advantages of PID controllers and their simplicity, there are some drawbacks. For example, if the controlled process is nonlinear and the setpoint changes, it will be necessary to retune the controller parameters in order to keep the desired performance. When the operating point changes frequently, it is advisable to have a table of PID control parameters where each set of data in the table corresponds to a particular operating point of the controlled process. This *gain scheduling* technique is widely used when highly nonlinear system with a frequent change of the operating point needs to be controlled. This technique will be described in more detail later on in this chapter. In addition, when the process has variable time delays, varying parameters, large nonlinearities and when the process is disturbed by noise, the use of PI and PID controllers may not result in the desired performance. In such cases, it is desirable to tune the controller parameters on-line as the process parameters change. For this purpose, the self-tuning version of the PID controller is suitable. This idea was employed by different researchers, and is beyond the scope of this chapter (Omatu *et al.*, 1996).

9.4 Adaptive control – an overview

The development of adaptive control started in the 1950s due to the need for more sophisticated controllers in the aerospace industry. Conventional controllers were not able to provide satisfactory control for high performance aircraft and rockets whose dynamics are highly nonlinear. Due to a lack of theoretical understanding and due to implementational difficulties, adaptive control at that time was more of academic than practical importance. In the 1970s, several parameter estimation techniques were proposed for various adaptive control schemes. That period was also characterized by the rapid development of microprocessors, which enabled implementation of more complex control algorithms. Theoretical results about the stability of adaptive control algorithms were mostly obtained in the early 1980s. Before that time, the stability analysis was based on some restrictive assumptions about controlled system dynamics. In the 1980s, considerable research efforts were directed towards robust control. This period was particularly important because several model predictive adaptive control techniques were developed. Ideas of robust and adaptive control theory merged together in the late 1980s and early 1990s providing suitable design tools for many different industrial processes (Ioannou and Sun, 1996). Applications of adaptive control for nonlinear process control often combine conventional control law, self-tuning, and neural networks for nonlinear parameter estimation. Such an example of speed control for an electric vehicle by using the self-tuning PID neuro controller can be found in Omatu *et al.* (1996). Another source of applications of neural networks in adaptive control of nonlinear systems is the monograph of Ng (1997). The adaptive backstepping algorithm for control of nonlinear processes has become a very powerful technique since the early the 1990s (Krstić *et al.*, 1995). Adaptive and learning techniques have been combined with PID controllers as convenient tools for design and tuning of these (PID) most widely used controllers. The PID control research community used extensively in the 1990 different adaptive and learning techniques for tuning and design of PID control parameters (Lelic and Gajic, 2000). The adaptive control techniques have found their way in to many industrial controllers (Wellstead and Zarrop, 1991; Äström and Wittenmark, 1995). Many of the adaptive control techniques, such as autotuning, self-tuning, gain scheduling, and adaptive feedforward are offered as standard features in industrial PID controllers. Especially popular adaptive controllers, used in process industries, are model predictive type techniques.

The formal definition of an adaptive controller can be found in Äström and Wittenmark (1995): ‘An adaptive controller is a controller with adjustable parameters and a mechanism for adjusting the parameters.’ It is intuitively clear that the role of adaptive controller is to modify its behaviour as a response to changes in the process dynamics and the changes in the dynamics of disturbances acting on the process. Adaptive control algorithms can be grouped into several categories: gain scheduling, model reference adaptive control, self-tuning control, suboptimal control, expert tuning systems, etc. In the following, we will introduce some of them.

9.4.1 Gain scheduling

This control technique is used when the process to be controlled is highly nonlinear and its operating point changes frequently. A typical example of such a process (system) is aircraft dynamics. Assume that the total operating range of the process dynamics is composed of N operating points. Each of the operating points has approximately linear dynamics. For each operating point i, $i = 1, 2, \ldots, N$, there exists a set of known linear parameters corresponding to that operating point. Every such linear subsystem has the corresponding feedback controller with a constant parameter vector, say θ_i, designed to meet performance requirements at that specific operating point, i. When the process moves from one operating point to another, the parameters of the process change so that it is necessary that the controller adjusts its parameters to meet the performance criteria for the new operating point. A common way to switch between different controller parameters is to use a lookup table, coupled with appropriate logic for detecting the change in the operating point. In some cases, the change between two different controller parameter sets may be large, causing the (unwanted) excessive change in the control signal. This can be handled by interpolation or by increasing the number of operating points. The gain scheduling control scheme is shown in Figure 9.1.

The advantage of gain scheduling is that the controller parameters (gains) can be changed very quickly. On the other hand, obtaining the controller parameters for different operating points is a time consuming task, because it is necessary to find the controller gains for each of N different operating points. In this case, the gains are precomputed off-line. If the number of the operating points does not cover the whole system dynamic range, no feedback action is available at such operation points and the control system may become unstable.

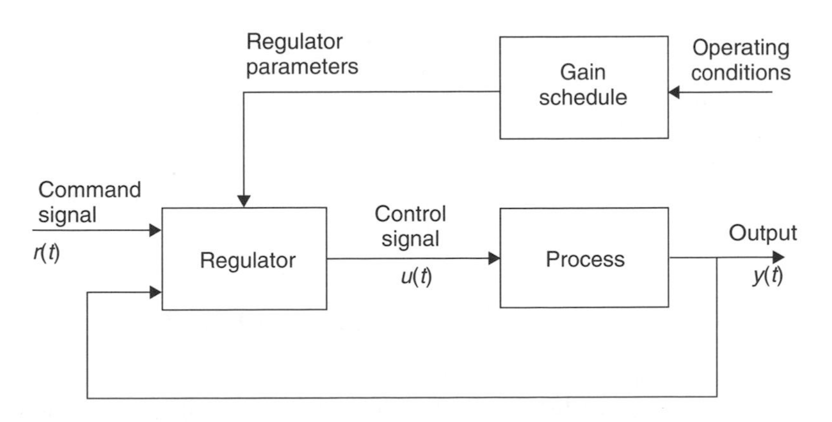

Fig. 9.1 Gain scheduling control scheme.

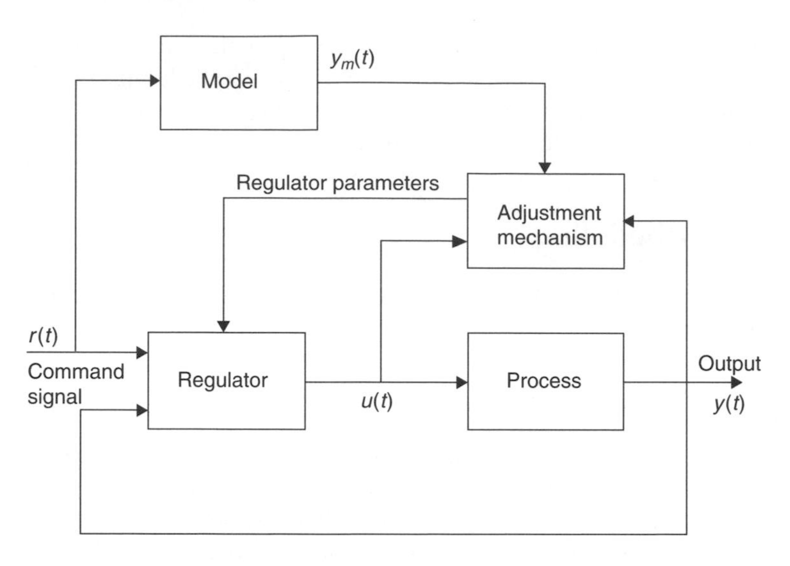

Fig. 9.2 Model reference adaptive control scheme.

9.4.2 Model reference adaptive control (direct adaptive control)

Model reference adaptive control is derived from the model following problem in which the desired performance is expressed in terms of reference model. This model gives the desired response to a set point (command) signal. The desired response is compared with the output of the system forming an error signal. The control objective is to minimize this error signal by adjusting the controller parameters. Mechanisms for adjusting parameters are mostly based either on the gradient type methods (the MIT rule) or on the Lyapunov stability techniques. Figure 9.2 shows the block diagram of the basic MRAC.

The original MIT rule has the following form:

$$\frac{d\theta}{dt} = -\alpha e \frac{de}{d\theta} \tag{9.7}$$

in which e is the model error and θ is a vector of adjustable parameters. The parameter α determines the adaptation rate. The MIT rule works well for small adaptation rate, but stability of the algorithm is not guaranteed. It is a heuristic method taken from nonlinear optimization. There are several modified adaptation rules based on using the Lyapunov stability theory, for which the controller is designed in such a way and the parameters are chosen in such a manner that the obtained closed-loop system is stable (a Lyapunov function for such a system exists).

9.4.3 Self-tuning control (indirect adaptive control)

The main philosophical difference between conventional design methods and self-tuning control is that in self-tuning we introduce control (or signal processing) algorithms with coefficients that can vary in time. Conventional control theory, on the other

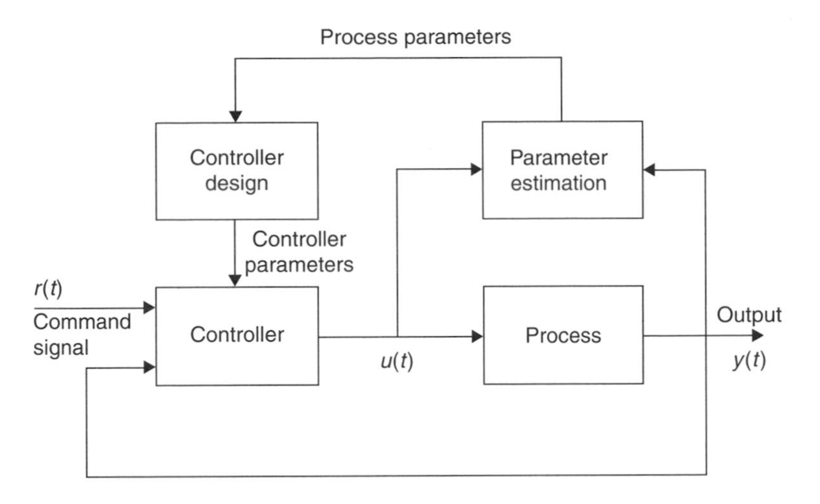

Fig. 9.3 Self-tuning control scheme.

hand, assumes that both the process and controller parameters are constant. Hence, the basic idea in self-tuning theory is to find an algorithm, which automatically adjusts its parameters in order to meet a particular control performance. Figure 9.3 shows a typical self-tuning control scheme. The controller, process, and feedback from the output to the controller, make the classical control system. The outer loop consists of a recursive parameter estimator and a controller design blocks. Self-tuning controllers use the *certainty equivalence principle* so that the process parameters are estimated in real time. The controller design block uses these parameters as if they were equal to the true values of the process parameters. Uncertainty of these parameters is not considered.

On-line parameter estimation is the key component of adaptive control. In the self-tuning control scheme in Figure 9.3, the recursive parameter estimator appears explicitly so that this self-tuning control scheme is called explicit self-tuning control. Another variant for this technique is indirect adaptive control, where the control law is computed indirectly, after the process model parameters are identified.

There are several potential problems with adaptive control techniques. For example, parameter convergence to their true values is not guaranteed since adaptive controllers are nonlinear; stability under adaptive control is proved only for restricted classes of systems.

9.5 System models for adaptive control

Adaptive control algorithms can be implemented either in continuous or discrete time forms. However, since digital computers are used for implementation of the outer control loop (parameter estimation plus controller design), the discrete form control implementation comes as a natural solution. The recursive nature of adaptive control algorithms makes the discrete form implementation even more appropriate. Most actual processes are nonlinear in nature and evolve continuously in time, but in most cases their models are linear both in parameters and data (Wellstead and Zarrop, 1991; Astrom and Wittenmark, 1995; Ljung, 1999).

9.5.1 System identification basics

System identification is a procedure for finding a mathematical model for a dynamic system from observed data. A typical procedure for system identification requires that a model structure is chosen and available process data is used to find the best model that fits given information according to a certain criterion. A common assumption in system identification is that the unknown system is linear. Linear system theory is well understood and there exist numerous papers and books covering this area (see, for example Ljung, 1999). Although this is almost never true in real applications, linear models are acceptable in cases when the operating point of the controlled process is fixed or slowly varying. Nonlinear model identification is much more complex than linear, because the choice of model structure for nonlinear model is difficult since there are many different types of static (and dynamic) nonlinearities. If a nonlinear system is linearized around a certain operating point, then the linear model obtained is in general time varying. If there exists enough physical understanding about the system, then it is possible to construct a model that has static nonlinearities. This *semi-physical modelling* can be realized by using *Wiener-Hammerstein* models (Ljung, 1999). The static nonlinear functions are parameterized in terms of physical parameters (saturation point, saturation level) or in black-box terms. Nonlinear black-box models can be used when there is not enough knowledge about the physics of the system. These models can be realized from input and output data by different modelling techniques, such as neural networks, wavelets, fuzzy logic, to name a few. There are three basic entities in the identification procedure (Ljung, 1999): (i) a data set; (ii) a set of candidate models – a model structure; (iii) a rule by which the candidate models can be assessed using the data, like the least squares selection rule.

9.5.2 Recursive parameter estimation

Assume that the system model is given in the discrete time form:

$$A(z^{-1})y(t) = B(z^{-1})u(t-1) + x(t) \tag{9.8}$$

where $A(z^{-1})$, $B(z^{-1})$ are polynomials in the backward shift operator form, such that $z^{-1}x(t) = x(t-1)$

$$\begin{aligned} A(z^{-1}) &= 1 + a_1 z^{-1} + \ldots + a_n z^{-n} \\ B(z^{-1}) &= b_0 + b_1 z^{-1} + \ldots + b_m z^{-m} \end{aligned} \tag{9.9}$$

and $y(t)$, $u(t-1)$, $x(t)$ are, respectively, process output, input, and disturbance. The disturbance $x(t)$ can contain several components:

$$x(t) = D(z^{-1})v(t) + d(t) + C(z^{-1})e(t) \tag{9.10}$$

where $C(z^{-1})$ and $D(z^{-1})$ are polynomials of known degrees. The term $D(z^{-1})v(t)$ is a measurable load disturbance. The second term in Equation 9.10, $d(t)$, is a drift, and $C(z^{-1})e(t)$ is coloured noise. In practical situations, it is not likely that the system has all three disturbance components as shown in Equation 9.10. In addition, the drift is usually equal to some constant offset $d(t) = d_0$, and it can be a part of the process or contributed by the sensor instruments. A measurable, but non-controllable load disturbance $v(t)$ can be compensated by feed-forward control. If noise is not stationary, then the last term on the right hand side of Equation 9.10 can be represented as $(C(z^{-1})/\Delta(z^{-1}))e(t)$, where $\Delta(z^{-1}) = 1 - z^{-1}$ is a discrete form of a differentiator. In this presentation, without loss of generality, it is assumed that the first two terms on the right-hand side of the disturbance vector $x(t)$ of Equation 9.10 are not present. Then, from (Equation 9.8 and Equation 9.10), we have the so-called, *controlled auto regressive moving average* (CARMA) system model:

$$A(z^{-1})y(t) = B(z^{-1})u(t-1) + C(z^{-1})e(t) \tag{9.11}$$

If noise is not stationary, this model can be represented by the *controlled auto regressive integrated moving average* (CARIMA) that has the form:

$$A(z^{-1})y(t) = B(z^{-1})u(t-1) + \frac{C(z^{-1})}{\Delta(z^{-1})}e(t) \tag{9.12a}$$

or

$$A(z^{-1})\Delta y(t) = B(z^{-1})\Delta u(t-1) + C(z^{-1})e(t) \tag{9.12b}$$

Note that in Equation 9.12a, the increments of input signal, $\Delta u(t) = u(t) - u(t-1)$, and the increments of the output signal, $\Delta y(t) = y(t) - y(t-1)$, are present.

For estimation purposes, the system model is written in the *regression model* form:

$$y(t) = \phi^T(t)\theta + e(t) \tag{9.13a}$$

where θ is the vector of unknown parameters, defined by

$$\theta^T = [a_1\, a_2\, \ldots\, a_n\, b_0\, b_1\, \ldots\, b_m\, c_1\, c_2\, \ldots\, c_p] \tag{9.13b}$$

and ϕ is a *regression vector* consisting of measured input and output variables and past values of unobservable disturbance $e(t)$.

$$\phi^T(t) = [-y(t-1)\ldots - y(t-n)\; u(t-1)\ldots u(t-m)\; e(t-1)\ldots e(t-p)] \tag{9.13c}$$

or (for CARIMA models)

$$\phi^T(t) = [-\Delta y(t-1)\ldots - \Delta y(t-n)\; \Delta u(t-1)\ldots \Delta u(t-m)\; e(t-1)\ldots e(t-p)] \tag{9.13d}$$

The vector $\phi(t)$ contains past disturbance values $e(t-1), e(t-2), \ldots, e(t-p)$, which are, generally, unknown. Note that, for the sake of simplicity, the measurable load disturbance and the drift terms (Equation 9.11) are not included in 9.11–9.13. Let us assume that Equation 9.11 is a description of a system. Our goal is to determine the vector of parameters, θ, by using available data. Assume that for the system of correct structure, we have:

$$y(t) = \phi^T(t)\hat{\theta} + \hat{e}(t) \tag{9.14}$$

where $\hat{\theta}$ is the vector of adjustable model parameters and $\hat{e}(t)$ is the corresponding modelling error at time t. It follows from Equations 9.11 and 9.14 that:

$$\hat{e}(t) = e(t) + \phi^T(t)(\theta - \hat{\theta}) \tag{9.15}$$

The modelling error $\hat{e}(t)$ depends on $\hat{\theta}$. In some cases, the minimized modelling errors is equal to the white noise sequence corrupting the system output data.

In order to be useful for adaptive control, the parameter estimation procedure should be iterative, thus allowing the estimated model 9.14 of the system to be updated in each sampling interval, as new data becomes available. There is a variety of recursive parameter estimation schemes – in this chapter we will only consider the *extended least square* algorithm. For more details about other recursive estimation techniques the reader is referred to Ljung (1999). Assume that the system model to be estimated is given in the CARMA form:

$$\hat{A}(z^{-1})y(t) = \hat{B}(z^{-1})u(t-1) + \hat{C}(z^{-1})\hat{e}(t) \tag{9.16}$$

Define the parameter and the regression vectors, respectively, as:

$$\theta^T = [\hat{a}_1 \ldots \hat{a}_n \; \hat{b}_0 \ldots \hat{b}_m \; \hat{c}_1 \ldots \hat{c}_p] \tag{9.17}$$

$$\phi^T(t) = [-y(t-1) \ldots - y(t-n) \; u(t-1) \ldots u(t-m) \; \varepsilon(t-1) \ldots \varepsilon(t-p)] \tag{9.18}$$

where

$$\varepsilon(t) = y(t) - \phi^T(t)\hat{\theta}(t-1) \tag{9.19}$$

is called the *prediction error* using the output prediction based on information up to the discrete time instant $t-1$. Below we present the complete extended recursive least square estimation algorithm.

Algorithm 1: Recursive extended least square estimation

At time step $t+1$:

(i) Form $\phi^T(t+1)$ using the new data, $u(t+1)$, $y(t+1)$, and find $\varepsilon(t+1)$ using

$$\varepsilon(t+1) = y(t+1) - \phi^T(t+1)\hat{\theta}(t)$$

(ii) Form $P(t+1)$ from

$$P(t+1) = P(t)\left[\mathrm{I}_m - \frac{\phi(t+1)\phi^T(t+1)P(t)}{1 + \phi^T(t+1)\mathrm{P}(t)\phi(t+1)}\right]$$

(iii) Update $\hat{\theta}(t)$

$$\hat{\theta}(t+1) = \hat{\theta}(t) + P(t+1)\phi(t+1)\varepsilon(t+1)$$

(iv) Wait for the next time step to elapse and loop back to step (i).

Note that $P(t)$ is the *covariance matrix* defined as $P(t) = [X^T(t)X(t)]^{-1}$, and

$$X(t) = [\phi\ (1)\ \phi\ (2) \ldots \phi(t)]^T$$

is the vector of past regression vectors (Equation 9.18).

9.5.3 Estimator initialization

The recursive algorithms require initial estimates for ϕ (0), $\hat{\theta}$ (0), and $P(0)$. A common procedure is to fill in the data vector ϕ (0) from the past input and output samples just before the estimation started. The prediction error values can be set to zero

$$\phi^T(t) = [-y(-1) \ldots - y(-n)\ u(-1) \ldots u(-m)\ 0 \ldots 0]$$

The initial estimate $\hat{\theta}(0)$ of the parameter vector can be set in a number of ways. The best approach is to make use of prior knowledge of the system whose parameters we want to estimate. If the prior information is not available, then we set $\hat{a}_1 = 1, \hat{a}_i = 0, i = 2, \ldots, n$ and $\hat{b}_i = 0, i = 0, \ldots, m$. The initial value $P(0)$ for the covariance matrix reflects our uncertainty concerning the unknown parameters. If we have no prior knowledge of the system parameters in the model, then a large initial covariance would reflect this. In the same way, if the initial parameters $\hat{\theta}$ (0) are known to be close to the true values, then a small covariance matrix should be used. A typical choice for $P(0)$ is the diagonal matrix of the form:

$$P(0) = rI \tag{9.20}$$

For large $P(0)$, the scale factor r is set between 100 and 1000 and for small $P(0)$ the value of r is set between 1 and 10.

9.6 Design of self-tuning controllers

9.6.1 Generalized minimum variance (GMV) control

The GMV control algorithm was originally proposed by Clarke and Gawtrop (1975). The objective of the standard minimum variance controller is to regulate the output of a stochastic system to a constant (zero) set point. In other words, it is required at each time t to find the control $u(t)$ which minimizes the output variance, expressed as a criterion:

$$J = E\{\phi^2(t+k)\} \tag{9.21}$$

where k is the pure time delay and $\phi(t+k)$ is a generalized (pseudo) output defined as:

$$\phi(t+k) = Py(t+k) + Qu(t) - Rr(t) \tag{9.22}$$

The variables in the above formula are as follows: P, R finite polynomials in z^{-1}; $y(t)$ system output at time t; $r(t)$, set point sequence; $u(t)$, control signal at time t. Note that the system model is given in CARMA form:

$$Ay(t) = z^{-k}Bu(t) + Ce(t) \tag{9.23}$$

Equations 9.22 and 9.23 yield:

$$\phi(t+k) = \frac{PB + QA}{A}u(t) - Rr(t) + \frac{PC}{A}e(t+k) \tag{9.24}$$

The last term on the right hand side represents the past and future noise values. Notice that $e(t+1), e(t+2), \ldots, e(t+k)$ are future values of noise that cannot be calculated. However, $e(t), e(t-1), \ldots$, can be calculated from Equation 9.23. Future noise values can be separated from the past values by introducing the following polynomial identity:

$$PA = E + z^{-k}G \tag{9.25}$$

in which E and G are polynomials in z^{-1} of the following orders $n_e = k - 1$ and $n_g = \max(n_a - 1, n_p + n_c - k)$. After some straightforward manipulations Equations 9.23–9.25 produce

$$\phi(t+k) = \frac{1}{C}[(BE + QC)u(t) + Gy(t) - CRr(t)] + Ee(t+k) \tag{9.26}$$

The error term $Ee(t+k)$ is uncorrelated with the remainder of the right-hand side of Equation 9.26. The criterion of Equation 9.21 has a minimum when:

$$(BE + QC)u(t) + Gy(t) - CRr(t) = 0 \tag{9.27}$$

or, in a more compact three-term controller form

$$\begin{aligned} &Fu(t) + Gy(t) + Hr(t) = 0 \\ &F = BE + QC, \quad H = -CR \end{aligned} \tag{9.27a}$$

From this formula we find the control signal $u(t)$ which minimizes the variance of $\phi(t+k)$ represented by Equation 9.21:

$$\begin{aligned} u(t) &= -\frac{G}{BE + QC}y(t) + \frac{CR}{BE + QC}r(t) \\ &= -\frac{G}{F}y(t) - \frac{H}{F}r(t) \end{aligned} \tag{9.28}$$

The closed-loop equation of the system controlled by the *generalized minimum variance* controller (Equation 9.27) is obtained by substituting Equation 9.28 in 9.23, that is

$$\begin{aligned} y(t) &= \frac{z^{-k}BR}{PB + QA}r(t) + \frac{BE + QC}{PB + QA}e(t) \\ &= \frac{z^{-k}BR}{T}r(t) + \frac{F}{T}e(t) \end{aligned} \tag{9.29}$$

Notes:

(i) If $P = 1$, $Q = 0$, $R = 0$, the classic minimum variance control law (Astrom and Wittenmark, 1973) is obtained as

$$u(t) = -\frac{G}{BE}y(t), \quad y(t) = Ee(t) \tag{9.30}$$

The minimum variance control is the oldest self tuning controller. It is stable only if the system is non-minimum phase (the polynomial B has all its zeros within the unit circle).

(ii) The zero steady-state tracking error is obtained if the following condition is satisfied

$$\left.\frac{BR}{PB + QA}\right|_{z=1} = 1 \tag{9.31}$$

(iii) By proper choice of polynomials P and Q, it is possible to choose the poles of the closed-loop system defined by polynomial T (Equation 9.29)

$$PB + QA = T \tag{9.32}$$

Example 1

Design a GMV controller for the system in Equation 9.23 with unit time delay ($k = 1$), and the system polynomials

$$A = 1 + 0.4z^{-1} - 0.45z^{-2}, \quad B = 1 + 0.8z^{-1}, \quad C = 1$$

Let $P = 1 = Q, R = r_0$. From Equation 9.25:

$$1 + 0.4z^{-1} - 0.45z^{-2} = e_0 + z^{-1}(g_0 + g_1z^{-1})$$

we find E and G as

$$E = e_0 = 1, \quad G = g_0 + g_1z^{-1} = 0.4 - 0.45z^{-1}$$

The GMV controller, calculated from Equation 9.27, has the form:

$$(2 + 0.8z^{-1})u(t) = r_0r(t) - (0.4 - 0.45z^{-1})y(t)$$

This equation and Equation 9.23 yield the closed-loop system:

$$y(t) = \frac{r_0(1 + 0.8z^{-1})}{2 + 1.2z^{-1} - 0.45z^{-2}}r(t-1) + \frac{2 + 0.8z^{-1}}{2 + 1.2z^{-1} - 0.45z^{-2}}e(t)$$

For the zero steady state tracking error, the coefficient r_0 has the following value:

$$r_0 = \frac{T(1)}{B(1)} = \frac{2.75}{1.8} = 1.5278$$

The performance of this system is shown in Figure 9.4.

Algorithm 2: Self-tuning generalized minimum variance controller

At time step t:

(i) Form the pseudo output $\phi(t)$:

$$\phi(t) = Py(t) + Qu(t-k) - Rr(t-k)$$

(ii) Estimate the controller polynomials $\hat{F}$, $\hat{G}$, $\hat{H}$ from:

$$\phi(t) = \hat{F}u(t-k) + \hat{G}y(t-k) + \hat{H}r(t-k) + \hat{e}(t)$$

by using the recursive estimation algorithm.

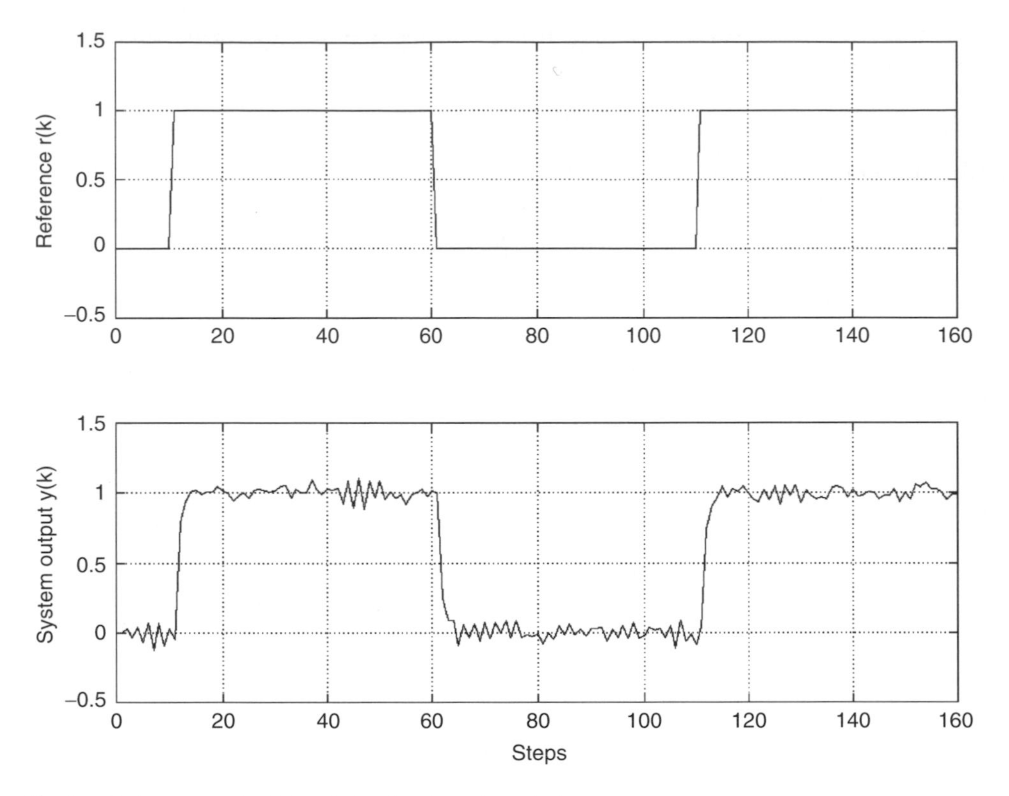

Fig. 9.4 Performance of GMV control system from Example 1.

(iii) Apply control $u(t)$ using the control law

$$\hat{F}u(t) = -\hat{G}y(t) - \hat{H}r(t)$$

(iv) Wait until the next sampling instant and go to Step 1.

Example 2

If the system is non-minimum phase (polynomial B has roots outside of the unit circle) the closed-loop system may become unstable. Consider the system model from Example 1, but with $B = 1 + 1.2z^{-1}$. The closed-loop system is

$$y(t) = \frac{r_0(1 + 1.2z^{-1})}{2 + 1.6z^{-1} - 0.45z^{-2}} r(t-1) + \frac{2 + 1.2z^{-1}}{2 + 1.6z^{-1} - 0.45z^{-2}} e(t)$$

Figure 9.5 indicates the unstable closed-loop response ($r_0 = (2 + 1.6 - 0.45)/(2 + 1.2) = 0.9844$).

The closed-loop response can be tailored by a proper choice of P and Q polynomials, resulting in a stable closed-loop control even for non-minimum phase system.

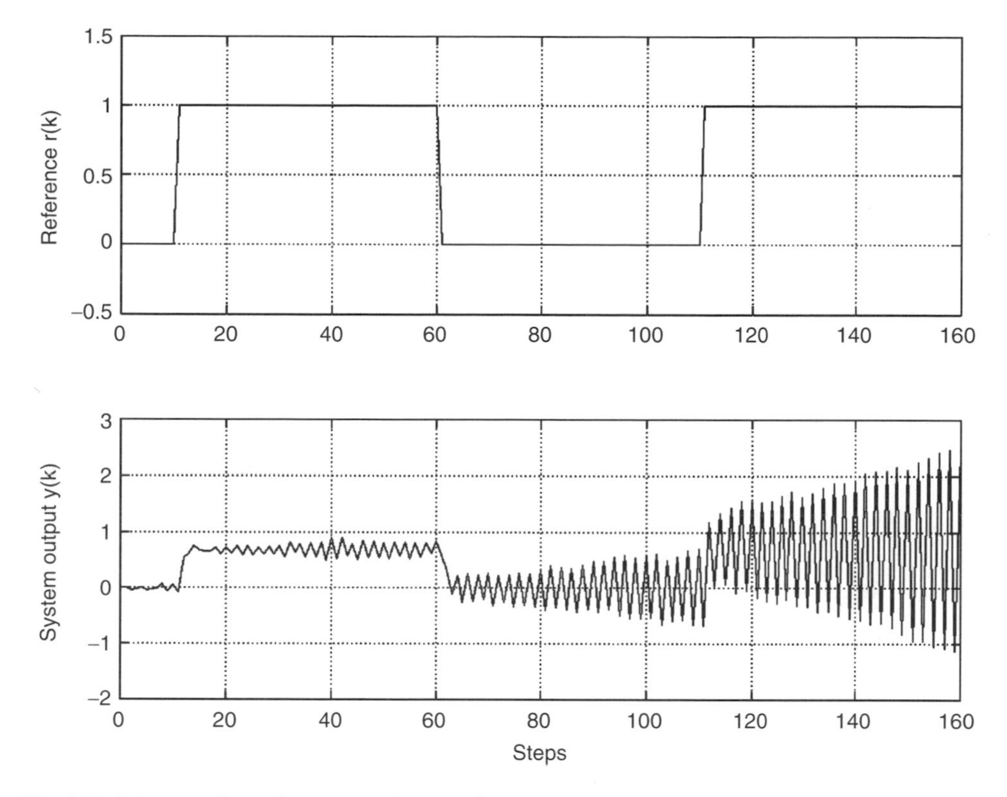

Fig. 9.5 GMV controller performance in the case of a non-minimum phase system.

Example 3

Design a GMV control for the system in Example 2 such that the closed-loop characteristic equation is $T = 1 - 0.8z^{-1}$. From Equation 9.29 we have:

$$PB + QA = T \Rightarrow (p_0 + p_1 z^{-1})(1 + 1.2z^{-1}) + q_0(1 + 0.4z^{-1} - 0.45z^{-2})$$
$$= 1 - 0.8z^{-1}$$

The above polynomial equation produces $P = -3.7059 + 1.7647z^{-1}$, $Q = 4.7059$. Polynomials F and G are obtained from Equation 9.25 as:

$$E = e_0 = -3.7059, \quad G = 0.2823 + 2.3735z^{-1} - 0.7953z^{-2}$$

leading to the stable closed-loop system (Equation 9.28):

$$y(t) = \frac{r_0(1 + 1.2z^{-1})}{1 - 0.8z^{-1}} r(t-1) + \frac{1 - 4.4471z^{-1}}{1 - 0.8z^{-1}} e(t)$$

With $r_0 = (1 - 0.8)/(1 + 1.2) = 0.0909$, the closed-loop system performance is shown in Figure 9.6.

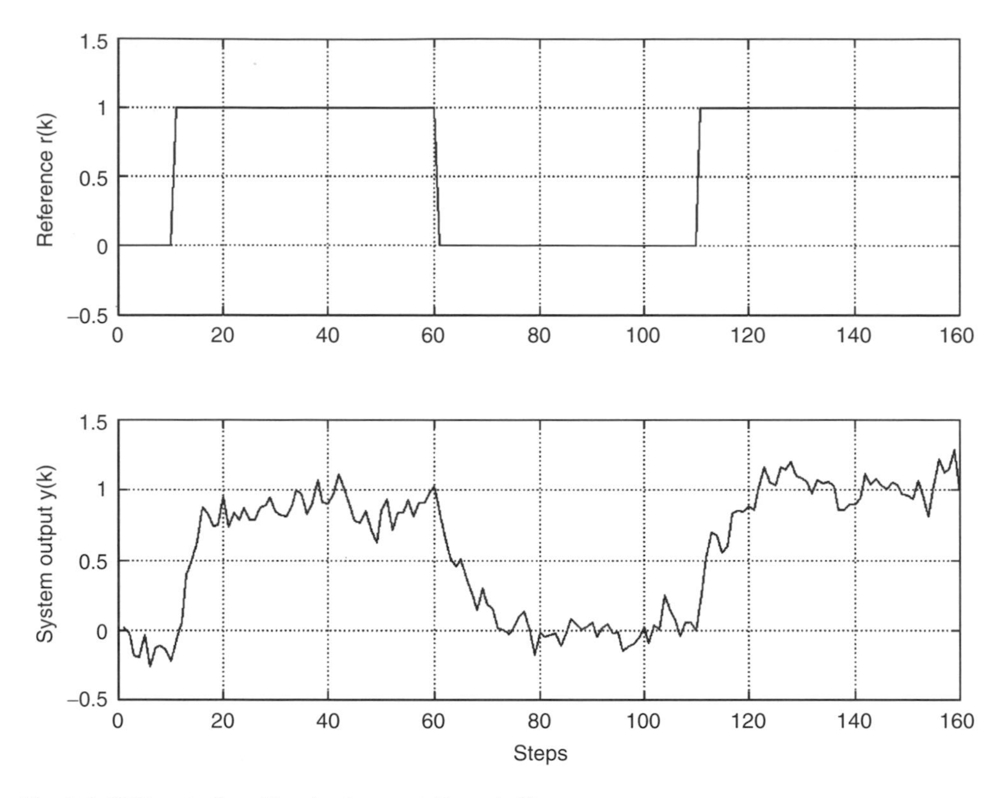

Fig. 9.6 GMV controller with pole-placement (Example 3).

9.6.2 Pole placement control

Pole placement self-tuning control originally was proposed by Wellstead *et al.* (1979) and Astrom and Wittenmark (1980). The basic idea behind this control algorithm is to push the closed-loop poles of the controlled system into prespecified positions defined by the closed-loop characteristic polynomial T. Assume that the system model is:

$$Ay(t) = Bu(t-1) + Ce(t) \tag{9.33}$$

and the control-loop is closed with the three-term controller:

$$Fu(t) = Hr(t) - Gy(t) \tag{9.34}$$

Combining the two above equations gives the closed-loop description

$$(AF + z^{-1}BG)y(t) = z^{-1}BHr(t) + CFe(t) \tag{9.35}$$

The polynomials F and G are chosen such that the following polynomial equation is satisfied:

$$AF + z^{-1}BG = CT \tag{9.36}$$

The polynomial Equation 9.36 can be solved if A and B do not have any other common zeros except for those of the stable polynomial C, and the orders of the polynomials

F and G satisfy:

$$n_f = n_b$$
$$n_g = n_a - 1 \quad (n_a \neq 0) \tag{9.37}$$

with an additional constraint $n_f \leq n_a + n_b - n_c$. The substitution of Equation 9.36 in Equation 9.35 yields

$$y(t) = \frac{NB}{TC} r(t-1) + \frac{F}{T} e(t) \tag{9.38}$$

Example 4
Calculate the pole placement controller for the system from Example 2 such that the closed-loop poles are defined by the polynomial T given in Example 3. The controller polynomials F and G are computed from Equation 9.36:

$$(1 + 0.4z^{-1} - 0.45z^{-2})(f_0 + f_1 z^{-1}) + z^{-1}(1 + 1.2z^{-1})(g_0 + g_1 z^{-1}) = 1 - 0.8z^{-1}$$

which leads to the linear system:

$$\begin{bmatrix} 1 & 0 & 0 & 0 \\ 0.4 & 1 & 1 & 0 \\ -0.45 & 0.4 & 1.2 & 1 \\ 0 & -0.45 & 0 & 1.2 \end{bmatrix} \begin{bmatrix} f_0 \\ f_1 \\ g_0 \\ g_1 \end{bmatrix} = \begin{bmatrix} 1 \\ -0.8 \\ 0 \\ 0 \end{bmatrix}$$

The solution of this system gives:

$$F = f_0 + f_1 z^{-1} = 1 - 4.4471z^{-1}, \quad G = g_0 + g_1 z^{-1} = 3.2471 - 1.6676z^{-1}$$

The polynomial $N = n_0$ is calculated such that the steady state error is equal to zero:

$$H = n_0 = \frac{T(1)}{B(1)} = \frac{1 - 0.8}{1 + 1.2} = 0.0909$$

The above calculations lead to the pole placement controller of Equation 9.34:

$$(1 - 4.4471z^{-1})u(t) = 0.0909r(t) - (3.2471 - 1.6676z^{-1})y(t)$$

and the closed-loop system Equation 9.38:

$$y(t) = \frac{0.0909(1 + 1.2z^{-1})}{1 - 0.8z^{-1}} r(t-1) + \frac{1 - 4.4471z^{-1}}{1 - 0.8z^{-1}} e(t)$$

This system is the same as GMV pole placement controller in Example 3. Pole placement controllers have higher output variance than regular GMV controllers.

For the system without noise, the closed-loop system looks like:

$$y(t) = \frac{0.0909(1 + 1.2z^{-1})}{1 - 0.8z^{-1}} r(t-1)$$

Figure 9.7 shows the performance of this system.

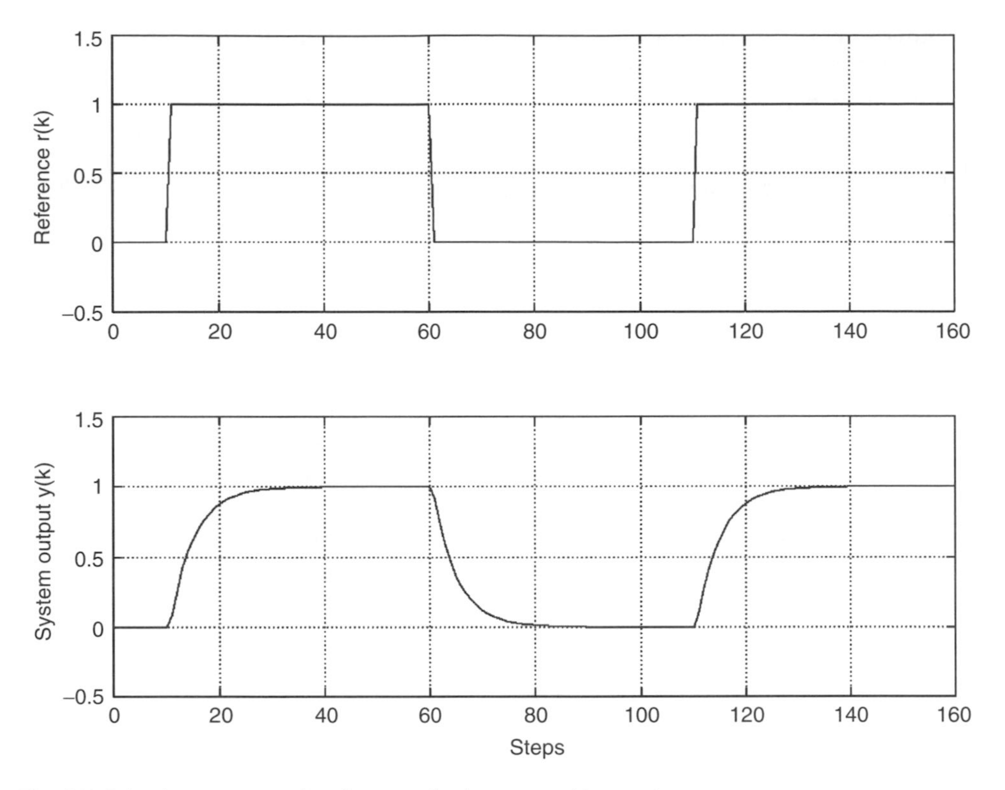

Fig. 9.7 Pole placement control performance for the system without noise.

Algorithm 3: Self-tuning pole placement control

At each sample time t:

(i) Using the recursive estimation algorithm estimate $\hat{A}$, $\hat{B}$, $\hat{C}$ in the model

$$\hat{A}y(t) = \hat{B}u(t-1) + \hat{C}e(t)$$

(ii) Calculate the controller coefficients using

$$\hat{F}\hat{A} + z^{-1}\hat{B}\hat{G} = \hat{C}T$$

$$\hat{H} = \frac{T(1)}{\hat{B}(1)}\hat{C}$$

(iii) Apply the control law

$$\hat{F}u(t) = \hat{H}r(t) - \hat{G}y(t)$$

(iv) Wait until the next sampling instant $t+1$ and go to Step 1.

9.6.3 Model predictive control

A lot of attention has been directed in recent years to self-tuning controllers based on model predictive control (MPC) (Richalet *et al.*, 1978; Cutler and Ramaker, 1980;

De Keyser and van Cauwenberge 1982, Mosca *et al.*, 1984; Peterka, 1984; Clarke *et al.*, 1987). Common to all these approaches is to use a multi-stage cost function whose minimization yields the required control law. These methods are also known as *long range predictive control*, *extended range predictive control*, *receding horizon*, or *rolling horizon* control algorithms. None of the above methods, however, easily encompasses prespecification of the closed-loop pole set. Based on the approach of Clarke and co-workers that led to the generalized predictive controller (GPC), Lelic and Zarrop (1987) proposed a new type of the pole placement controller – generalized pole placement (GPP). The MPC controllers are computationally more demanding than the minimum variance and pole placement self-tuning controllers, but they have numerous advantages over other adaptive control techniques. Some of the advantages are:

- They can deal with a variety of processes including unstable, non-minimum phase, long and variable time delay processes.
- The feed-forward control is introduced in a natural way to compensate for measurable disturbances.
- Most of the MPC methods have a set of parameters 'tuning knobs' which allows tailoring of the closed-loop performance. By changing these parameters, the MPC algorithm can result in some other control techniques (PID, pole-placement, MRAC).
- When the future reference trajectories are known, such as in robotics and batch control, 'program control mode' can be used, allowing smoother transient response.
- Multivariable versions of these controllers are easily derived in most cases.
- They are more robust (if there are constrains in the control signal) than the other classic adaptive controllers.

MPC algorithms are characterized by the following strategy, shown in Figure 9.6:

(i) At each moment t (presence) and for a given horizon N (called prediction horizon) a forecast of the future outputs $\hat{y}(t+k)$, $k = 1 \ldots N$ is obtained. These predicted output values depend on the known values up to time t (past inputs and outputs) and on the future control signals $u(t+k)$, $k = 0, \ldots, N-1$, which have to be calculated.

(ii) The set of future control signals is calculated by optimizing a determined criterion in order to keep a process output as close as possible to the reference trajectory $r(t+k)$.

(iii) The control signal $u(t)$ is applied to the process and all other future signals are discarded.

In the following, two MPC algorithms are described – GPC and GPP.

9.6.4 Generalized predictive control

The generalized predictive control (GPC) can be obtained by minimization of the following cost function

$$J(N1, N) = E_t \left[\sum_{i=N1}^{N} (Py(t+i) - Rr(t+i))^2 + \sum_{i=1}^{Nu} \lambda (Qu(t+i-1))^2 \right] \tag{9.39}$$

$E_t[\cdot]$ is the expectation operator conditioned to data up to time t, P, R and Q are finite polynomials in z^{-1}. Q contains the factor $\Delta(z^{-1})$, and polynomial arguments are omitted for brevity. The first sum on the right hand side represents the tracking error between the set point sequence $\{r(t)\}$ and the system output sequence $\{y(t)\}$. The second term on the right-hand side with the weighting parameter λ imposes a constraint on the future control signal increments. The parameter N is called the output horizon and it defines the number of future outputs in the cost function. NU is the control horizon and defines the number of the control signal increments which minimize the cost function. The control increments $\Delta u(t), \ldots, \Delta u(t+NU-1)$ are calculated at each time t, to minimize the cost function Equation (9.39), but only $\Delta u(t)$ is applied to the system. Note that the control increments can be replaced by $u(t), \ldots, u(t+NU-1)$, if the CARMA model (Equation 9.11) of the process is used, as reported in Lelic and Zarrop (1987). The CARIMA model (Equation 9.12b) has a term $\Delta(z^{-1})$, which serves as an integrator in the closed loop and therefore eliminates steady state errors.

Example 5
Given the system

$$(1 - az^{-1})y(t) = bu(t-1) + e(t) \tag{9.40}$$

Chose the parameter of the cost function (Equation 9.39) as

$$NU = N = 2, \quad N1 = 1, \quad P = 1 + z^{-1}, \quad Q = 1 = R \tag{9.41}$$

From (Equation 9.40) we find $y(t+1)$, $y(t+2)$:

$$y(t+1) = ay(t) + bu(t) + e(t+1) \tag{9.42}$$

$$\begin{aligned} y(t+2) &= ay(t+1) + bu(t+1) + e(t+2) \\ &= a^2y(t) + abu(t) + bu(t+1) + ae(t+1) + e(t+2) \end{aligned} \tag{9.43}$$

Therefore, the cost function is

$$\begin{aligned} J(1,2) &= E_t \sum_{i=1}^{2} \{[y(t+i) + py(t+i-1) - w(t+i)]^2 + \lambda u(t+i-1)^2\} \\ &= [(a+p)y(t) + bu(t) - w(t+1)]^2 \\ &\quad + [a(a+p)y(t) + b(a+p)u(t) + u(t+1) - w(t+2)]^2 \\ &\quad + \lambda u(t)^2 + \lambda u(t+1)^2 + \text{constant} \end{aligned} \tag{9.44}$$

Minimizing Equation 9.44 with respect to $u(t)$, $u(t+1)$ leads to

$$\begin{bmatrix} b^2 + b^2(a+p)^2 + \lambda & b(a+p) \\ b(a+p) & 1+\lambda \end{bmatrix} \begin{bmatrix} u(t) \\ u(t+1) \end{bmatrix} = \begin{bmatrix} b & b(a+p) \\ 0 & 1 \end{bmatrix} \begin{bmatrix} r(t+1) \\ r(t+12 \end{bmatrix} - (a+p) \begin{bmatrix} b + ab(a+p) \\ a \end{bmatrix} y(t) \tag{9.45}$$

The 'rolling horizon' controller is obtained from Equation 9.45 by extracting $u(t)$, so the GPC controller has the form:

$$[\lambda^2 + \lambda + \lambda b^2 + \lambda b^2(a+p)^2 + b^2]u(t) = (1+\lambda)br(t+1) + \lambda b(a+p)r(t+2) - b(a+p)[(1+\lambda) + \lambda a(a+p)]y(t) \qquad (9.46)$$

The controller coefficients are nonlinear in p, λ and in general involve increasing powers as the output horizon is extended. It is not a simple task to assign the closed-loop poles of the system nor to tell what the transient response will look like. The cost function proposed in the next section overcomes this problem and allows the assignment of closed-loop poles in a straightforward manner whatever the choice of output horizon or control horizon.

9.6.5 Generalized pole placement control

The cost function (Equation 9.39) is replaced by either of the following two functionals

$$J_1(N1, N) = E_t\left[\sum_{i=N1}^{N} \phi_i^2(t) + \sum_{i=1}^{NU} \lambda \Delta u^2(t+i-1)\right] \qquad (9.47)$$

with

$$\phi_i(t) = Py(t+i) - Rr(t+i) + (Q/S)\Delta u(t-1) \qquad (9.48)$$

or

$$\phi_i(t) = Py(t+i) - Rr(t+i) + Q\Delta u(t-1) + Sy(t) \qquad (9.49)$$

P, Q, R, S are finite polynomials in z^{-1} and S is a monic polynomial ($S(0) = 1$). For simplicity, it is assumed that $N1 = 1$. In the following, the GPP controller for the CARMA model and generalized output (Equation 9.49) is derived. A very similar procedure, when applied to the cost function (Equation 9.39) results in the GPC controller. In this derivation the CARIMA model (Equation 9.12b) of the system is used (note that $\overline{A} = \Delta A$):

$$\overline{A}(z^{-1})y(t) = B(z^{-1})\Delta u(t-1) + C(z^{-1})e(t)$$

In order to derive the GPP controller two polynomial partitions are needed. First partition splits the term $\Phi_i(t)$ into a sum of two uncorrelated terms – its predicted value at time t and a prediction error, that is:

$$PC = \overline{A}F_i + z^{-i}G_i, \quad \deg(F_i) = i-1, \quad \deg(G_i) = \max(n_a, \deg(P) + n_c - i) \qquad (9.50)$$

The partition shown above leads via standard manipulation to:

$$\phi_i(t) = \frac{BF_i}{C}\Delta u(t+i-1) + Q\Delta u(t-1) + \left(\frac{G_i}{C} + S\right)y(t) - Rr(t+i) + F_i e(t+1) \qquad (9.51)$$

The first term on the right-hand side can be partitioned in the standard way into future (unknown) signal increments $\Delta u(t+N-1), \ldots, \Delta u(t)$ and past, already applied control signal increments $\Delta u(t-1), \Delta u(t-2), \ldots$, by invoking a polynomial identity (provided that the first coefficient of polynomial B, $b_0 \neq 0$) of the form

$$BF_i = CE_i + z^{-i}\Gamma_i, \quad \deg(E_i) = i-1, \deg(\Gamma_i) = \max(n_b - 1, n_c - i),$$
$$(\Gamma_i = 0 \text{ if } n_b = 0) \tag{9.52}$$

From Equations 9.51 and 9.52, we have

$$\phi_i(t) = E_i\Delta u(t+i-1) + C^{-1}[(\Gamma_i + CQ)\Delta u(t-1)$$
$$+ (G_i + CS)y(t) - Rr(t+i)] + F_i e(t+1) \tag{9.53}$$

These equations, when put together for $i = 1, 2, \ldots, N$, give a compact, vector form

$$\mathbf{\Phi} = \mathbf{E}\Delta\mathbf{u} + \mathbf{\Gamma}\Delta u(t-1) + \mathbf{G}y(t) - \mathbf{r} + \mathbf{f} \tag{9.54}$$

where

$$\mathbf{\Phi}^T = [\phi_1(t)\phi_2(t)\ldots\phi_N(t)] \tag{9.55}$$

$$\mathbf{u}^T = [\Delta u(t)\Delta u(t+1)\ldots\Delta u(t+NU-1)] \tag{9.56}$$

$$\mathbf{E} = \begin{bmatrix} e_0 & 0 & \cdots & 0 \\ e_1 & e_0 & \cdots & 0 \\ \vdots & \vdots & \ddots & \vdots \\ e_{N-1} & e_{N-2} & \cdots & e_{N-NU} \end{bmatrix}, \quad E_i(z) = \sum_{j=0}^{i-1} e_j z^j \tag{9.57}$$

$$\mathbf{f}^T = [F_1 e(t+1)\ldots F_N(t+N)] \tag{9.58}$$

$$\mathbf{r}^T = [Rr(t+1)\ldots Rr(t+N)] \tag{9.59}$$

$$\mathbf{G}^T = [G_1\ldots G_N] \tag{9.60}$$

$$\mathbf{\Gamma}^T = [\Gamma_1 + Q/S\ldots\Gamma_N + Q/S] \tag{9.61}$$

With this notation the cost function (Equation 9.47) has vector matrix form:

$$J(1, N) = E_t[\mathbf{\Phi}^T\mathbf{\Phi} + \lambda\Delta\mathbf{u}^T\Delta\mathbf{u}] \tag{9.62}$$

Minimization of $J(1, N)$ with respect to $\Delta\mathbf{u}$ gives the control law:

$$\Delta\mathbf{u} = (\mathbf{E}^T\mathbf{E} + \lambda\mathbf{I})^{-1}\mathbf{E}^T[\mathbf{r} - \mathbf{G}y(t) - \mathbf{\Gamma}\Delta u(t-1)] \tag{9.63}$$

This formula defines a vector of future controls between t and $t + \text{N} - 1$. However, only the first component $\Delta u(t)$ of the vector $\Delta\mathbf{u}$ is applied to the system. The remaining elements are discarded so that the control law can be written in the standard controller form:

$$F\Delta u(t) + Gy(t) - Nr(t+N) = 0 \tag{9.64}$$

The polynomials F, G, and N are computed from:

$$F = C + z^{-1}(\mathbf{k}^T \mathbf{\Gamma\prime}) + z^{-1}\lambda_N CQ \tag{9.65}$$

$$G = (\mathbf{k}^T \mathbf{G\prime}) + \lambda_N CS \tag{9.66}$$

$$N = RCK \tag{9.67}$$

and $\mathbf{k}^T$ is the first row of the matrix $(\mathbf{E}^T\mathbf{E} + \lambda\mathbf{I})^{-1}\mathbf{E}^T$, that is

$$\mathbf{k}^T = [k_1 k_2 \ldots k_N] \tag{9.68}$$

$$\lambda_N = \sum_{i=1}^{N} k_i \tag{9.69}$$

$$K = k_N + k_{N-1}z^{-1} + \ldots + k_N z^{-N+1} \tag{9.70}$$

$$\mathbf{\Gamma}^T = [\Gamma_1 \Gamma_2 \ldots \Gamma_N] \tag{9.71}$$

$$\mathbf{G}^T = [G_1 G_2 \ldots G_N] \tag{9.72}$$

Note that the control law (Equation 9.64) includes the *integral action* because the CARIMA model contains an integrator. This assures the elimination of the steady state disturbances.

The closed-loop system equation is obtained from Equations 9.12b and 9.64, and takes the form:

$$y(t) = \frac{BN}{\overline{A}F + z^{-1}BG} r(t + N - 1) + \frac{FC}{\overline{A}F + z^{-1}BG} e(t) \tag{9.73}$$

We select the characteristic closed-loop polynomial H so that

$$\overline{A}F + z^{-1}BG = CT \tag{9.74}$$

where the polynomial T defines the required closed-loop poles.

By some algebra with Equations 9.65–9.67, Equation 9.74 can be written as:

$$z^{-1}\lambda_N C(\overline{A}Q + BS) = C(T - \overline{A}) - z^{-1}\mathbf{k}^T(\overline{A}\mathbf{\Gamma}' + B\mathbf{G}') \tag{9.75}$$

Since $D = z(\overline{A} - T)$ and $\overline{A}\mathbf{\Gamma}' + B\mathbf{G}' = C\mathbf{X}$ (Wellstead and Zarrop, 1991), Equation 9.75 is simplified into the following equation:

$$\lambda_N(\overline{A}Q + BS) = D - (\mathbf{k}^T\mathbf{X}) \tag{9.75}$$

The unique solution of Equation 9.75 exists for Q and S if $\overline{A}$ and B are coprime and the degrees of the system and design polynomials satisfy the following conditions

$$\begin{aligned} \deg(Q) &= \max(0, n_b) \\ \deg(S) &= n_a) \\ n_t &\leq n_a + n_b + 1 \\ \deg(P) &\leq n_a + 1 \end{aligned} \tag{9.76}$$

An explicit GPP CARIMA self-tuning controller

Employing the certainty equivalence principle, the unknown A and B CARMA model polynomials are replaced by their estimated values $\hat{A}$ and $\hat{B}$. Estimation of these polynomials is performed from the model in incremental form as follows:

$$\hat{A}\Delta y(t) = \hat{B}\Delta u(t-1) + \hat{C}e(t) \tag{9.77}$$

The explicit GPP self-tuning algorithm for CARIMA system models is outlined below. It requires certain prior information concerning the system to be controlled. The orders of the polynomials A and B are assumed to be known. In addition, certain additional parameters are required in order to have the algorithm running:

(i) The polynomials P, R in the cost function (9.39).
(ii) The output horizon N and the control horizon NU.
(iii) The control weighting factor λ in the cost function (9.39).
(iv) The desired closed-loop pole set determined by the characteristic polynomial T. Note that the closed-loop pole set T determines the polynomials Q, S through the identity Equation 9.75.

The algorithm steps at each sample interval are:

Step 1: Estimate $\hat{A}$, $\hat{B}$, and $\hat{C}$ from (Equation 9.77). Compute $\hat{\bar{A}} = \Delta\hat{A}$.
Step 2: Compute D and $\mathbf{X}$ from the corresponding equations.
Step 3: Solve the polynomial equation

$$\bar{A}Q + z^{-1}BS = D - (\mathbf{k}^T\mathbf{X})$$

for the polynomials Q and S.
Step 4: Derive F, G, and N from Equations 9.65–9.67.
Step 5: Compute the control increment $\Delta u(t)$ from

$$F\Delta u(t) = Gy(t) - Nr(t+N)$$

Step 6: Apply the input $u(t) = u(t-1) + \Delta u(t)$ to the system and return to Step 1.

9.7 Concluding remarks

In this chapter we have presented the essence of the adaptive control techniques that can be used by researchers and practitioners dealing with automated highways and general problems of automatic control of moving vehicles. A brief survey of the corresponding results that appear in the journal papers during the last decade is given. The main adaptive control techniques are presented in sufficient details so that the corresponding part of the chapter can be used as a self-study guide to adaptive control for control engineers interested in applying the adaptive control theory results in practice.

References

Ackermann, J. (1997). 1996 Bode Prize Lecture: Robust control prevents car skidding. *IEEE Control Systems Magazine*, **17**, 23–31.

Alvarez, L., Horowitz, R. and Li, P. (1999). Traffic flow control in automated highway systems. *Control Engineering Practice*, **7**, 1071–8.

An, P. and Harris, C. (1996). An intelligent driver warning system for vehicle collision avoidance, *IEEE Transactions on Systems, Man, and Cybernetics–Part A: Systems and Humans*, **26**, 254–61.

Astrom, K.J. and Hagglund, T. (1995). *PID Controllers – Theory, Design, and Tuning*, 2nd ed., Instrument Society of America.

Astrom, K.J. and Wittenmark, B. (1973). On self-tuning regulators. *Automatica*, **9**, 185–89.

Astrom, K.J. and Wittenmark, B. (1980). Self-tuning regulator based on pole-zero placement. *IEEE Proceedings Part D: Control Theory and Applications*, **127**, 120–30.

Astrom, K.J. and Wittenmark, B, (1995). *Adaptive Control* 2nd edn., Addison Wesley.

Bakhtiari-Nejad, F. and Kamari-Mohammadi, A. (1998). Active vibration control of vehicles with elastic body using adaptive control. *Journal of Vibration and Control*, **4**, 463–79.

Blochl, B. and Tsinas, L. (1994). Automated road following using fuzzy control. *Control Engineering Practice*, **2**, 305–11.

Byrne, R., Abdallah, C. and Dorato, P. (1998). Experimental results in robust lateral control of highway vehicles. *IEEE Control Systems Magazine*, **18**, 70–76.

Camacho, Bordons, (1999). *Model Predictive Control*. Springer.

Chakroborty, P. and Kikuchi, S. (1999). Evaluation of General Motors based car-following models and proposed fuzzy inference model. *Transportation Research*, **C**, 209–35.

Chien, C., Zhang, Y. and Ioannou, P. (1997). Traffic density control for automated highway systems. *Automatica*, **33**, 1273–85.

Chu, K. (1974). Optimal decentralized control of a string of coupled systems. *IEEE Transactions on Automatic Control*, **13**, 243–46.

Clarke, D.W. and Gawtrop, P.G, (1975). Self-Tuning Controller. *IEEE Proceedings Part D: Control Theory and Applications*, **122**(9), 929–934.

Clarke, D.W., Mohtadi, C. and Tuffs, P.S. (1987). Generalized Predictive Control. *Automatica*, **23**, 137–60.

Cutler, C.R. and Ramaker, B.L. (1980). Dynamic Matrix Control – a computer control algorithm. *Joint Automatic Control Conference*, San Francisco, California.

De Keyser, R.M.C. and Van Cauwenberge, A.R. (1982). Typical application possibilities of self-tuning predictive control. *Proceeding of the 6th IFAC Symposium on Identification and System Parameter Estimation*, Washington, D.C., USA.

Fenton, R. and Bender, G. (1969). A study of automatic car following. *IEEE Transactions on Vehicular Technology*, **18**, 134–40.

Fritz, H. (1996). Neural speed control for autonomous road vehicle. *Control Engineering Practice*, **4**, 507–12.

Hauksdottir, A. and Fenton, R. (1985). On the design of a vehicle longitudinal controller. *IEEE Transactions on Vehicular Technology*, **34**, 182–87.

Huang, S. and Ren, W. (1997). Design for vehicle following control systems with actuator delays. *International Journal of Systems Science*, **28**(2), 145–51.

Holzmann H., Halfmann, Ch., Germann S., Würtenberger M. and Isermann R. (1997). Longitudinal and lateral control and supervision of autonomous electric vehicles. *Control Engineering Practice*, **5**(11), 1599–605.

Ioannou, P. and Chien, C. (1993). Autonompus intelligent cruise control. *IEEE Transactions on Vehicular Technology*, **42**, 657–72.

Ioannou, P.A. and Sun J. (1996). *Robust Adaptive Control.* Prentice Hall.

Ishida, A., Takada, M., Narazaki, K. and Ito, O. (1992). A self-tuning automotive cruise control system using the time delay controller. *SAE paper, No. 920159.*

Krstić M., Kanellakopoulos I. and Kokotović P. (1995). *Nonlinear and Adaptive Control Design.* John Wiley & Sons, Inc., New York.

Lelic, M. (1998). *Understanding PID Controllers: Algorithms, Implementation, and Tuning.* Corning Inc. Technical Report R-13998, Corning, New York.

Lelic, M. and Gajic, Z. (2000). Reference guide to PID control in nineties. *Proceedings of IFAC Workshop PID'00,* Terrasa, Spain.

Lelic, M.A. and Zarrop, M.B. (1987). Generalized pole placement self-tuning controller – Part 1: basic algorithm. *International Journal of Control*, **46**(2), 547–68.

Levine J. and Athans, M. (1966). On the optimal error regulation of a string of moving vehicles. *IEEE Transactions on Automatic Control*, **11**, 355–61.

Ljung, L. (1999). *System Identification – Theory for the User*, (2nd edn), Prentice Hall PTR.

Mosca, E., Zappa, G. and Manfredi, C. (1984). Multistep horizon self-tuning controller: the MUSMAR approach. *Proc. IFAC 9th World Congress*, Budapest, Hungary.

Murray, R. and Sastry, S. (1993). Nonholonomic motion planning: steering using sinusoids. *IEEE Transactions on Automatic Control*, **38**, 700–16.

Ng, G.W. (1997). *Application of Neural Networks to Adaptive Control of Nonlinear Systems.* UMIST Control Systems Centre Series, Research Studies Press, John Willey & Sons.

Omatu, S. Khalid, M. and Yusof, R., (1996). *Neuro-Control and its Applications.* Springer-Verlag, London.

Peppard, L. (1974). String stability of a relative motion PID vehicle control systems. *IEEE Transactions on Automatic Control*, **19**, 259–31.

Peterka, V. (1984). Predictor-based self-tuning control. *Automatica*, **20**, 39–50.

Pfefferl, J. and Färber, G. (1998). Applying load adaptive real-time algorithms to a vehicle control systems. *Control Engineering Practice*, **6**, 541–46.

Raza, H. and Ioannou, P. (1996). Vehicle following control design for automated highway systems. *IEEE Control Systems Magazine*, **16**, 43–60.

Raza, H., Xu, Z., Yang, B. and Ioannou, P.A. (1997). Modelling and control design for a computer-controlled brake system. *IEEE Transactions on Control System Technology* **5**(3), 279–96.

Richalet, J., Rault, A., Testud, J.L. and Papan, J. (1978). Model predictive heuristic control: application to industrial process. *Automatica*, **14**, 413–28.

Sheikholeslam, Desoer, S. (1992). A system level study of the longitudinal control of a platoon of vehicles. *ASME Journal of Dynamics, Measurements and Control*, **114**, 286–92.

Shladover, S., (1991). Longitudinal control of automotive vehicles in close-formation platoons. *ASME Journal of Dynamics, Measurements and Control*, **113**, 231–41.

Shoureshi, R. and Knurek, T. (1996). Automotive applications of a hybrid active noise vibration control. *IEEE Control Systems Magazine*, **16**, 72–78.

Sugisaka, M., Wang X. and Lee J. (1998). Intelligent control strategy for a mobile vehicle. *Applied Mathematics and Computation*, **91**, 91–98.

Swaroop, D. and Hedrick, J.K. (1999). Constant spacing strategies for platooning in automated highway systems. *Transactions of ASME*, **121**, 462–470.

Tagasaki, G. and Fenton, R. (1977). On the identification of vehicle longitudinal dynamics. *IEEE Transactions on Automatic Control*, **22**, 606–615.

Tsugawa, S., Aoki, M., Hosaka A. and Seki K. (1997). A survey of present IVHS activities in Japan. *Control Engineering Practice*, **5**, 11, 1591–597.

Unsal, C., Kachroo, P. and Bay, J. (1999). Multiple stochastic learning automata for vehicle path control in an automated highway system. *IEEE Transactions on Systems, Man, and Cybernetics – Part A: Systems and Humans*, **29**, 120–128.

Varaiya, P. (1993). Smart cars on smart roads. *IEEE Transactions on Automatic Control*, **38**, 195–207.

Wellstead, P.E., Prager, D.L. and Zanker, P.M. (1979). A pole-assignment self tuning regulator. *Proc. IEEE*, **126**(8), 781–87.

Wellstead, P.E. and Zarrop, M.B. (1991). *Self-Tuning Systems: Control and Signal Processing*, Wiley.

Zhang, Y., Kosmatopoulos, E.B., Ioannou, P.A and Chen, C.C. (1999). Autonomous intelligent cruise control using front and back information for tight vehicle following manoeuvers. *IEEE Trans. Vehicular Technology*, **48**(1), 319–28.

Ziegler, J. and Nichols, N. (1942). *Transactions of ASME*, **64**, 759–768.

Fuzzy control

Mark Hitchings, Ljubo Vlacic
Intelligent Control Systems Laboratory, School of Microelectronic Engineering, Griffith University, Brisbane, Australia

and

Vojislav Kecman
The University of Auckland, Auckland, New Zealand

This chapter gives a background on intelligent control techniques, neural networks and fuzzy logic, and explains how these techniques, fuzzy logic in particular, can be implemented in solving a distance and tracking control – two very important tasks of intelligent autonomous (driver-less) vehicles. The described fuzzy control system design process is derived from the authors' recent experience in prototyping the distance and tracking control system for cooperative autonomous vehicles.

10.1 Introduction

10.1.1 Intelligent control techniques

This chapter is about learning (from experimental data) and about transferring human knowledge into analytical models. These techniques are often named as intelligent ones. Performing such tasks belongs to soft computing too (see for example Gupta and Sinha, 2000; Kecman, 2000). Neural networks (NNs) and support vector machines (SVMs) are the mathematical structures (models) that stand behind the idea of learning, while fuzzy logic (FL) systems are aimed at embedding structured human knowledge into workable algorithms. However, there is no clear boundary between these two modelling approaches. The notions, basic ideas, fundamental approaches and concepts common to these two fields are introduced here, as well as the differences between them, and then are discussed in some detail. Also, at the very beginning it should be noted that there is no difference in representational capacity between the NNs and SVMs. In short, the structure and the mathematical model of NNs and SVMs are the same. However, there is a difference in learning. NNs learn by applying some form of gradient optimization while the learning of the weights in SVMs boils down to solving quadratic (or linear) programming problem. Very detailed presentation of SVMs can be found in Kecman (2000). Here we will discuss NNs and FLMs primarily as a basic intelligent tool for modelling nonlinearities.

Recently many new products, equipped with 'intelligent' controllers, have been launched on the market, a lot of efforts have been shown in R&D departments around the world and numerous papers have been written on how to apply NNs, SVMs and FL models, and related ideas of learning (in classic control terms of adaptation), for solving control problems of both linear and nonlinear systems. NN based algorithms have been recognized as attractive alternatives to the standard and well established adaptive control schemes. Due to their well known ability to be a universal approximator of multivariate functions, NN and FLM are of particular interest for controlling highly nonlinear, partially known and complex plants or processes. Many good and promising results have been reported and the whole field is developing rapidly but it is still in the initial phase. This is mainly due to a lack of firm theory despite the very good and seemingly far-reaching and important experimental results.

Before plunging into a description of basic adaptive control schemes for linear systems, and into presentation of NN and FLM based adaptive control structures, let us clarify the differences and similarities between the concepts of *learning* and *adaptation*. These two concepts form the basis of intelligent control systems today. This might be important because the present intelligent control is a mixture and an overlapping of classic *adaptive* control and standard NN based *learning* techniques.

In the simplest of colloquial terms we might say that learning is an adaptation from scratch, where scratch means ignorance of the system. Thus, learning usually means acquiring knowledge of a previously unknown system or concept, and adaptation connotes change or refinement of the existing knowledge of the system. This process of knowledge acquisition is naturally achieved through learning, which is, in the field of control, usually called adjustment or tuning. Historically, adaptive techniques originated from attempts to control *time-varying* and *linear* systems. NNs, SVMs and FLMs are by their very nature nonlinear modelling tools. Thus an alternative way to think about the learning may also be that it represents an adaptation of the nonlinear NN and/or FLM.

In recent years neural networks and fuzzy logic models have been used in many different fields. There is practically no area of human activity left untouched by NN and/or FLM. These two primary *modelling tools* are the two sides of the same coin. When and why one should try to solve a given problem by NNs (i.e., SVMs), and when the more appropriate tool would be FL model, depends upon availability of previous knowledge (expertise) about the system to be controlled, as well as upon the amount of measured process data. More about the answer to this question may be found in Table 10.1 that is self explanatory. In short, the less previous knowledge there is, the more neural and less fuzzy our model will be and vice versa. When applied in a system control area, neural networks can be regarded as *nonlinear identification* tools. This is the closest connection with the standard and well developed field of estimation or identification of linear control systems. If the problem at hand is a linear one, NN would degenerate into one single linear neuron and in that case the weights of the neuron correspond to the parameter of the plant's discrete transfer function $G(z)$.

In order to avoid too high an expectation of these new concepts in computing (and particularly after they have been connected with intelligence), as well as avoiding under-estimating these computing paradigms, it might be of use to mention the basic advantages

Table 10.1 Neural networks, support vector machines and fuzzy logic modelling as examples of modelling tools or approaches at extreme poles

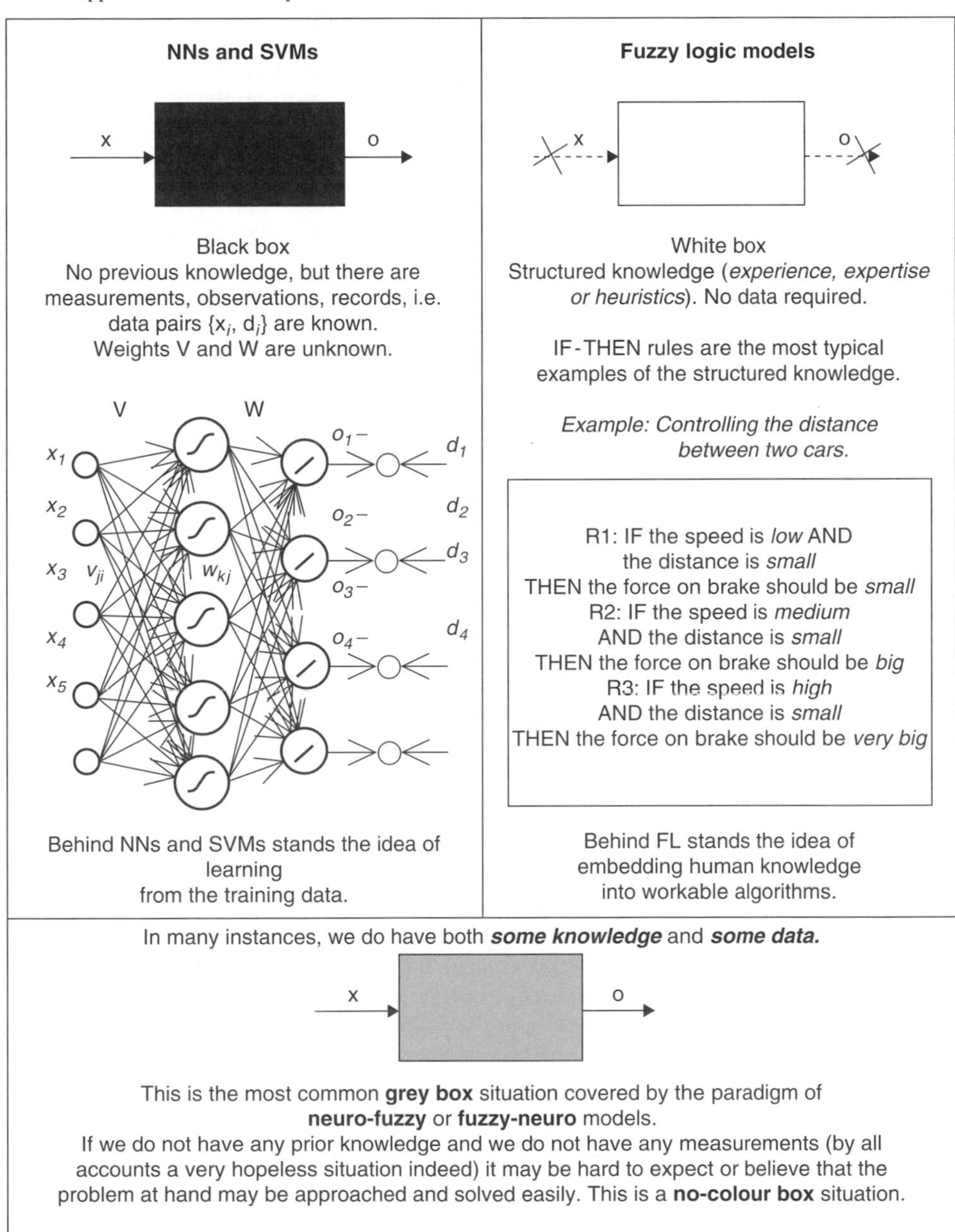

claimed for NN and FLM (Table 10.2) and the counter-claims advanced by their discouraged users or cautious beginners in the field (Table 10.3). Because of the wide range of applications of these two new modelling tools it is hard to disprove or object to some of these claims and counter-claims. It is certain that everyone working with NN and FLM will find his or her own answers to at least some or most of these claims.

Table 10.2 Claimed advantages of NN and FLM

Neural networks	Fuzzy logic models
(i) have the property of learning from the data mimicking human learning ability.	(i) are an efficient tool for embedding human (structured) knowledge into useful algorithms.
(ii) can approximate any multivariate nonlinear function.	(ii) are efficient for modelling of substantially nonlinear processes.
(iii) do not require deep understanding of the process or the problem being studied.	(iii) are applicable when a mathematical model is unknown or impossible to obtain.
(iv) are robust to the presence of noisy data.	(iv) operate successfully despite a lack of precise sensor information.
(v) have parallel structure and can be easily implemented in hardware.	(v) are useful at the higher levels of hierarchical control systems.
(vi) the same NN can cover a broad range of tasks.	(vi) are the appropriate tool in generic decision making processes.

Table 10.3 Claimed disadvantages of NN and FLM

Neural networks	Fuzzy logic models
(i) Need extremely long training or learning time (problems with local minima or multiple solutions) with little hope of many real time applications.	(i) There must be human solutions to the problem and this knowledge must be structured. Experts may have problems with structuring the knowledge.
(ii) do not uncover basic internal relatins of physical variables and do not increase our knowledge of the process under consideration.	(ii) Experts sway between extreme poles–too much aware of their knowledge.
(iii) are prone to bad generalization. (When a large number of weights are present, there is a tendency to overfit the data with poor performance on previously unseen data during the test phase.)	(iii) The number of rules increases exponentially with the increase in the number of inputs, as well as the number of fuzzy subsets per input variable.
(iv) There is very little or no guidance at all about the NN structure, optimization procedure or even the type of NN for a particular problem.	(iv) Learning (changing of membership functions' shapes and positions and/or the rules) is highly constrained and typically a more complex task than in NN.

But the fast growing number of companies and products (software and hardware) in the NN and FLM fields with applications in almost all areas of human activity and the growing number of new neural and fuzzy computing theories and paradigms, are proof that despite the many still open questions and problems, NN and FLM are already well established engineering tools which are overcoming their 'teething problems' and are becoming common computational means for many everyday real life tasks and problems.

10.1.2 Distance and tracking control – problem definition

Distance control for autonomous vehicles relates to the automatic maintenance of the longitudinal distance between two successive vehicles travelling in the same direction in the same environment. It is the goal of the vehicles to maintain the distance without intervention from external supervisory systems or humans (Hitchings, 1999).

Tracking control for autonomous vehicles relates to the automatic maintenance of coincident trajectories of two successive vehicles travelling in the same direction in the same environment. It is the goal of the vehicles, using lateral control systems, to maintain their coincident paths automatically without intervention from external supervisory systems or humans (Hitchings, 1999).

Distance control and tracking control are separate control problems, however the goal of distance and tracking control is to automatically control both parameters under the one control scheme. By performing both distance and tracking control the solution becomes a multi-variable control system. It is important to note that the intelligent vehicles are defined as *autonomous*, which means they require no intervention from humans or external supervisory systems. In this context *autonomous* refers to the fact that the vehicles are capable of *perception*, *cognition* and *action* (Vlacic *et al.*, 1998). *Perception* is achieved using sensory systems such as distance and tracking sensors. *Cognition* refers to the ability of the vehicles for intelligent decision making which is the output of the control system on board the vehicle. The *action* is the response of the system as a result of the perception and cognition. All hardware and software to run the systems must reside on board the vehicles. The vehicles must also be travelling in the same environment. In the application of autonomous vehicles, the *environment* is typically a restricted pathway or lane in which the vehicle must follow a set of rules relating to its navigation. These rules may include:

- The vehicle must maintain its trajectory within boundaries established to define the environment.
- The direction of travel of the vehicles is usually fixed.
- The speed of the vehicles is restricted to an upper limit.
- The vehicle must not collide with other vehicles or the environment boundaries.

The last point suggests that the vehicles must have a high level of autonomy in order to successfully detect and avoid collisions with vehicles and other objects in the environment. The sensor systems required for this autonomy typically consume considerable computing resources in order to successfully implement the control strategy and may be beyond the scope of solutions with cost restrictions.

The minimum longitudinal distance, d_{min} that may be safely maintained between road vehicles is given by Palmquist (1993):

$$d_{min} = vT + v^2/2r \qquad (10.1)$$

where: d_{min} is the minimum longitudinal distance between the vehicles.
v is the individual vehicle velocity.
T is reaction time of a human driver.
The term $v^2/2r$ is the braking distance required when applying a retardation value r.

The stopping distance for vehicles under human control on typical roads ranges between 150 and 300 metres (Palmquist, 1993). Clearly, removing the driver reaction time will reduce the stopping distance and thus decrease the distance, *d* between the vehicles required. This reduction will result in an increase in road traffic capacity. Removal of driver reaction time is only possible through automation of distance control of the vehicle.

This chapter will discuss the automation of distance and tracking control of a vehicle.

A solution to the distance and tracking control problem for autonomous vehicles consists of the following:

1. Distance sensors to determine the distance between vehicles.
2. Tracking error sensors for the following vehicle/s to assume a position in line with the trajectory of the leading vehicle.
3. A method of determining the speed of the vehicle/s.
4. Vehicle motion control hardware and algorithm resident in software to control – both speed and direction of travel.
5. On board computing device such as a microprocessor or microcontroller.
6. Distance and tracking control algorithm resident in software.
7. Communications link between the vehicles.
8. Integration with other vehicle behaviours.

During the last ten years, an autonomous, driver-less motion of the vehicles has become a very popular and challenging research area. For example, odometric vehicle positioning and traced control using communication is discussed in Tsumura *et al.* (1992); fuzzy control of distance in autonomous vehicle applications is addressed in Dickerson *et al.* (1997) and Holve and Protzel (1995a); conventional (PI, PID etc.) control of distance in autonomous vehicle applications is discussed in Raza and Ioannou (1996), Palmquist (1993) and Zielke *et al.* (1992); a comparison between PID and fuzzy control of distance for autonomous vehicle applications is given in Tso *et al.* (1994) and Schone and Sun (1995); general issues relating to distance control and platooning of intelligent vehicles are given in Chira-Chavala and Yoo (1993), Variaya (1993) and Tso (1994); sensors for the measurement of distance and tracking applications are discussed in Grosch (1995), Rohling and Lissel (1995), Everett (1995), Borenstein (1996) and Yamaguchi (1996). A theory to underpin operation of cooperative vehicles can be found in Vlacic *et al.* (2000).

10.2 Fuzzy control systems – theoretical background

10.2.1 Overview

This section will first provide an overview of fuzzy logic-based control systems and the standard additive model for fuzzy systems. Following this, sections detailing fuzzification, fuzzy inference and defuzzification methods commonly used in fuzzy control systems will be provided. Finally a brief overview of the fuzzy control system design process will complete this section.

Fuzzy logic enables the designer to describe the desired system behaviour using descriptive language. Fuzzy logic also allows the designer to express a certain level

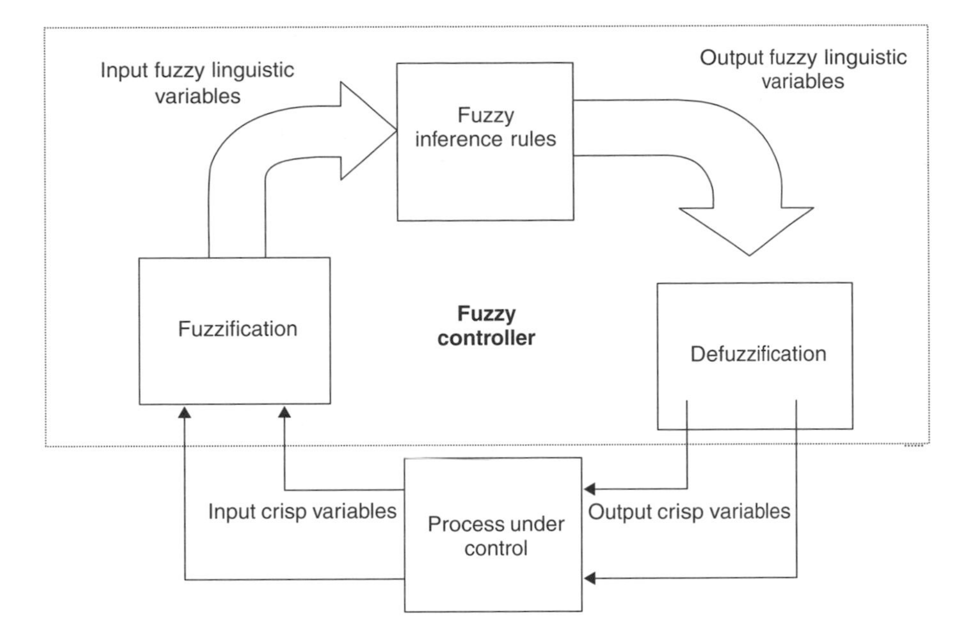

Fig. 10.1 Typical fuzzy controller structure.

of vagueness associated with an item under consideration. Fuzzy control systems take real input values, such as those obtained from sensor measurements and associate them with language descriptions of their magnitude. The fuzzy control system then assigns a language output description based on the input using if-then type rules. Finally the system takes the descriptive output and converts it back into a real output value as a control signal. A typical fuzzy control system structure is shown in Figure 10.1.

Fuzzy logic is gaining widespread use by control engineers for the following general reasons (von Altrock, 1995):

- Fuzzy control systems may be designed to match any set of input to output data. This makes it ideal for nonlinear, multi-variable control systems that are difficult to control using conventional linear control techniques like PID.
- Fuzzy logic techniques offer flexibility to the designer, whereby entire fuzzy systems may be changed quickly and easily.
- Fuzzy logic allows the designer the freedom to design complex systems modelled on natural human thought. This allows the encoding of expert systems of control.
- Imprecise data is tolerated by fuzzy systems.
- Conventional control techniques may be embedded into a fuzzy control system.

Inputs to a fuzzy control system are described using *linguistic variables*. A linguistic variable is a grouping of *terms*, i.e. the attributes, that associate a degree of membership (DoM) of a real, *crisp* value to the linguistic variable (Zadeh, 1996). A typical linguistic variable is expressed as:

$$\textbf{Linguistic\_Variable}\ (\text{term } 1, \text{term } 2, \ldots \text{term } n) \qquad (10.2)$$

where: n is the number of terms in the linguistic variable.

A crisp value in a fuzzy control system may be an input or output physical variable such as input sensor readings or output voltage levels. The degree of membership μ, to which a crisp value belongs to a linguistic variable, is simply an expression between zero and one of how true the membership is. A DoM of zero means the crisp value has no membership and a DoM of one means the crisp value fully belongs to a term.

The fuzzification stage takes the crisp values from the process under control and maps them to the terms by assigning a degree of membership, μ. A membership basis function (MBF), which defines each of the terms in a linguistic variable, is used for this mapping. The input crisp values are usually obtained from sensor readings and the designer defines the scope of the inputs and the shape of the MBFs. An example MBF is shown in Figure 10.2.

The input crisp value x, is mapped to the linguistic variable as follows:

$$\mu_{MFn}(x) \in [0, 1] \qquad x \in X$$

where: x is the input, crisp value.
X is called the *universe* and is the range of all possible values the crisp input may take.

Using the membership functions in Figure 3.2 as an example, the input $x = 20$ will yield:

$\mu_{MF1}(20) = 0$
$\mu_{MF2}(20) = 0.8$
$\mu_{MF3}(20) = 0.2$
$\mu_{MF4}(20) = 0$ and so on ....

It can be seen that it is possible to have more than one non-zero term in a linguistic variable. The fuzzy inference rules map fuzzy linguistic inputs to fuzzy linguistic

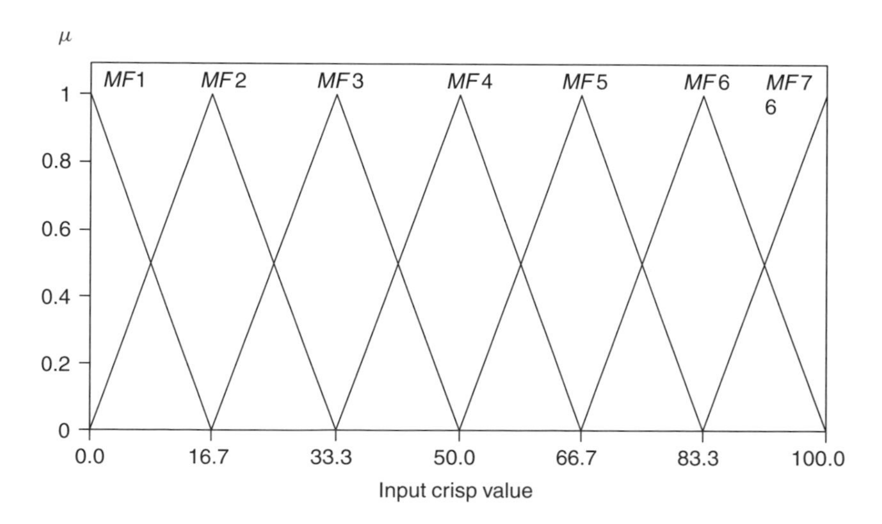

Fig. 10.2 Example fuzzification membership basis function.

outputs. An example fuzzy inference rule is:

$$\textbf{If DISTANCE is FAR then SPEED is FAST} \tag{10.3}$$

where: DISTANCE is the input linguistic variable.
FAR is a term, i.e. the attribute, of DISTANCE.
SPEED is the output linguistic variable.
FAST is a term, i.e. the attribute, of SPEED

The fuzzy inference rules usually take the form of a set of rules that describe the complete system behaviour. The IF part of a rule is called the *aggregation* and the THEN part is called the *composition*. A rule that has a non-zero output degree of truth is termed as having *fired*.

Defuzzification takes the output of the fuzzy inference rules and converts them into crisp values to send to the process under control. Due to the fact that more than one term in each input fuzzy variable may be non-zero, it is possible for more than one rule from the inference stage to fire. The defuzzification stage must resolve the contributions from all the fired rules and make a compromise between, or plausible resolution of, the results of these rules. There are several methods of defuzzification, which will be discussed in later sections. Defuzzification applies the reverse process to fuzzification – as the name suggests. The fuzzy output from the inference rules is mapped from a fuzzy, ambiguous quantity to a crisp, real valued output. A membership function similar to that used in Figure 10.2 may also be used for this purpose.

10.2.2 Additive fuzzy systems – the standard additive model

Additive fuzzy systems add the then-parts of the if-then rules when the rules fire (Kosko, 1997; Kecman, 2000). An additive fuzzy system F, stores m fuzzy rules of the form: 'If $X = A_j$ then $Y = B_j$' and computes the output $F(x)$ as the centroid of the summed and partially fired then part fuzzy sets B'_j. Each real input x fires all of the rules to some degree (most to zero degree) and in parallel. The system then assigns weights to the then-parts to give the new fuzzy sets B'_j and sums these to form the output set B. The system defuzzifies B (maps it to a real value) to give the output $y = F(x)$ by taking the centroid of B or by using other means which will be detailed later. This concept may be expressed as a structure diagram shown in Figure 10.3 (Kosko, 1997).

The rule weights w_j scale each rule term in the sum to reflect rule creditability, frequency or 'usuality'. These rule weights may be used by a learning fuzzy system for tuning and are, in many non-learning cases (including the system presented in this chapter), all treated as equal and unity where:

$$w_1 = w_2 = w_3 = \ldots = w_m = 1 \tag{10.4}$$

The sum is taken over m rules that map fuzzy subsets $A_j \subset R^n$ to fuzzy subsets $B_j \subset R^p$ with R^n the set of real inputs and R^p the set of real outputs. The *fit* (fuzzy unit) value $a_j(x)$ states the degree to which the input x belongs to the if-part fuzzy set A_j. The additive fuzzy system $F{:}R^n \to R^p$ acts as an associative processor or fuzzy

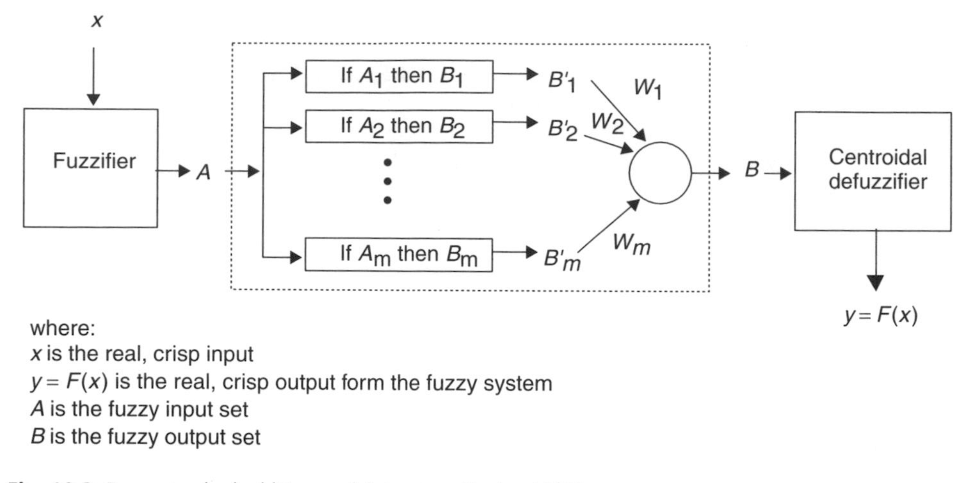

Fig. 10.3 Fuzzy standard additive model structure (Kosko, 1997).

associative mapping (FAM):

$$
\begin{aligned}
a_j &: R^n \longrightarrow [0, 1] \\
b_j &: R^p \longrightarrow [0, 1]
\end{aligned}
\tag{10.5}
$$

The set function a_j is often termed the membership function (MBF).

Each fuzzy rule defines a fuzzy 'patch' or subset of the input-output state space $X \times Y$. The additive fuzzy system $F{:}X \rightarrow Y$ approximates a function $f{:}X \rightarrow Y$ by covering its graph with patches and averaging (adding) those patches that overlap (Kosko, 1997). Figure 10.4 shows an example of the patch defined by the rule:

$$\text{If } X = \text{IN}_4 \text{ then } Y = \text{OUT}_2 \tag{10.6}$$

All fuzzy systems suffer from the 'curse of dimensionality'. This simply means that in order to approximate most functions completely, they require a large set of rules. The number of rules grows exponentially as the dimensions of X and Y increases.

The SAM theorem (Kosko, 1997) states that if a fuzzy system $F{:}R^n \rightarrow R^p$ is a standard additive model (SAM):

$$
\begin{aligned}
F(x) &= \text{centroid}(B) = \text{Centroid} \\
&\left(\sum_{j=1}^{m} wj \times aj(x) \times Bj \right)
\end{aligned}
\tag{10.7}
$$

where: w_j is the weights of the rules.
$a_j(x)$ is the degree that the input vector $x \in X$ belongs to the fuzzy set $A_j \subset R^n$ (also called the membership function).
B_j is the then-part fuzzy set.

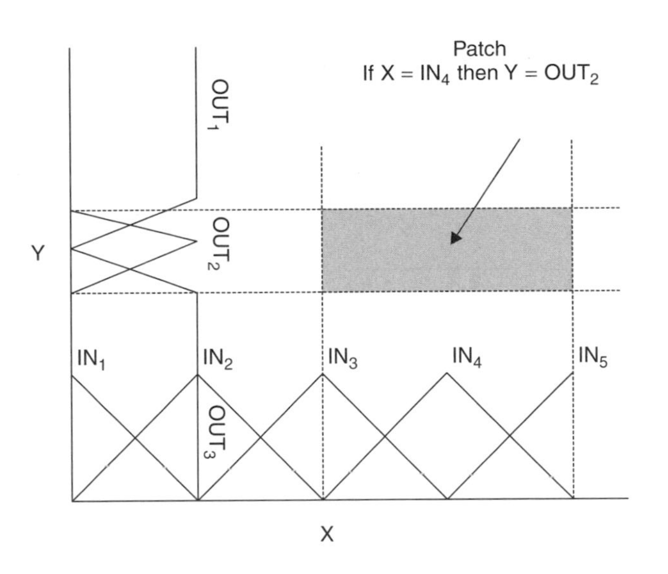

Fig. 10.4 A fuzzy rule as a state space patch.

Then $F(x)$ is a convex sum of the m then part set centroids:

$$F(x) = \frac{\sum_{j=1}^{m} w_j \times a_j(x) \times V_j \times C_j}{\sum_{j=1}^{m} w_j \times a_j(x) \times V_j} \tag{10.8}$$

where: V_j is the volume
C_j is the centroid of the then-part fuzzy set B_j.

If the then-part sets B_j all have the same volumes V_j and rule weights w_j and the centroids c_j coincide with the peaks P_j of the sets B_j then the SAM model above simplifies to the centre of gravity (COG) fuzzy model:

$$F(x) = \frac{\sum_{j=1}^{m} a_j(x) \times P_j}{\sum_{j=1}^{m} a_j(x)} \tag{10.9}$$

The COG/SAM model is used in the fuzzy controller for distance and tracking control presented in this chapter.

10.2.3 Fuzzification methods

All fuzzification methods determine the degree of membership (DoM) a real, crisp input has to a linguistic variable. For a continuous variable, the DoM is expressed by a

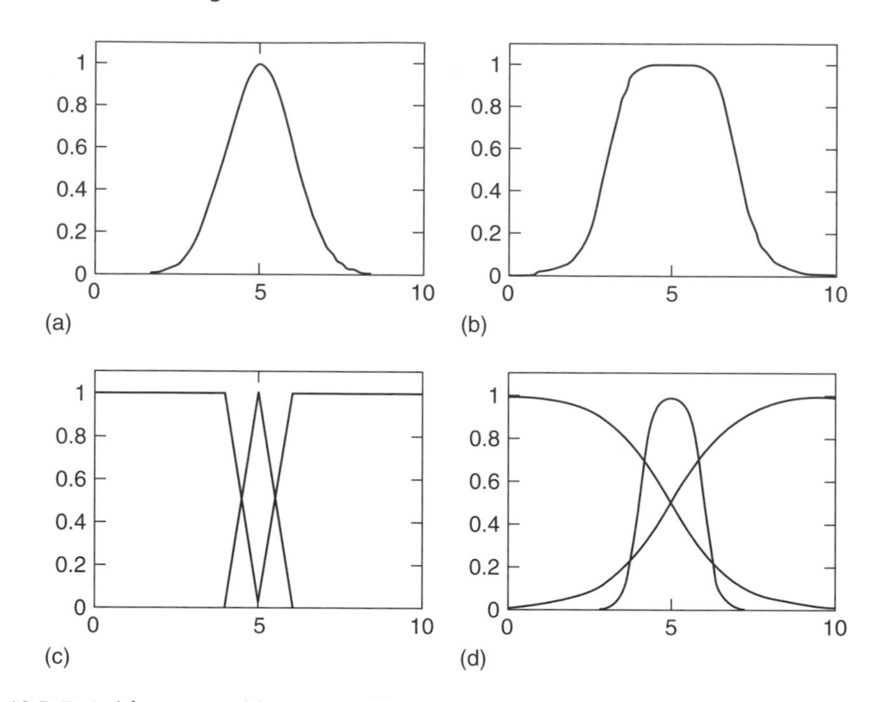

Fig. 10.5 Typical fuzzy MBFs: (a) Gaussian; (b) Bell; (c) Triangular (lambda), *Z*-type and *S*-type trapezoidal; (d) Sigmoidal and differential sigmoidal.

mathematical function called a membership function (MBF). The membership functions map each value of crisp inputs (x) to the membership degree (μ) in linguistic terms. Usually membership functions for all terms in a linguistic variable are expressed in one diagram – an example of which is given in Figure 10.2. The shape of the membership functions is many and varied – as is the reasoning behind their use. Examples of some typically used MBFs are given in Figure 10.5.

The expressions for the triangular (also called *lambda-type*), Z-type and S-type trapezoidal functions are derived in a later sections of this chapter. The general normalized expressions for each of the remaining functions in Figure 10.5 are as used in (Mathworks, 1998):

Gaussian:

$$\mu(x) = e^{[-(x-c)^2]/2\times\varepsilon^2} \tag{10.10}$$

where: ε is the width.
c is the centre.
x is the input value.

General Bell function (extension of the Cauchy probability distribution function):

$$\mu(x) = \frac{1}{1 + \left|\left(\frac{(x-c)}{a}\right) 2b\right|} \tag{10.11}$$

where: a is the width.
b is the rise of the slope.
c is the centre.
x is the input value.

Sigmoid curve membership function:

$$\mu(x) = \frac{1}{1 + e^{-a \times (x-c)}} \tag{10.12}$$

where: a is the rise of the slope.
c is the mid-point of the rise.
x is the input value.

Membership functions are intended to approximate the way humans linguistically interpret real values. Psycholinguistic studies (von Altrock, 1995) have shown that membership functions should follow a set of axioms:

Axiom 1: $\mu(x)$ is continuous over X.
Axiom 2: $d(\mu(x))/\mathrm{d}x$ is continuous over X.
Axiom 3: $d^2(\mu(x))/\mathrm{d}x^2$ is continuous over X.
Axiom 4: $\min \mu(\max_x(\mathrm{d}^2(\mu(x))/\mathrm{d}x^2))$ is continuous over X.

where: μ is the degree of membership and X is the universe of the base input variable x.

10.2.4 Fuzzy inference methods

The fuzzy inference rules map fuzzy linguistic inputs to fuzzy linguistic outputs. The inference rules consist of two parts:

- the aggregation which resolves multiple pre-conditions of a rule into one degree of truth using fuzzy logic operators;
- the composition part which assigns an outcome for each rule fired.

The aggregation and composition parts are discussed separately in this section. There are two commonly used types of fuzzy inference systems – Mamdani and Sugeno type systems. The Mamdani-type system is the most commonly used in control applications and is used in the application of fuzzy control to the distance and tracking control problem presented in this chapter. The Mamdani method expects the output membership functions to be fuzzy sets (von Altrock, 1995). After the aggregation process there is a fuzzy set output variable that requires defuzzification to output a crisp value from the system. It is possible, and in most cases more efficient, to use a single spike as the output membership function rather than use a distributed fuzzy set. This is known as a Singleton output membership function and it can be thought of as a pre-defuzzified fuzzy set. The Sugeno method, rather than integrating across the two-dimensional function to find a centroid, uses a weighted average of a few data points. In general Sugeno-type systems are used to model inference systems which have linear or constant membership functions (Mathworks, 1998).

Fuzzy inference aggregation

Each fuzzy inference rule may have multiple pre-conditions (antecedents) which must be resolved into one degree of truth for each rule. Examples of multiple antecedent rules are:

IF distance is far ***AND*** *Alignment is Left THEN ....*

IF height is tall ***OR*** *length is as long THEN ....* (10.13)

It can be seen that there are two pre-conditions in the rules above which may be resolved by using the AND, OR and NOT operators. It must be noted that the fuzzy AND, OR and NOT operators differ from their Boolean counterparts. Boolean operators cannot cope with degrees of truth to set membership and therefore cannot be applied to fuzzy operations. It has been proposed (von Altrock, 1995; Mathworks, 1998) that the following fuzzy operators be used in place of the standard Boolean operators:

Operator	***Fuzzy equivalent***	
AND	$\mu_{A\wedge B} = \min\{\mu_A, \mu_B\}$	(14)
OR	$\mu_{A\vee B} = \max\{\mu_A, \mu_B\}$	
NOT	$\mu_{\neg A} = 1 - \mu_A$	

There are several other methods that have been proposed for fuzzy AND, OR and NOT operators but these methods are not detailed here. Any fuzzy operator used must conform to certain properties in order for it to be a valid method. The rules for the fuzzy AND and OR operators will now be given.

The intersection (fuzzy AND) of two fuzzy sets A and B is specified in general by a binary mapping T, which aggregates two membership functions as follows (Mathworks, 1998):

$$\mu_{A\cap B} = T(\mu_A, \mu_B)$$

These fuzzy intersection operators are usually referred to as *T-norm* (triangular-norm) operators and must conform to certain requirements.
A *T-norm* operator is a binary mapping $T(.,.)$ satisfying:

- *boundary*: $T(0, 0) = 0,\ T(a, 1) = T(1, a) = a$. This requirement ensures that the T-norm operator is a correct generalization of crisp sets.
- *monoticity*: $T(a, b) \Leftarrow T(c, d)$ if $a \Leftarrow c$ and $b \Leftarrow d$. This requires that an increase in the membership values of A or B cannot produce and increase in the membership value of $A \cap B$.
- *commutativity*: $T(a, b) = T(b, a)$. The operator must be indifferent to the order of the fuzzy sets to be combined.
- *associativity*: $T(a, T(b, c)) = T(T(a, b), c)$. The ordering of the intersection must be able to be done in any pair-wise groupings.

The union (fuzzy OR) of two fuzzy sets A and B is specified in general by a binary mapping S, which aggregates two membership functions as follows (Mathworks, 1998):

$$\mu_{A\cup B} = S(\mu_A, \mu_B)$$

These fuzzy union operators are usually referred to as *T-conorm* (triangular-conorm) or *S-norm* operators and must conform to certain requirements.

A *T-conorm* or *S-norm* operator is a binary mapping $S(.,.)$ satisfying:

- *boundary*: $S(1, 1) = 1$, $S(a, 0) = S(0, a) = a$. This requirement ensures that the *S-norm* operator is a correct generalization of crisp sets.
- *monoticity*: $S(a, b) \Leftarrow S(c, d)$ if $a \Leftarrow c$ and $b \Leftarrow d$. This requires that an increase in the membership values of A or B cannot produce an increase in the membership value of $A \cup B$.
- *commutativity*: $S(a, b) = S(b, a)$. The operator must be indifferent to the order of the fuzzy sets to be combined.
- *associativity*: $S(a, S(b, c)) = S(S(a, b), c)$. The ordering of the union must be able to be done in any pair-wise groupings.

The fuzzy inference method chosen for the distance and tracking control was MAX/MIN. This is so called due to the aggregation part being computed using the MIN (fuzzy AND) operator and the composition part being computed using the MAX (fuzzy OR) operator. Another method called MAX/PROD uses the MAX operator for the composition part and multiplies (PROD) the degrees of truth of the inputs to the IF-part in the aggregation.

Fuzzy inference composition

The composition is the second step to fuzzy rule inference and links the validity of a rule of the entire condition with the degree of support (DoS). The DoS is the weighting of the rule w_j, which scales the rule according to a desired 'importance' of its outcome. Learning systems use this method of rule weighting in order to adjust the contribution the rule makes to the whole solution. The contribution each rule makes to an overall solution, may be expressed as a two-dimensional matrix called a fuzzy associative map (FAM). This map shows the degree of support each rule has for a given input pre-condition. Figure 10.6 shows an example of a fuzzy associative map. The matrix squares depicted in black represent that the rule has a zero degree of support (DoS) – meaning the rule has no effect. Various shades of grey from black to white represent an increasing degree of support up to the point where white is full ($\mu = 1$) support.

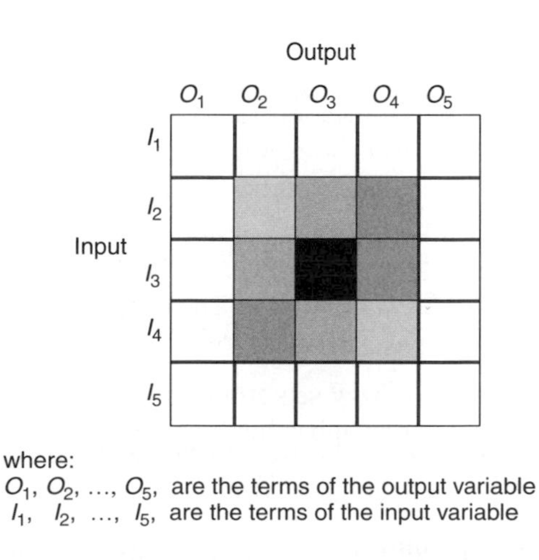

Fig. 10.6 Hypothetical fuzzy associative map (FAM).

In general, non-learning MAX/MIN or MAX/PROD methods do not weight the rules and treat them all as unity and equal. A particular outcome is not unique to each rule. Consider the two rules below:

$$Rule_1 : \textit{IF Distance is OK AND Alignment is Far\_Left then LWS is Fast and RWS is Slow}$$

$$Rule_2 : \textit{IF Distance is Close AND Alignment is Left then LWS is Fast and RWS is Slow} \quad (10.15)$$

The same outcome (*LWS is Fast and RWS is Slow*) is desired for both rules. If both of these rules fire to some non-zero degree, the outcome must be resolved into one fuzzy set. In MAX/MIN and MAX/PROD systems, fuzzy rules are defined alternatively – either $Rule_1$ is true OR $Rule_2$ is true OR ... OR $Rule_n$ is true. The MAX operator in both cases is used to resolve one fuzzy value for the same outcome. As an example, consider the case where $Rule_1$ is true to a degree of $\mu = 0.6$ and $Rule_2$ is true to a degree of $\mu = 0.1$. The fuzzy OR operator (MAX) will determine that the overall output, *LWS is Fast and RWS is Slow* is true to:

$$\max\{\mu_1, \mu_2\} = \max\{0.6, 0.1\} = 0.6$$

10.2.5 Defuzzification methods

The input to the defuzzification stage is a fuzzy set B and the output from the defuzzification stage is a single crisp value $y = F(x)$ which is a function of the fuzzy system. The fuzzy set consists of one or more non-zero values that are generally combined and defuzzified using one of the following methods:

- Centroidal – centre of area or gravity (CoA or CoG)
- Bisector or centre of maximum (CoM)
- Mean of maximum (MoM)
- Left of maximum (LoM)
- Right of maximum (RoM).

The fuzzy set input to a defuzzification stage may be interpreted as a number of non-zero degrees of truth (DoT) of the membership function for the output variable. The DoT for each term in the variable can be expressed in graphical format as shown in Figure 10.7.

It can be seen that the terms MF3 and MF4 are non-zero outputs from the inference stage where:

$$\mu_{MF3} = 0.2$$

$$\mu_{MF3} = 0.8$$

There are, in general two methods used to defuzzify a linguistic variable (von Altrock, 1995):

- Determine the 'best compromise'.
- Determine the 'most plausible result'.

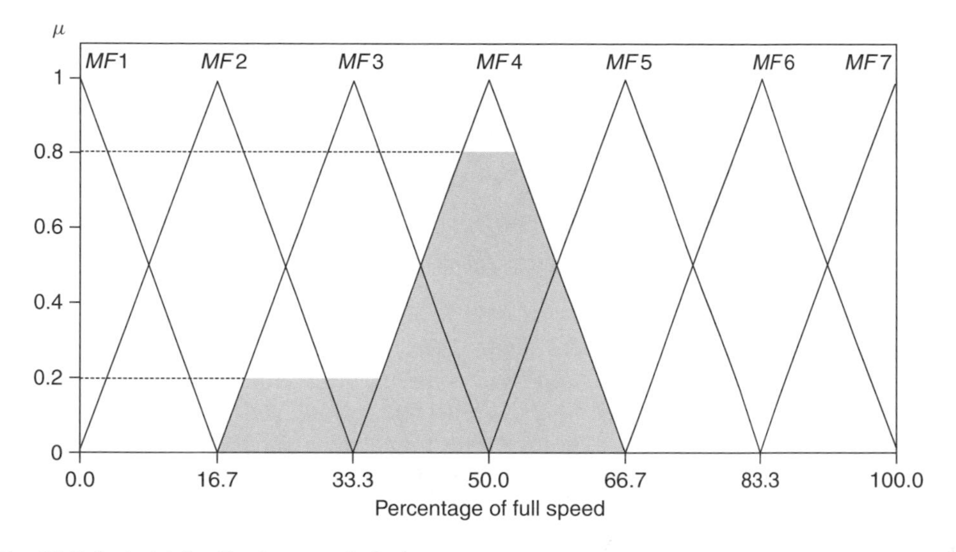

Fig. 10.7 Typical defuzzification scenario for fuzzy output set.

The 'best-compromise' method selects a typical value for each non-zero term and forms a compromise between them in order to obtain a single crisp value. Examples of the 'best compromise' method are CoM, CoG and CoA. The 'most plausible result' method of defuzzification selects the typical value of the term that is most valid (has the largest DoT). LoM, RoM and MoM are examples of 'most plausible result' methods. Each of the defuzzification methods mentioned earlier will now be discussed. Figure 10.8 shows the graphical interpretations of each method.

LoM, RoM methods are derived from the mean of maximum (MoM) method and simply select the middle, left or right of the term that has the greatest degree of truth associated with it. The centre of maximum (CoM) method selects the 'best compromise' between the non-zero terms. The maximum or centre of each non-zero term is firstly resolved to their crisp 'typical' value. A 'weight' equal to the DoT of the term is assigned to each typical value. The balance between these weighted typical values is found by:

$$y(x) = \frac{\sum_{j=1}^{m} \mu_{MFj} \times t_j}{\sum_{j=1}^{m} \mu_{MFj}} \tag{10.16}$$

where: $y(x)$ is the defuzzified value.
μ_{MFj} is the degree of truth of the jth term.
t_j is the typical value of the jth term.

The centre of area (CoA) defuzzification method proceeds as follows:

- Each term is cut at the degree of truth of the respective term. This is shown in Figure 10.7 as the lines horizontally truncating the non-zero terms.
- The areas under each of the resulting functions of all terms are then superimposed.

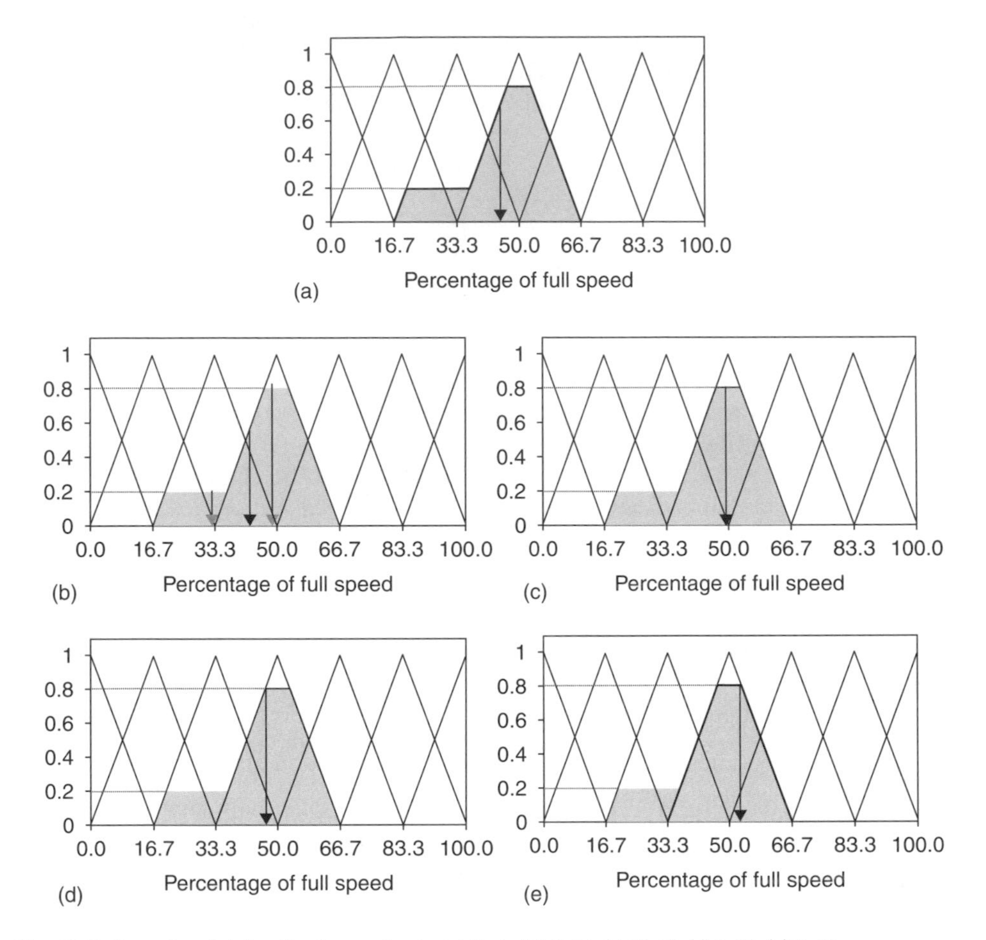

Fig. 10.8 Examples of defuzzification methods: (a) CoA; (b) CoM; (c) MoM; (d) LoM; (e) RoM.

- The area under the total function created by the superimposed, truncated terms is found and a centre of balance (or gravity) for this area is found and is the resulting crisp value.

Of the methods mentioned above, the MoM method is the most computationally efficient; however, it presents a problem of discontinuity of its output (von Altrock, 1995). The continuity property is important for most closed loop control applications. If the output of a fuzzy control system directly controls a variable of the process, jumps in the output variable of a fuzzy controller can cause instabilities and oscillations. Hence, most closed loop control systems use CoM defuzzification. Only when the output from a fuzzy controller is passed to an integrator before output, is a MoM system a viable alternative. The CoA/CoG method of defuzzification is the most computationally expensive method. The sheer number of multiplications and divisions required to perform the CoA method make it difficult to implement in small microprocessor systems. A fast CoA algorithm has been developed which improves its computational efficiency, however this method still requires more computational resources than the other methods mentioned here. The CoM method offers a slight trade-off between computational efficiency and accuracy of the output in terms of the 'best compromise' value. For this reason it is recommended for closed loop control systems to be implemented on microprocessor/microcontroller – based systems.

10.3 Fuzzy control systems – design steps

To successfully design and implement a fuzzy control system the following steps are recommended:

1. System design
   - the functions and requirements of the system to be defined
   - the linguistic variables of the system to be created
   - the structure of the system to be designed
   - the control strategy as the fuzzy inference rules are to be defined
   - an appropriate defuzzification method is to be selected and carried out.
2. Off-line optimization
   This step typically consists of simulation methods with real process data, mathematical modelling, transfer characteristics analysis and/or time response analysis.
3. On-line optimization
   On-line optimization may be an alternative or in addition to off-line optimization. Circuit emulation techniques that support on-the-fly modifications on a running system are typically used in this step. Due to a lack of availability of circuit emulation tools for the chosen test-bed, the on-line optimization stage was not used in the solution presented in this chapter.
4. Implementation
   After completion the fuzzy logic system was implemented on the target hardware platform. This involved converting the system design into assembler code.
5. Testing and verification
   The implemented system was tested against requirements and verified for correct operation.

Section 10.4 will detail the implementation issues associated with applying fuzzy control methods to distance and tracking control for autonomous vehicles. The section will also outline the design process followed during the design of the fuzzy control system that was described in Hitchings (1999).

10.4 Fuzzy control of distance and tracking

This section discusses issues and methods relevant to the application of fuzzy control to the distance and tracking control problem for autonomous vehicles. For the benefit of readers, the issues that are related to the design and development of fuzzy control systems are addressed by way of describing the real design and development process that took place in conjunction with the prototyping of the distance and tracking control algorithm for cooperative autonomous vehicles. The developed prototype has been tested in experimenting with five laboratory-scaled test vehicles, autonomous mobile robots – Figure 10.9. The details can be found in Hitchings (1999). Once implemented,

Fig. 10.9 The ICSL cooperative mobile robots.

the distance and tracking control system increases both the traffic safety and traffic capacity of pre-existing road infrastructure.

After some design considerations are identified, the design process undertaken to develop the fuzzy controller for distance and tracking control is described. Next, the system requirements and functions are identified. The fuzzification, fuzzy inference and defuzzification methods used in the fuzzy-control based solution are then detailed.

10.4.1 Considerations

It was considered important to identify the existing hardware test-bed constraints prior to designing an upgrade distance and tracking control function. Several practical system considerations were identified as having a limiting effect on the implementation of a fuzzy control scheme for distance and tracking control on the test-beds. These limitations are imposed on the design by physical constraints in hardware and are briefly described here. Among these considerations are:

1. Distance measurement update time. An existing time-of-flight ultrasonic ranging subsystem interface is used to calculate the distance from the following vehicle to the

leader. The time between obtaining sensor readings from the ultrasonic transducer and its microcontroller interface is at least 125 milliseconds. This means that a maximum of eight readings per second is available to the system. This is a restriction imposed by the ultrasonic transducer and cannot be reduced on the current hardware test-beds. However, it was noted that the vehicles are slow moving (less than 0.1 metres per second) and this sensor update time was thought to be sufficient for this application.

2. Microcontroller platform. The microcontroller platform used on the test-beds is an 8-bit, 1 MHz bus speed processor. The data memory is limited to 192 bytes and program memory limited to 4 kbytes. While it was thought that this platform would be sufficient to implement the fuzzy control system, the use of computationally expensive methods of fuzzification and defuzzification were ruled out. These methods were investigated in the simulation stage, however the large number of calculations required made them impractical for the processing platform to be used. Literature recommends that only the standard S-type, Z-type and triangular (lambda) membership functions be used on an 8-bit microcontroller for computational efficiency as well as the use of fast fuzzification algorithms. The computationally more demanding basis functions such as spline or Gaussian functions were not considered possible for implementation on the target hardware platform.
3. The test-bed used is based on a distributed processing platform. Separate processors locally control each of the major system elements such as sensor interface and motor control modules. The processors are physically separated but logically connected using an I^2C data communications bus. This bus, like any communications medium, introduces transmission delays and noise to data. It is possible that the sensor input or motor drive output data may be corrupted, lost or may introduce additional noise to the overall solution. The transmission delays introduced by the communications bus are dependent on data or request commands but is of the order of 160 microseconds for 1 byte of data. Execution times for system functions range from 180 microseconds for a motor velocity set-point change to 125 milliseconds for an ultrasonic sensor update.

10.4.2 Design strategy

The design strategy of the distance and tracking controller followed the design strategy presented in Section 10.3. The following sections outline the major design steps undertaken.

System design

The results of the system design stage are described in Sections 10.4.3 to 10.4.8. Broadly speaking, the system design stage entailed the following:

- definition of the functions and requirements of the system;
- creation of the linguistic variables of the system;
- design of the structure of the system;
- formulation of the control strategy as the fuzzy inference rules;
- selection of an appropriate defuzzification method.

Off-line optimization

Off-line optimization was performed by developing a simulation model of the test-beds and applying a fuzzy control algorithm based around the design obtained from the system design stage. The goal of the off-line optimization stage was to test the performance of the fuzzy controller in the simulation environment and manually tune the fuzzy controller parameters in order to obtain satisfactory performance from the system. It was not the goal of the off-line optimization stage to attempt to obtain an optimal controller for distance and tracking control. The fuzzy controller parameters were manually tuned (in software) until it was thought that the response of the fuzzy controller to a range of initial errors and disturbances was satisfactory. It should be noted that the developed fuzzy control system does not improve the parameters of the system automatically. Such a system would involve the application of learning capabilities to the system such as those used in neuro-fuzzy systems. Such systems integrate neural networks and fuzzy control principles to improve the parameters of the system automatically – as explained in the introductory section of this chapter.

The ease of manual tuning of the fuzzy control parameters proved the off-line optimization stage invaluable. The steps undertaken in the off-line optimization stage included:

- development of a kinematic model of the test-bed vehicles;
- design of the fuzzy control algorithm in the Matlab Fuzzy Logic Toolbox environment;
- performing simulation of the fuzzy control of distance and tracking on the developed model by applying the fuzzy control algorithm to the test-bed model. The resulting system was tested under a number of initial conditions, set-point values and disturbances. Results were recorded by plotting the system input/output relationships and adjusting fuzzification, inference and defuzzification parameters for best system response.

On-line optimization

Due to the limited availability of in-circuit emulator platforms for the hardware test-bed, the on-line optimization stage was omitted.

Implementation

The fuzzy control algorithm, its fuzzification and defuzzification parameters and rule base were used as a guide in the implementation of the controller on the test-beds. The entire fuzzy control algorithm was written in assembler language without the use of any tools to aid this process. The implementation of the designed system proceeded as follows:

- Design of the distance and tracking fuzzy controller module circuitry. This involved an initial 'breadboard-based' hardware prototype.
- Design of the distance and tracking fuzzy controller module printed circuit board.
- Procurement of components and printed circuit board from suppliers.
- Assembly and testing of the developed distance and tracking fuzzy controller hardware module.
- Integration of the distance and tracking fuzzy controller module onto the test-beds.

- Coding of the fuzzy control algorithm and basic functionality testing in a processor-specific simulator environment.
- Integration of code into the distance and tracking fuzzy controller hardware module.
- Testing for basic functionality of the solution.

Testing and verification

Once the system was implemented successfully on the test-beds, the performance of the overall solution was tested. This was done in order to verify the effectiveness of the approach and compare the implementation results to those obtained in the simulation stage. This stage involved some final manual tuning of the fuzzy controller parameters in order to obtain satisfactory distance and tracking control. The purpose of the testing and verification stage was not to attempt to obtain an optimal controller for distance and tracking control. The testing and verification stage involved the following:

- Fitting the test-beds with a pen plotting apparatus so that the trajectories of the vehicles could be recorded on plotting paper as they moved. This was important in order to compare the following vehicles' trajectory with that of the leader in order to obtain a measure of the tracking control performance of the solution.
- Testing the solution under a number of initial conditions, set-point values and disturbances. Errors in tracking and distance were recorded and graphed.
- Testing the developed prototype under the same conditions as the simulation model. This was done in order to compare the two sets of results and the accuracy of the simulation model.

10.4.3 System requirements and functions

Overview

The system requirements and functions stage involved the identification of the controlled variables, input and output variables, and scope of the input and output variables of the system. The distance and tracking fuzzy controller is intended to operate in the follower vehicle. The lead vehicle has to navigate itself around its environment and inform the follower of any major travel path changes beyond the capabilities of the follower's distance and tracking controller using a radio-communications channel. The follower has to navigate so that the desired set-point distance is maintained as well as ensuring its trajectory is in alignment with that of the leader. It follows that there are two main controlled variables of the system:

1. Distance – where the following vehicle must control its speed such that a set distance from the leader is maintained while they are in motion.
2. Tracking – where the following vehicle must mimic the travel path of the lead vehicle. This also requires that the following vehicle align itself with the lead vehicle so that it follows the travel path of the leader with as little error in their trajectories as possible. The error in the trajectory of the follower vehicle with respect to that of the leader at any time instant will be referred to as its *alignment* in this chapter.

Input and output variables

In order to ensure that the distance and tracking control goals are met, the fuzzy controller must be supplied the following input data:

1. Distance to the leader.
2. Alignment in trajectories.

An ultrasonic transducer and a time-of-flight ranging algorithm were employed on the follower test-beds to measure the distance to the leader. The following vehicle measures its alignment with respect to the leader. In order for the follower to perform this function, the leader emits a laser signal directly behind its centre, which is used by the follower to measure its alignment. The follower has 39 infrared detectors arranged in a linear array, which measures the received strength of the laser signal transmitted by the leader. These detectors are configured to provide a measure of the signal strength of the laser signal incident on them by converting it from light intensity to a voltage level. When the following robot is exactly in line with the leader, the sensor in the middle of the array shows an output voltage much higher than the other sensors in the array. When the following robot drifts from the travel path of the leader, the voltage readings from the sensors to the left or right of the middle will increase – giving the follower a measure of its alignment with respect to that of the leader.

The follower vehicle must control both its speed and orientation in order to align its trajectory with that of the leader while maintaining the separation distance. The test-beds feature a differential drive system, which allows both the speed and orientation of the vehicle to be controlled by controlling the left and right wheel velocities. Because of this configuration, there are two output variables from the fuzzy controller as follows:

1. Left wheel velocity.
2. Right wheel velocity.

These controller outputs are sent from the fuzzy controller module to the testbed motion controller module, which uses these values as set-point values for its left and right wheel PI velocity controllers. Each PI wheel velocity controller performs the single task of keeping the velocity parameter at a constant level quite well. However the operating points derived from the overall process of distance and tracking control represent a nonlinear, multi-variable control task. The fuzzy controller serves this task.

Scope of input and output variables

The range of possible values for the input variables must be defined before defining the input fuzzy membership functions. Physical sensor constraints and the range of useful values obtained from them determine this scope of input variables. The hardware test-beds employ an ultrasonic ranging system to measure distance to the leader. These sensors have a physical measurement range of 20 centimetres up to a maximum of 11 metres. For the purposes of the implementation on the slow moving vehicles, it was decided to set the upper bounds of distance measurements to 2 metres. The distance is then converted to a normalised ratio between 0 and 2, representing the distance of the follower with respect to that of the set-point distance. For example, a ratio of 1 means the distance measured is exactly equal to the set point. A ratio of 2 represents that the distance between the leader and follower is twice the set-point value.

The alignment is provided to the fuzzy controller as a scalar value between 1 and 39, representing the sensor number that has detected the most incident laser light from the leader. An input alignment value of 1 indicates that the far-left sensor is active and an input value of 39 indicates that the far right sensor is active. Sensor number 20 is the centre sensor and indicates the follower trajectory is in line with the leader.

In summary, the scope of input signals to the fuzzy controller is:

Distance to leader	20 to 200 centimetres converted to a normalized ratio between 0 to 2 with respect to the set-point distance.
Alignment	A real number between 1 and 39 where 1 is the far-left sensor and 39 is the far right sensor. Sensor number 20 is the centre sensor and indicates the follower trajectory is in line with the leader.

The output wheel velocities must be given to the test-bed motor control module in a specific format. The formatting dictates that the wheel speeds be provided separately for each wheel and should specify a percentage of full speed.

In summary, the scope of output signals from the controller is:

Left wheel velocity	0 to 100 per cent
Right wheel velocity	0 to 100 per cent

10.4.4 Definitions of linguistic variables

The initial definitions of the input and output linguistic variables were based on the following conditions:

- Only Z-type, lambda-type and S-type basis functions be used to define the linguistic variable terms.
- The input and output linguistic variables contain at most seven terms. This is a restriction imposed by the memory capacity of the test-beds but is considered quite sufficient for the application.
- The linguistic variables are symmetric. This means that the number of terms in the variables must be odd with one term describing the middle of the upper and lower extremes of the variable input space.

Equally spaced linguistic variable terms were used in the initial definitions, with the simulation stage intended to refine their size and shape in order to obtain better controller performance. The terms for the input and output linguistic variables were defined as follows:

Input linguistic variables:

Distance ∈ {Too_Far, Far, Bit_Far, OK, Bit_Close, Close, Too_Close}
Alignment ∈ {Far_Left, Left, Bit_Left, Centre, Bit_Right, Right, Far_Right} (10.17)

Output linguistic variables:

Left_Wheel_Speed ∈ {Very_Slow, Slow, Bit_Slow, Medium, Bit_Fast, Fast, Very_fast}
Right_Wheel_Speed ∈ {Very_Slow, Slow, Bit_Slow, Medium, Bit_Fast, Fast, Very_fast} (10.18)

The input linguistic variables, 'Distance' and 'Alignment' are given below in Figures 10.10 and 10.11 respectively.

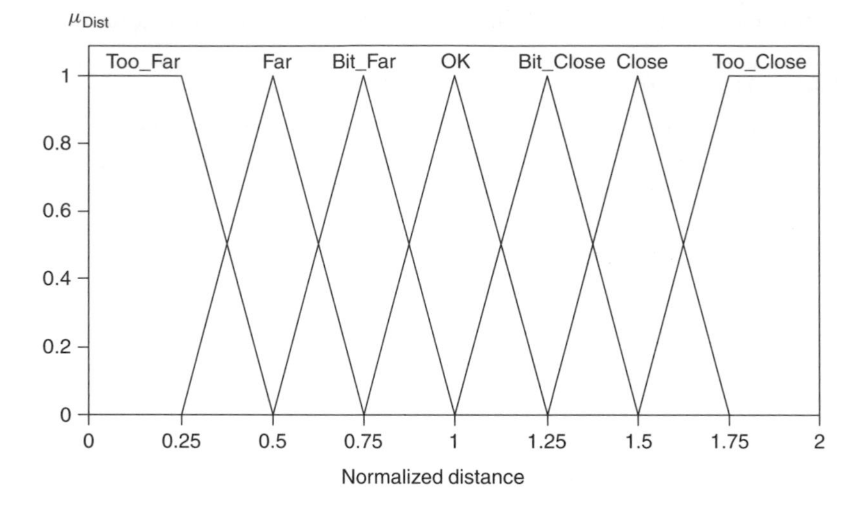

Fig. 10.10 Distance linguistic variable membership functions.

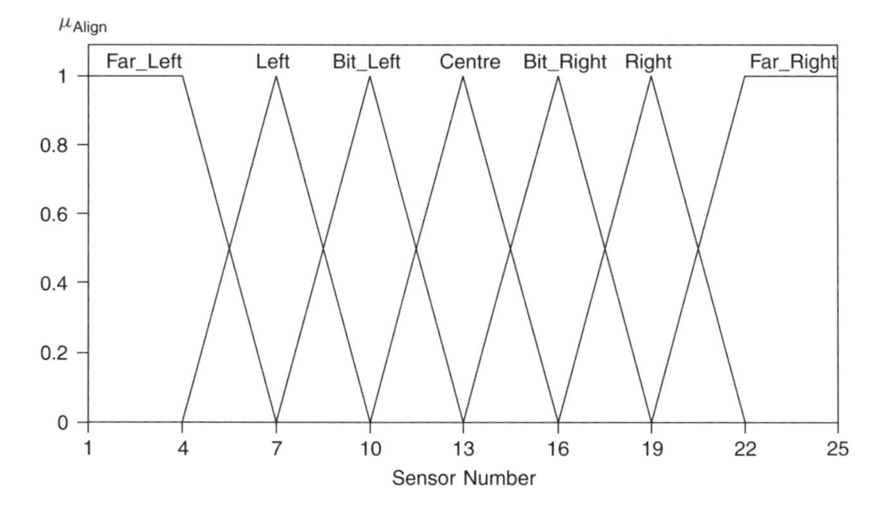

Fig. 10.11 Alignment linguistic variable membership functions.

The output linguistic variables, 'Left_Wheel_Speed (LWS)' and 'RightWheel_Speed (RWS)' are given below in Figures 10.12 and 10.13 respectively.

10.4.5 System structure

The fuzzy control system for distance and tracking takes inputs from the distance and alignment sensors and fuzzifies these input crisp values. The inference stage is then used to produce a number of rule outputs, which are then defuzzified into single crisp output values to be sent to the motor drive module of the test-bed. The two outputs from the fuzzy controller are used as set-point values by the motor drive module,

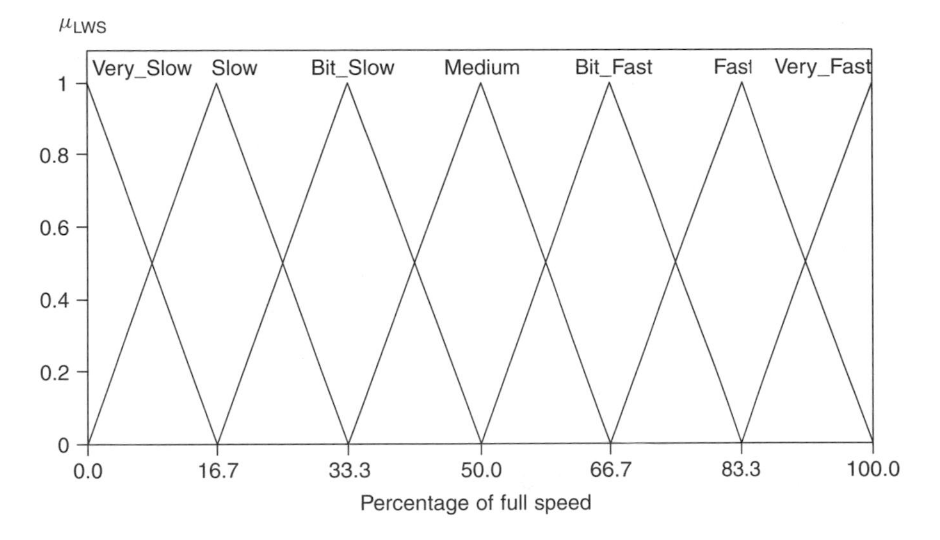

Fig. 10.12 Left wheel speed linguistic variable membership functions.

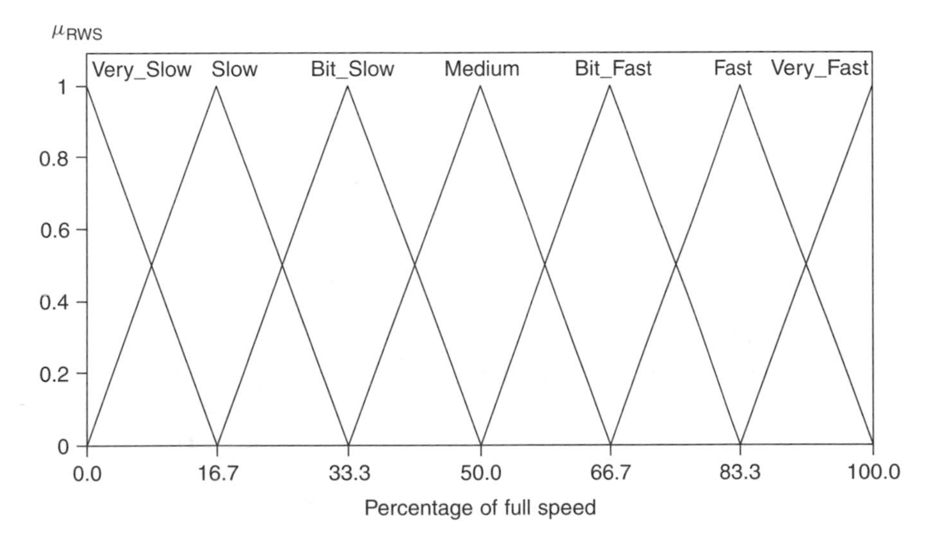

Fig. 10.13 Right wheel speed linguistic variable membership functions.

which has two separate PI controllers – one for each wheel velocity control. Each PI wheel velocity controller performs the single task of keeping the velocity parameter at a constant level quite well. However the operating points derived from the overall process of distance and tracking control represent a nonlinear, multi-variable control task. The fuzzy controller serves this task. The implementation of the distance and tracking control system represents an example of a 'hybrid' control system – using conventional and fuzzy control in the one solution.

Pre-processing of the sensor data occurs prior to its use in the fuzzy controller. This involves the following:

- Normalizing the distance sensor reading to the set-point distance. This yields an input universe for the distance variable of [0, 2], where 0 represents no separation distance between the two vehicles and 2 represents twice the set-point distance between them.
- Deriving a single alignment result. This involves reading the input laser signal strength from all 39 sensors, determining a threshold value for ambient light effects and normalizing each sensor reading to the threshold value. More than one sensor may have a non-zero, normalized reading and therefore a single alignment reading must be resolved from all non-zero inputs. A centre of gravity (COG) algorithm is used to produce the alignment sensor value to be passed to the fuzzy controller.

The complete distance and tracking control system structure is shown in Figure 10.14.

10.4.6 Fuzzification method

The method of choice for fuzzification in 8-bit microcontroller platforms with limited memory resources is the *fast-fuzzification method for standard MBFs*. This method can implement fuzzification of standard membership basis functions (MBFs) in as little as a few comparisons, subtractions and one multiplication per linguistic variable term. This is a vast advantage in terms of computing complexity when compared to other methods of fuzzification such as the use of Gaussian and Spline functions. It is for this reason that the fast fuzzification algorithm was chosen for the distance and tracking fuzzy controller. It is difficult to find a comprehensive treatment of the fast-fuzzification algorithm, however an attempt will now be given here. The following description will detail S-type, Z-type and lambda-type standard MBFs only. Pi-type membership functions are merely an extension of this work and were not used in the implementation of the distance and tracking fuzzy controller.

Lambda-type MBFs

Consider the lambda-type MBF shown in Figure 10.15 in an input universe X forming the closed interval $[a, b]$. The fuzzification stage of a fuzzy controller requires that an input signal x be mapped to each term MF_n, in a linguistic variable. This simply means that an input signal is assigned a degree of membership, $\mu_{MFn}(x)x \in X$ for each term, MF_n of a linguistic variable. For example, the linguistic variable, distance contains the following terms:

$$\mathbf{Distance(MF_n)} \in \{\mathbf{Too\_Far,\ Far,\ Bit\_Far,\ OK,\ Bit\_Close,\ Close,\ Too\_Close}\}$$

where $n = [1{:}7]$ (10.19)

An input normalized distance value is assigned a degree of membership between 0 and 1 for each of these seven terms. The assumption is that an input x, is defined for all X and the term is continuous over the closed interval (universe) $X = [a, b]$. The degree of membership of the input x, for a lambda-type MBF in a fast fuzzification algorithm is found by applying a piece-wise approach. Let us divide the universe X, into three distinct intervals (Xa_1, Xa_2 and Xa_3) where,

$$X = Xa_1 \cup Xa_2 \cup Xa_3$$

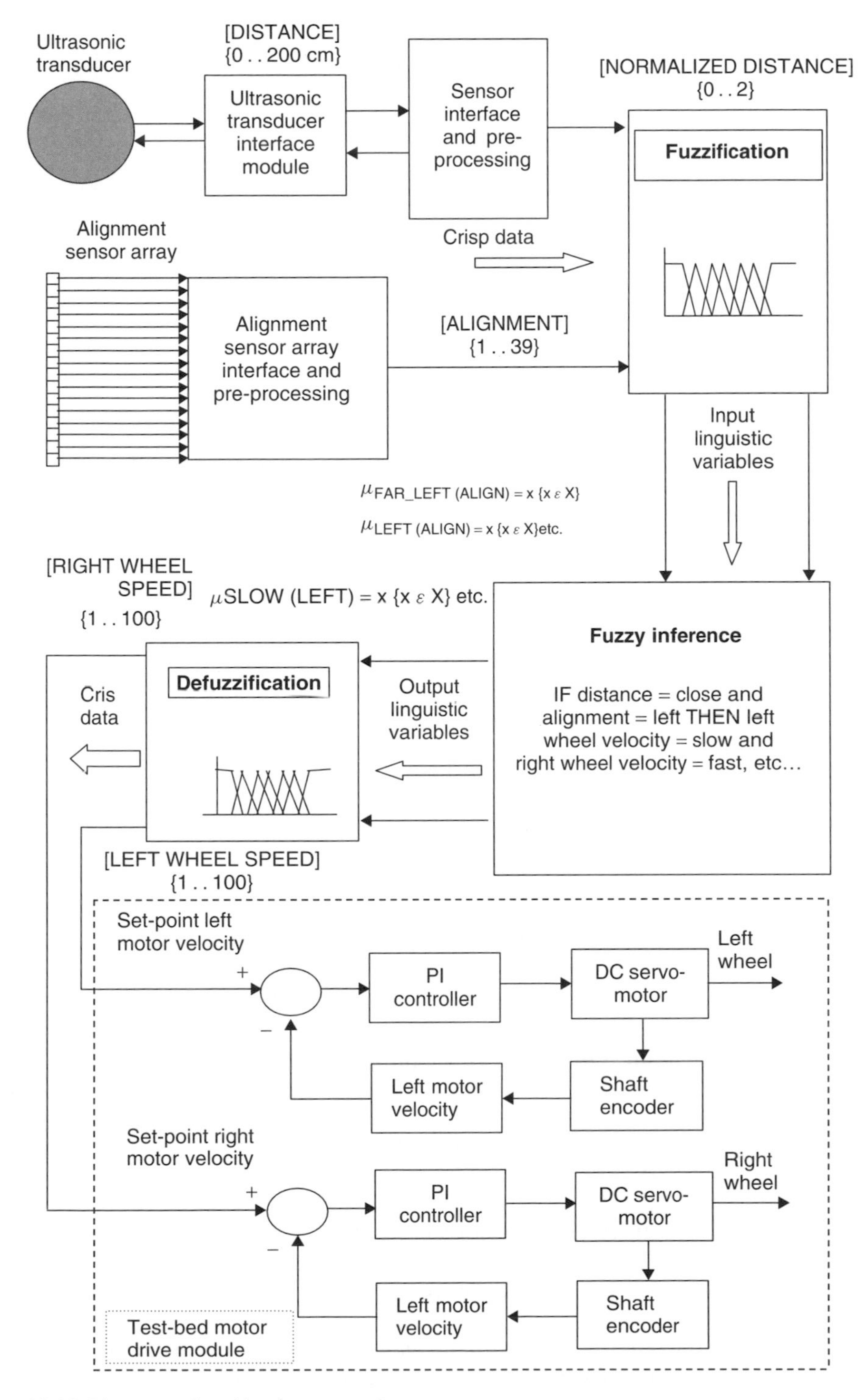

Fig. 10.14 Distance and tracking fuzzy control system structure.

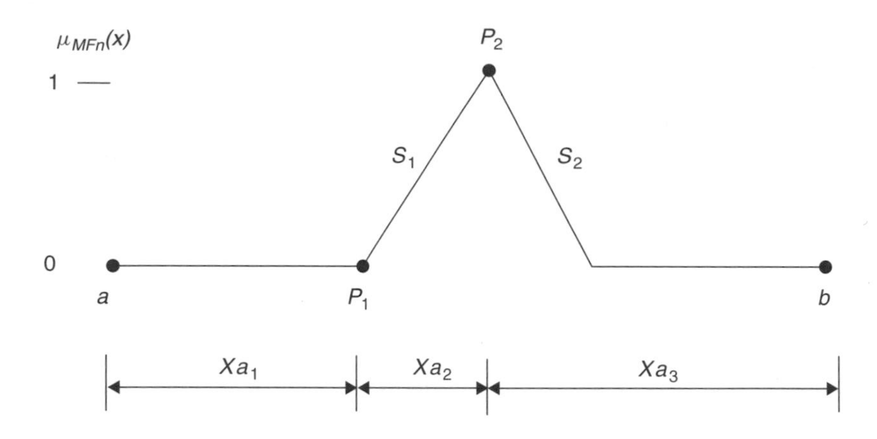

Fig. 10.15 Lambda-type membership basis function.

where Xa_1, Xa_2 and Xa_3 are mutually exclusive sub-sets of X. Any input signal $x \in X$, may therefore fall into only one of these areas – each with its own mapping relationship.

Define:
Area 1 as the half-closed interval between points P_1 and a so that

$$Xa_1 = [a, P_1]$$

Therefore, with reference to Figure 10.15, an input signal x, occurring in the half-closed interval Xa_1 has a degree of membership to a lambda-type MBF of:

$$\mu_{MFn}(x) = 0 \qquad x \in Xa_1$$

Define:
Area 2 as the half-closed interval between points P_2 and P_1 so that

$$Xa_2 = [P_1, P_2]$$

Therefore, with reference to Figure 10.15, an input signal x, occurring in the half-closed interval Xa_2 has a degree of membership to a lambda-type MBF of:

$$\mu_{MFn}(x) = (x - P_1) \times S_1 \qquad x \in Xa_2$$

Define:
Area 3 as the closed interval between points b and P_2 so that

$$Xa_3 = [P_2, b]$$

Therefore, with reference to Figure 10.15, an input signal x, occurring in the closed interval Xa_3 has a degree of membership to a lambda-type MBF of:

$$\mu_{MFn}(x) = \max\{0, 1 - ((x - P_2) \times S_2)\} \qquad x \in Xa_3$$

In summary, for lambda-type MBFs, the degree of membership of an input signal $x \in X$ is as follows:

$$\mu_{MFn}(x) = \left\{ \begin{array}{cc} 0 & x \in Xa_1 \\ (x - P_1) \times S_1 & x \in Xa_2 \\ \max\{0, 1 - ((x - P_2) \times S_2)\} & x \in Xa_3 \end{array} \right\} \tag{10.20}$$

where: $Xa_1 \cup Xa_2 \cup Xa_3 = \mathrm{X}$

Z-Type MBFs

Consider the Z-type MBF shown in Figure 10.16 in an input universe X forming the closed interval $[a, b]$. Once again, the fuzzification stage of a fuzzy controller requires that an input signal x be mapped to each term MF_n, in a linguistic variable. This assumes that an input x is defined for all X and the term is continuous over the closed interval (universe) $X = [a, b]$. The degree of membership of the input x, for a Z-type MBF in a fast fuzzification algorithm is also found by applying a piece-wise approach. Let us divide the universe X, into two distinct intervals (Xa_1 and Xa_2) where,

$$X = Xa_1 \cup Xa_2$$

where Xa_1 and Xa_2 are mutually exclusive sub-sets of X. Any input signal $x \in X$, may therefore fall into only one of these areas – each with its own mapping relationship.

Define:

Area 1 as the half-closed interval between points P_1 and a so that

$$Xa_1 = [a, P_1]$$

Therefore, with reference to Figure 10.16, an input signal x, occurring in the half-closed interval Xa_1 has a degree of membership to a Z-type MBF of:

$$\mu_{MFn}(x) = 1 \qquad x \in Xa_1$$

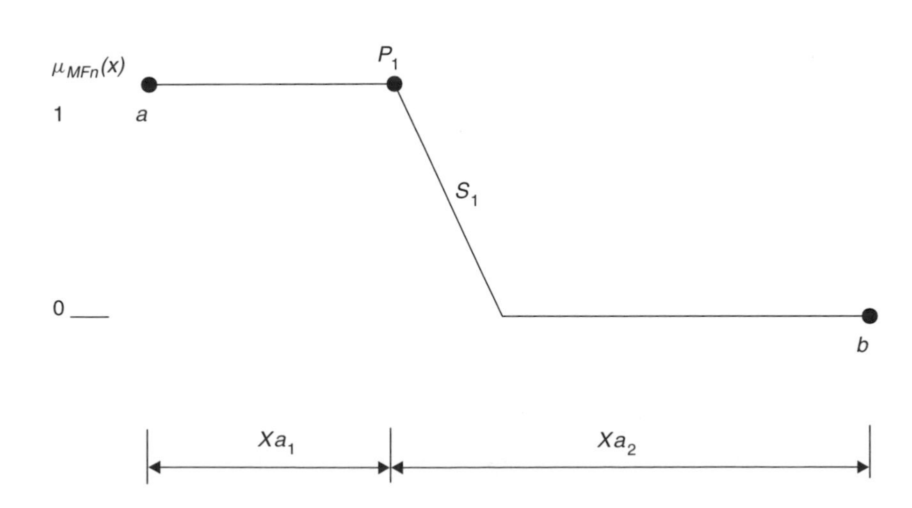

Fig. 10.16 *Z*-type membership basis function.

Define:
Area 2 as the closed interval between points b and P_1 so that

$$Xa_2 = [P_1, b]$$

Therefore, with reference to Figure 10.16, an input signal x, occurring in the closed interval Xa_2 has a degree of membership to a Z-type MBF of:

$$\mu_{MFn}(x) = \max\{0, 1 - ((x - P_1) \times S_1)\} \qquad x \in Xa_2$$

In summary, for Z-type MBFs, the degree of membership of an input signal $x \in X$ is as follows:

$$\mu_{MFn}(x) = \begin{Bmatrix} 1 & x \in Xa_1 \\ \max\{0, 1 - ((x - P_1) \times S_1)\} & x \in Xa_2 \end{Bmatrix} \tag{10.21}$$

where: $Xa_1 \cup Xa_2 = \mathrm{X}$

Consider the S-type MBF shown in Figure 10.17 in an input universe X forming the closed interval $[a, b]$. Once again, the fuzzification stage of a fuzzy controller requires that an input signal x be mapped to each term MF_n, in a linguistic variable. This assumes that x is defined for all X and the term is continuous over the closed interval (universe) $X = [a, b]$. The degree of membership of the input x, for an S-type MBF in a fast fuzzification algorithm is also found by applying a piece-wise approach. Let us divide the universe X, into two distinct intervals-(Xa_1 and Xa_2) where,

$$X = Xa_1 \cup Xa_2$$

Where Xa_1 and Xa_2 are mutually exclusive sub-sets of X. Any input signal $x \in X$, may therefore fall into only one of these areas – each with its own mapping relationship.

Define:
Area 1 as the half-closed interval between points P_1 and a so that

$$Xa_1 = [a, P_1]$$

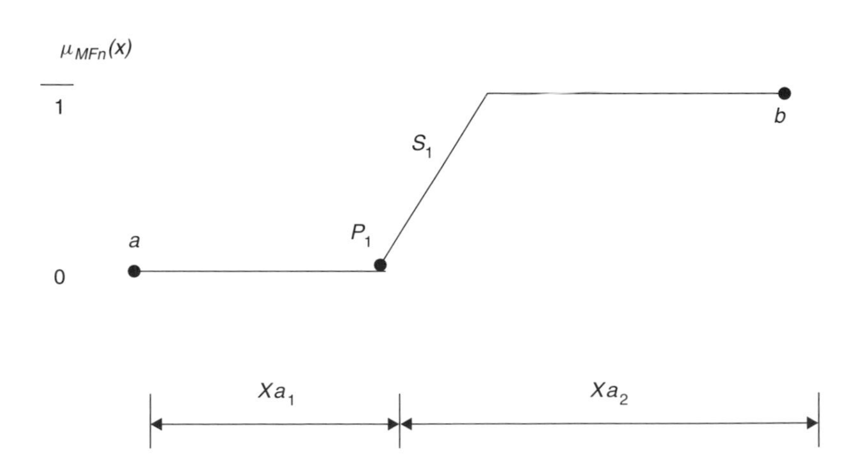

Fig. 10.17 *S*-type membership basis function.

Therefore, with reference to Figure 10.17, an input signal x, occurring in the half-closed interval Xa_1 has a degree of membership to an S-type MBF of:

$$\mu_{MFn}(x) = 0 \qquad x \in Xa_1$$

Define:
Area 2 as the closed interval between points b and P_1 so that

$$Xa_2 = [P_1, b]$$

Therefore, with reference to Figure 10.17, an input signal x, occurring in the closed interval Xa_2 has a degree of membership to an S-type MBF of:

$$\mu_{MFn}(x) = \min\{1, 1 - ((x - P_1) \times S_1)\} \qquad x \in Xa_2$$

In summary, for S-type MBFs, the degree of membership of an input signal $x \in X$ is as follows:

$$\mu_{MFn}(x) = \left\{ \begin{array}{cc} 0 & x \in Xa_1 \\ \min\{1, 1 - ((x - P_1) \times S_1)\} & x \in Xa_2 \end{array} \right\} \tag{10.22}$$

where: $Xa_1 \cup Xa_2 = \mathrm{X}$

10.4.7 Fuzzy inference rules

Computation of the fuzzy inference rules consists of two components:

- Aggregation: computation of the IF parts of the rules.
- Composition: computation of the THEN parts of the rules.

The fuzzy inference method chosen for the distance and tracking control was MAX/MIN. This terminology arises due to the aggregation part being computed using the MIN (fuzzy AND) operator and the composition part being computed using the MAX (fuzzy OR) operator. It must be noted that the fuzzy AND, OR and NOT operators differ from their Boolean counterparts. Boolean operators cannot cope with degrees of truth to set membership and therefore cannot be applied to fuzzy operations. It has been proposed (von Altrock, 1995; Yager and Rybalov, 1997) that the following fuzzy operators be used in place of the standard Boolean operators:

Operator	***Fuzzy equivalent***
AND	$\mu_{A \wedge B} = \min\{\mu_A, \mu_B\}$
OR	$\mu_{A \vee B} = \max\{\mu_A, \mu_B\}$
NOT	$\mu_{\neg A} = 1 - \mu_A$

Consider the example inference rule below:

IF Distance = Far AND Alignment = Left THEN Left_Wheel_Speed = Fast

AND Right_Wheel_Speed = Slow (10.23)

It can be seen that the aggregation part combines the two pre-conditions of Distance = Far and Alignment = Left. The aggregation part of the inference stage defines whether the rule is valid or not. When a rule is valid it is often referred to as having 'fired'. The degrees to which the statements 'Distance = Far' and 'Alignment = Left' are true are assigned during the fuzzification stage and are combined during the aggregation stage using the fuzzy AND operator.

The distance and tracking controller has two input variables – distance and alignment, each with seven variable terms. This yields a fuzzy inference rule base of 49 dual-output rules. The terminology 'dual-output rules' refers to the left and right wheel velocities as forming two separate outputs asserted on the same inference condition. The initial fuzzy inference rule base designed for the distance and tracking controller is provided in Table 10.4.

Once a fuzzy inference rule base has been defined, it may be optimized to remove redundant rules or rules that never fire. Redundant rules are defined as rules that have no impact on the system. It may be concluded from Table 10.4 that there are no redundant rules and that all rules have a chance of firing. The execution of the fuzzy inference may be greatly reduced by using the assertion.

If the MIN operator is used, just one invalid condition in the aggregation stage deactivates the entire rule – removing the need for further calculation in the composition stage which will arrive at a zero result. The savings in execution and memory resources can range from 5 to 50 times using this method. It can be seen from the terms in the distance and alignment variables in Figures 10.10 and 10.11 that there can be at most two non-zero terms for any input x. That is, $\mu_{MFn}(x) \neq 0$ can only be true for at most two terms in the distance variable and two terms in the alignment variable. Therefore, from the aggregation stage there can be at most four rules that can possibly fire. This assertion is only true for the membership functions such as those presented in Section 10.4.4 where at most two terms overlap to a non-zero degree. If the shape or position of these membership function terms is changed, there may be more or less possible numbers of non-zero terms presented to the fuzzy inference stage and hence it may be possible that more than four rules may fire.

The degree of truth (DoT) of a rule is calculated by the aggregation stage and the composition stage assigns the action to be taken. The fuzzy sets from the consequent of each rule are combined through the aggregation operator and the resulting fuzzy set is defuzzified to yield the system output. This type of fuzzy inference is referred to as a Mamdani-type inference. A fuzzy rule base defines the rules alternatively – either rule 1 is true OR rule 2 is true OR rule 3 is true and so on. Rules that have the same outcome are resolved using the MAX operator.

10.4.8 Defuzzification method

The result of the fuzzy inference is both fuzzy and ambiguous in that two different actions may have non-zero DoTs. The resulting actions must be combined to form one crisp, real valued output for the left and right wheel velocities. These velocities are used as a set-point value by two separate PI controllers – one for each wheel. The method chosen for defuzzification was the centre of maximum (CoM) method. The defuzzification method consists of a two-step approach for defuzzification:

Table 10.4 Fuzzy inference rule base used in the distance and tracking controller

Fuzzy rule	IF	Distance =	Alignment =	THEN	LWS =	RWS =
1	IF	TOO_CLOSE	FAR_LEFT	THEN	MED	VERY_SLOW
2	IF	CLOSE	FAR_LEFT	THEN	BIT_FAST	SLOW
3	IF	BIT_CLOSE	FAR_LEFT	THEN	FAST	BIT_SLOW
4	IF	OK	FAR_LEFT	THEN	VERY_FAST	MED
5	IF	BIT_FAR	FAR_LEFT	THEN	VERY_FAST	MED
6	IF	FAR	FAR_LEFT	THEN	VERY_FAST	MED
7	IF	TOO_FAR	FAR_LEFT	THEN	VERY_FAST	MED
8	IF	TOO_CLOSE	LEFT	THEN	BIT_SLOW	VERY_SLOW
9	IF	CLOSE	LEFT	THEN	MED	SLOW
10	IF	BIT_CLOSE	LEFT	THEN	BIT_FAST	BIT_SLOW
11	IF	OK	LEFT	THEN	FAST	MED
12	IF	BIT_FAR	LEFT	THEN	VERY_FAST	BIT_FAST
13	IF	FAR	LEFT	THEN	VERY_FAST	BIT_FAST
14	IF	TOO_FAR	LEFT	THEN	VERY_FAST	BIT_FAST
15	IF	TOO_CLOSE	BIT_LEFT	THEN	SLOW	VERY_SLOW
16	IF	CLOSE	BIT_LEFT	THEN	BIT_SLOW	SLOW
17	IF	BIT_CLOSE	BIT_LEFT	THEN	MED	BIT_SLOW
18	IF	OK	BIT_LEFT	THEN	BIT_FAST	MED
19	IF	BIT_FAR	BIT_LEFT	THEN	FAST	BIT_FAST
20	IF	FAR	BIT_LEFT	THEN	VERY_FAST	FAST
21	IF	TOO_FAR	BIT_LEFT	THEN	VERY_FAST	FAST
22	IF	TOO_CLOSE	CENTRE	THEN	VERY_SLOW	VERY_SLOW
23	IF	CLOSE	CENTRE	THEN	SLOW	SLOW
24	IF	BIT_CLOSE	CENTRE	THEN	BIT_SLOW	BIT_SLOW
25	IF	OK	CENTRE	THEN	MED	MED
26	IF	BIT_FAR	CENTRE	THEN	BIT_FAST	BIT_FAST
27	IF	FAR	CENTRE	THEN	FAST	FAST
28	IF	TOO_FAR	CENTRE	THEN	VERY_FAST	VERY_FAST
29	IF	TOO_CLOSE	BIT_RIGHT	THEN	VERY_SLOW	SLOW
30	IF	CLOSE	BIT_RIGHT	THEN	SLOW	BIT_SLOW
31	IF	BIT_CLOSE	BIT_RIGHT	THEN	BIT_SLOW	MED
32	IF	OK	BIT_RIGHT	THEN	MED	BIT_FAST
33	IF	BIT_FAR	BIT_RIGHT	THEN	BIT_FAST	FAST
34	IF	FAR	BIT_RIGHT	THEN	FAST	VERY_FAST
35	IF	TOO_FAR	BIT_RIGHT	THEN	FAST	VERY_FAST
36	IF	TOO_CLOSE	RIGHT	THEN	VERY_SLOW	BIT_SLOW
37	IF	CLOSE	RIGHT	THEN	SLOW	MED
38	IF	BIT_CLOSE	RIGHT	THEN	BIT_SLOW	BIT_FAST
39	IF	OK	RIGHT	THEN	MED	FAST
40	IF	BIT_FAR	RIGHT	THEN	BIT_FAST	VERY_FAST
41	IF	FAR	RIGHT	THEN	BIT_FAST	VERY_FAST
42	IF	TOO_FAR	RIGHT	THEN	BIT_FAST	VERY_FAST
43	IF	TOO_CLOSE	FAR_RIGHT	THEN	VERY_SLOW	MED
44	IF	CLOSE	FAR_RIGHT	THEN	SLOW	BIT_FAST
45	IF	BIT_CLOSE	FAR_RIGHT	THEN	BIT_SLOW	FAST
46	IF	OK	FAR_RIGHT	THEN	MED	VERY_FAST
47	IF	BIT_FAR	FAR_RIGHT	THEN	MED	VERY_FAST
48	IF	FAR	FAR_RIGHT	THEN	MED	VERY_FAST
49	IF	TOO_FAR	FAR_RIGHT	THEN	MED	VERY_FAST

- Compute one typical value for each term in the output linguistic variable.
- The degrees of truth from the rule outputs of the inference stage are assigned to the typical values for each term and balanced out using a best compromise method.

The output linguistic variables were firstly defined, appearing in Section 10.4.4. Each term of the output variables has a maximizing interval and so the median of the

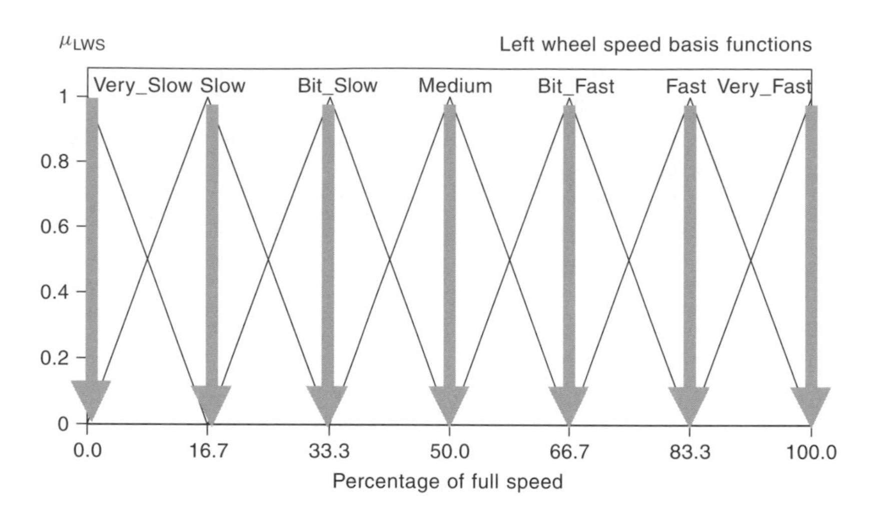

Fig. 10.18 Typical values for the left wheel speed linguistic variable membership function.

maximizing set was chosen. Figure 10.18 graphically depicts the typical values for the left wheel speed output.

The CoM method of defuzzification is identical to the centre of gravity method using Singleton membership functions. The fuzzy system, $F{:}R^n \rightarrow R^p$ has m rules of the form 'If X is a fuzzy set A, then Y is the fuzzy set B'. Each input x fires all of the rules in parallel in this additive fuzzy system (Kosko, 1997). The *standard additive model* or SAM defines a function $F{:}R^n \rightarrow R^p$ as follows:

$$F(x) = \frac{\sum_{j=1}^{m} w_j \times a_j(x) \times V_j \times C_j}{\sum_{j=1}^{m} w_j \times a_j(x) \times V_j} \tag{10.24}$$

The SAM stores m rules of the 'patch' form $A_j \times B_j$ or the word form 'If input variable X equals fuzzy set A_j, then output Y equals fuzzy set B_j. The SAM uses the volume or area V_j of set B_j and its centroid or centre of gravity c_j. Each real input x in the universe of inputs X, fires the IF part of all m rules in parallel and to some degree (most to zero degree) $a_j(x)$ in [0,1]. Each rule in the fuzzy inference set may be assigned a *weight* $w_j > 0$. This will have the effect of increasing or decreasing the effect of a fired rule. This is a simple form of *fuzzy associative mapping* or FAM. This SAM fuzzy system may be reduced to the centre of gravity (CoG) form if the centres of gravity are equal to the peaks P_j of the then part of the inference stage and the rule weights w_j and volumes V_j are equal. This reduces the SAM system to:

$$F(x) = \frac{\sum_{j=1}^{m} a_j(x) \times P_j}{\sum_{j=1}^{m} a_j(x)} \tag{10.25}$$

The fuzzy controller for distance and tracking was designed assuming that all rule weights were equally important and the standard triangular (lambda) MBFs assured that the centroids were the peaks or 'typical values'.

A fuzzy control system can typically be analysed by plotting a contour surface mapping of input to output variables. Such a transfer characteristic plot shows linear regions, non-monotonous regions, and regions of instability (if any), where no rules fire. Surface plots for the initial design of the distance and tracking controller were produced showing the transfer characteristics of the system with the variables, distance and alignment as the system inputs and the left wheel speed and right wheel speed as the system outputs. The surface texture is determined by the fuzzy inference rules that plot the transfer function for each of the two outputs as a function of the inputs. The surface plots for the left wheel speed and right wheel speed system outputs are shown as Figures 10.19 and 19.20 respectively. It can be seen from these plots that both outputs are nearly identical and feature a complex surface with most linear, monotonous regions small in size.

10.5 Conclusion

This chapter has outlined the relevant issues and theory related to intelligent control systems techniques focusing on a fuzzy logic which enables the control system designer to describe the desired system behaviour using descriptive language and also to express a certain level of vagueness associated with an item under consideration. A fuzzy control system takes real input values, such as those obtained from sensor measurements, and associates them with language descriptions of their magnitude. Then, it assigns a

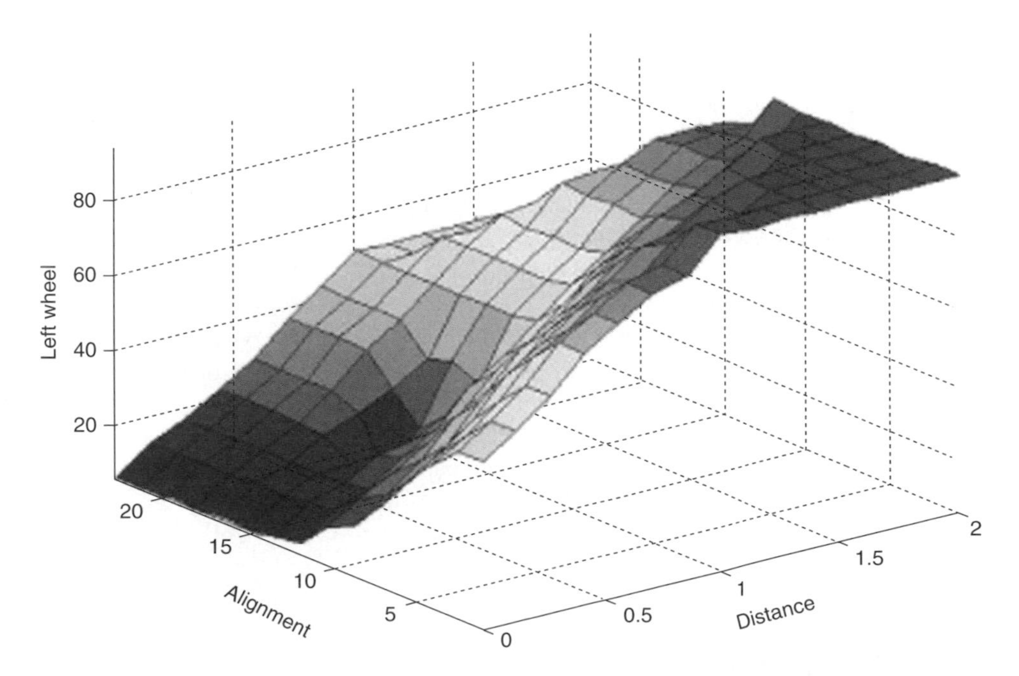

Fig. 10.19 Surface plot for the left wheel speed output of the fuzzy controller for dist[illegible]ce and tracking.

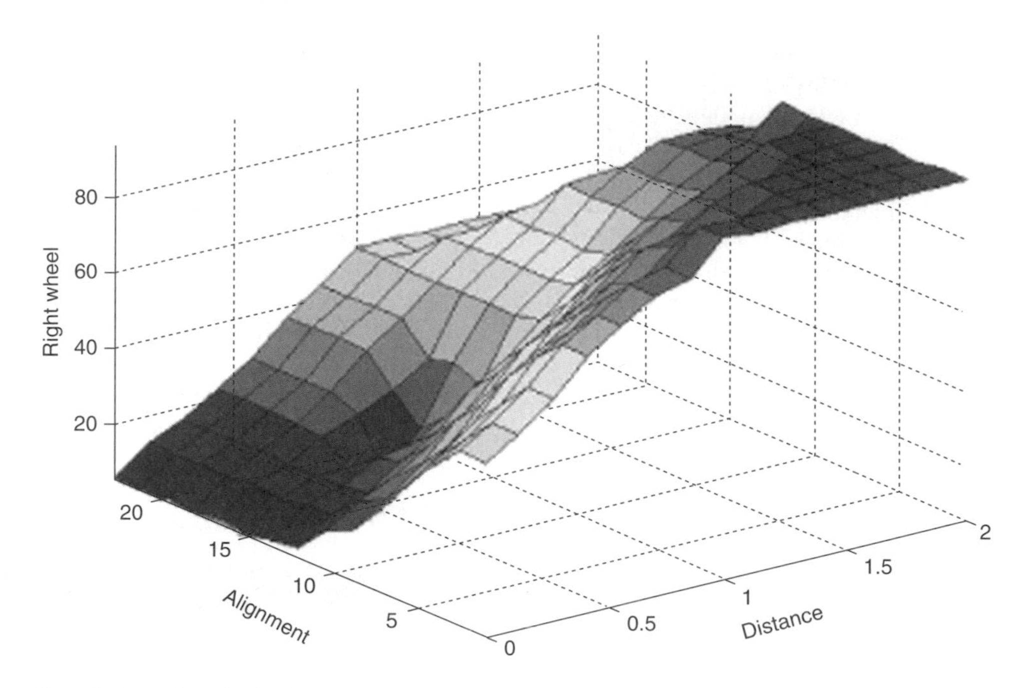

Fig. 10.20 Surface plot for the right wheel speed output of the fuzzy controller for distance and tracking.

language output description based on the input using if-then type rules. Finally, a fuzzy control system takes the descriptive output and converts it back into a real output value as a control signal. A fuzzy control system design process is illustrated in the details by presenting all the design steps that were carried out in conjunction with prototyping of the distance and tracking fuzzy control system for autonomous, driver-less vehicle.

A distance between two successive vehicles, if maintained by a control system, may be significantly less than the minimum safe distance that a human driver can maintain. This factor leads to increased road capacity and reduction in the accidents caused by human error. If this longitudinal distance is reduced towards its minimum point without reducing safety, the greater road capacity becomes. Further, if the trajectory control of the vehicle is removed from human control, the chances for human error are limited. Of course, there are many and varied issues that must be addressed before the full deployment of total vehicle automation. Some of them are addressed elsewhere in this book (for example, see Chapters 6, 12, 13 and 14).

10.6 Abbreviations

CoA	centre of area
CoG	centre of gravity
CoM	centre of maximum
DoM	degree of membership
DoS	degree of support
DoT	degree of truth

FAM	fuzzy associative mapping
FL	fuzzy logic
FLM	fuzzy logic models
LoM	left of maximum
MAX	maximum
MBF	membership function
MIN	minimum
MoM	mean of maximum
NN	neural networks
PI	proportional-integral
PID	proportional-integral-derivative
PROD	multiplier
RoM	right of maximum
SAM	standard additive model
SVMs	support vector machines

References

Borenstein, J. (1996). Navigating Mobile Robots: Systems and Techniques. A.K. Peters Publishing Ltd, Wellesley, USA.

Chira-Chavala, T. and Yoo, S.M. (1993). Characteristics of Automated HOV Lanes. *Transportation Quarterly*, **47**, No. 4, October, pp. 545–560, Virginia, USA.

Dickerson, J.A., Kim, H. and Kosko, B. (1997). Fuzzy Control for Platoons of Smart Cars. In: *Fuzzy Engineering*, Prentice-Hall, pp. 177–209, New Jersey, USA.

Du, Y. and Papanikolopoulos, N. (1997). Real-Time Vehicle Following Through a Novel Symmetry-Based Approach. *Proceedings of the 1997 IEEE International Conference on Robotics and Automation*, pp. 3160–66, IEEE, USA.

Everett, H.R. (1995). *Sensors for Mobile Robots: Theory and Applications*. A. K. Peters Publishing, Wellesley, Massachusetts, USA.

Grosch, T. (1995). Radar Sensors for Automotive Collision Warning and Avoidance, *SPIE*, **2463**, pp. 239–45.

Gupta, M.M. and Sinha, N.K. (eds) (1995). *Intelligent Control: Theory and Practice*. IEEE Press, Piscataway, NJ.

Haugen, J. (1993). Smart Cruise: A Deployment Issue. *Automotive Industries Magazine*, May, 5–6.

Hitchings, M. (1999). Distance and Tracking Control for Autonomous Vehicles. *MphilChapter*, Griffith University.

Holve, R. and Protzel, P. (1995). Adaptive Fuzzy Control for Driver Assistance in Car-Following. *Proceedings of the 3rd European Congress in Intelligent Techniques and Soft Computing*, Aachen, Germany, August, pp. 1149–53.

Holve, R. and Protzel, P. (1995b). Generating Fuzzy Rules for the Acceleration Control of an adaptive Cruise Control System. Proceedings of the NAFIPS Conference, Berkeley, California, USA, June.

Kecman, V. (2000). *Learning and Soft Computing, with Support Vector Machines, Neural Networks and Fuzzy Logic Models*. The MIT Press, Cambridge, MA.

Kosko, B. (1997). *Fuzzy Engineering*, Prentice-Hall Publishing, New Jersey, USA.

Mathworks Inc. (1998). *Fuzzy logic toolbox for use with Matlab – User's guide*. Mathworks Inc., Massachusetts, USA.

Palmquist, U. (1993). Intelligent Cruise Control and Roadside Information. *IEEE Micro*, **13**, February, pp. 20–28.
Raza, H. and Ioannou, P. (1996). Vehicle Following Control Design for Automated Highway Systems. *IEEE Control Systems*, December, pp. 43–60.
Rohling, H. and Lissel, E. (1995). 77 GHz Radar Sensor for Car Applications. *Proceedings of the IEEE International Conference on Radar Technologies*, pp. 373–9.
Schone, A. and Sun, X. (1995). Different In-Vehicle Control Concepts for the Distance Control of Cars Moving in Convoys. *Proceedings of the 2nd International Conference on Road Vehicle Automation*, Bolton, UK, September, pp. 50–69.
Tso, S.K. and Fung, Y.H. (1994). Real-time Fuzzy Logic Design and Implementation of Autonomous Vehicle Control. *Proceedings of the 3rd International Conference on Automation, Robotics and Computer Vision*, Singapore, November, pp. 2102–6.
Tsumura, T., Okubu, H. and Komatsu, N. (1992). Directly Follow-up and/or Traced Control System of Multiple Ground Vehicles. *Proceedings of the 1992 IEEE/RSJ International Conference on Intelligent Robots and Systems*, pp. 1769–76.
Variaya, P. (1993). Smart Cars on Smart Roads –Problems of Control. *IEEE Transactions on Automatic Control*, **38**, No. 2, 195–203.
Vlacic, L., Hitchings, M., Kajitani, M. and Kanamori, C. (1998). Cooperative Autonomous Road Robots. *Proceedings of the ICARCV98 conference*, Singapore.
Vlacic, L., Engwirda, A. and Kajitani, M. (2000). Cooperative Behaviour of Intelligent Agents: Theory and Practice. In *Soft Computing and Intelligent Systems: Theory and Applications*, N.N. Sinha and M.M. Gupta (eds) Academic Press, pp. 279–307.
von Altrock, C. (1995). Fuzzy Logic and Neuro-Fuzzy Applications Explained, Prentice-Hall Publishing, New Jersey. f
Yager, R. and Rybalov, A. (1997). Noncommutative Self-Identity Aggregation. *International Journal of Soft Computing-Fuzzy Sets and Systems*, **85**, No. 1, January, pp. 73–82.
Yamaguchi, J. (1996). Honda Prototype Vehicle for Japanese Automated Highway System. *Automotive Engineering Magazine*, April, pp. 35–40.
Zadeh, L.A. (1996). Fuzzy Control: Issues, Contentions and Perspectives. *Proceedings of IFAC 13th Triennial World Congress*, San Francisco, pp. 257–62.
Zielke, T., Brauckmann, M. and Von Seelan, W. (1992). CARTRACK: Computer Vision-Based Car Following. *Proceedings of the First IEEE Workshop on Applications of Computer Vision*, Palm Springs, California, pp. 156–63.

Recommended reading

The interested reader may wish to consult several readily available sources for more material on this broad and interdisciplinary field

Books on neural networks

Bishop, C.M. (1995). *Neural Networks for Pattern Recognition*. Clarendon Press, Oxford, UK.
Cichocki, A. and Unbehauen, R. (1993). *Neural Networks for Optimization and Signal Processing*. B.G. Teubner, Stuttgart and John Wiley and Sons, Chichester.
Haykin, S. (1994). *Neural Networks: A Comprehensive Foundation*. MacMilan College Publishing Co., New York.
Zurada, J. (1992). *Introduction to Artificial Neural Systems*. West Publishing Co, St Paul, MN.

Books on fuzzy logic

Kahlert, J. and Frank, H. (1994). *Fuzzy-Logik und Fuzzy-Control* (in German), Vieweg-Verlag, 2 Auflage, Wiesbaden.

Klir, J.G. and Folger, T.A. (1988). *Fuzzy Sets, Uncertainity, and Information.* Prentice Hall, Englewood Cliffs, NJ.

Zimmermann, H.J. (1991). *Fuzzy Set Theory and Its Applications*, 2nd edn. Kluwer, Boston.

Special books devoted to the intelligent control

Antsaklis, P. and Passino, K. (1993). *An Introduction to Intelligent and Autonomous Control.* Kluwer Academic Publishers, Massachusetts, USA.

Gupta, M.M. and Sinha, N.K. (eds) (1995). *Intelligent Control: Theory and Practice.* IEEE Press, Piscataway, NJ.

Sinha, N.K. and Gupta, M.M. (eds) (2000). Soft Computing & Intelligent Systems: Theory and Applications. Academic Press.

Miller, T.W., Sutton, III, R.S. and Werbos, P.J. (eds) (1990). *Neural Networks for Control.* The MIT Press, Cambridge, MA.

White, D.A. and Sofge, D.A. (eds) (1992). *Handbook of Intelligent Control.* Van Nostrand Reinhold, New York.

Worth consulting on intelligent control

Special Issues on Neural Networks in Control Systems of the IEEE *Control System Magazine*, **10**, No. 3, April (1990); **12**, No. 3, April (1992); and **15**, No. 3, June (1995).

Additional sources on the related topics

Arkin, R.C. (1998). Behaviour Based Robotics. MIT Press Publishing, Massachusetts, USA.

Borenstein, J. and Koren, Y. (1987). Motion Control Analysis of a Mobile Robot. *Journal of Dynamic Systems, Measurement and Control*, **109**, No. 2, The American Society of Mechanical Engineers, USA, pp. 73–9.

Brookes, R.A. (1991). Intelligence Without Representation. *Artificial Intelligence*, **47**, Elsevier Science Publishers, USA, pp. 139–59.

Chapuis, R., Potelle, A., Brame, J. and Chausse, F. (1995). Real-time Vehicle Trajectory Supervision on the Highway. *The International Journal of Robotics Research*, **6**, December, MIT, pp. 531–42.

Congress, N. (1996). Smart Road, Smart Car – The Automated Highway System. *Public Roads Magazine*, Autumn, USA, pp. 46–51.

Dorf, R.C. (1992). *Modern Control Systems.* 6th Edition, Addison-Wesley Publishing, Massachusetts, USA, pp. 75–9.

Engwirda, A., Vlacic, Lj.B., Hitchings, M.R. and Kajitani, M. (1998). Cooperation Among Autonomous Robots: Unsignalised Road Intersection Problem. *Proceedings of the Distributed Autonomous Robotic Systems Conference*, Karlsruhe, Germany.

Franklin, G.F., Powell, J.D. and Emami-Naeini, A. (1994). *Feedback Control of Dynamic Systems*, 3rd edn, Addison-Wesley Publishing, Massachusetts, USA, pp. 664–9.

Franssila, J. and Heikki, K. (1992). *Fuzzy Control of an Industrial Robot in a Transputer Environment.* Tampere University of Technology, Control Engineering Laboratory, Tampere, Finland.

Gajic, Z. and Lelic, M. (1996). *Modern Control Systems Engineering.* Prentice-Hall Publishing, Hertfordshire, UK.

Hamel, T. and Meizel, D. (1996). Robust Control Laws for Wheeled Mobile Robots. *Proceedings IFAC 13th Triennial World Congress*, San Francisco, USA, pp. 175–80.

Hitchings, M.R. (1998). TR-014 Lane-Following Obstacle Avoidance for Autonomous Mobile Robots. Technical Report, Intelligent Control Systems Laboratory, Griffith University, Australia.

Hitchings, M.R. (1998). TR-015 Corridor Traversal for Autonomous Mobile Robots. Technical Report, Intelligent Control Systems Laboratory, Griffith University, Australia.

Hitchings, M.R. (1998). TR-013 Users Manual and Developers Guide for the ICSL CAMRG2 Robots. Technical Report, Intelligent Control Systems Laboratory, Griffith University, Australia.

Hitchings, M.R., Engwirda, A.E. and Vlacic, L.B. (1997). Distance Control Applied to Cooperative Autonomous Mobile Robots. *Proceedings of the Australian Control Conference CONTROL'97*, Institution of Engineers, Australia, pp. 275–80.

Hitchings, M.R., Engwirda, A.E., Vlacic, L.B. and Kajitani, M. (1997). Two Sensor Based Obstacle Avoidance for Autonomous Vehicles. *Proceedings of the 3rd IFAC Symposium on Intelligent Autonomous Vehicles*, Spain, March (1998).

Jiang, Z. and Nijmeijer, H. (1997). Tracking Control of Mobile Robots: A Case Study in Backstepping. *Automatica*, **33**, No. 7, Elsevier Science, UK, pp. 1393–9.

Jones, J. and Flynn, A. (1993). Mobile Robots: Inspiration to Implementation. A.K. Peters Publishing Ltd, Wellesley, USA.

Kolodko, J. (1997). TR-007 ICSL Technical Report – Ultrasonic Ranging Subsystem for the ICSL Robots. *ICSL*, Griffith University.

Kuhn, T. and Wernstedt, J. (1996). Controlling of Autonomous Robot Systems with Fuzzy Techniques. *Proceedings of IFAC 13th Triennial World Congress*, San Francisco, pp. 257–62.

Kuo, B. (1995). *Automatic Control Systems*, 7th edn, Prentice-Hall International Publications, New Jersey, USA.

Langer, D. and Thorpe, C. (1997). Sonar-Based Outdoor Vehicle Navigation. In: *Intelligent Unmanned Ground Vehicles – Autonomous Navigation Research at Carnegie Mellon University.* Kluwer Academic Publishers, Massachusetts, USA, pp. 159–85.

Ming, A., Masek, V., Kanamori, C. and Kajitani, M., (1996). Cooperative Operation of Two Mobile Robots. *Proceedings of Distributed Autonomous Robotic Systems 2*, Asama, Fukuda, Arai and Endo (eds), Springer-Verlag, pp. 339–49.

Noreils, F.R. (1993). Toward a Robot Architecture Integrating Cooperation Between Mobile Robots: Application to Indoor Environment. *The International Journal of Robotics Research*, **12**, No. 1, February, Massachusetts, USA, pp. 79–98.

Olney, R., Wragg, R., Schumacher, R. and Landau, F. (1996). Automotive Collision Warning System. *Delco Electronics Corporation – Automotive Electronics Development*, Malibu, California.

Rausch, W.A. and Levi, P., (1996). Asynchronous and Synchronous Cooperation – Demonstrated by Deadlock Resolution in a Distributed Robot System. University of Stuttgart, Applied Computer Science – Image Understanding, Stuttgart, Germany.

Rosenblatt, J. and Thorpe, C. (1997). Behaviour Based Architecture for Mobile Navigation. In *Intelligent Unmanned Ground Vehicles – Autonomous Navigation Research at Carnegie Mellon University.* Kluwer Academic Publishers, Massachusetts, USA, pp. 19–32.

Shladover, S. (1993). Potential Contributions of Intelligent Vehicle/Highway Systems (IVHS) to Reducing Transportation's Greenhouse Gas Productions. *Transportation Research*, **27A**, No. 3, Pergamon Press, UK, pp. 207–16.

Stevens, W.B. (1996). The Automated Highway System Program: A Progress Report. *IFAC 13th Triennial Congress*, San-Francisco, USA, pp. 25–33.

Toyota Motor Corporation, (1996). *Breaking with the Past, Heading for the Future – Toyota ITS Vision*. Toyota Motor Corporation, Japan.

Van den Bogaert, T., Lemoine, P., Vacherand, F. and Do, S. (1993). Obstacle Avoidance in Panorama Espirit II Project. *Proceedings of the IFAC Intelligent Autonomous Vehicles Conference*, UK, pp. 51–56.

11

Decisional architectures for motion autonomy

Christian Laugier

and

Thierry Fraichard

INRIA, Rhône-Alpes and Gravir, France

11.1 Introduction

Autonomy in general and motion autonomy in particular has been a long standing issue in robotics. In the late 1960s and early, 1970s, Shakey (Nilsson, 1984) was one of the first robots able to move and perform simple tasks autonomously. Ever since, many authors have proposed control architectures to endow robot systems with various autonomous capabilities. Some of these architectures are reviewed in Section 11.2 and compared to the one presented in Section 11.3. These approaches differ in several ways, however it is clear that the control structure of an autonomous robot placed in a dynamic and partially known environment must have both *deliberative* and *reactive* capabilities. In other words, the robot should be able to decide which actions to carry out according to its goal and current situation; it should also be able to take into account events (expected or not) in a timely manner.

The control architecture presented in this chapter aims at meeting these two requirements. It is designed to endow a car-like vehicle moving on the road network with motion autonomy. It was initially developed within the framework of the French Praxitèle programme aimed at the development of a new urban transportation system based on a fleet of electric vehicles with autonomous motion capabilities (Daviet and Parent, 1996); more recent work on this topic is done within the framework of the LaRA (Automated Road) French project. The road network is a complex environment; it is partially known and highly dynamic with moving obstacles (other vehicles, pedestrians, etc.) whose future behaviour is not known in advance. However the road network is a structured environment with motion rules (the highway code) and it is possible to take advantage of these features in order to design a control architecture that is efficient, robust and flexible.

This chapter is organized as follows: in Section 11.2, an overview of the existing approaches to implementing a control architecture is presented and discussed; this section describes the three main classes of existing approaches (deliberative, reactive and hybrid), along with their main implementation alternatives and their respective advantages and drawbacks.

Section 11.3 describes the hybrid control architecture developed by Sharp. The rationale of the architecture and its main features are overviewed in Section 11.3.1. This section introduces the key concept of the *sensor-based manoeuvre*, i.e. general templates that encode the knowledge of how a specific motion task is to be performed. The models of the experimental car-like vehicles that are used throughout the chapter are then described in Section 11.3.2. Afterwards the concept of the *sensor-based manoeuvre* is explored in Section 11.3.3 and three types of manoeuvres are presented in detail (Sections 11.3.4, 11.3.5 and 11.3.6). These manoeuvres have been implemented and successfully tested on our experimental vehicles; the results of these experiments are finally presented in the Section 11.4.

One important component of the architecture is the *motion planner* whose purpose is to determine the trajectory leading the vehicle to its goal. Motion planning for car-like vehicles in dynamic environments remains an open problem and a practical solution to this intricate problem is presented in Section 11.5.

11.2 Robot control architectures and motion autonomy

11.2.1 Definitions and taxonomy

The development of robot control architectures constitutes for engineers and scientists one of the most challenging frameworks for integrating and testing *intelligent* systems, inspired from attributes of living beings such as perception, interaction and reasoning. Robot control architectures are rather understood in terms of software architectures, and consequently are closer to domains related to computer science and control engineering. A basic definition for a robot control architecture can be found in (Arkin, 1998): 'Robotic architecture is the discipline devoted to the design of highly specific and individual robots from a collection of common software building blocks.'

The state-of-the-art in this domain includes a large number of approaches, sometimes guided by research work on ethology and cognitive sciences. One of the most challenging domains for testing and evaluating these approaches, particularly when real-time constraints have to be verified, is mobile robotics. This is why most of the significant contributions in this research field come from work on mobile robot and autonomous guided vehicles. The next sections outline the state-of-the-art in mobile robot architectures, using a commonly agreed taxonomy. This taxonomy is based on three main paradigms on which a large number of control architectures have been developed:

- *The deliberative paradigm*. In this approach, the system uses a model of the world – an *a priori* known model, or a model reconstructed from sensory data – in order to plan the actions that the robots have to execute. This approach leads to a sequential decomposition of the whole process, and to highly hierarchical systems.

- *The reactive paradigm.* This approach is based on a tight coupling between sensors and actuators, for continuously producing the required controls. This approach usually relies on a decomposition of the system into elementary behaviours which can be combined and executed concurrently.
- *The hybrid paradigm.* This approach consists in combining the deliberative and reactive paradigms, in order to try to exploit the advantages of the two previous approaches. Most of the current approaches are of this type.

11.2.2 Deliberative architectures

This approach relies on traditional paradigms of artificial intelligence. It tries to implement a simplified view of human reasoning, and it is often referred as the 'sense-model-plan-act' (SMPA) scheme. In practice, this concept is implemented into robotic control systems using a hierarchical architecture made of three main components: perception (which includes sensing and modeling functions), decision and action.

- *Perception.* Considered as a key feature in a robotic system, the *perception function* may be seen as the first basic function of a deliberative architecture. The main purpose of this first stage of the control process is to construct a model of the environment from sensory data and a *a priori* knowledge (e.g. topological or a grid-based models). This model is subsequently used for planning robot actions and for checking that the robot actions have correctly been executed. However, world reconstruction from sensory data is a complete active research domain, having already motivated a great number of research works and approaches; this research domain is still open.
- *Decision.* The second processing phase of a deliberative architecture is referred as the *decision* module. It consists in 'reasoning' about the task model and the environment model, in order to decide what is the more appropriate sequence of actions to execute. In practice, this reasoning phase is often implemented as an *off-line motion planning* task. This is why motion planning has been a very active research domain for about 20 years.
- *Action.* The last processing phase of a deliberative architecture is to control the robot actuators in order to execute the planned actions. Recent research work in this domain addresses robust control techniques and sensor-based control approaches.

The first robot control architectures reported in the literature are based on such an approach. In particular, this type of architecture has been used for controlling the first mobile robot having a partial autonomy: the robot Shakey (Nilsson, 1984). This robot, designed at the beginning of the 1970s at the Stanford Research Institute, used a video-camera as a sensor and was theoretically able to move in a highly constrained environment. Its reasoning capabilities were derived from problem-solving techniques developed in the field of artificial intelligence. The typical tasks that could be achieved by Shakey consisted in finding a known object (i.e. an object described by its shape and its colour) in a room, and in pushing this object up to a given point. Unfortunately, each simple movement of Shakey required more than one hour of external computing, and it had a strong probability of failure at execution time.

The architecture of the Stanford Cart developed at the University of Carnegie-Mellon (Moravec, 1983) is also representative of the deliberative paradigm. In this approach, a 3D-vision system provided the robot with the positions of the objects located in its environment. Then, the motion planner generated a collision-free path allowing the robot to reach the desired placement. Finally, the system controlled the robot actuators in order to move it along the planned path, for a distance of about 1 metre. Unfortunately, the complete SMPA process had to be carried out after each motion of this type until the goal had been reached, and each iteration required to wait for about 15 minutes (mainly because of the image processing time).

The main serious problems related to purely deliberative architectures are the following:

- The main drawback of this type of approach relies in its intrinsic incapacity to cope with unpredicted events (mainly because of the large reaction time which is required for processing the whole (SMPA) cycle). Consequently, it is almost impossible to take into account dynamic objects or obstacles detected while the robot is moving. The main reason for these limitations comes from the slowness of the modelling and planning phases, which cannot be carried out in real time, even when they are implemented on large computers external to the robot.
- The second difficulty is related to the modelling phase itself, which is in charge of reconstructing a model of the robot environment from sensory data. Indeed, this problem in its whole generality represents a complete research domain which is still open (even if impressive results have already been obtained by researchers in the field of computer vision). It is well known, that relating sensory data to real objects is a difficult task because of the noisy, inaccurate, and often spread nature of the information to process. This difficulty is related to the fact that there is a great difference between *sensing* and *perception*.
- The intrinsic differences which exist between the model and the real world introduce strong uncertainties on the positions/orientations of the robot and of the obstacles. Taking into account these uncertainties is obviously necessary for obtaining a robust system. Unfortunately, this requirement makes the planning phase much more complex, and poses several modelling and algorithmic problems which are still open.

Consequently, the use of such an approach seems to be limited to the case of a robot evolving in a static, strongly constrained, and *a priori* known environment. This is why the purely deliberative approach is not used anymore in recent robot control architectures (even if it has raised key research issues that are still actively studied). However, it should be stressed that the deliberative approach has a significant advantage: it includes the property to apply high-level reasoning capabilities at the planning phase, and consequently to be potentially able to cope with complex missions combining several goals and task constraints. As it will be shown in Section 11.2.3, this property doesn't hold when applying purely reactive approaches.

11.2.3 Reactive architectures

The basic idea

Because of the above-mentioned strong limitations of the purely deliberative architectures, some researchers have developed in the 1980s a new approach inspired by ethology and entomology. Nowadays, we know that many live beings, in particular the insects, have very few capacities for 'modelling' and 'reasoning'. Despite this, they are able to achieve quite complex tasks; to an external observer, they exhibit a global behaviour which seems to be the result of intelligence. On the basis of this observation, some researchers have proposed an alternative to deliberative architectures consisting in making use of purely 'reactive behaviours'. This approach basically consists in removing the 'modelling' and 'planning' phases from the decisional loop, and in trying to produce 'intelligent behaviours' driven by sensory data: the robot *reacts intelligently* to what it *senses*. Brooks (Brooks, 1990) justifies the use of such an approach, by claiming that the 'best model of the world is the world itself'.

The implementation of such an approach is based upon the combination of several elementary modules implementing very simple (reactive) behaviours. Such approaches are often referred as *behavioural based architectures*: the observed behaviour of the robot is the result of the combination of some various elementary behaviours; it *emerges* from the interaction of the involved elementary behaviours with themselves and with the environment. Each elementary behaviour (e.g. avoiding an obstacle or the heading to a goal) performs a close coupling between the sensors and the effectors of the robot. The intrinsic low complexity of the involved processing, along with the parallel structure of the behaviours, leads to a high speed execution property.

Note: The previous type of system is usually referred as a 'reactive control system'. However, one can find two different definitions of reactivity in the literature: for peoples in the fields of computer science and robotics, a reactive system is 'a system able to react continuously to a physical environment, at a speed determined by this environment' (Harel and Pnueli, 1985); for peoples in the field of cognitive sciences, an agent is said to be reactive 'if it does not have an explicit representation of the world' (Ferber, 1995). Even if they look different, these two definitions are not contradictory, since they imply that the involved controllers react directly to the stimuli coming from the physical world; hence avoiding the construction of a 'non-correct' model of the world, and enabling high execution speeds. The next subsections illustrate some typical implementations of this type of architecture.

Subsumption architecture

R. Brooks is one of the pioneers in the domain of reactive architectures for robots, and his work on this topic is certainly the most known in our scientific community (Brooks, 1990). His behavioural architecture is based on a vertical decomposition into several *levels of competence* (see Figure 11.1). Each of these levels represents an

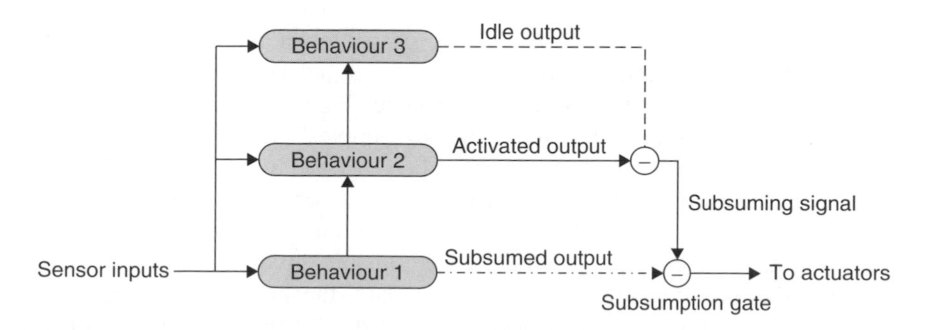

Fig. 11.1 Principle of the subsumption architecture (Brooks).

independent behaviour, receiving data from sensors and acting on the robot actuators. The first implementation of this architecture has been realized using augmented finite state machine (AFSM) models, a communication mechanism based on messages, and an inhibition/suppression mechanism for modifying dynamically the stimulus or response signals of some active modules.

It should be noticed that such an architecture doesn't have any central decisional kernel for selecting the required behaviours: this choice is continuously done at execution time using the structure of the implemented layers and communication network.

Eight levels of competence have been defined by Brooks as a guide for his work (see Brooks (1990) for more details): the lower levels of competence consist in (1) 'avoiding obstacles', (2) 'wandering aimlessly around', and (3) 'exploring the world'; the higher levels consist in (7) 'formulating and executing plans', and (8) 'reasoning about the behaviour of objects and modifying plans accordingly'. This approach has been used to develop several small autonomous robots at MIT. However, only the first three levels of competence have been implemented.

Such an architecture exhibit some interesting properties underlined by Brooks: (1) *a short response time* which provides the robot with the capacity to move in a dynamic environment, (2) *a robust controller* which can potentially work even if one of the modules doesn't work correctly (this is due to the parallel structure and the relative independence of the behaviours), (3) *an incremental structure* which potentially allows the implementation of new higher levels of competence, and (4) *a simple implementation* which drastically reduces the production cost and makes miniaturization possible (see for instance the proposal of Brooks for exploration robots (Brooks and Flynn, 1989)).

However, the experience has proven that such an architecture can hardly be fully implemented (except the lower levels of competence), and consequently it exhibits strong limitations coming from its lack of high-level reasoning. Indeed, even if this capacity is theoretically present in the highest levels of competence, these levels have never been implemented and the feasibility still remains doubtful. Consequently, only simple behaviours can be implemented in the current state-of-the-art.

Other reactive architectures

The supposed potentialities of the behavioural approach have motivated various developments and theories.

As has been previously mentioned, the subsumption architecture uses a hierarchy of behaviours, in which only one module at a time 'controls' the robot (by possibly integrating the results of some other behaviours located lower in the hierarchy). Another possible approach, as the one proposed by Anderson and Donath (1990), consists in combining the controls simultaneously 'recommended' by several behaviours. In such an approach, each reactive module implements a simple 'reflex behaviour' having no memory capacity. Such a type of behaviour has been defined by McFarland (McFarland, 1987): 'a reflex behaviour is the simplest form of reaction to an external stimulation; stimuli such as a sudden change in the level of illumination or a contact arising at a given point of the body, generate an automatic, involuntary and stereotyped response.' In the approach of Anderson and Donath, the involved modules generate artificial potential fields whose composition is used to determine the motion direction to be followed by the robot (by using the classical gradient technique). The set of the behaviours to be considered at a given time is selected according to criteria taking into the account the task to be achieved and the characteristics of the environment. In a similar way, Rosenblatt and Payton (Rosenblatt and Payton, 1989) have proposed to use an artificial neural network to combine the controls recommended by the selected behaviours; however, their system has been developed using an empirical approach, and the way the controls have to be appropriately combined is rather difficult to define.

In order to overcome the previous difficulty, some authors have tried to use training techniques for realizing the selection mechanism. For instance, Humphrys (1995) has proposed to apply a learning phase to each individual behavioural module, in order to construct a function for evaluating the 'quality' of the proposed actions; then this function is used on-line to select the best ranked behaviours. Lin (1993) developed a similar approach by using Q-learning techniques for implementing the training phase.

Although some improvements have been proposed for the behaviours selection and composition mechanisms, all these approaches exhibit the same general limitations as the subsumption architecture (see Section 11.2.2).

11.2.4 Hybrid architectures

How to hybridize?

The purely reactive and purely deliberative approaches represent two extremes that many authors naturally tried to combine. The objective is to try to preserve the potential high-level reasoning capacity of the deliberative approaches, while ensuring the robustness and the short response time of reactive approaches. Such approaches are referred to as *hybrid architectures*. However, several types of hybrid architecture can be distinguished depending on the way the deliberative and reactive components have been combined. A commonly used classification (with some minor variations) in the literature consists in considering three main types of approaches:

- *Deliberative-based hybrid approaches*. In such approaches, the *planning function* has a predominant role (i.e. motion is driven by planning using the SMPA paradigm), and the *reactive functions* are only added for dealing with some exceptions. The related elementary reactive actions can either be integrated as part of the architecture, or inserted into the generated motion plans by the planner.
- *Reactive-based hybrid approaches*. In this type of approach, the robot motions are executed under the supervision of the reactive functions, and the planning functions are mainly used as a 'guidance resource' for the reactive component of the system.
- *Three-layered hybrid approaches*. Most of the current developments on robot control architectures are based on this type of approach. Such an approach may be seen as the implementation of an 'adaptable planning-reacting scheme'. The basic idea consists in adding an intermediate layer for appropriately interfacing the deliberative and reactive functions. Then, one can consider that this type of architecture relies on the three following basic logical functions: *planning, sequencing* and *reacting*. In such a paradigm, the behaviours of the reactive layer are conditionally instantiated by the sequencing module, according to some sensing conditions and constraints defined by the planning module.

Deliberative-based hybrid architectures

JPL exploratory robot architecture (Gat) This architecture has been proposed by Gat (Gat *et al.*, 1990) at JPL (Jet Propulsion Laboratory), for providing a planet exploratory mobile robot with some partial autonomy capabilities (since such a robot cannot be directly teleoperated from the earth because of communication delays). As shown in Figure 11.2, this architecture is based on four main modules: the Perception Module, the Path Planner, the Execution Monitoring Planner, and the System Executive.

The main task of the Perception Module is to build a local map of the environment, using sensory data and global data provided by the orbiter. Then, this map is used by the Path Planner and by the Execution Monitoring Planner to respectively plan a collision-free path (of roughly 10 metres) and a complete plan including execution monitoring parameters. This motion plan is obtained by simulating the displacement of the robot along the planned collision-free path, and by anticipating the possible failures and the sensory data to monitor. The parameterized motion plan is finally used by the System Executive to monitor the execution of the robot task. In practice, only some predefined and simple 'Reflex Actions' can be introduced into the motion plan (e.g. stopping the robot and moving back to a safe position). Such an architecture basically applies the sequential SMPA paradigm, while executing some predefined reflex manoeuvres when dangerous situations have been detected. This approach represents to some extent the minimum level of integration of a reactive component into a deliberative architecture.

Payton's architecture This architecture (Payton, 1986) is based on a hierarchical decomposition, in which each layer is characterized by a type of sensory data processing (modelling). As shown in Figure 11.3, this architecture is composed of a layered perception system and of four main decisional modules: (1) the 'Mission planner' defines a sequence of geographical goals to reach along with their associated motion

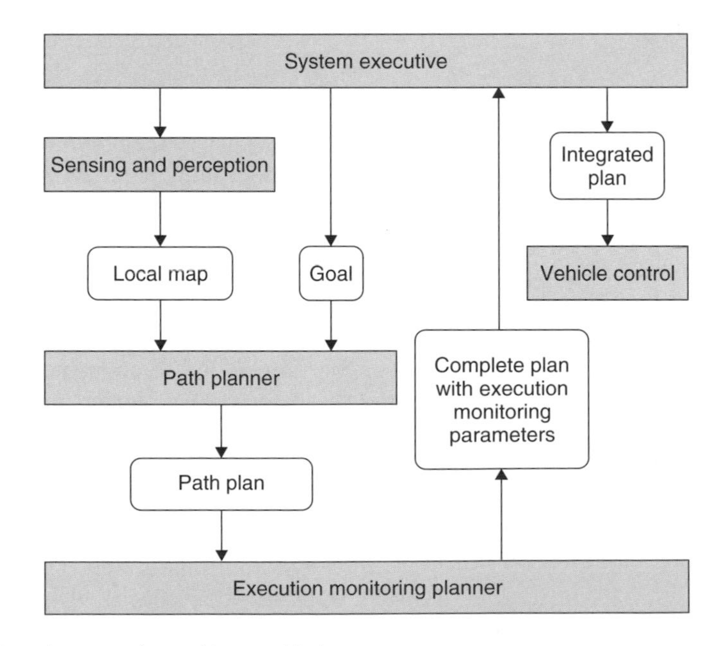

Fig. 11.2 JPL exploratory robot architecture (Gat).

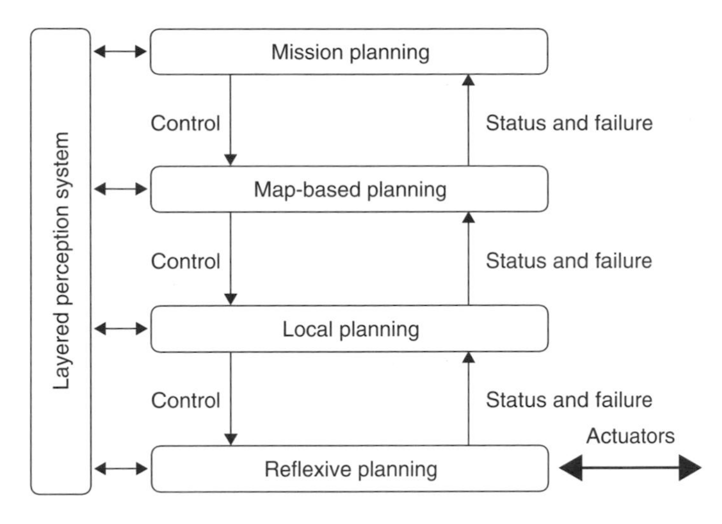

Fig. 11.3 Payton's architecture.

constraints; (2) the 'Map-based planner' uses the global world model to generate paths connecting the previous geographical goals (the response time of this planner is of a few minutes); (3) the 'Local planner' determines the details of the motions which are required for moving the robot along the planned paths (the response time of this planner is of a few seconds); (4) the 'Reflexive planner' controls in real time the execution of the motion task.

From the implementation point of view, this architecture has been developed using expert agents communicating between them using a blackboard technique. Using this approach, the activity of a particular module can theoretically be controlled by a higher layer through the selection of the expert agents to be activated. However, only the implementation of some expert agents belonging to the 'Reflexive planner' (e.g. follow a wall, or avoid an obstacle) has been deeply described. In this implementation, the related reflex behaviours are associated with some virtual sensors in charge of providing a specialized information (e.g. obstacle detection, object recognition, localization). Then, the activation of the appropriate reflex behaviours is done using a blackboard technique and some predefined priorities.

Quite complex missions have been planned and executed with a significant level of reactivity using this approach. The main limitations of the system come from both the limited communication mechanism existing between the different layers, and the predefined combination of behaviours. Later on, Payton *et al.* (1990) and Rosenblatt (1997) have improved the reactivity of the system by using a distributed control arbitration technique, allowing them to combine controls coming from both the reactive behaviours and the planning layers.

TCA (Simmons) The TCA (task control architecture) proposed by Simmons (1994) represents a new alternative to traditional hierarchical approaches. This architecture is composed of an arbitrary number of specialized modules, communicating through messages with a central management module. The specialized modules carry out the tasks which are specific to the robot to control, whereas the central management module supervises the functioning of the whole system and controls the routeing of

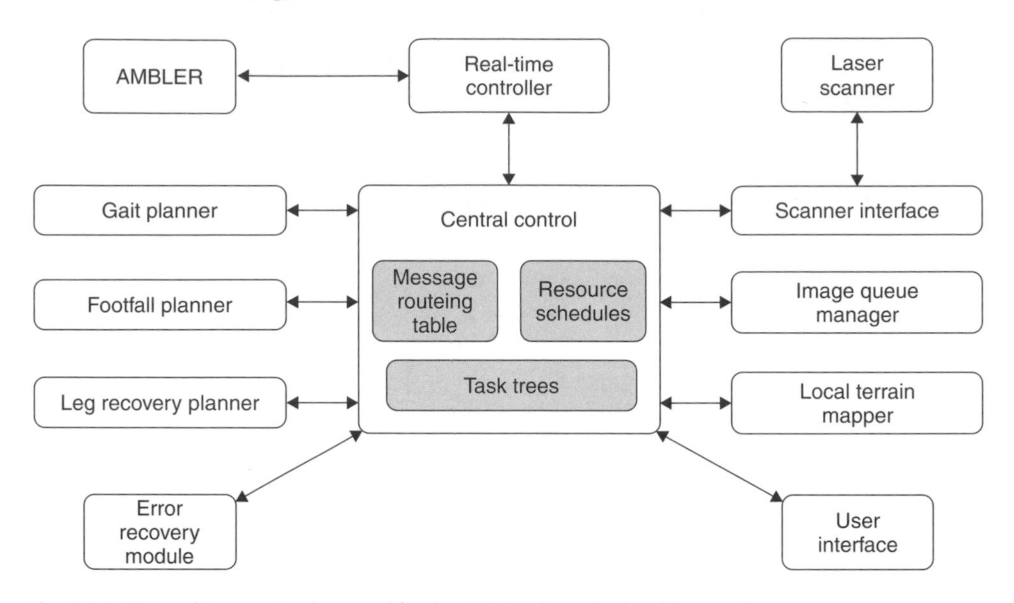

Fig. 11.4 TCA architecture implemented for the AMBLER legged robot (Simmons).

the messages between the various modules; messages can be used for an information request, for sending a command, or for asking for a task decomposition to the planners. The TCA architecture makes use of a hierarchical representation of the tasks/sub-tasks relationships (called the 'task tree') for maintaining an internal representation of the robot task to execute.

Figure 11.4 shows how the TCA architecture has been implemented for controlling the Ambler legged robot. However, this implementation put the emphasis onto the planning functions (e.g. gait planner, footfall planner, etc.), and reduces the reactivity to the processing of some exceptions (for stabilizing the robot). The main drawback of this architecture relies on the centralized processing schema and its associated communication mechanism, which often implies rather long response times incompatible with fast robots. This is why the author has also implemented additional (i.e. apart from the TCA architecture) some 'emergency reflexes' for quickly stabilizing the Ambler robot when a problem arose.

Reactive-based hybrid architectures

AuRA architecture (Arkin) The AuRA architecture proposed by Arkin (1987; 1989; 1990) is mainly based on the concepts of 'motor schema' and 'perceptive schema', which are used for describing the links existing between action and perception (like in reactive approaches). A motor schema specifies a generic behaviour which can be instantiated under some conditions, for generating a particular type of robot motion (e.g. moving along a straight line, moving towards a goal position, or avoiding a given obstacle); each motor schema is associated to a perceptive schema (*action-oriented perception*) in charge of providing the required information.

The AuRA architecture is mainly composed of two components: a hierarchical component in charge of the modelling and planning tasks, and a reactive component inhabited by the motor and perceptual schemas (see Figure 11.5). The hierarchical component is composed of three classical layers: the *Mission planner* which generates

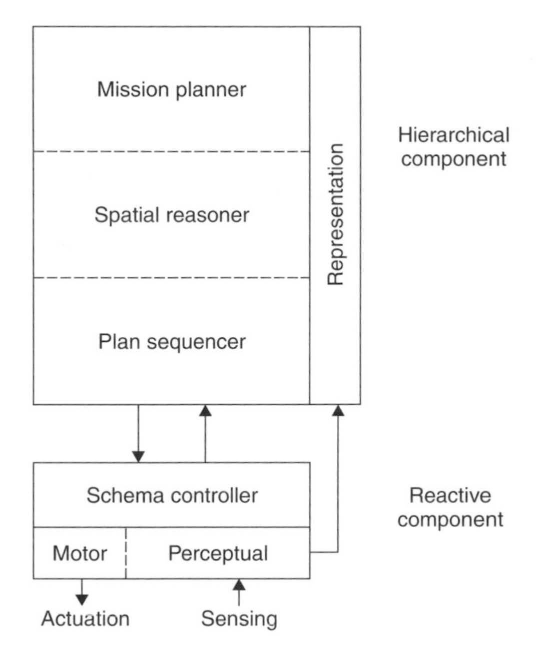

Fig. 11.5 AuRA architecture (Arkin).

a sequence of sub-goals to achieve, the *spatial reasoner (or navigator)* which constructs executable paths using cartographic data stored in a long-term memory, and the *plan sequencer (or pilot)* which selects and instantiates the appropriate behaviours. The reactive component makes use of a vector field approach to combine the movements proposed by the activated motor schemas, and to generate the required controls. In practice, the deliberative part of the system mainly produces way-points and associated behaviours; it is reactivated only when a fatal failure has been detected (no more motion or timeout). This approach doesn't allow the processing of more complex missions combining several manoeuvres; it also suffers from the classical drawback of reactive approaches: the combination of behaviours generates motions which can hardly been predicted, and conflicts may appear when 'opposite' behaviours have to be considered.

SSS architecture (Connell) The SSS (Symbolic, Subsumption, Servo) architecture proposed by Connell (1992) is composed the three layers (see Figure 11.6), associated with three levels of discretization of the robot state space and of the time. The lower layer (Servo) operates using continuous space and time representations for controlling the robot and the sensing functions. The intermediate layer (Subsumption) works using a continuous time representation and a discrete state space model for generating specialized behaviours (e.g. wall following, or crossing a door). The higher layer (Symbolic) operates under discrete space and time representations for selecting the behaviours to apply according to the task to achieve and to the arising events; this layer makes use of a map of the environment containing 'landmarks' connected by paths (straight line segments).

In practice, this architecture mainly behaves as an improved subsumption architecture, having a more elaborated mechanism for selecting or inhibiting behaviours (the

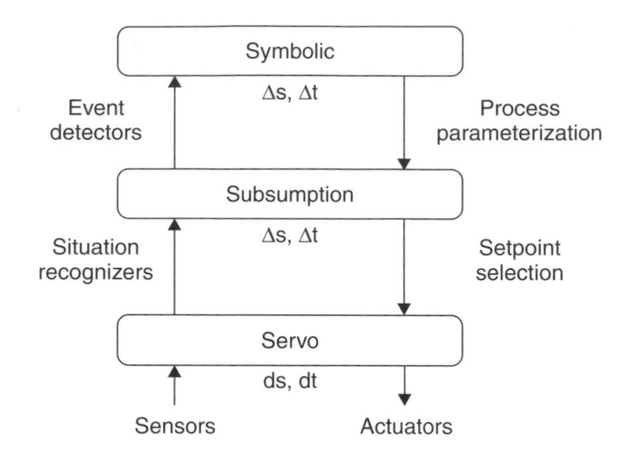

Fig. 11.6 SSS architecture (Connel).

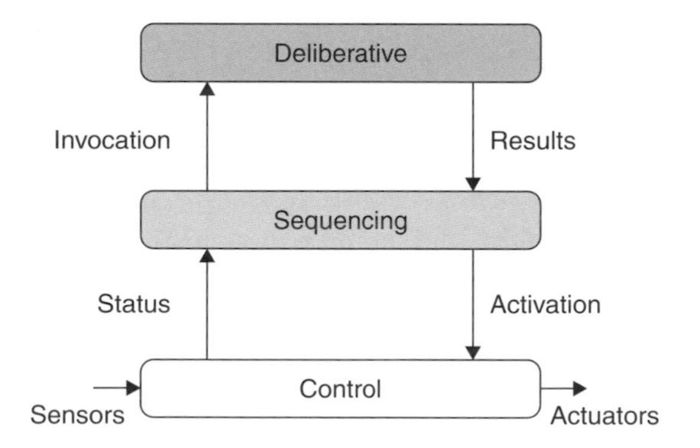

Fig. 11.7 ATLANTIS architecture (Gat).

role of the symbolic layer is mainly to construct a contingency table indicating under which conditions the behaviours of the subsumption layer have to be activated).

Three-layered hybrid architectures

ATLANTIS (Gat) and 3T (Bonasso) architectures The ATLANTIS (Gat, 1992; 1997) and 3T (Bonasso *et al.*, 1996) architectures are both based on three main layers: (1) the higher layer which includes the deliberative functions (planners), (2) the intermediate layer whose purpose is to manage the various sequences of actions to execute, and (3) the lower layer which includes the reactive control mechanisms. These three functional layers are generally represented in all the contemporary mobile robot control architectures. The intermediate layer, which can be seen as an 'advanced interface' between the deliberative and reactive components, has an important role to play for appropriately integrating high-level task representations and real-time reactive capabilities. The three previous layers are respectively called *Deliberator, Sequencer* and *Controller* by Gat (Figure 11.7), and *Planner, Sequencer* and *Skill Manager* by Bonasso (Figure 11.8).

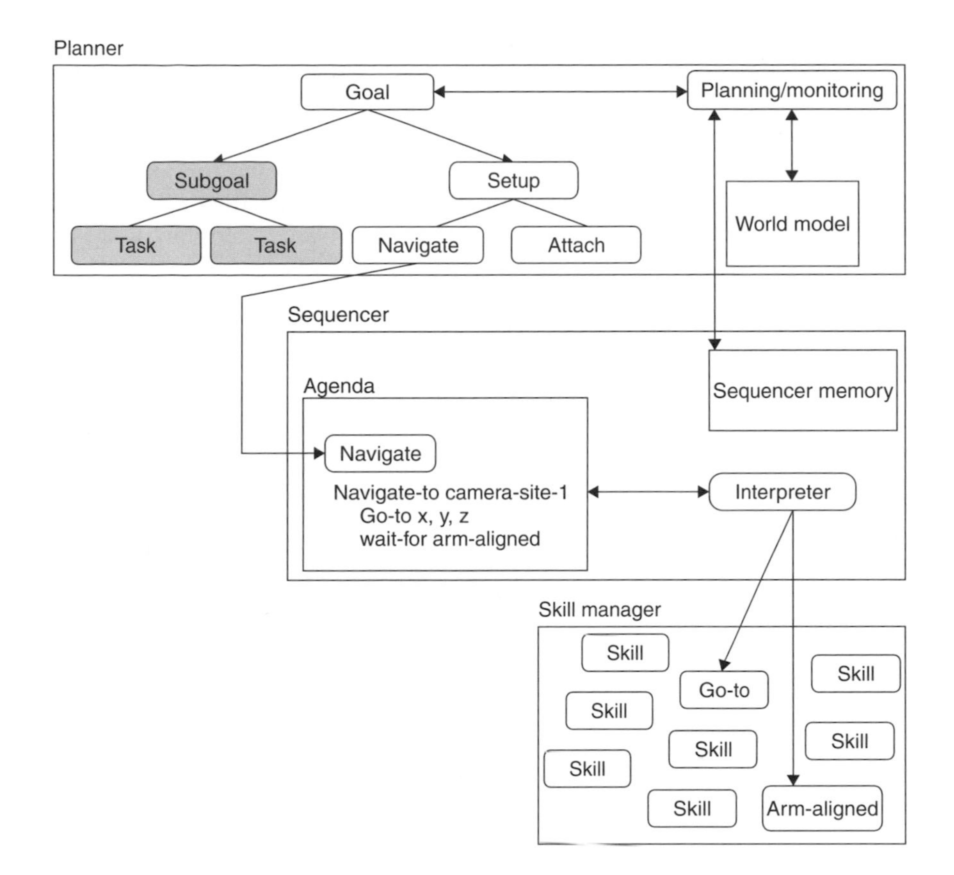

Fig. 11.8 3T architecture (Bonasso).

In these architectures, the lower layer gathers together several functions implementing simple behaviours such as wall following or obstacle avoidance; it constitutes a library of *skills*, used at the request of the higher levels. The intermediate layer (the *Sequencer*) selects and parameterizes the set of behaviours to apply in the current state of the robot task; in order to authorize the concurrent execution of several alternative action plans (the appropriate solution being chosen at execution time according to some identified internal an external events), this layer makes use of the 'conditional sequencing' principle (Gat, 1997). Using this approach, the role of the higher layer is to produce action plans for 'guiding' the robot movements, rather than generating a single sequence of actions. In the 3T architecture, the action plans are previously sent to the *Sequencer*, whereas they are produced in response to the requests from the *Sequencer* in the ATLANTIS system.

The implementation of these three basic layers varies from one system to another. The way the *Sequencer* is implemented may obviously have strong consequences to the scope and the robustness of the whole system. In the 3T and ATLANTIS architectures, this layer has respectively been implemented using the RAP and ESL languages. These two languages have similar characteristics, and they both rely on the same basic principle (Noreils, 1990): 'Rather than trying to build algorithms that never fail (which is impossible when dealing with real robots), it is better to build algorithms that never fail to detect a failure'; Gat (1997) calls such a type of failure, a *'cognizant failure'*. The RAP (Reactive Action Packages (Firby, 1989)) language, basically allows the specification of a set of procedures (sequences of actions) which have to be activated when some predefined conditions are verified. A particular RAP specifies the different strategies which are known to achieve a given goal according to some contextual conditions; in this description, any strategy may in turn be specified using some other RAPs. Then, the appropriate RAPs are successively activated according to an agenda constructed and updated by an interpreter (Figure 11.8).

HILARE architecture (LAAS) In the HILARE architecture from LAAS[1] (Alami *et al.*, 1998), the three previous layers are roughly represented by the *Decisional, Execution Control* and *Functional layers* (Figure 11.9); this architecture is completed by a *Logical*

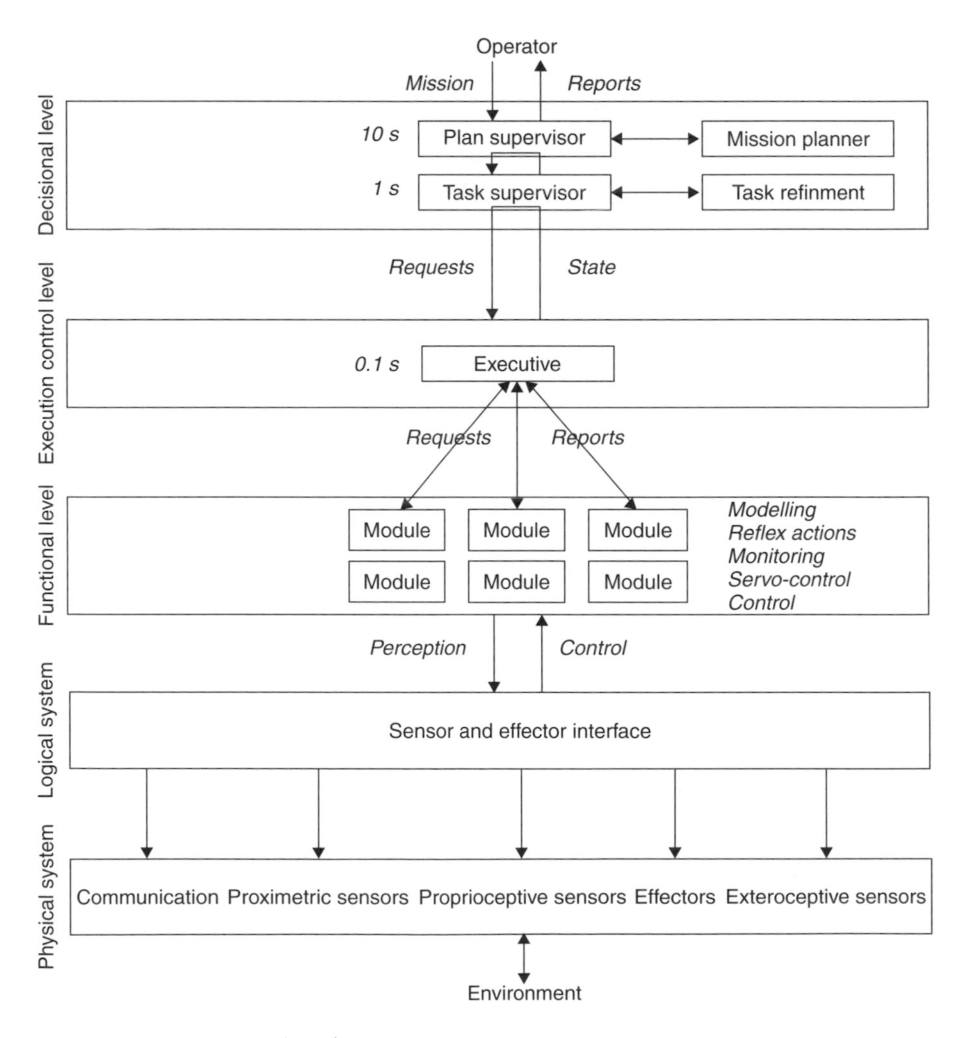

Fig. 11.9 HILARE architecture (LAAS).

[1] Laboratoire d'Analyse et d'Architecture des Systèmes, Toulouse

robot layer whose purpose is to interface the system with the physical resources (in order to increase the portability).

The *Decisional layer* is in charge of processing all the requests requiring a global knowledge of the task and of the execution context (i.e. planning and decision-making). In order to be compatible with the constraints of the reactive functions of the architecture, this layer has been divided into two components having different response times: the *Plan supervisor*, which generates appropriate action plans from a description of the tasks to achieve, and the *Task supervisor*, which is in charge of supervising the work of the intermediate layer and of producing (when it is needed) the task refinements. The action plans that are produced include the required modalities of execution (mainly events to monitor and their associated reflex actions).

The *Execution control layer* is in charge of executing in a reactive way the action plans produced by the *Decisional layer*. For that purpose, it continuously selects, parameterizes, and activates the appropriate modules of the *Functional layer*. This is done using an automation automatically generated from a set of logical rules.

The *Functional layer* is composed of all the functions which are required for processing the perception and robot actions (e.g. sensory data processing, events observers, motion controls, etc.). These functions have been encapsulated into modules communicating between themselves and with the *Execution control layer* using a client/server protocol. Then, a particular module is activated through a request (including the involved execution parameters) sent by the *Execution control layer* or by an other module; this module remains active until the task has been completed or a failure has been detected. From the implementation point of view, this layer is organized as a network of interacting modules.

Sharp architecture (INRIA) The Sharp architecture developed at INRIA[2] (Laugier *et al.*, 1998) exhibits the same basic functional structures as the previously described three-layered systems (i.e. the Deliberator, the Sequencer, and the Controller). The main objective of this architecture is to take into account the particular constraints and properties of a car-like vehicle, in order to be able to *efficiently and safely* control the motions of a car moving on the road network; another constraint is to obtain *smooth motions*, i.e. to appropriately control velocities and accelerations.

In the Sharp architecture, the three previous functional layers are respectively represented by the *Planner*, the *Mission Scheduler*, and the *Motion Controller* (see Section 11.3.1 for more details). However, the efficiency and robustness of the system has been improved by introducing a new concept for constructing on-line the action plans: the concept of '*sensor-based manoeuvre*' (SBM) which can be seen as *meta-skill*. Using this approach, the *Mission Scheduler* can efficiently construct motion plans (i.e. with the required response time) by combining previously planned trajectories with appropriate instantiations of these generic skills (SBM). This approach has been both motivated and justified by the fact that in the considered application (i.e. an intelligent vehicle moving on the road network), the number of different types of manoeuvres to execute is finite and not too large. In particular, it is useless to fully re-plan a similar sequence of actions (i.e. which mainly differ from one to the other by the execution

[2] Institut National de Recherche en Informatique et Automatique

parameters) each time that the vehicle is going to execute an overtaking or a parallel parking manoeuvre. This approach is described in more detail in Section 11.3.

11.2.5 Conclusion

Nowadays, the three-layer paradigm represents a consensus about the conception of a hybrid control architecture. Yet, the level of abstraction, of reactivity, and of capacity of these layers are still a design problem and a motivation for many research works. This is probably due to the fact that the nature of the integration of deliberative and reactive features is not yet well understood; it is also probably due to the fact that no universal solution exists and that a particular type of control architecture responds to a particular niche of application. The control architecture presented in Section 11.3 has specially been designed and implemented for the intelligent vehicle application.

11.3 Sharp control and decisional architecture for autonomous vehicles

11.3.1 Overview of the Sharp architecture

The Sharp control and decisional architecture is depicted in Figure 11.10. As has previously been mentioned, this architecture is a three-layer based hybrid architecture that has been specially designed and implemented for controlling an autonomous road vehicle. This architecture takes advantages of the particular features of the considered application domain (and in particular the fact that the road network is a structured environment with motion rules), for making use of appropriate constructions (the *sensor-based manoeuvres* or *SBM*) for improving the efficiency, the robustness and the flexibility of the system. The SBM concept is a key concept of our control and decisional architecture; it is derived from the artificial intelligence paradigm of *script* (Rich and Knight, 1983). A script is a general template that encodes procedural knowledge of how a specific type of task is to be performed. A script is fitted to a specific task through the instantiation of variable parameters in the template; these parameters can come from a variety of sources (*a priori* knowledge, sensor data, output of other modules, etc.). Script parameters fill in the details of the script steps and allow it to deal easily with the current task conditions.

The introduction of SBM was motivated by the observation that the kind of motion task that a vehicle has to perform can usually be described as a series of simple steps (a script). An SBM is a script, it combines *control* and *sensing skills*. Skills are elementary functions with real-time abilities: sensing skills are functions processing sensor data whereas control skills are control programs (open or closed loop) that generate the appropriate commands for the vehicle. Control skills may use data provided directly by the sensors or by the sensing skills.

As it has already been mentioned, the idea of combining basic real-time skills to build a plan in order to perform a given task can be found in some other robot control architectures (see Section 11.2); this approach permits the authors to obtain robust, flexible and reactive behaviours. From the conceptual point of view, an SBM can be seen

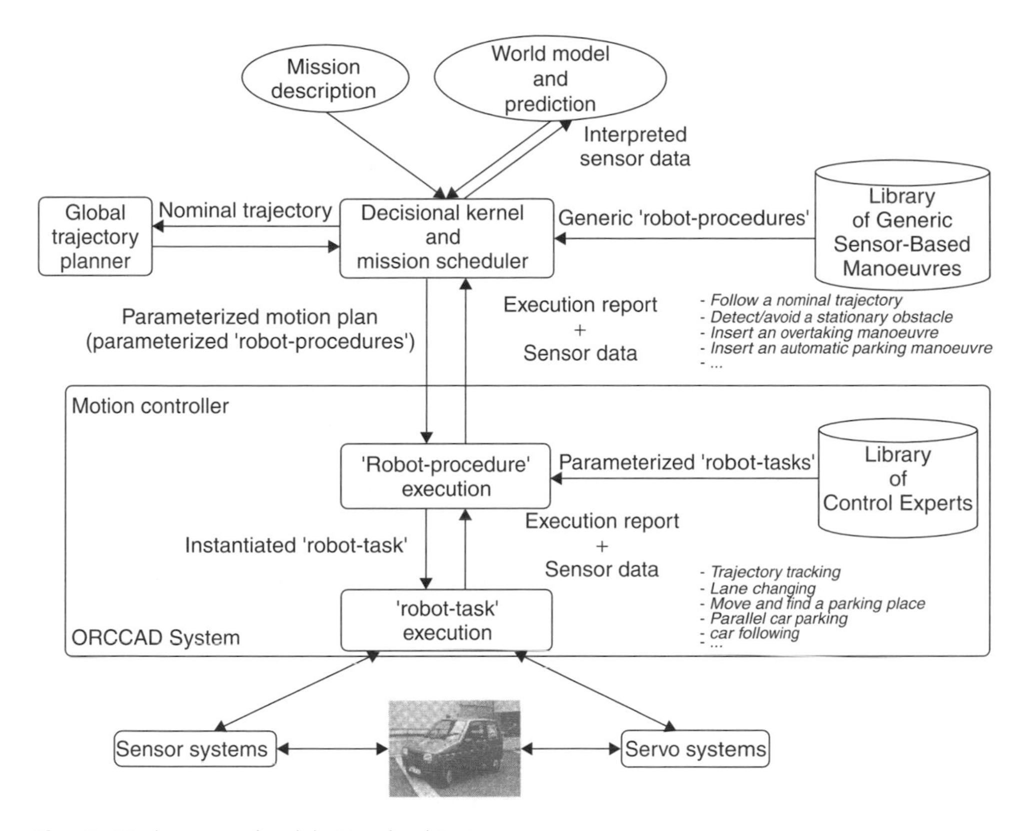

Fig. 11.10 Sharp control and decisional architecture.

as 'meta-skill', which encapsulates high-level expert human knowledge and heuristics about how to perform a specific motion task (see Section 11.3.3). Accordingly they allow a reduction in the planning effort required to address a given motion task, thus improving the overall response-time of the system, while retaining the good properties of a skill-based architectures, i.e. robustness, flexibility and reactivity.

Our control and decisional architecture features three main components, the *Mission scheduler*, the *Motion planner*, and the *Motion controller*, which are described below.

Mission scheduler

When given a mission description, e.g. 'park at location l', the *Mission scheduler* generates a *parameterized motion plan (PMP)* which is an ordered set of generic sensor-based manoeuvres (SBM) possibly completed with nominal trajectories. The SBMs are selected from a SBM library, according to the current execution context. An SBM may require a nominal trajectory (as is the case for instance of the 'follow trajectory' SBM). A nominal trajectory is a continuous time-ordered sequence of (position, velocity) of the vehicle that represents a theoretically safe and executable trajectory, i.e. a collision-free trajectory which satisfies the kinematic and dynamic constraints of the vehicle. When they are needed, such trajectories are computed by the *Motion planner*, under the request of the *Mission scheduler*.

The involved SBMs, along with their associated nominal trajectories, are passed to the *Motion controller* for their reactive executions.

Motion planner

The *Motion planner* is in charge of generating collision-free trajectories which satisfy the kinematic and dynamic constraints of the vehicle. Such trajectories are computed using:

- an *a priori known* or *acquired model* of the vehicle environment,
- the current *sensor data*, e.g. position and velocity of the moving obstacles,
- a *world prediction* that gives the most likely behaviours of the moving obstacles.

Motion planning is detailed in the Section 11.5.

Motion controller

The goal of the *motion controller* is to execute in a reactive way the current SBM of the PMP. For that purpose, the current SBM is instantiated according to the current execution context, i.e. the variable parameters of the SBM are set by using the *a priori* known or sensed information available at the time, e.g. road curvature, available lateral and longitudinal space, velocity and acceleration bounds, distance to an obstacle, etc. As mentioned above, an SBM combines control and sensing skills that are either parameterized control programs or sensor data processing functions. It is up to the *Motion controller* to control and coordinate the execution of the different skills required. The sequence of *control skills* that is executed for a given SBM is determined by the events detected by the *sensor skills*. When an event that cannot be handled by the current SBM happens (e.g. the intrusion of an unexpected obstacle which cannot be avoided using the current *control skills*), the *Motion controller* reports a failure to the *mission scheduler* which updates the current PMP either by applying a re-planning procedure (time permitting), or by selecting in real-time an SBM adapted to the new situation.

11.3.2 Models of the vehicles

The Sharp control and decisional architecture has been tested on two experimental vehicles with slightly different kinematic characteristics. The first one is a commercial *Ligier* electrical vehicle (Figure 11.11(a)). The second one is a special prototype especially designed for the purpose of the *Automated Public Car* project (Laugier and Parent, 1999) (Figure 11.11(b)). The kinematics of the Ligier is that of a regular car whereas the Cycab has four wheels that can be steered (a steering angle ϕ on the front wheels induces a steering angle $-k\phi$ on the rear wheels). Accordingly, its kinematics is slightly different.

The kinematic properties of a car-like vehicle are explored in detail in Section 11.5. From a control point of view, the respective models of the Ligier (left) and the Cycab (right) are:

$$\begin{cases} \dot{x} = v\cos(\theta + \phi) \\ \dot{y} = v\cos(\theta + \phi) \\ \dot{\theta} = \dfrac{v}{L}\sin\phi \end{cases} \qquad \begin{cases} \dot{x} = v\cos(\theta + \phi) \\ \dot{y} = v\cos(\theta + \phi) \\ \dot{\theta} = v\dfrac{\sin(\phi + k\phi)}{L\cos(k\phi)} \end{cases} \tag{11.1}$$

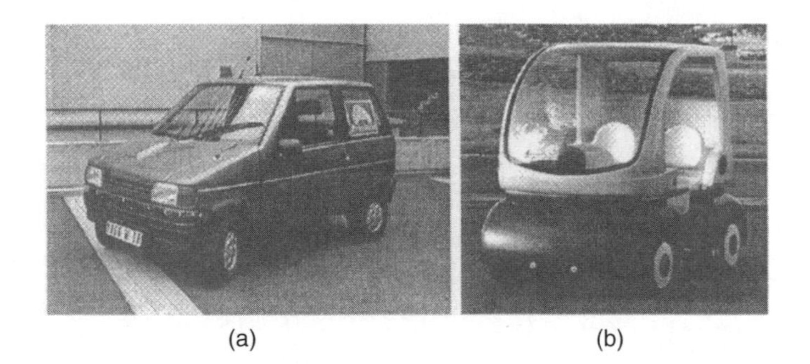

(a) (b)

Fig. 11.11 The *Ligier* vehicle (a), and the *Cycab* (b).

where x and y are the coordinates of the front axle midpoint, θ is the orientation of the vehicle and L is the wheel base. The controls are ϕ the steering angle and ν the velocity of the front wheels.

11.3.3 Concept of sensor-based manoeuvre

As has previously been mentioned, our control and decisional architecture strongly relies upon the concept of sensor-based manoeuvre (SBM) for providing the system with the required reactivity and robustness properties, while being able to generate smart motion controls for the vehicle. At a given time instant, the vehicle is carrying out a particular SBM that has been instantiated to fit the current execution context (see Section 11.3.1). SBMs are general templates encoding the knowledge of how a given motion task is to be performed. They combine real-time functions, control and sensing skills, that are either control programs or sensor data processing functions. From the practical point of view, a SBM can be seen as a specialized controller which generates *safe and smooth motions* for executing in a reactive way a given type of manoeuvre (i.e. by combining some predefined sensory modalities and controls).

In the sequel, we will use two particular types of SBM for illustrating this concept and for showing how it works in practice: the 'trajectory following' SBM, and the 'parallel parking' SBM. These two types of SBM have been developed and integrated in our control and decisional architecture; they have also been implemented and successfully tested on a real automatic vehicle; the results of these experiments are presented in Section 11.4. The Orccad tool (Simon *et al.*, 1993) has been selected to implement both SBMs and *skills*: robot procedures (in the Orccad formalism) are used to encode SBM's, while 'robot-tasks' encode *skills*; robot procedures and robot tasks can both be represented as finite automata or transition diagrams. The 'trajectory following' and 'parallel parking' SBMs are depicted in Figure 11.12 as transition diagrams. The control skills are represented by square boxes, e.g. 'find parking place', whereas the sensing skills appear as predicates attached to the arcs of the diagram, e.g. 'parking place detected', or conditional statements, e.g. 'obstacle overtaken?'. The control skills are used to control the motions of the vehicle and to activate the selected sensors; the task of the sensing skills is to evaluate the involved perception-based predicates or conditional statements.

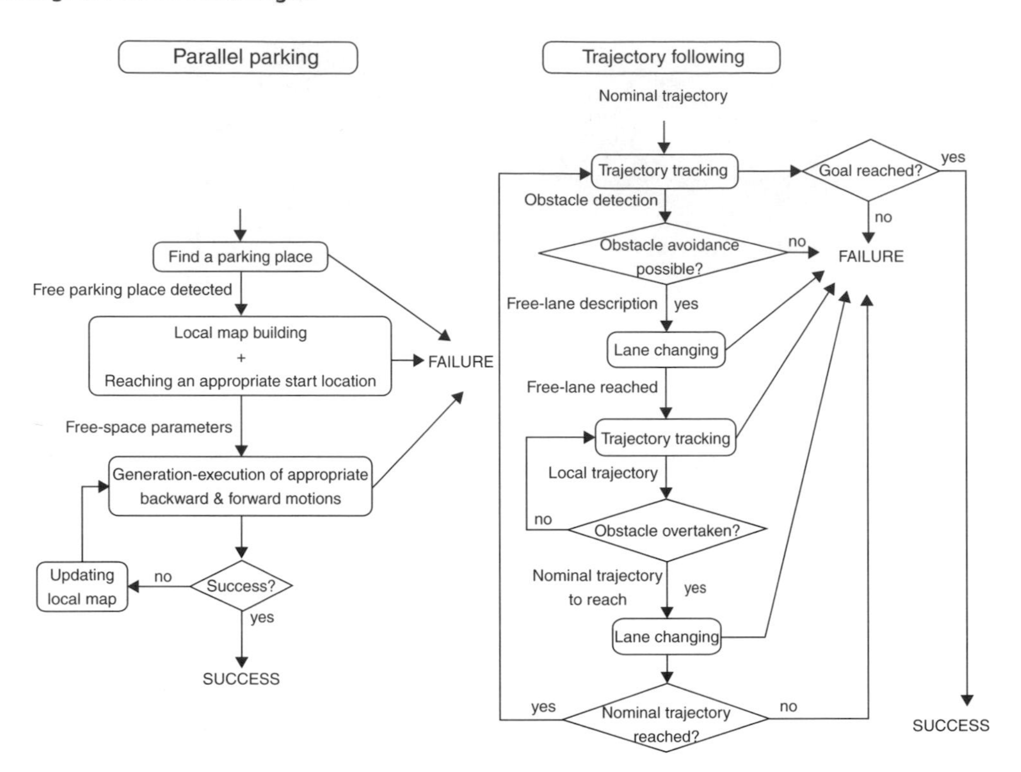

Fig. 11.12 The 'parallel parking' and 'trajectory following' SBM.

The next two sections describe how the two SBMs illustrated in Figure 11.12 operate. Section 11.3.6 presents an other type of SBM involving a specialized sensing device.

11.3.4 Reactive trajectory following

Outline of the SBM

The purpose of the trajectory following SBM is to allow the vehicle to follow a given nominal trajectory as closely as possible, while reacting appropriately to any unforeseen obstacle obstructing the way of the vehicle. Whenever such an obstacle is detected, the nominal trajectory is locally modified in real time, in order to avoid the collision. This local modification of the trajectory is done in order to satisfy a set of different motion constraints: collision avoidance, time constraints, kinematic and dynamic constraints of the vehicle. In a previous approach, a fuzzy controller combining different basic behaviours (trajectory tracking, obstacle avoidance, etc.) was used to perform trajectory following (Garnier and Fraichard, 1996). However this approach proved unsatisfactory: it yields oscillating behaviours, and does not guarantee that all the aforementioned constraints are always satisfied.

The trajectory following SBM makes use of *smooth local trajectories* to avoid the detected obstacles. These local trajectories allow the vehicle to move away from the obstructed nominal trajectory, and to catch up this nominal trajectory when the (stationary or moving) obstacle has been overtaken. All the local trajectories verify the motion constraints. This SBM relies upon two control skills, *trajectory tracking* and *lane changing* (see Figure 11.12), that are detailed now.

Trajectory tracking

The purpose of this control skill is to issue the control commands that will allow the vehicle to track a given nominal trajectory. Several control methods for non-holonomic robots have been proposed in the literature. The method described in (Kanayama *et al.*, 1991) ensures stable tracking of a feasible trajectory by a car-like robot. It has been selected for its simplicity and efficiency. The vehicle's control commands are of the following form:

$$\dot{\theta} = \dot{\theta}_{ref} + v_{R,ref}(k_y y_e + k_\theta \sin\theta_e), \tag{11.2}$$

$$v_R = v_{R,ref}\cos\theta_e + k_x x_e, \tag{11.3}$$

where $q_e = (x_e, y_e, \theta_e)^T$ represents the error between the reference configuration q_{ref} and the current configuration q of the vehicle ($q_e = q_{ref} - q$), $\dot{\theta}_{ref}$ and $v_{R,ref}$ are the reference velocities, $v_R = v\cos\phi$ is the rear axle midpoint velocity, k_x, k_y, k_θ are positive constants (the reader is referred to Kanayama *et al.* (1991) for full details about this control scheme).

When the reference trajectory is considered as too far from the current vehicle configuration (i.e. out of the range of validity of the error parameters of the Kanayama control law), a smooth local trajectory is generated and tracked in order to appropriately catch up the reference trajectory (Figure 11.13). These local trajectories are generated using second degree polynomial functions.

Lane changing

This control skill is applied to execute a lane changing manoeuvre. The lane changing is carried out by generating and tracking an appropriate smooth local trajectory. Let T be the nominal trajectory to track, d_T be the distance between T and the middle line of the free lane to reach, s_T be the curvilinear distance along T between the vehicle and the obstacle (or the selected end point for the lane change), and $s = s_t$ be the curvilinear abscissa along T since the starting point of the lane change (see Figure 11.14).

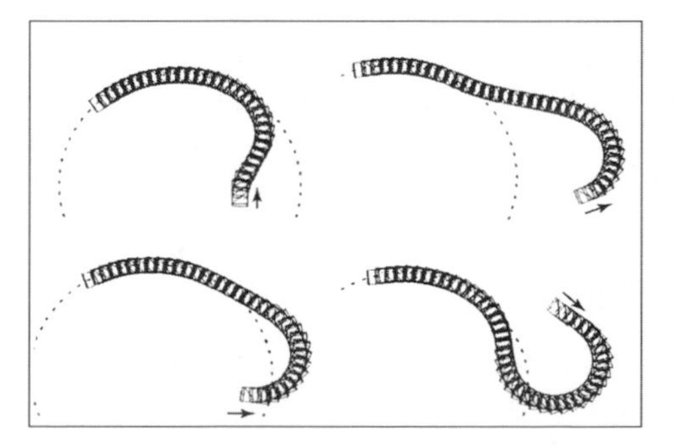

Fig. 11.13 Examples of local 'catching up' trajectories.

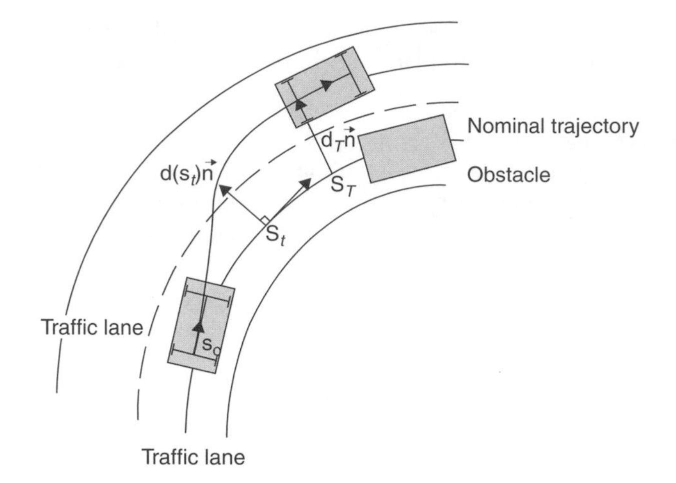

Fig. 11.14 Generation of smooth local trajectories to avoid an obstacle.

A feasible smooth trajectory for executing a lane change can be obtained using the following quintic polynomial (see Nelson (1989)):

$$\mathrm{d}(s) = \mathrm{d}_T\left(10\left(\frac{s}{s_T}\right)^3 - 15\left(\frac{s}{s_T}\right)^4 + 6\left(\frac{s}{s_T}\right)^5\right), \tag{11.4}$$

In this approach, the distance d_T is supposed to be known beforehand. Then, the minimal value required for s_T can be estimated as follows:

$$S_{T,\min} = \frac{\pi\sqrt{kd_T}}{2C_{\max}}, \tag{11.5}$$

where $C_{\max}$ stands for the maximum allowed curvature:

$$C_{\max} = \mathbf{min}\left\{\frac{\tan(\phi_{\max})}{L}, \frac{\gamma_{\max}}{V_{R,ref}^2}\right\}, \tag{11.6}$$

$\gamma_{\max}$ is the maximum allowed lateral acceleration, and $k > 1$ is an empirical constant, e.g. $k = 1.17$ in our experiments.

At each time t from the starting time T_0, the reference position p_{ref} is translated along the vector $d(S_t).\vec{n}$, where $\vec{n}$ represents the unit normal vector to the nominal velocity vector along $\mathcal{T}$; the reference orientation θ_{ref} is converted into $\theta_{ref}+$ arctan $(\partial d/\partial s(s_t))$, and the reference velocity $V_{R,ref}$ is obtained using the following equation:

$$V_{R,ref}(t) = \frac{dist(p_{ref}(t), p_{ref}(t+\Delta t))}{\Delta t}, \tag{11.7}$$

where *dist* stands for the Euclidean distance. As shown in Figure 11.12, this type of control skill can also be used to avoid a stationary obstacle, or to overtake another vehicle. As soon as the obstacle has been detected by the vehicle, a value $s_{T,\min}$ is computed according to Equation 11.5 and compared with the distance between the vehicle and the obstacle. The result of this computation is used to decide which behaviour to apply: avoid the obstacle, slow down or stop. In this approach, an obstacle avoidance or overtaking manoeuvre consists of a lane changing manoeuvre towards a collision-free 'virtual' parallel trajectory (see Figure 11.14). The lane changing skill operates the following way:

1. Generate a smooth local trajectory τ_1 which connects τ with a collision-free local trajectory τ_2 'parallel' to $\mathcal{T}$ (τ_2 is obtained by translating appropriately the involved piece of $\mathcal{T}$).
2. Track τ_1 and τ_2 until the obstacle has been overtaken.
3. Generate a smooth local trajectory τ_3 which connects τ_2 with $\mathcal{T}$, and track τ_3.

11.3.5 Parallel parking

The purpose of the parallel parking SBM is to automatically park the vehicle within an unknown parking area. This SBM comprises three main steps (Figure 11.12): (1) localizing a free parking place, (2) reaching an appropriate start location with respect to the selected parking place, and (3) performing the parallel parking manoeuvre using iterative backward and forward motions until the vehicle is parked.

Finding a parking place

During this step, the vehicle moves slowly along the traffic lane and uses its range sensors to build a local map of the environment and detect obstacles. The local map is used to determine whether free parking space is available to park the vehicle. If an obstacle is detected during the motion of the vehicle, another SBM e.g. the trajectory following SBM is activated for avoiding this obstacle.

Reaching an appropriate start location

A typical situation at the beginning of a parallel parking manoeuvre is depicted in Figure 11.15. The autonomous vehicle A1 is in the traffic lane. The parking lane with parked vehicles B1, B2 and a parking place between them is on the right-hand side of A1. L1 and L2 are respectively the length and width of A1, and D1 and D2 are the distances available for longitudinal and lateral displacements of A1 within the place. D3 and D4 are the longitudinal and lateral displacements of the corner A13 of A1 relative to the corner B24 of B2.

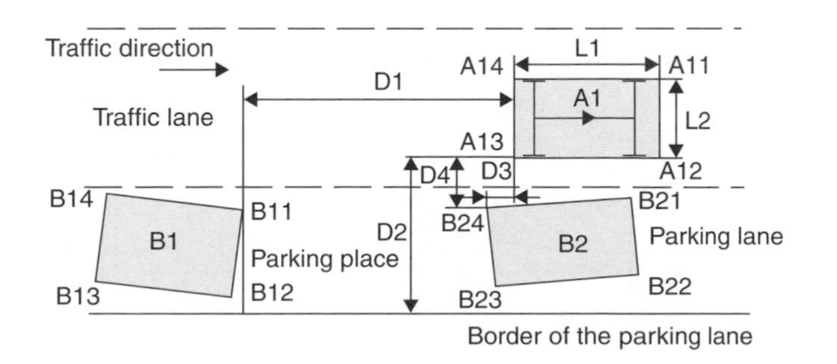

Fig. 11.15 Situation at the beginning of a parallel parking manoeuvre.

Distances D1, D2, D3 and D4 are computed from data obtained by the sensor systems. The length (D1 – D3) and width (D2 – D4) of the free parking place are compared with the length L1 and width L2 of A1 in order to determine whether the parking place is sufficiently large.

Performing the parking manoeuvre

During parallel parking, iterative low-speed backward and forward motions with coordinated control of the steering angle and locomotion velocity are performed to produce a lateral displacement of the vehicle into the parking place. The number of such motions depends on the distances D1, D2, D3, D4 and the necessary parking depth (that depends on the width L2 of the vehicle A1). The start and end orientations of the vehicle are the same for each iterative motion.

For the i-th iterative motion (but omitting the index 'i'), let the start coordinates of the vehicle be $x_0 = x(0)$, $y_0 = y(0)$, $\theta_0 = \theta(0)$ and the end coordinates be $x_T = x(T)$, $y_T = y(T)$, $\theta_T = \theta(T)$, where T is duration of the motion. The 'parallel parking' condition means that

$$\theta_0 - \delta_\theta < \theta_T < \theta_0 + \delta_\theta, \tag{11.8}$$

where $\delta_\theta > 0$ is a small admissible error in orientation of the vehicle.

The following control commands of the steering angle ϕ and locomotion velocity v provide the parallel parking manoeuvre (Paromtchik and Laugier, 1996b):

$$\phi(t) = \phi_{\max} k_\phi A(t), \quad 0 \le t \le T, \tag{11.9}$$

$$v(t) = v_{\max} k_v B(t), \quad 0 \le t \le T, \tag{11.10}$$

where $\phi_{\max} > 0$ and $v_{\max} > 0$ are the admissible magnitudes of the steering angle and locomotion velocity respectively, $k_\phi = \pm 1$ corresponds to a right side $(+1)$ or left side (-1) parking place relative to the traffic lane, $k_v = \pm 1$ corresponds to forward $(+1)$ or backward (-1) motion,

$$A(t) = \begin{cases} 1, & 0 \le t < t', \\ \cos \dfrac{\pi(t - t')}{T^*}, & t' \le t \le T - t', \\ -1, & T - t' < t \le T, \end{cases} \tag{11.11}$$

$$B(t) = 0.5\left(1 - \cos \frac{4\pi t}{T}\right), \quad 0 \le t \le T, \tag{11.12}$$

where $t' = \dfrac{T - T^*}{2}$, $T^* < T$. The shape of the type of paths that corresponds to the controls (Equations 11.11 and 11.12) is shown in Figure 11.16.

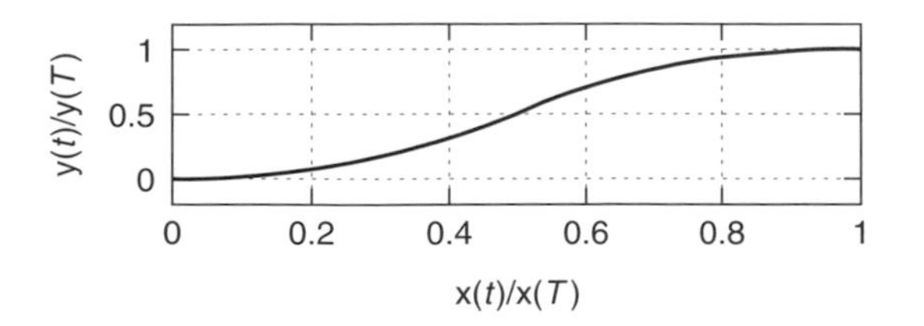

Fig. 11.16 Shape of a parallel forward/backward motion.

The commands (Equations 11.9 and 11.10) are open-loop in the (x, y, θ)-coordinates. The steering wheel servo-system and locomotion servo-system must execute the commands (Equations 11.9 and 11.10), in order to provide the desired (x, y)-path and orientation θ of the vehicle. The resulting accuracy of the motion in the (x, y, θ)-coordinates depends on the accuracy of these servo-systems. Possible errors are compensated by subsequent iterative motions.

For each pair of successive motions $(i, i+1)$, the coefficient k_v in Equation 11.10 has to satisfy the equation $k_{v,i+1} = -k_{v,i}$ that alternates between forward and backward directions. Between successive motions, when the velocity is null, the steering wheels turn to the opposite side in order to obtain a suitable steering angle $\phi_{\max}$ or $-\phi_{\max}$ to start the next iterative motion.

In this way, the form of the commands in Equations 11.9 and 11.10 is defined by Equations 11.11 and 11.12 respectively. In order to evaluate (Equations 11.9–11.12 for the parallel parking manoeuvre, the durations T^* and T, the magnitudes $\phi_{\max}$ and $v_{\max}$ must be known.

The value of T^* is lower-bounded by the kinematic and dynamic constraints of the steering wheel servo-system. When the control command (Equations 11.9) is applied, the lower bound of T^* is:

$$T^*_{\min} = \pi \ \mathbf{max} \left\{ \frac{\phi_{\max}}{\dot{\phi}_{\max}}, \sqrt{\frac{\phi_{\max}}{\ddot{\phi}_{\max}}} \right\}, \tag{11.13}$$

where $\dot{\phi}_{\max}$ and $\ddot{\phi}_{\max}$ are the maximal admissible steering rate and acceleration respectively for the steering wheel servo-system. The value of $T^*_{\min}$ gives duration of the full turn of the steering wheels from $-\phi_{\max}$ to $\phi_{\max}$ or vice versa, i.e. one can choose $T^* = T^*_{\min}$.

The value of T is lower-bounded by the constraints on the velocity $v_{\max}$ and acceleration $\dot{v}_{\max}$ and by the condition $T^* < T$. When the control command (11.10) is applied, the lower bound of T is:

$$T_{\min} = \mathbf{max} \left\{ \frac{2\pi v'(\mathrm{D1})}{\dot{v}_{\max}}, T^* \right\}, \tag{11.14}$$

where $v'(\mathrm{D1}) \leq v_{\max}$, empirically-obtained function, serves to provide a smooth motion of the vehicle when the available distance D1 is small.

The computation of T and $\phi_{\max}$ aims to obtain the maximal values such that the following 'longitudinal' and 'lateral' conditions are still satisfied:

$$|(x_T - x_0)\cos\theta_0 + (y_T - y_0)\sin\theta_0| < \mathrm{D1}, \tag{11.15}$$

$$|(x_0 - x_T)\sin\theta_0 + (y_T - y_0)\cos\theta_0| < \mathrm{D2}. \tag{11.16}$$

Using the maximal values of T and $\phi_{\max}$ assures that the longitudinal and, especially, lateral displacement of the vehicle is maximal within the available free parking space. The computation is carried out on the basis of the model (Equation 11.1) when the commands (Equations 11.9 and 11.10) are applied. In this computation, the value of $v_{\max}$ must correspond to a safety requirement for parking manoeuvres, e.g. $v_{\max} = 0.75$ m/s was found empirically.

At each iteration i the parallel parking algorithm is summarized as follows:

1. Obtain available longitudinal and lateral displacements D1 and D2 respectively by processing the sensor data.
2. Search for maximal values T and ϕ_{max} by evaluating the model (Equation 11.1) with controls (Equations 11.9 and 11.10) so that the conditions (Equations 11.15 and 11.16) are still satisfied.
3. Steer the vehicle by controls (Equations 11.9 and 11.10) while processing the range data for collision avoidance.
4. Obtain the vehicle's location relative to environmental objects at the parking place. If the 'parked' location is reached, stop; otherwise, go to step 1.

When the vehicle A1 moves backwards into the parking place from the start location shown in Figure 11.15, the corner A12 (front right corner of the vehicle) must not collide with the corner B24 (front left corner of the place). The start location must ensure that the subsequent motions will be collision-free with objects limiting the parking place. To obtain a convenient start location, the vehicle has to stop at a distance D3 that will ensure a desired minimal safety distance D5 between the vehicle and the nearest corner of the parking place during the subsequent backward motion. The relation between the distance D1, D2, D3, D4 and D5 is described by a function $\mathcal{F}(\mathrm{D1, D2, D3, D4, D5}) = 0$.

This function can not be expressed in closed form, but it can be estimated for a given type of vehicle by using the model (Equation 11.1) when the commands (Equations 11.9 and 11.10) are applied. The computations are carried out off-line and the results are stored in a look-up table which is used on-line, to obtain an estimate of D3 corresponding to a desired minimal safety distance D5 for given D1, D2 and D4 (Paromtchik and Laugier, 1996a). When the necessary parking 'depth' has been reached, clearance between the vehicle and the parked ones is provided, i.e. the vehicle moves forwards or backwards so as to be in the middle of the parking place between the two parked vehicles.

11.3.6 Platooning

The platooning SBM allows the controlled vehicle to automatically follow an other vehicle (this leading vehicle can either have been moved autonomously, or driven by a human driver). This SBM takes as input the current (velocity, position, orientation) parameters of the vehicle to control,[3] and it generates in real-time the required lateral and longitudinal controls. The platooning SBM operates in two phases (Daviet and Parent, 1996): (1) determining the relative velocity and position/orientation parameters, and (2) generating the required longitudinal and lateral controls.

Determining the state parameters

The assessment of the velocity and of the position/orientation parameters of the leading vehicle has to be performed at a rate consistent with the servo-loop frequency (50 Hz in practice). In our implementation, these parameters are evaluated using a linear camera

[3] The (velocity, position, orientation) parameters of the following vehicle are computed in real-time from the sensory data; they are expressed relatively to the leading vehicle reference frame.

(equipped with appropriate optical lenses) located in the automatic vehicle, and an infrared target located at the rear side of the leading vehicle (see Section 11.4). The position/orientation parameters are represented by the longitudinal and lateral distances DX and DY between the two vehicles, and by the angle $D\psi$ between the main axes of the two vehicles; the velocity parameter is obtained by derivating the position parameters.

Generating the required controls

Following the leading vehicle is performed by controlling, at the servo-loop frequency, the acceleration/deceleration of the automated vehicle along with the angular velocity of its steering wheel.

As for the *longitudinal control*, the basic idea is to set a linear relation between the distance and the speed of the two vehicles:

$$X_l - X_f = d_{\min} + hV_f \tag{11.17}$$

where X_l, X_f, and V_f are respectively the position of the leading vehicle, the position of the following vehicle, and the velocity of the following vehicle, $d_{\min}$, is the minimum distance between the two vehicles, and h is a time constant ($d_{\min} = 1$ m and $h = 0.35$ s in the reported experiments). This approach has led us to make use of the following controller (see (Daviet and Parent, 1996) for more details):

$$A_f = C_v \Delta V + C_p(\Delta X - hV_f - d_{\min}) \tag{11.18}$$

where A_f is the acceleration of the following vehicle, $\Delta V = V_l - V_f$, and $\Delta X = X_l - X_f$; the control gains C_p and C_v have been chosen as follows: $C_v = 1/h$, and $C_p = \min(1/h, A_{\max}/V_f)$. The fact that the position gain factor is variable allows the controller to take into account the acceleration saturation and to deal with large initial errors (since C_p decreases when the speed increases).

As for the *lateral control*, we have applied a simple approach based onto the classical 'tractor model'. This approach leads the controller to always set the orientation of the steering wheel in a direction parallel to the orientation of the leading vehicle. This approach generate stable behaviours, but it leads the following vehicle to weakly cut the turns (this might be a problem for controlling a platoon of several vehicles in a constrained area).

In a more recent work, we have slightly modified the longitudinal and lateral controls in order to avoid the above-mentioned problem and to increase the robustness of the 'target tracking' behaviour. The chosen approach mainly consists in coupling the controller with an on-line 'local trajectory generator' which tries to continuously evaluate the sequences of states–i.e. the (position, orientation, velocity) parameters–of the leading vehicle on a short time interval (instead of only using an instantaneous approach). This approach allows us to still control the motions of the following vehicle, when the target has been lost for a short time period. Current work deals with the processing of exceptions and of the car entrance and exit procedures.

11.4 Experimental results

11.4.1 Experimental vehicles

The approach described in this chapter has been implemented and tested on our experimental automatic vehicles (a modified Ligier electric car, and the Cycab electric vehicle designed and developed at INRIA (Baille *et al.*, 1999). These vehicles are equipped with the main following capabilities:

1. a *sensor unit* to measure relative distances between the vehicle and environmental objects,
2. a *servo unit* to control the steering angle and the locomotion velocity,
3. a *control unit* that processes data from the sensor and servo units in order to 'drive' the vehicle by issuing appropriate servo commands.

These vehicles can either be manually driven, or they can move autonomously using the *control unit*; this unit is based on a Motorola VME162-CPU board and a transputer network for the Ligier, and on a distributed control architecture implemented using a CAN bus and micro-controllers for the Cycab (Baille *et al.*, 1999). The *sensor unit* of the vehicle makes use of a belt of ultrasonic range sensors (Polaroid 9000) and of a linear CCD-camera. The *servo unit* consists of a steering wheel servo-system, a locomotion servo-system for forward and backward motions, and a braking servo-system to slow down and stop the vehicle. The steering wheel servo-system is equipped with a direct current motor and an optical encoder to measure the steering angle; the locomotion servo-system is equipped with an asynchronous motor and two optical encoders located onto the rear wheels (for odometry data); the Ligier vehicle is also equipped with a hydraulic braking servo-system.

The ultrasonic range sensors used in the described experiments have a measurement range of 0.5–10.0 m, and a sampling rate is 60 ms. The sensors are activated sequentially in order to make more robust measurements in the different regions defined by the vehicle i.e. four sensors are emitting/receiving signals at each time step for sensing each side of the car). The minimal number of ultrasonic sensors required by the parallel parking SBM is eight: three for looking in the forward direction, two located on each side of the vehicle and one for looking in the backward direction. This ultrasonic sensor system is intended to test the control algorithms for low-speed motion only; a more complex sensor system (e.g. a combination of vision and ultrasonic sensors) should be used to ensure reliable operation in a more dynamic environment. This is the purpose of current work.

The CCD-camera has a resolution of 2048 pixels, and it operates at a rate of 1000 Hz; it is equipped with a cylindrical lens and an infrared polarized filter. This device operates in relation with an infrared target for localizing the leading vehicle in the platooning SBM; this target is made of three pulsing sets of LED organized along vertical lines, as shown in Figure 11.22.

The *motion controller* of our control and decisional architecture monitors the current steering angle, locomotion velocity, travelled distance, coordinates of the vehicle and range data from the environment, calculates appropriate local trajectories, and issues the required servo commands. It has been implemented using the Orccad software tools

(Simon *et al.*, 1993) running on a workstation; the compiled code is transmitted via Ethernet to the *control unit* operating under the VxWorks real-time operating system.

11.4.2 Experimental run of the trajectory following manoeuvre

An experimental run of the trajectory following SBM with obstacle avoidance on a circular road (roundabout) is shown in Figure 11.17. In this experiment, the Ligier vehicle follows a nominal trajectory along the curved traffic lane, and it finds on its way another vehicle moving at a lower velocity (see Figure 11.17(a). When the moving obstacle is detected, a local trajectory for a right lane change is generated by the system, and the Ligier performs the lane changing manoeuvre, as illustrated in Figure 11.17(b). Afterwards, the Ligier moves along a trajectory parallel to its nominal trajectory, and a left lane change is performed as soon as the obstacle has been overtaken (Figure 11.17(c)). Finally the Ligier catches up its nominal trajectory, as illustrated in Figure 11.17(d)).

The corresponding motion of the vehicle is depicted in Figure 11.18(a). The steering and velocity controls applied during this manoeuvre are shown in Figure 11.18(b) and Figure 11.18(c). It can be noticed in this example that the velocity of the vehicle has increased when moving along the local 'parallel' trajectory (Figure 11.18(c)); this is due to the fact that the vehicle has to satisfy the time constraints associated with its nominal trajectory.

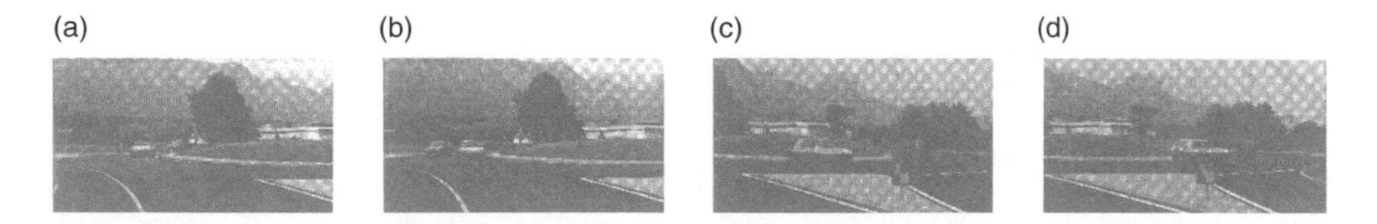

Fig. 11.17 Snapshots of trajectory following with obstacle avoidance in a roundabout: (a) following the nominal trajectory, (b) lane changing to the right and overtaking, (c) lane changing to the left, (d) catching up with the nominal trajectory.

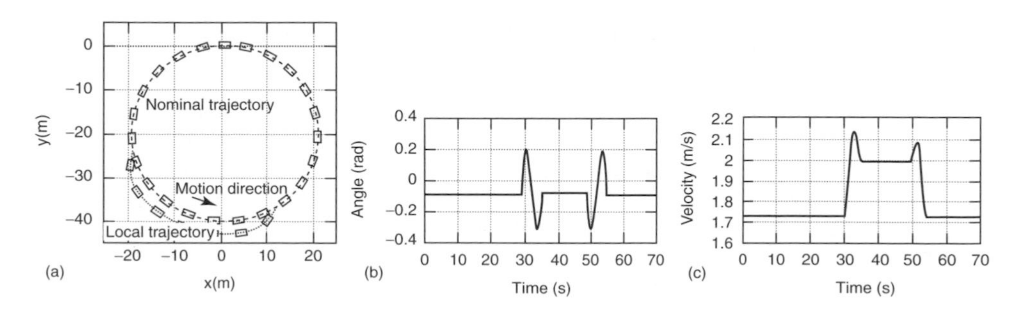

Fig. 11.18 Motion and control commands in the "roundabout" scenario: (a) motion, (b) steering angle and (c) velocity controls applied.

11.4.3 Experimental run of the parallel parking manoeuvre

An experimental run of the parallel parking SBM in a street is shown in Figure 11.19. This manoeuvre can be carried out in environments including moving obstacles, e.g. pedestrians or some other vehicles (see the video Paromtchik and Laugier (1997)). In this experiment, the Ligier was manually driven to a position near the parking place, the driver started the autonomous parking mode and left the vehicle. Then, the Ligier moved forward autonomously in order to localize the parking place, obtained a convenient start location, and performed a parallel parking manoeuvre. When, during this motion a pedestrian crosses the street in a dangerous proximity to the vehicle, as shown in Figure 11.19(a), this moving obstacle is detected, the Ligier slows down and stops to avoid the collision. When the way is free, the Ligier continues its forward motion. Range data is used to detect the parking space. A decision to carry out the parking manoeuvre is made and a convenient start position for the initial backward movement is obtained, as shown in Figure 11.19(b). Then, the Ligier moves backwards into the parking space, as shown in Figure 11.19(c). During this backward motion, the front human-driven vehicle starts to move backwards, reducing the length of the bay. The change in the environment is detected and taken into account. The range data shows that the necessary 'depth' in the bay has not been reached, so further iterative motions are carried out until it has been reached. Then, the Ligier moves to the middle between the rear and front vehicles, as shown in Figure 11.19(d). The parallel parking manoeuvre is completed.

The corresponding motion of the vehicle is depicted in Figure 11.20(a) where the motion of the corners of the vehicle and the midpoint of the rear wheel axle are plotted. The control commands (Equations 11.9 and 11.10) for parallel parking into a parking place situated at the right side of the vehicle are shown in Figure 11.20(b) and (c) respectively. The length of the vehicle is $L1 = 2.5$ m, the width is $L2 = 1.4$ m, and

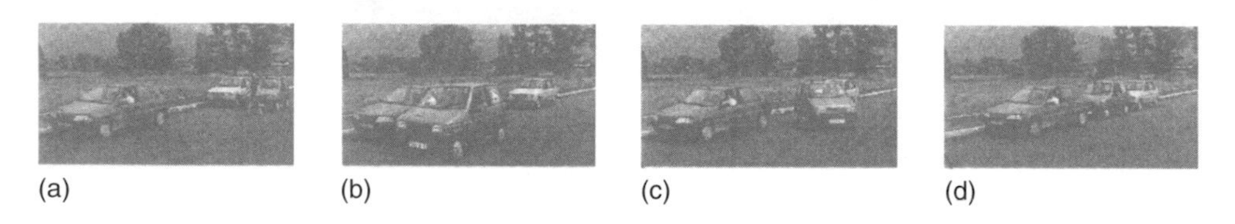

Fig. 11.19 Snapshots of parallel parking: (a) localizing a free parking place, (b) selecting an appropriate start location, (c) performing a backward parking motion; (d) completing the parallel parking.

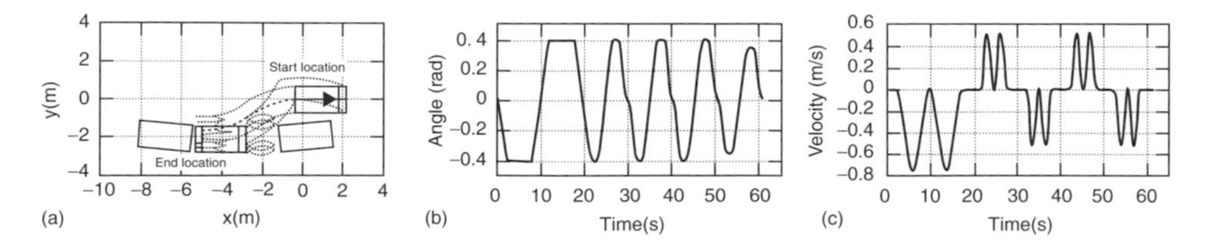

Fig. 11.20 Motion and control commands in the parallel parking scenario: (a) motion, (b) steering angle and (c) velocity controls applied.

the wheelbase is $L = 1.785$ m. The available distances are D1 = 4.9 m, D2 = 2.7 m relative to the start location of the vehicle.

The lateral distance D4 = 0.6 m was measured by the sensor unit. The longitudinal distance D3 = 0.8 m was estimated so as to ensure the minimal safety distance D5 = 0.2 m. In this case, five iterative motions are performed to park the vehicle. As seen in Figure 11.20, the duration T of the iterative motions, magnitudes of the steering angle ϕ_{max} and locomotion velocity v_{max} correspond to the available displacements D1 and D2 within the parking place (e.g. the values of T, ϕ_{max} and v_{max} differ for the first and last iterative motion).

11.4.4 Experimental run of the platooning manoeuvre

An experimental run of the platooning SBM in a street is shown in Figure 11.21. The linear camera and the infrared target is shown in Figure 11.22. During the execution of a platooning manoeuvre, the linear camera operates at a frequency of 1000 Hz for providing the relative position/orientation parameters of the two vehicles; the accuracy of the measurement has been estimated at a value of 1 mm for a distance of 10 m. It has experimentally been shown that the system is robust according to various lighting and light reflecting conditions (thanks to the camera characteristics, to the pulsing infrared target, and to the used filters). Experiments have been conducted at speeds up to 60 km/h, with decelerations up to 2 m/s^2. The distance between the vehicles is proportional to the speed (see Section 11.3.6), with a gap of 0.3 s.

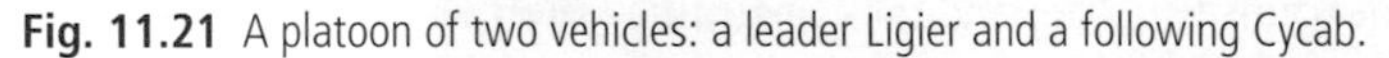

Fig. 11.21 A platoon of two vehicles: a leader Ligier and a following Cycab.

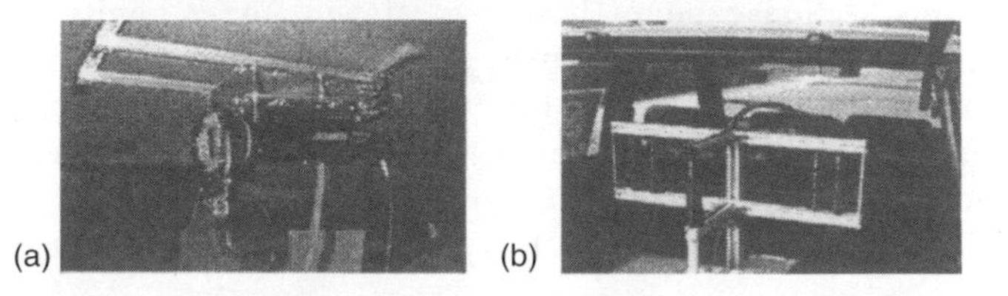

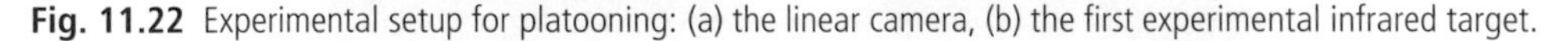

Fig. 11.22 Experimental setup for platooning: (a) the linear camera, (b) the first experimental infrared target.

11.5 Motion planning for car-like vehicles

11.5.1 Introduction

The purpose of every robot is to perform actions in its workspace (grasping and mating parts, moving around to explore or survey, etc.). Carrying out a given action usually implies that a motion be made by the robot hence the importance, in robotics, of motion planning, i.e. the determination of the motion that is to be performed in order to achieve a given task. This importance is naturally reflected in the number and variety of research works that have dealt with motion planning in the past 30 years.

Latombe's (1991) book is undoubtedly the reference book for robot motion planning. Its table of contents reveals the importance of what Latombe refers to as the *basic motion planning problem*. Six out of ten chapters are dedicated to this problem, which is to plan a collision-free path for a robot moving freely amidst stationary obstacles. The basic motion planning problem is readily illustrated with the concept of *configuration space* that was introduced in robotics in the late 1970s by Udupa (1977) and Lozano-Perez and Wesley (1979a). The *configuration* of a robot is a set of independent parameters representing the position and orientation of every part of the robot. In its configuration space, a robot is represented as a point, stationary obstacles are represented as forbidden regions[4] and motion planning between a start and a goal configuration is reduced to finding a path, i.e. a continuous sequence of configurations, that avoids the forbidden regions.

The basic motion planning problem is essentially geometric, it deals with collision avoidance of stationary obstacles and it computes a path, i.e. a geometric curve in the configuration space. However there is much more to motion planning than that, especially when the robot considered is a car-like vehicle. For a start, such a vehicle cannot move freely: it is subject to *nonholonomic constraints* that restrict its motion capabilities (a car cannot make sidewise motions for instance). Then it usually moves in a workspace that contains *moving obstacles* (they should be avoided too!). In addition to that, it may also be necessary to take into account *dynamic constraints*, e.g. bounded accelerations, that further affect the vehicle's motion capabilities (they cannot be ignored when the vehicle is moving fast).

In summary, physical and temporal constraints as well as geometrical ones must be considered when planning the motions of a car-like vehicle. Such additional constraint yields extensions to the basic motion planning problem that raise new problems and further complicate motion planning. In this case, the output of motion planning is a trajectory, i.e. a path parameterized by time. Trajectory planning with its time dimension permits to take into account time-dependent constraints such as moving obstacles and the dynamic constraints of the vehicle.

The rest of this section explores how to deal with these kinds of constraints. It reviews the main approaches that have been developed in order to deal with all or part of the constraints considered (Section 11.5.2). Then it introduces a method that, unlike most of the methods developed before, attempts to take into account all the aforementioned constraints simultaneously; it is based upon the concept of *state-time space* (Section 11.5.3). For the sake of clarity, this general method is presented in the case of

[4] The set of configurations yielding a collision between the robot and the obstacle.

a car-like vehicle moving along a given path (Sections 11.5.4 and 11.5.5). Such a path is collision-free and respects the vehicle's nonholonomic constraints. The particular problem of planning a nonholonomic path is explored afterwards (Section 11.5.6).

11.5.2 Main approaches to trajectory planning

Nonholonomic constraints

Path planning with nonholonomic constraints is a research field in itself and has motivated a large number of research works in the past 15 years. The review of the relevant literature is made in Section 11.5.6, which is dedicated to this topic.

Dynamic constraints

There are several results for time-optimal trajectory planning for Cartesian robots subject to bounds on their velocity and acceleration (Canny *et al.*, 1990; Ó'Dúnlaing, 1987). Besides optimal control theory provides some exact results in the case of robots with full dynamics and moving along a given path (Bobrow *et al.*, 1985(b); Shiller and Dubowsky, 1985; Shiller and Lu, 1990). Using these results, some authors have described methods that compute a local time-optimal trajectory (Shiller and Dubowsky, 1989; Shiller and Chen, 1990). The key idea of these works is to formulate the problem as a two-stage optimization process: optimal motion time along a given path is used as a cost function for a local path optimization (hence local time-optimality). However the difficulty of the general problem and the need for practical algorithms led some authors to develop approximate methods. Their basic principle is to define a grid which is searched in order to find a near-time-optimal solution. Such grids are defined either in the workspace (Shiller and Dubowsky, 1988), the configuration space (Sahar and Hollerbach, 1985), or the state space of the robot (Canny *et al.*, 1988; Donald and Xavier, 1990; Jacobs *et al.*, 1989).

Moving obstacles

A general approach that deals with moving obstacles is the configuration-time space approach which consists of adding the time dimension to the robot's configuration space (Erdmann and Lozano-Perez, 1987). The robot maps in this configuration-time space to a point moving among stationary obstacles. Accordingly the different approaches developed in order to solve the path planning problem in the configuration space can be adapted in order to deal with the specificity of the time dimension and used (see Latombe (1991). Among the existing works are those based upon extensions of the visibility graph (Erdmann and Lozano-Perez, 1987; Fujimura and Samet, 1990; Reif and Sharir, 1985) and those based upon cell decomposition (Fujimura and Samet, 1989; Shih *et al.*, 1990).

Few research works take into account moving obstacles and dynamic constraints simultaneously, and they usually do so with far too simplifying assumptions e.g. Fujimura and Samet (1989) and Ó'Dúnlaing (1987). More recently (Fiorini and Shiller, 1996), has presented a two-stage algorithm that computes a local time-optimal trajectory for a manipulator arm with full dynamics and moving in a dynamic workspace: the solution is computed by first generating a collision-free path using the concept of velocity obstacle, and then by optimizing it thanks to dynamic optimization.

11.5.3 Trajectory planning and state-time space

State-time space is a tool to formulate problems of trajectory planning in dynamic workspaces. In this respect, it is similar to the concept of configuration space (Lozano-Perez and Wesley, 1979b) which is a tool to formulate path planning problems. State-time space permits to study the different aspects of dynamic trajectory planning, i.e. moving obstacles and dynamic constraints, in a unified way. It seems from two concepts which have been used before in order to deal respectively with moving obstacles and dynamic constraints, namely the concepts of *configuration-time space* (Erdmann and Lozano-Perez, 1987), and *state space*, i.e. the space of the configuration parameters and their derivatives. Merging these two concepts leads naturally to state-time space, i.e. the state space augmented of the time dimension. In this framework, the constraints imposed by both the moving obstacles and the dynamic constraints can be represented by static forbidden regions of state-time space. In addition, a trajectory maps to a curve in state-time space hence trajectory planning in dynamic workspaces simply consists in finding a curve in state-time space, i.e. a continuous sequence of state-times between the current state of the robot and a goal state. Such a curve must obviously respect additional constraints due to the fact that time is irreversible and that velocity and acceleration constraints translate to geometric constraints on the slope and the curvature along the time dimension. However it is possible to extend previous methods for path planning in configuration space in order to solve the problem at hand.

In particular, a method to solve trajectory planning in dynamic workspaces problems when cast in the state-time space framework is presented (Section 11.5.5). It is derived from a method originally presented in Canny *et al.* (1988), and extended to take into account the time dimension of the state-time space. It follows the paradigm of near-time-optimization: the search for the solution trajectory is performed over a restricted set of *canonical trajectories* hence the near-time-optimality of the solution. These canonical trajectories are defined as having piecewise constant acceleration that changes its value at given times. Moreover, the acceleration is selected so as to be either minimum, null or maximum (bang controls). Under these assumptions, it is possible to transform the problem of finding the time-optimal canonical trajectory to finding the shortest path in a directed graph embedded in the state-time space.

11.5.4 Case study

State-time space was first introduced in Fraichard and Laugier (1992) to plan the motion of a point robot subject to simple velocity and acceleration bounds and moving along a given path amidst moving obstacles. Later, a mobile robot subject to full dynamic constraints was considered; first, in the case of a one-dimensional motion along a given path Fraichard (1993), and then, in the case of a two-dimensional motion on a planar surface (Fraichard and Scheuer, 1994).

It has been decided to focus herein on the case of a car-like vehicle with full dynamics and moving along a given path. The main reason for this choice is that, in this particular case, the state-time space is three-dimensional thus permitting a clear presentation of the concept of state-time space. Besides, although one-dimensional only, this motion planning problem does feature all the key characteristics of trajectory planning in dynamic workspaces, i.e. full dynamics and moving obstacles, and the concepts presented hereafter can easily be extended to problems of higher dimension (see for instance Fraichard and Scheuer (1994). Accordingly the method presented here could readily be used within a path-velocity decomposition scheme (Kant and Zucker, 1986) to plan motions along a given path taking into account the vehicle dynamics (as in Bobrow *et al.* (1985a) or Shin and McKay (1985)) or moving obstacles (as in Kyriakopoulos and Saridis (1991) or Ó'Dúnlaing (1987)).

In summary, this section addresses trajectory planning for a car-like vehicle A which moves along a given path S on a planar workspace W cluttered up with stationary and moving obstacles. It is assumed that S is collision-free with the stationary obstacles of W and that it is feasible, i.e. that it respects the kinematic constraints that restricts the motion capabilities of A. The problem then is to compute a trajectory for A that follows S, is collision-free with the moving obstacles of W and satisfies the dynamic constraints of A.

Model of the path S

As mentioned earlier, the car-like vehicle A moves along a given path S which is collision-free with the stationary obstacles of W and which is feasible, i.e. it respects the kinematic constraints of A. Those kinematic constraints are studied in more detail in Section 11.5.6. It appears there that good feasible paths for a car-like vehicle should be planar curve made up of straight segments and circular arcs of radius $1/\kappa_{\max}$ connected with clothoid arcs. S is defined accordingly. Our main concern is that in planning 'high' speed and forward motions only, S should also be of class C^2. The C^2 property ensures that the path is manoeuvre-free and that A can follow it without having to stop to change its direction. Assuming that A moves along S, it is possible to reduce a configuration of A to the single variable s which represents the distance travelled along S.

Model of the Vehicle A

In this section, we start by presenting the dynamic model of $\mathcal{A}$ that is used.[5] Then we describe the dynamic constraints that are taken into account.

Dynamic model of $\mathcal{A}$ $\mathcal{A}$ is modelled as a rigid body supported by four wheels with rigid suspensions. Without loss of generality, it is assumed that the $\vec{t}$ axis of the frame attached to $\mathcal{A}$ coincides with the unit vector tangent to the path S at point R (Figure 11.23). The $\vec{b}$ axis points in the positive direction normal to the plane. The $\vec{n}$ axis is chosen so that $(\vec{t}, \vec{n}, \vec{b})$ is right-handed. Note that the line of the radius of curvature at point R coincides with $\vec{n}$.

The motion of $\mathcal{A}$ along S obeys Newtonian dynamics. The external forces acting on $\mathcal{A}$ are the gravity force $\vec{G}$ and the ground reaction $\vec{R}$ which can be decomposed into their perpendicular components:

$$\vec{G} = -mg\vec{b} \tag{11.19}$$

$$\vec{R} = R_t\vec{t} + R_n\vec{n} + R_b\vec{b} \tag{11.20}$$

where m is the mass of $\mathcal{A}$ and g the gravity constant. The equation of motion of $\mathcal{A}$ can be expressed in terms of the tangential velocity $\dot{s}$ and the tangential acceleration

[5] This model is the two-dimensional instance of the model presented in Shiller and Chen (1990).

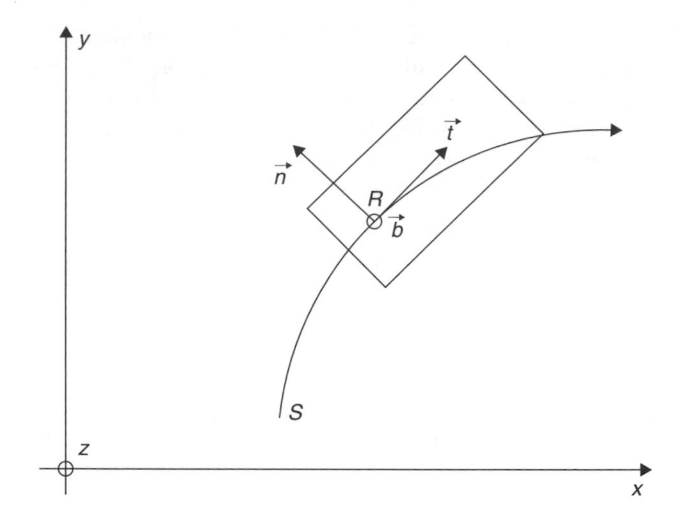

Fig. 11.23 The frame attached to $\mathcal{A}$.

$\ddot{s}$, namely:

$$\vec{G} + \vec{R} = m\ddot{s}\,\vec{t} + m\kappa_s\dot{s}^2\,\vec{n}$$

where κ_s is the signed curvature of the path at position s (κ_s is positive if the radial direction coincides with $\vec{n}$ and negative otherwise, $-\kappa_{\max} \leq \kappa_s \leq \kappa_{\max}$). Using Equations 11.19 and 11.20, this equation can be rewritten in the following set of equations:

$$R_t = m\ddot{s} \tag{11.21}$$

$$R_n = m\kappa_s\dot{s}^2 \tag{11.22}$$

$$R_b = mg \tag{11.23}$$

Equations 11.21 to 11.23 represent the forces required to maintain the velocity $\dot{s}$ and the acceleration $\ddot{s}$ of $\mathcal{A}$ at a given position s along the path. Although simple, this model is rich enough in the sense that the constraints associated are truly dynamic (they lead to state-dependence of the set of allowable accelerations).

Dynamic constraints of $\mathcal{A}$ Three dynamic constraints are taken into account (engine force, sliding and velocity constraints). They are presented in the next three sections. Afterwards they are transformed into constraints on the tangential velocity $\dot{s}$ and the tangential acceleration $\ddot{s}$.

Engine force constraint

When the vehicle is moving, the torque applied by the engine on the wheels translates into a planar force F whose direction is $\vec{t}$ and whose modulus is $m\ddot{s}$. This force is bounded by the maximum (resp. minimum) equivalent engine force:

$$F_{\min} \leq F \leq F_{\max} \tag{11.24}$$

These bounds are assumed to be constant and independent of the speed.

Sliding constraint

The component of $\vec{R}$ in the plane $\vec{t} \times \vec{n}$ represents the friction that is applied from the ground to the wheels. This friction is constrained by the following relation:

$$\sqrt{R_t^2 + R_n^2} \leq \mu R_b \tag{11.25}$$

where μ is the friction coefficient between the wheels and the ground. If this constraint is violated then $\mathcal{A}$ will slide off the path.

Velocity constraint

Our main constraint being in planning forward motions, the velocity $\dot{s}$ is constrained by the following relation:

$$0 \leq \dot{s} \leq \dot{s}_{\max} \tag{11.26}$$

where $\dot{s}_{\max}$ is the highest velocity allowed.

Tangential acceleration constraints

The engine force constraint (Equation 11.24) yields the following feasible acceleration range:

$$\frac{F_{\min}}{m} \leq \ddot{s} \leq \frac{F_{\max}}{m} \tag{11.27}$$

Besides substituting Equations 11.21, 11.22 and 11.23 in Equation 11.25 and solving it for $\ddot{s}$ yields the following relation which expresses the feasible acceleration range due to the sliding constraint:

$$-\sqrt{\mu^2 g^2 - \kappa_s{}^2 \dot{s}^4} \leq \ddot{s} \leq \sqrt{\mu^2 g^2 - \kappa_s{}^2 \dot{s}^4} \tag{11.28}$$

The final feasible acceleration range is therefore given by the intersection of Equations 11.27 and 11.28:

$$\max\left(\frac{F_{\min}}{m}, -\sqrt{\mu^2 g^2 - \kappa_s{}^2 \dot{s}^4}\right) \leq \ddot{s} \leq \min\left(\frac{F_{\max}}{m}, \sqrt{\mu^2 g^2 - \kappa_s{}^2 \dot{s}^4}\right) \tag{11.29}$$

Tangential velocity constraints

Velocity $\dot{s}$ must respect Equation 11.26. In addition, the argument under the square roots in Equation 11.28 should be positive. When $\kappa_s \neq 0$, $\dot{s}$ must respect the following constraint:

$$-\sqrt{\frac{\mu g}{|\kappa_s|}} \leq \dot{s} \leq \sqrt{\frac{\mu g}{|\kappa_s|}} \tag{11.30}$$

The final feasible velocity range is therefore given by the intersection of Equations 11.26 and 11.30:

$$0 \leq \dot{s} \leq \min\left(\dot{s}_{\max}, \sqrt{\frac{\mu g}{|\kappa_s|}}\right) \tag{11.31}$$

The latter constraint can be expressed as a set of forbidden states, i.e. points of the $s \times \dot{s}$ plane. Let $\mathcal{TV}$ be this set of states, it is defined as:

$$\mathcal{TV} = \left\{(s, \dot{s}) | 0 > \dot{s} \text{ or } \dot{s} > \min\left(\dot{s}_{\max}, \sqrt{\frac{\mu g}{|\kappa_s|}}\right)\right\}$$

Moving obstacles

$\mathcal{A}$ moves in a workspace $\mathcal{W} \in \mathbb{R}^2$ which is cluttered up with stationary and moving obstacles. The path $\mathcal{S}$ being collision-free with the stationary obstacles, only the moving obstacles have to be considered when it comes to planning $\mathcal{A}$'s trajectory.

Let $\mathcal{B}_i$, $i \in \{1, \ldots, b\}$, be the set of moving obstacles. Let $\mathcal{B}_i(t)$ denote the region of $\mathcal{W}$ occupied by $\mathcal{B}_i$ at time t and $\mathcal{A}(s)$ the region of $\mathcal{W}$ occupied by $\mathcal{A}$ at position s along $\mathcal{S}$. If, at time t, $\mathcal{A}$ is at position s and if there is an obstacle $\mathcal{B}_i$ such that $\mathcal{B}_i(t)$ intersects $\mathcal{A}(s)$ then a collision occurs between $\mathcal{A}$ and $\mathcal{B}_i$. Accordingly the constraints imposed by the moving obstacles on $\mathcal{A}$'s motion can be represented by a set of forbidden points of the $s \times t$ plane. Let $\mathcal{TB}$ be this set of forbidden points, it is defined as:

$$\mathcal{TB} = \{(s, t) | \exists i \in \{1, \ldots, b\}, \mathcal{A}(s) \cap \mathcal{B}_i(t) \neq \emptyset\}$$

State-time space of $\mathcal{A}$

As mentioned earlier, the configuration of $\mathcal{A}$ is reduced to the single variable s which represents the distance travelled along $\mathcal{S}$. A state of $\mathcal{A}$ is therefore represented by a pair $(s, \dot{s}) \in [0, s_{\max}] \times [0, \dot{s}_{\max}]$ where $s_{\max}$ is the arc-length of $\mathcal{S}$.

A **state-time** of $\mathcal{A}$ is defined by adding the time dimension to a state hence it is represented by a triple $(s, \dot{s}, t) \in [0, s_{\max}] \times [0, \dot{s}_{\max}] \times [0, \infty)$. The set of every state-time is the **state-time space** of $\mathcal{A}$; it is denoted by $\mathcal{ST}$.

A state-time is admissible if it does not violate the no-collision and velocity constraints presented earlier. Before defining an admissible state-time formally, let us define $\mathcal{TB}'$, the set of state-times which entail a collision between $\mathcal{A}$ and a moving obstacle. $\mathcal{TB}'$ is simply derived from $\mathcal{TB}$:

$$\mathcal{TB}' = \{(s, \dot{s}, t) | \exists i \in \{1, \ldots, b\}, \mathcal{A}(s) \cap B_i(t) \neq \emptyset\}$$

Similarly we define $\mathcal{TV}'$, the set of state-times which violate the velocity constraint Equation 11.31. $\mathcal{TV}'$ is simply derived from $\mathcal{TV}$:

$$\mathcal{TV}' = \left\{(s, \dot{s}, t) | 0 > \dot{s} \text{ or } \dot{s} > \min\left(\dot{s}_{\max}, \sqrt{\frac{\mu g}{|\kappa_s}}\right)\right\}$$

Accordingly a state-time q is **admissible** if and only if:

$$q \in \mathcal{ST} \backslash (\mathcal{TB}' \cup \mathcal{TV}')$$

where $E \backslash F$ denotes the complement of F in E. The set of every admissible state-time is the **admissible state-time space** of $\mathcal{A}$, it is denoted by $\mathcal{AST}$ and defined as:

$$\mathcal{AST} = \mathcal{ST} \backslash (\mathcal{TB}' \cup \mathcal{TV}')$$

Figure 11.24 depicts the state-time space of $\mathcal{A}$ in a simple case where there is only one moving obstacle which crosses $\mathcal{S}$.

In this framework, a *trajectory* Γ for $\mathcal{A}$ between an initial state $(s_i, \dot{s}_i)$ and a final state $(s_f, \dot{s}_f)$ can be represented by a curve of $\mathcal{ST}$, i.e. a continuous sequence of state-times between the initial state-time $(s_i, \dot{s}_i, 0)$ and a final state-time $(s_f, \dot{s}_f, t_f)$.t_f is the duration of the trajectory Γ. The acceleration profile of Γ is a continuous map

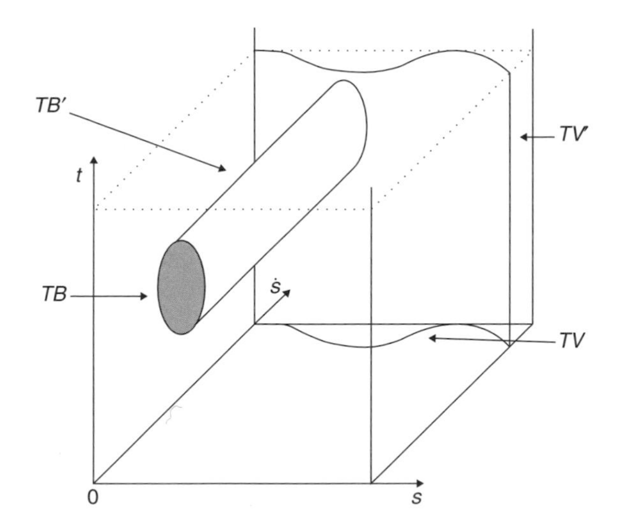

Fig. 11.24 $\mathcal{ST}$, the state-time space of $\mathcal{A}$.

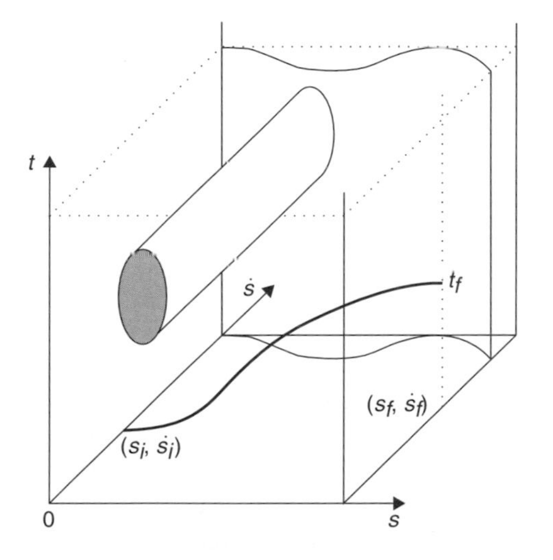

Fig. 11.25 A trajectory between $(s_i, \dot{s}_i)$ and $(s_f, \dot{s}_f)$.

$\ddot{s}: [0, t_f] \rightarrow \mathbb{R}$. $\ddot{s}(t)$ represents the acceleration which is applied to $\mathcal{A}$ at time t. Note that the velocity $\dot{s}$ and position s of $\mathcal{A}$ along $\mathcal{S}$ are respectively defined as the first and second integral of $\ddot{s}$ subject to an initial position and velocity. In order to be feasible, Γ has to verify the different constraints presented in the previous sections, i.e. it must be collision-free with the moving obstacles and respect Equations 11.29 and 11.31. Figure 11.25 depicts an example of trajectory between $(s_i, \dot{s}_i)$ and $(s_f, \dot{s}_f)$.

Statement of the problem

Finally, we can formally state the problem which is to be solved. Let $(s_i, \dot{s}_i)$ be the state of $\mathcal{A}$ and $(s_f, \dot{s}_f)$ its goal state. A trajectory $\Gamma : [0, 1] \rightarrow \mathcal{ST}$ is a solution to the problem at hand if and only if:

1. $\Gamma(0) = (s_i, \dot{s}_i, 0)$ and $\Gamma(1) = (s_f, \dot{s}_f, t_f)$.
2. $\Gamma \subset \mathcal{AST}$.
3. Γ's acceleration profile respects Equation 11.29.

Naturally, we are interested in finding a time-optimal trajectory, i.e. a trajectory such that t_f should be minimal.

11.5.5 Solution algorithm

Outline of the approach

The method that we have developed in order to solve the problem at hand, i.e. to find a curve Γ of the state-time space $\mathcal{ST}$ which respect the various constraints presented in the previous section, was initially motivated by the work described in Canny *et al.* (1988). For reasons which will be discussed later in Section 11.5.5, we follow the paradigm of near-time-optimization, i.e. instead of trying to find out the exact time-optimal trajectory between an initial and a final state, we compute an approximate time-optimal solution by performing the search over a restricted set of *canonical trajectories*. These canonical trajectories are defined as having piecewise constant acceleration $\ddot{s}$ that can only change its value at given times $k\tau$ where τ is a time-step and k some positive integer. Besides $\ddot{s}$ is selected so as to be either minimum, null or maximum. Under these assumptions, it is possible to transform the problem of finding the time-optimal canonical trajectory to finding the shortest path in a directed graph $\mathcal{G}$ embedded in $\mathcal{ST}$. The vertices $\mathcal{G}$ form a regular grid embedded in $\mathcal{ST}$ while the edges corresponds to canonical trajectory segments that each takes time τ. The next sections respectively present the canonical trajectories, the graph $\mathcal{G}$, the search algorithm and experimental results. Finally we discuss the interest of such an approach.

Canonical trajectories

The definition of the canonical trajectories depends on discretizing time – a time-step τ is chosen – and selecting an acceleration $\ddot{s}$ that respects the acceleration constraint (Equation 11.29) and which is either minimum, null or maximum. From a practical point of view, the set of accelerations is discretized – an acceleration-step δ is chosen – and the acceleration applied to $\mathcal{A}$ at each time-step, i.e. the minimum, null or maximum one, is selected from this discrete set. As we will see further down, this discretization yields a regular grid in $\mathcal{ST}$.

First let us determine the minimum (resp. maximum) acceleration $\ddot{s}_{\min}$ (resp. $\ddot{s}_{\max}$) that can be applied to $\mathcal{A}$. $\ddot{s}_{\min}$ and $\ddot{s}_{\max}$ are derived from (Equation 11.29 by noting that the acceleration that can be applied to $\mathcal{A}$ is maximum (resp. minimum) when the curvature κ_s is null, in other words:

$$\ddot{s}_{\min} = \max\left(\frac{F_{\min}}{m}, -\sqrt{\mu^2 g^2}\right)$$

$$\ddot{s}_{\max} = \min\left(\frac{F_{\max}}{m}, \sqrt{\mu^2 g^2}\right)$$

The interval $[\ddot{s}_{\min}, \ddot{s}_{\max}]$ is therefore the overall range of accelerations that can be applied to $\mathcal{A}$. Now, when $\mathcal{A}$ follows a given path $\mathcal{S}$, at each time instant, it can withstand a range of accelerations that is a subset of $[\ddot{s}_{\min}, \ddot{s}_{\max}]$, this subset is derived from Equation 11.29 and depends on the current curvature of $\mathcal{S}$. Given the acceleration step δ, Δ, the overall discrete set of accelerations that can be applied to $\mathcal{A}$ is defined as:

$$\Delta = \left\{ i\delta | i \in \mathbb{N}, \left\lceil \frac{\ddot{s}_{\min}}{\delta} \right\rceil \leq i \leq \left\lfloor \frac{\ddot{s}_{\max}}{\delta} \right\rfloor \right\}$$

Let $\Gamma : [0, 1] \rightarrow \mathcal{ST}$ be a trajectory and $\ddot{s} : [0, t_f] \rightarrow \Delta$ its acceleration profile. Γ is a **canonical trajectory** if and only if:

- $\ddot{s}$ only changes its value at times $\kappa\tau$ where $\kappa \in \mathbb{N}$, $0 \leq \kappa \leq \lfloor t_f/\tau \rfloor$.
- Let $\ddot{s}^{\kappa\tau}_{\min}$ (resp. $\ddot{s}^{\kappa\tau}_{\max}$) be the minimum (resp. maximum) acceleration allowed w.r.t. the state of $\mathcal{A}$ at time $\kappa\tau$. $\ddot{s}(\kappa\tau)$ is chosen from Δ so as to be either null or as close as possible to $\ddot{s}^{\kappa\tau}_{\min}$ and $\ddot{s}^{\kappa\tau}_{\max}$. Thus we have:

$$\ddot{s}(\kappa\tau) \in \left\{ \delta \left\lceil \frac{\ddot{s}^{\kappa\tau}_{\min}}{\delta} \right\rceil, 0, \delta \left\lfloor \frac{\ddot{s}^{\kappa\tau}_{\max}}{\delta} \right\rfloor \right\}$$

As we will see later in this section, $\ddot{s}^{\kappa\tau}_{\min}$ and $\ddot{s}^{\kappa\tau}_{\max}$ are computed so as to ensure that the acceleration constraint (Equation 11.29) is respected along the trajectory until the next acceleration change. Note that such a trajectory is very similar to the so-called 'bang-bang' trajectory of the control literature except that, in our case, the acceleration switches occur at regular time intervals.

State-time graph $\mathcal{G}$

Let q be a state-time, i.e. a point of $\mathcal{ST}$. It is a triple $(s, \dot{s}, t)$. It can equivalently be represented by $q(t) = (s(t), \dot{s}(t))$. Let $q(\kappa\tau) = (s(\kappa\tau), \dot{s}(\kappa\tau))$ be a state-time of $\mathcal{A}$ and $q((\kappa + 1)\tau)$ one of the state-times that $\mathcal{A}$ can reach by a canonical trajectory of duration τ. $q((\kappa + 1)\tau)$ is obtained by applying an acceleration $\ddot{s} \in \Delta$ to $\mathcal{A}$ for the duration τ. Accordingly we have:

$$\dot{s}((\kappa + 1)\tau) = \dot{s}(\kappa\tau) + \ddot{s}\tau \quad (11.32)$$

$$s((\kappa + 1)\tau) = s(\kappa\tau) + \dot{s}(\kappa\tau)\tau + \tfrac{1}{2}\ddot{s}\tau^2 \quad (11.33)$$

By analogy with (Canny *et al.*, 1988), the trajectory between $q(\kappa\tau)$ and $q((\kappa + 1)\tau)$ is called a $(\ddot{s}, \tau)$-**bang**. The state-time $q((\kappa + 1)\tau)$ is reachable from $q(\kappa\tau)$. Obviously a canonical trajectory is made up of a sequence of $(\ddot{s}, \tau)$-bangs.

Let $q(m\tau)$, $m \geq k$, be a state-time reachable from $q(k\tau)$. Assuming that $\dot{s}(k\tau)$ is a multiple of $\delta\tau$, it can be shown that the following relations hold for some integers α_1 and α_2:

$$s(m\tau) = s(k\tau) + \alpha_1 \frac{1}{2}\delta\tau^2$$

$$\dot{s}(m\tau) = \dot{s}(k\tau) + \alpha_2\delta\tau$$

Thus all state-times reachable from one given state-time by a canonical trajectory lie on a regular grid embedded in $\mathcal{ST}$. This grid has spacings of $\delta\tau^2/2$ in position, of $\delta\tau$ in velocity and of τ in time.

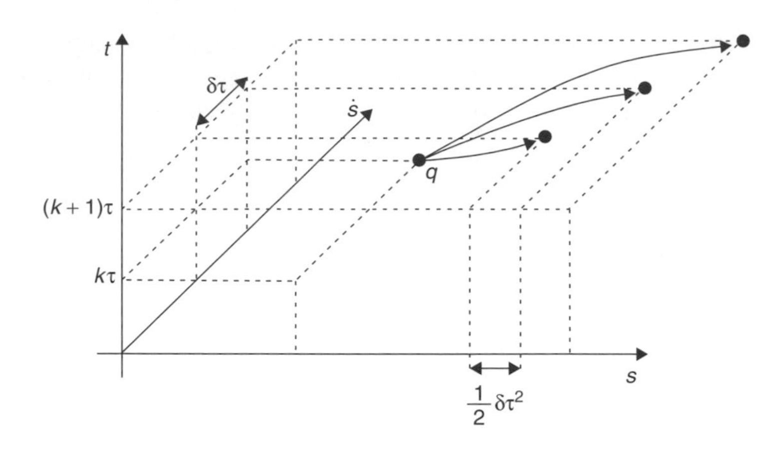

Fig. 11.26 $\mathcal{G}$, the graph embedded in $\mathcal{ST}$.

Consequently it becomes possible to define a directed graph $\mathcal{G}$ embedded in $\mathcal{ST}$. The nodes of $\mathcal{G}$ are the grid-points while the edges of $\mathcal{G}$ are $(\ddot{s}, \tau)$-bangs between pairs of nodes. $\mathcal{G}$ is called the **state-time graph**, Let η be a node in $\mathcal{G}$, the state-times reachable from η by a $(\ddot{s}, \tau)$-bang lie on the grid, they are nodes of $\mathcal{G}$ (Figure 11.26). An edge between η and one of its neighbours represents the corresponding $(\ddot{s}, \tau)$-bang. A sequence of edges between two nodes defines a canonical trajectory. The time of such a canonical trajectory is trivially equal to τ times the number of edges in the trajectory. Therefore the shortest path between two nodes (in term of number of edges) is the *time-optimal canonical trajectory* between these nodes.

Let $\mathbf{s} = (s_i, \dot{s}_i)$ be the initial state of $\mathcal{A}$ and $\mathbf{g} = (s_f, \dot{s}_f)$ be its goal state. Without loss of generality it is assumed that the corresponding initial state-time $s^* = (s_i, \dot{s}_i, 0)$ and the corresponding set of goal state-times $\mathbf{G}^* = \{(s_f, \dot{s}_f, k\tau) \text{ with } k \geq 0\}$ are grid-points. Accordingly searching for a time-optimal canonical trajectory between **s** and **g** is equivalent to searching a shortest path in $\mathcal{G}$ between the node $\mathbf{s}^*$ and a node in $\mathbf{G}^*$.

From a practical point of view, the state-time graph $\mathcal{G}$ is embedded in a compact region of $\mathcal{ST}$. More precisely, the time component of the grid-points is upper bounded by a certain value $t_{\max}$ which can be viewed as a time-out. The number of grid-points is therefore finite and so is $\mathcal{G}$. Accordingly the search for the time-optimal canonical trajectory can be done in a finite amount of time.

Searching the state-time graph

Search algorithm We use an A^* algorithm to search $\mathcal{G}$ (Nilsson, 1980). Starting with $\mathbf{s}^*$ as the current node, we expand this current node, i.e. we determine all its neighbours, then we select the neighbour which is the best according to a given criterion (a cost function) and it becomes the current node. This process is repeated until the goal is reached or until the whole graph has been explored. The time-optimal path is returned using back-pointers. In the next two sections, we detail two key-points of the algorithm, namely the cost function assigned to each node and the node expansion.

Cost function A^* assigns a cost $f(\eta)$ to every node η in $\mathcal{G}$. Since we are looking for a time-optimal path, we have chosen $f(\eta)$ as being the estimate of the time-optimal path in $\mathcal{G}$ connecting $\mathbf{s}^*$ to $\mathbf{G}^*$ and passing through η. $f(\eta)$ is classically defined as the sum of two components $g(\eta)$ and $h(\eta)$:

- $g(\eta)$ is the duration of the path between $\mathbf{s}^*$ and η, i.e. the time component of η.
- $h(\eta)$ is the estimate of the time-optimal path between η and an element of $\mathbf{G}^*$, i.e. the amount of time it would take $\mathcal{A}$ to reach $\mathbf{g}$ from its current state with a 'bang-coast-bang' acceleration profile, i.e. maximum overall acceleration $\ddot{s}_{\max}$, null acceleration and minimum overall acceleration $\ddot{s}_{\min}$. When such an acceleration profile does not exist,[6] $h(\eta)$ is set to $+\infty$.

The heuristic function $h(\eta)$ is trivially admissible, thus A^* is guaranteed to generate the time-optimal path whenever it exists (Nilsson, 1980). Moreover the fact that $f(\eta)$ is locally consistent improves the efficiency of the algorithm.

Expansion of a node The neighbours of a given node $\eta = (s, \dot{s}, k\tau)$ are the nodes which can be reached from η by a $(\ddot{s}, \tau)$-bang. As mentioned earlier, $\ddot{s} \in \{\lfloor \ddot{s}^{k\tau}_{\min} + \delta \rfloor, 0, \lceil \ddot{s}^{k\tau}_{\max} - \delta \rceil\}$. $\ddot{s}^{k\tau}_{\min}$ and $\ddot{s}^{k\tau}_{\max}$ have to be computed so as to ensure that the acceleration constraint (Equation 11.29) is respected along the corresponding $(\ddot{s}, \tau)$-bang. This computation is done in a conservative way. First the farthest position, say s^+, that $\mathcal{A}$ can reach from its current state is determined. It is the position reached after a $(\ddot{s}_{\max}, \tau)$-bang. Then the maximum curvature between s and s^+ is determined and substituted into (Equation 11.29) so as to yield the desired acceleration bounds $\ddot{s}^{k\tau}_{\min}$ and $\ddot{s}^{k\tau}_{\max}$. Finally it remains to check that the $(\ddot{s}, \tau)$-bang associated with each of the candidate neighbours does not violate the velocity and collision avoidance constraints, i.e. that the $(\ddot{s}, \tau)$-bang is included in $\mathcal{AST}$.

As we will see now, it is not necessary to compute $\mathcal{AST}$ to check these two points. Velocity checking is done by using the motion equation of $\mathcal{A}$, while collision checking is performed directly in $\mathcal{W}$. Let us consider a $(\ddot{s}, \tau)$-bang taking place between time instants $k\tau$ and $(k+1)\tau$. $\forall t \in [0, \tau]$ and according to Equations 11.32 and 11.33, we have:

$$\dot{s}(k\tau + t) = \dot{s}(k\tau) + \ddot{s}t \tag{11.34}$$

$$s(k\tau + t) = s(k\tau) + \dot{s}(k\tau)t + \frac{1}{2}\ddot{s}t^2 \tag{11.35}$$

Velocity constraint Using Equation 11.34, it is straightforward to check that a $(\ddot{s}, \tau)$-bang does not violate the velocity bonds of Equation 11.30 stated in Section 11.5.4.

Collision avoidance Recall that a $(\ddot{s}, \tau)$-bang between $k\tau$ and $(k+1)\tau$ is collision-free if and only if:

$$\forall t \in [k\tau, (k+1)\tau], \forall i \in \{1, \ldots, b\}, \mathcal{A}(s(t)) \cap \mathcal{B}_i(t) = \emptyset$$

Equation 11.35 provides the position of $\mathcal{A}$ at every time along the $(\ddot{s}, \tau)$-bang and collision checking can efficiently be performed by computing the intersection between the two planar regions $\mathcal{A}(s(t))$ and $\mathcal{B}_i(t)$.

---

[6] In this case, η is no longer reachable.

It might be desirable to add a safety margin to the collision checking procedure so as to incorporate the uncertainty on the motions of $\mathcal{A}$ and of the moving obstacles. In this case, the collision avoidance condition becomes:

$$\forall t \in [k\tau, (k+1)\tau], \forall i \in \{1, \ldots, b\}, \mathcal{G}(\mathcal{A}(s(t)), sm) \cap \mathcal{B}_i(t) = \emptyset$$

where $\mathcal{G}(X, sm)$ denotes the planar region X isotropically grown of the safety margin sm. The safety margin can integrate both a fixed and a velocity-dependent term, e.g. $sm = c_0 + c_1\dot{s}(t)$ with c_0 and $c_1 \in \mathbb{R}$.

Complexity issues Expanding a node of the graph $\mathcal{G}$ can be done efficiently in constant time (recall that the admissible state-time space $\mathcal{AST}$ is not computed, collision checking is performed directly in the two-dimensional workspace $\mathcal{W}$). The heuristic function used for the A^* search is both admissible and locally consistent (see Section 11.5.5), the time complexity of the A^* algorithm is therefore $O(n)$ where n is the number of vertices in $\mathcal{G}$ (Farreny and Ghallab, 1987). n is defined as:

$$n = \frac{2s_{\max}}{\delta\tau^2}\frac{\dot{s}_{\max}}{\delta\tau}\frac{t_{\max}}{\tau}$$

The acceleration step and, to a greater extent, the time step are key factors as far as the running time of the algorithm is concerned. Experimental running times and a discussion about the choice of the discretization steps are given later in Section 11.5.5.

Implementation and experiments

The algorithm presented earlier has been implemented in C on a Sparc station. Two examples of trajectory planning are depicted in Figure 11.27 and 11.28. In each case, there are two windows: a trace window showing the part of the graph which has been explored and a result window displaying the final trajectory. Any such window represents the $s \times t$ plane (the position axis is horizontal while the time axis is vertical; the frame origin is at the upper-left corner). The thick black segments represent the trails left by the moving obstacles and the little dots are points of the underlying grid. Note that the vertical spacing of the dots corresponds to the time-step τ. In both examples, $\mathcal{A}$ starts from position 0 (upper-left corner) with a null velocity, it is to reach position $s_{\max}$ (right border) with a null velocity.

The values of the different parameters and discretization steps in these experiments are selected in order to simulate a car-like vehicle moving in the road network: $\dot{s}_{\max} = 20\,\text{m/s} - \ddot{s}_{\min} = \ddot{s}_{\max} = 1\,\text{m/s}^2$. The idea is to plan the motion of the vehicle for the next 500 m ($s_{\max} = 500\,m$). The time horizon $t_{\max}$ is set to 25 s and the obstacles are assumed to keep a constant velocity over the time horizon. For a value of τ set of 0.5 s, the running time is of the order of 1 s.

Discussion on the proposed solution

As mentioned above in this section, the running time of the search algorithm depends on the size of the graph $\mathcal{G}$ which is to be explored (number of nodes). In turn this size is directly related to the value of the time-step τ – the smaller τ, the higher the number of vertices in $\mathcal{G}$. On the other hand, we intuitively[7] feel that the quality of the solution

[7] This intuition is confirmed in (Canny *et al.*, 1988) where it is shown that, for a correct choice of τ, any safe trajectory can be approximated to a tolerance ε by a safe canonical trajectory.

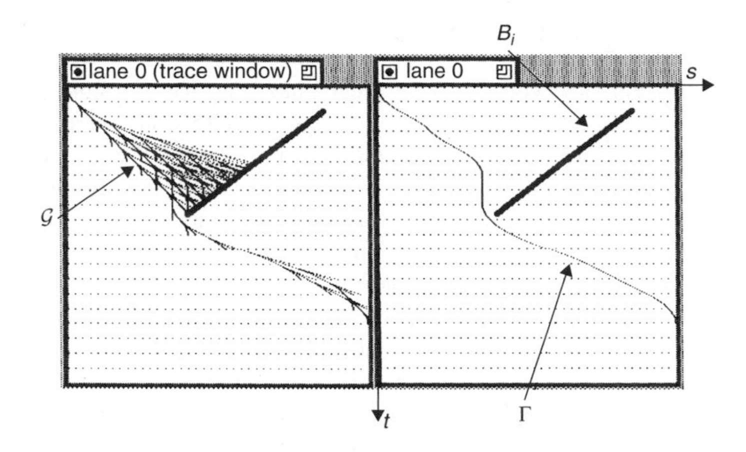

Fig. 11.27 Experimental results.

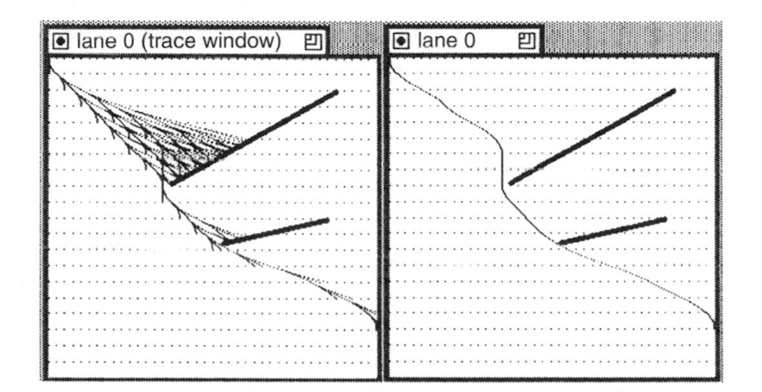

Fig. 11.28 Experimental results.

trajectory is also related to the value of τ – the smaller τ, the better the approximation. Thus it is possible to trade off the computation speed against the quality of the solution.

This property is very important and we would like to advocate this type of approach when dealing with an actual dynamic workspace. In such a workspace, it is usually impossible to have a full *a priori* knowledge of the motion of the moving obstacles. It is more likely that the knowledge that we have of their motions be restricted to a certain time interval, i.e. a time horizon. This time horizon may represent the duration over which an estimation of the motions of the moving obstacles is sound. The main consequence of this assumption is to set an upper bound on the time available to plan the motion of our vehicle (in a highly dynamic workspace, this upper bound may be very low). In this case, an approach such as the one we have presented is most interesting because its average running time can be tuned w.r.t. the time horizon considered.

11.5.6 Nonholonomic path planning

Nonholonomy is a classical concept from mechanics that was introduced in robotics by Laumond (1986). A nonholonomic system is subject to non-integrable equations involving the time derivative of its configuration parameters. These equations express constraints in the tangent space of the system at a given configuration, i.e. on the allowable velocities of the system. Nonholonomy usually arises when the system has less control parameters than configuration parameters. A car-like vehicle for instance has three configuration parameters (*xy* position and orientation) but only two control parameters (acceleration and steering). Thus it cannot change its orientation without also changing its position. As a consequence, any given path in the configuration space is not necessarily admissible which means that, even in the absence of obstacles, planning the motion of a nonholonomic system is not straightforward.

In the basic motion planning problem, the existence of a path between two configurations is characterized by the fact that these two configurations lie in the same connected component of the collision-free configuration space of the robot. In other words, a holonomic robot can reach any configuration within the connected component of the configuration space where it is located. This property no longer holds in the presence of nonholonomic constraints. Nonholonomy therefore raises a first problem which is: what is the reachable configuration space? The second problem is of course how to compute an admissible path, i.e. a path that respects the nonholonomic constraints of the robot.

Nonholonomy appears in systems as different as multifingered hands (Murray, 1990), hopping robots (Wang, 1996), or space robots (Nakamura and Mukherjee, 1989). However it concerns primarily wheeled mobile robots and most of the results obtained since 1986 have been obtained for wheeled vehicles such as unicycles, bicycles, two wheel-drive robots, cars, cars with one or several trailers, fire trucks, etc. This section presents the main results regarding path planning for the archetypal nonholonomic system represented by a car-like vehicle. The reader interested to know more about nonholonomy in general is referred to Li and Canny (1992) and Laumond (1998).

This section comprises two parts: the first part considers the 'classical' car-like robot, i.e. the one whose model is equivalent to that of an oriented particle moving in the plane. Henceforth, this car is called the *Reeds and Shepp car*.[8] The Reeds and Shepp car has been extensively studied in the literature and key results have been obtained as far as path planning is concerned. However, as will be seen below, the properties of the Reeds and Shepp car model restricts its applicability, hence the definition of a more complex model for the car. This new car is henceforth called the *Continuous-curvature car*; it is considered in the second part of this section.

Reeds and Shepp car

As mentioned earlier, the Reeds and Shepp car, or RS car, denotes a car-like vehicle whose mathematical model corresponds to that of an oriented particle moving in the plane. This model is by far the one that has been most widely used. The case where the car can move forward only was first addressed by Dubins (1957) who, among other things, gave a characterization of the shortest paths. Later, Reeds and Shepp

[8] After Reeds and Shepp (1990) who established its main properties.

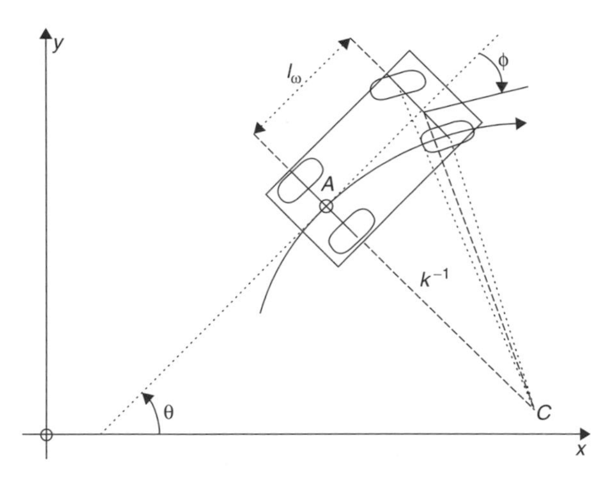

Fig. 11.29 A car-like robot.

(1990) considered the case where the car can change its direction of motion and extended Dubins' results. This section first presents the model of the RS car. Then it summarizes its main properties and overviews the main path planning techniques that were developed.

Model of the RS car Let $\mathcal{A}$ represent a RS car-like robot; it moves on a planar workspace $\mathcal{W} \equiv \mathbb{R}^2$ cluttered up with a set of stationary obstacles $\mathcal{B}_i, i \in \{1, \ldots, b\}$, modelled as forbidden regions of $\mathcal{W}$. $\mathcal{A}$ is modelled as a rigid body moving on the plane supported by four wheels making point contact with the ground: two rear wheels and two directional front wheels. It is designed so that the front wheels' axles intersect the rear wheels' axle at a given point C which is the rotation centre of $\mathcal{A}$. It takes three parameters to characterize the position and orientation of $\mathcal{A}$. A configuration of $\mathcal{A}$ is then defined by the triple $q = (x, y, \theta) \in \mathbb{R}^2 \times S^1$ where (x,y) are the coordinates of the rear axle midpoint A and θ the orientation of $\mathcal{A}$ (Figure 11.29).

Under perfect rolling assumption, a wheel moves in a direction normal to its axle. Therefore A must move in a direction normal to the rear wheels' axle and the following constraint holds accordingly (*perfect rolling constraint*):

$$\begin{cases} \dot{x} & = v\cos\theta \\ \dot{y} & = v\sin\theta \end{cases} \tag{11.36}$$

where v is the linear velocity of A, $|v| \leq v_{\max}$ ($\mathcal{A}$ moves forward when $v > 0$, stands still when $v = 0$, and moves backward when $v < 0$).

Let ϕ denote the steering angle of $\mathcal{A}$, i.e. the average orientation of the front wheels, and let κ denote the curvature of the xy-curve traced by A. κ is the inverse of the distance between C and A: $\kappa = {l_w}^{-1} \tan\phi$, where l_w is the wheelbase of $\mathcal{A}$. Since ϕ is mechanically limited, $|\phi| \leq \phi_{\max}$, the following constraint holds (*bounded curvature constraint*):

$$|\kappa| \leq \kappa_{\max} = {l_w}^{-1} \tan\phi_{\max} \tag{11.37}$$

According to Equation 11.36, θ is always tangent to the xy-curve traced by A and its derivative, i.e. the angular velocity ω, therefore satisfies $\dot{\theta} = \omega = v\kappa$. Selecting v and

ω as (coupled) control parameters, the model of $\mathcal{A}$ can be described by the following differential system:

$$\begin{pmatrix} \dot{x} \\ \dot{y} \\ \dot{\theta} \end{pmatrix} = \begin{pmatrix} \cos\theta \\ \sin\theta \\ 0 \end{pmatrix} v + \begin{pmatrix} 0 \\ 0 \\ 1 \end{pmatrix} \omega \qquad (11.38)$$

with $|v| \leq v_{\max}$, $\omega = v\kappa$, and $|\kappa| \leq \kappa_{\max}$. Because path planning is generally interested in computing shortest paths, it is furthermore assumed that $|v| = 1$ (thus the time and the arc length of a path are the same). The system (Equation 11.38) under the different control constraints defines the Reeds and Shepp car.

Admissible paths for the RS car Let C denote the configuration space of the RS car $\mathcal{A} : \mathcal{C} \equiv \mathbb{R}^2 \times [0, 2\pi]$. Let Π denote a path for $\mathcal{A}$, it is a continuous sequence of configurations: $\Pi(t) = (x(t), y(t), \theta(t))$. An admissible path must satisfy both constraints (Equations 11.36 and 11.37), it is a solution to the differential system (Equation 11.38); it is such that:

$$\begin{cases} x(t) = x(0) + \int_0^t v(\tau)\cos\theta(\tau)d\tau \\ y(t) = x(0) + \int_0^t v(\tau)\sin\theta(\tau)d\tau \\ \theta(t) = \theta(0) + \int_0^t \omega(\tau)d\tau \end{cases} \qquad (11.39)$$

with $|v(\tau)| = 1$, $\omega(\tau) = v(\tau)\kappa(\tau)$ and $|\kappa(\tau)| \leq \kappa_{\max}$.

Reachable configuration space for the RS car As mentioned earlier, the first question raised by nonholonomy is to determine whether the presence of nonholonomic constraints reduces the set of configurations that the RS car can reach. This question was first answered by Laumond (1986). Through an *ad hoc* geometric reasoning, Laumond established that the RS car could reach any configuration within the same connected component of the collision-free configuration space.

In fact, it turns out that this question is directly related to the controllability of differential systems. The small-time controllability[9] of a differential system implies that the existence of an admissible collision-free path is equivalent to the existence of a collision-free path (Laumond *et al.*, 1998, Theorem 3.1). Using tools from differential geometric control theory, it proved possible to show the small-time controllability of the RS car and therefore to redemonstrate Laumond's result (Barraquand and Latombe, 1990).

In summary, in spite of the presence of nonholonomic constraints, the RS car can reach any configuration within the connected component of the collision-free configuration space where it is located.

Optimal paths for the RS car Once the small-time controllability of the RS car has been established, it is interesting to find out the shortest path between two configurations in the absence of obstacles. Reeds and Shepp (1990) used differential calculus tools to

[9] A differential system is *locally controllable* if the set of configurations reachable from any configuration q by an admissible path contains a neighbourhood of q. It is *small-time controllable* if the set of configurations reachable from q before a given time t contains a neighbourhood of q for any t.

give a first characterization of the shortest paths for the RS car. Later, Boissonnat *et al.* (1991), and Sussmann and Tang (1991) used optimal control theory to refine Reeds and Shepp's result.

The optimal path for the RS car is made up of line segments and circular arcs of radius $1/\kappa_{\max}$; it is the shortest among a set of 46 paths that belong to one of the nine following families:

$$\begin{array}{ll} (i) & l^+l^-l^+ \text{ or } r^+r^-r^+ \\ (ii)(iii) & A|AA \text{ or } AA|A \\ (iv) & AA|AA \\ (v) & A|AA|A \\ (vi) & A|ASA|A \\ (vii)(viii) & A|ASA \text{ or } ASA|A \\ (ix) & ASA \end{array} \qquad (11.40)$$

where A (resp. S) denotes a circular arc (resp. line segment). | denotes a change of direction of motion (a cusp point). A may be replaced by r or l to specify a right (clockwise) or left (counterclockwise) turn. A + or − superscript indicates a forward or backward motion. Figure 11.30 depicts two examples of optimal paths for the RS car.

This result enables definition of what Laumond *et al.* (1998) call a *steering method* for the RS car, i.e. an algorithm that computes an admissible path between two configurations in the absence of obstacles. Let $Steer_{\mathrm{RS}}$ denote the steering method returning the optimal path for the RS car, i.e. the shortest path among the set defined by Equation 11.40.

Collision-free path planning for the RS car The complete path planning problem for the RS car must consider the constraints imposed both by the nonholonomic constraints and the obstacles of the environment. Although the results on the controllability and the optimal paths presented above did not consider the obstacles of the environment, it will be seen further down that they proved useful in the design of path planners for the RS car.

In the past 15 years, several solutions have been proposed to solve the full path planning problem for both the RS car (e.g. Barraquand and Latombe, 1989; Laumond *et al.*, 1989; Pommier, 1991), and the RS car that can move forward only (e.g. Jacobs and Canny, 1989; Fraichard, 1991; Laumond, 1987; Wilfong, 1988). This paragraph focuses on three of them that all share common features. For a start, they are generic, i.e. they can be applied to other types of robotics systems (the first two are completely

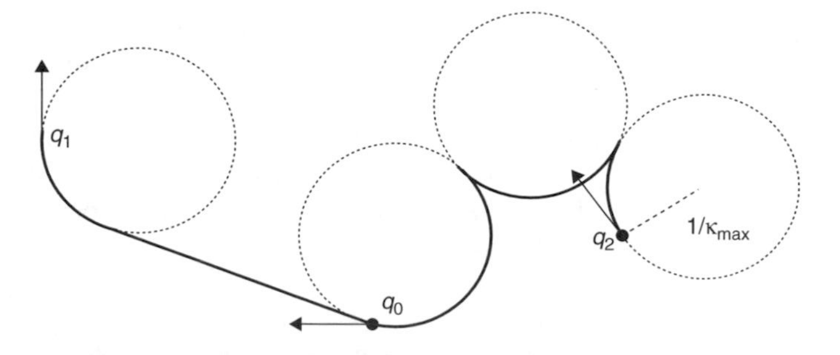

Fig. 11.30 Optimal paths for the Reeds and Shepp car.

general, they can deal with holonomic and nonholonomic systems). Then, all of them make use of a steering method such as $Steer_{RS} \cdot Steer_{RS}$ computes the optimal path between two configurations for the RS car but other steering methods could be used instead (see the review made in Laumond *et al.* (1998)).

- *Probabilistic path planning:* this generic path planning scheme was developed by Svestka and Overmars (1998) and Kavraki *et al.* (1996). It operates in two phases:
  - **Learning phase:** build a roadmap reflecting the connectivity of the collision-free configuration space. The roadmap is a graph whose nodes are randomly selected configurations and whose edges are admissible collision-free paths computed with a steering method and a collision checker.
  - **Query phase:** given a start and a goal configurations, use $Steer_{RS}$ to connect them to the roadmap. Search the roadmap for a solution path.
- *Ariadne's Clew algorithm:* this generic path planning scheme developed by Ahuactzin-Larios (1994) is slightly different from the probabilistic path planning approach. It incrementally builds a tree rooted at the start configuration. At each step, a new configuration is added to the tree; it is the farthest configuration reachable from the tree by the steering method considered (optimization tools such as genetic algorithms are used). In parallel, the steering method is used to determine whether the goal configuration is reachable from the new configuration.
- *Holonomic path approximation:* the controllability result on the existence of an admissible collision-free path is at the origin of a two-step nonholonomic path planning scheme that was introduced in Laumond *et al.* (1994):

  Step 1: compute a holonomic[10] collision-free path.

  Step 2: approximate the holonomic path by a sequence of admissible collision-free paths.

  Step 2 recursively subdivides the holonomic path and tries to connect the endpoints by using a steering method along with a collision checker.

Continuous-curvature car

As mentioned earlier, most path planning techniques for the RS car compute paths made up of line segments connected with tangential circular arcs of minimum radius. Reeds and Shepp's (1990) result concerning the shortest path for the RS car is a reason that explains this situation. No doubt that another reason for this situation is that they are easy to deal with from a computational point of view. However the curvature of this type of path is discontinuous: discontinuities occur at the transitions between segments and arcs. The curvature is directly related to the orientation of the front wheels of the car. Accordingly, if a car were to track precisely such a type of path, it would have to stop at each curvature discontinuity so as to reorient its front wheels. It is therefore desirable to plan continuous-curvature paths.[11] To address this issue, a new model for the car-like vehicle is introduced: the Continuous-curvature car, or CC car. This model is presented in the next paragraph. Unlike the RS car, the CC car has been little studied; several results have been obtained however, they are presented afterwards.

---

[10] It does not satisfy the nonholonomic constraints.

[11] As a matter of fact, it is emphasized in De Luca *et al.* (1998) that feedback controllers for car-like robots require this property in order to guarantee the exact reproducibility of a path.

Model of the CC car Let $\mathcal{A}$ now represent a CC car-like robot. As per Boissonnat *et al.* (1994), a configuration of $\mathcal{A}$ is now defined by the quadruple $q = (x, y, \theta, \kappa) \in \mathbb{R}^2 \times S^1 \times \mathbb{R}$. κ is introduced to characterize the orientation of the front wheels of $\mathcal{A}$. Considering κ as a configuration parameter ensures that it will vary continuously.

$\mathcal{A}$ is subject to the perfect rolling constraint (Equation 11.36) and the bounded curvature constraint (Equation 11.37). Let σ denote the derivative of $\kappa : \sigma = \phi / \cos^2 \phi$. The steering velocity of $\mathcal{A}$ is physically limited, $|\dot{\phi}| \leq \dot{\phi}_{\max}$, and the following constraint is added (*bounded curvature derivative constraint*):

$$|\sigma| \leq \sigma_{\max} = \dot{\phi}_{\max} \tag{11.41}$$

Accordingly the model of $\mathcal{A}$ can now be described by the following differential system:

$$\begin{pmatrix} \dot{x} \\ \dot{y} \\ \dot{\theta} \\ \dot{\kappa} \end{pmatrix} = \begin{pmatrix} \cos\theta \\ \sin\theta \\ \kappa \\ 0 \end{pmatrix} v + \begin{pmatrix} 0 \\ 0 \\ 0 \\ 1 \end{pmatrix} \sigma \tag{11.42}$$

with $|\kappa| \leq \kappa_{\max}$, $|v| = 1$ and $|\sigma| \leq \sigma_{\max}$. The system (Equation 11.42) under the different control constraints defines the Continuous-curvature car.

Admissible paths for the CC car The configuration space of $\mathcal{A}$ is now 4-dimensional: $\mathcal{C} = \mathbb{R}^2 \times [0, 2\pi] \times \mathbb{R}$. A path for $\mathcal{A}$ is a continuous sequence of configurations: $\Pi(t) = (x(t), y(t), \theta(t), \kappa(t))$. An admissible path must satisfy the constraints (Equations 11.36, 11.37 and 11.41); it is a solution of the differential system (Equation 11.42); it is such that:

$$\begin{cases} x(t) = x(0) + \int_0^t v(\tau)\cos\theta(\tau)d\tau \\ y(t) = x(0) + \int_0^t v(\tau)\sin\theta(\tau)d\tau \\ \theta(t) = \theta(0) + \int_0^t v(\tau)\kappa(\tau)d\tau \\ \kappa(t) = \kappa(0) + \int_0^t \sigma(\tau)\tau \end{cases} \tag{11.43}$$

with $|\kappa(\tau)| \leq \kappa_{\max}$, $|v(\tau)| = 1$ and $|\sigma(\tau)| \leq \sigma_{\max}$.

Reachable configuration space for the CC car The small-time controllability of the CC car has been established in Scheuer and Laugier (1998, Theorem 1). Accordingly, the existence of an admissible collision-free path is equivalent to the existence of a collision-free path. Furthermore, if a path exists between two configurations then an optimal path exists as well (Scheuer and Laugier, 1998, Theorem 2).

Optimal paths for the CC car The nature of the optimal paths for the CC car is more difficult to establish than for the RS car. However Scheuer (1998) demonstrates that, for the forward CC car, i.e. the CC car moving forward only, the optimal paths are made up of: (a) line segments, (b) circular arcs of radius $1/\kappa_{\max}$, and (c) clothoid arcs[12] of sharpness $\pm\sigma_{\max}$. Unfortunately, it appears that, whenever the shortest path includes a line segment, it is irregular and contains an infinite number of clothoid arcs that accumulate towards each endpoint of the segment (Boissonnat *et al.*, 1994). Furthermore, when the distance between two configurations is large enough, the shortest path contains a line segment hence an infinite number of clothoid arcs (Degtiariova-Kostova and Kostov, 1998).

In summary, although the exact nature of the optimal paths for the CC car has not been established yet, it seems reasonable to conjecture that they will (at least) be made up of line segments, circular arcs and clothoid arcs, and that they will be irregular in most cases.

Collision-free path planning for the CC car The main path planning techniques presented earlier are generic: they can be used for a wide variety of robotics systems. To some extent, they mainly rely upon the existence of a steering method for the robotics system considered.

As for the CC car and to the best of the authors' knowledge, there is only one steering method that has been developed so far. It has been proposed in Fraichard *et al.* (1999). Since the conjectured irregularity of the optimal paths for the CC car leaves little hope to ever have a steering method for the CC car that finds out the optimal path between two configurations, Fraichard *et al.* have developed a steering method $Steer_{CC}$ that computes admissible paths derived from the optimal paths for the RS car.

The paths computed by $Steer_{CC}$ are similar to those computed by $Steer_{RS}$ but, in order to ensure curvature continuity, the circular arcs are replaced by transitions called *CC turns* whose curvature varies continuously from 0 up and then down back to 0. A CC turn is made up of three parts: (a) a clothoid arc of sharpness $\sigma = \pm\sigma_{\max}$ whose curvature varies from 0 to $\kappa_{\max}$, (b) a circular arc of radius $1/\kappa_{\max}$, and (c) a clothoid arc of sharpness $-\sigma$ whose curvature varies from $\kappa_{\max}$ to 0 (Figure 11.31). The paths computed by $Steer_{CC}$ are not optimal but, based upon the result already established in Scheuer and Laugier (1998) for the forward CC car, it is conjectured that they are

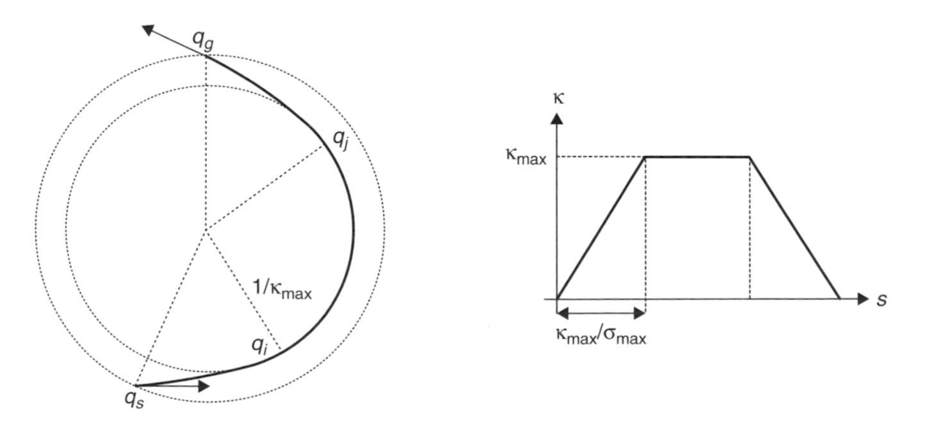

Fig. 11.31 A CC turn and its curvature profile.

[12] A clothoid is a curve whose curvature varies linearly with its arc length.

suboptimal, i.e. longer than the optimal path of no more than a given constant. This result is yet to be demonstrated however.

Reeds and Shepp car versus the Continuous-curvature car

$Steer_{CC}$ and $Steer_{RS}$ have both been has been implemented and compared. Figure 11.32 illustrates the results obtained. It appears that, for a given pair of (initial, goal) configurations, the resulting RS and CC paths may belong to the same family of path (Figure 11.32, top left), or to different families. CC paths may have the same number of back-up manoeuvres (Figure 11.32, top right), more (Figure 11.32, bottom left) or less (Figure 11.32, bottom right).

Further comparisons were made regarding the respective length of the paths and the time required for their computation. The ratio of CC over RS paths' lengths were computed for 100 pairs of (initial, goal) configurations. The results obtained are summarized in Table 11.1. In most cases (82 per cent), CC paths are only about 10 per cent longer than RS paths. Similar experiments were carried out for the computation time. The running time of both $Steer_{CC}$ and $Steer_{RS}$ are of the same order of magnitude

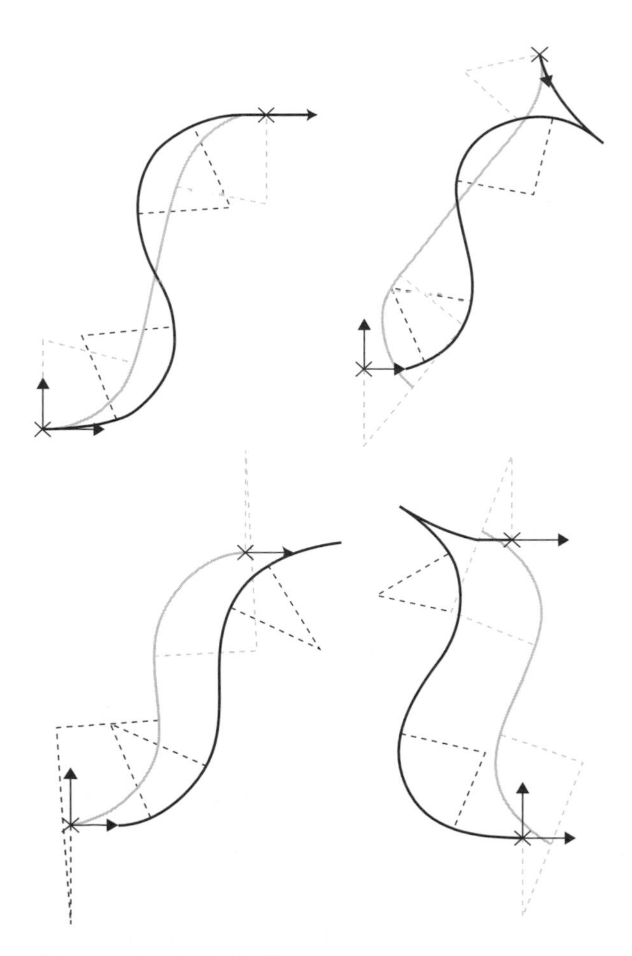

Fig. 11.32 RS (in grey) versus CC paths (in black).

Table 11.1 RS *vs.* CC paths' length

	min.	average	max.	deviation
ratio	1.00253	1.1065	2.45586	0.172188

Table 11.2 RS *vs.* CC paths' computation time

RS (1000 paths)	CC (1000 paths)	average ratio
3.466586 s.	4.483492 s.	1.33

(Table 11.2). Given that continuous curvature paths can be tracked with a much greater accuracy by a real car-like vehicle (see the experimental results obtained in Scheuer and Laugier, 1998), the results reported herein demonstrate the interest of CC paths (about the same computation time and same length).

Acknowledgements

This work was partially supported by the Eureka EU-45 European project *Prometheus*, the Inria-Inrets French programme *Praxitèle*, the Inco-Copernicus European project *Multi-agent robot systems for industrial applications in the transport domain* and the French research programme *La route automatisée*.

The authors would like to thank J. Hermosillo, F. Large, E. Gauthier, Ph. Garnier and I. Paromtchik for their invaluable contributions.

References

Ahuactzin-Larios, J.-M. (1994). Le Fil d'Ariane: Une Méthode de Planification Génerale. Application à la Planification Automatique de Trajectoires. Thèse de doctorat Inst. Nat. Polytechnique de Grenoble, Grenoble, France.

Alami, R., Chatila, R., Fleury, S., Ghallab, M. and Ingrand, F. (1998). An architecture for autonomy. *International Journal of Robotics Research*, **17**(4), 315–37. Special issue on integrated architectures for robot control and programming.

Anderson, T.L. and Donath, M. (1990). Autonomous Robots and Emergent Behavior: A Set of Primitive Behaviors for Mobile Robot Control. In *Proceedings of the IEEE-RSJ International Workshop on Intelligent Robots and Systems*, **2**, Tsuchiura, Japan, pp. 723–30.

Arkin, R.C. (1987). Motor Schema Based Navigation for a Mobile Robot. In *Proceedings of the IEEE International Conference on Robotics and Automation*, Raleigh, NC, USA, pp. 264–71.

Arkin, R.C. (1989). Motor schema-based mobile robot navigation. *International Journal of Robotics Research*, **8**, 92–112.

Arkin, R.C. (1990). Integrating Behavioral, Perceptual, and World Knowledge in Reactive Navigation. *Robotics and Autonomous Systems*, **6**, 105–22.

Arkin, R.C. (1998). *Behavior-based robotics – Intelligent robots and autonomous agents*. The MIT Press.

Baille, G., Garnier, Ph., Mathieu, H. and Pissard-Gibollet, R. (1999). Le Cycab de l'Inria Rhône-Alpes. Technical Report 229 Inst. Nat. de Recherche en Informatique et en Automatique. Montbonnot, France.

Barraquand, J. and Latombe, J.-C. (1989). On nonholonomic mobile robots and optimal maneuvering. *Revue d'Intelligence Artificielle*, **3**(2), 77–103.

Barraquand, J. and Latombe, J.-C. (1990). Controllability of Mobile Robots with Kinematic Constraints. Research Report STAN-CS-90–1317. Stanford University, Stanford, CA, USA.

Bobrow, J.E., Dubowsky, S. and Gibson, J.S. (1985). Time-optimal control of robotic manipulators along specified paths. *International Journal of Robotics Research*, **4**(3), 3–17.

Boissonnat, J.-D., Cérézo, A. and Leblond, J. (1991). Shortest paths of bounded curvature in the plane. Research Report 1503. Inst. Nat. de Recherche en Informatique et en Automatique. Rocquencourt, France.

Boissonnat, J.-D., Cérézo, A. and Leblond, J. (1994). A note on shortest paths in the plane subject to a constraint on the derivative of the curvature. Research Report 2160. Inst. Nat. de Recherche en Informatique et en Automatique, Rocquencourt, France.

Bonasso, R.P., Kortenkamp, D., Miller, D.P. and Slack, M. (1996). Experiences with an Architecture for Intelligent, Reactive Agents. *Lecture Notes in Computer Science*, **1037**, 187–202.

Brooks, R.A. (1990). A Robust Layered Control System for a Mobile Robot. In *Readings in Uncertain Reasoning*, G. Shafer and J. Perl (eds). Morgan Kaufmann, pp. 204–13.

Brooks, R.A. and Flynn, A.M. (1989). Fast, Cheap and Out of Control: A Robot Invasion of the Solar System. *Journal of the British Interplanetary System*, **42**(10), 478–85.

Canny, J., Donald, B., Reif, J. and Xavier, P. (1988). On the complexity of kinodynamic planning. In *Proceedings of the IEEE Symposium on the Foundations of Computer Sciences.* White Plains, NY, USA, pp. 306–16.

Canny, J., Rege, A. and Reif, J. (1990). An exact algorithm for kinodynamic planning in the plane. In *Proceedings of the ACM Symposium on Computational Geometry*. Berkeley, CA, USA, pp. 271–80.

Connell, J.H. (1992). SSS: A Hybrid Architecture Applied to Robot Navigation. In *Proceedings of the IEEE International Conference on Robotics and Automation*, Vol. 2, Nice, France, pp. 2719–24.

Daviet, P. and Parent, M. (1996). Longitudinal and Lateral servoing of vehicles in a platoon. In *Proceedings of the IEEE International Symposium on Intelligent Vehicles*, Tokyo, Japan, pp. 41–6.

De Luca, A., Oriolo, G. and Samson, C. (1998). Feedback control of a nonholonomic car-like robot. In *Robot motion planning and control*, (J.-P. Laumond, ed.). *Lecture Notes in Control and Information Science*, **229**, pp. 171–253. Springer.

Degtiariova-Kostova, E. and Kostov, V. (1998). Irregularity of Optimal Trajectories in a Control Problem for a Car-like Robot. Research Report 3411. Inst. Nat. de Recherche en Informatique et en Automatique.

Donald, B. and Xavier, P. (1990). Probably good approximation algorithms for optimal kinodynamic planning for cartesian robots and open-chain manipulators. In *Proceedings of the ACM Symposium on Computational Geometry*, Berkeley, CA, USA, pp. 290–300.

Dubins, L.E. (1957). On curves of minimal length with a constraint on average curvature, and with prescribed initial and terminal positions and tangents. *American Journal of Mathematics*, **79**, 497–517.

Erdmann, M. and Lozano-Perez, T. (1987). On multiple moving objects. *Algorithmica*, **2**, 477–521.

Farreny, H. and Ghallab, M. (1987). *Eléments d'Intelligence Artificielle*. Hermès.

Ferber, J. (1995). *Les systemes multi-agents, vers une intelligence collective*. Intereditions.

Fiorini, P. and Shiller, Z. (1996). Time optimal trajectory planning in dynamic environments. In *Proceedings of the IEEE International, Conference on Robotics and Automation*, Vol. 2, Minneapolis, MN, USA, pp. 1553–8.

Firby, R.J. (1989). Adaptive Execution in Complex Dynamic Domains. PHD thesis YALEU/CSD RR 672 Yale University.

Fraichard, Th. (1991). Smooth trajectory planning for a car in a structured world. In *Proceedings of the IEEE International Conference on Robotics and Automation*, Vol. 1, Sacramento, CA, USA, pp. 318–23.

Fraichard, Th. (1993). Dynamic Trajectory Planning with Dynamic Constraints: a 'State-Time Space' Approach. In *Proceedings of the IEEE-RSJ International Conference on Intelligent Robots and Systems*, Vol. 2, Yokohama, Japan, pp. 1394–1400.

Fraichard, Th. and Laugier, C. (1992). Kinodynamic planning in a structured and time-varying 2D workspace. In *Proceedings of the IEEE International Conference on Robotics and Automation*, Vol. 2, Nice, France, pp. 1500–1505.

Fraichard, Th. and Scheuer, A. (1994). Car-Like Robots and Moving Obstacles. In *Proceedings of the IEEE International Conference on Robotics and Automation*, Vol. 1, San Diego, CA, USA, pp. 64–9.

Fraichard, Th., Scheuer, A. and Desvigne, R. (1999). From Reeds and Shepp's to continuous-curvature paths. In *Proceedings of the IEEE International Conference on Advanced Robotics*, Tokyo, Japan, pp. 585–90.

Fujimura, K. and Samet, H. (1989). A hierarchical strategy for path planning among moving obstacles. *IEEE Transactions on Robotics and Automation*, **5**(1), 61–9.

Fujimura, K. and Samet, H. (1990). Motion planning in a dynamic domain. In *Proceedings of the IEEE International Conference on Robotics and Automation*, Cincinnatti, OH, USA, pp. 324–30.

Garnier, Ph. and Fraichard, Th. (1996). A Fuzzy Motion Controller for a Car-Like Vehicle. In *Proceedings of the IEEE-RSJ International Conference on Intelligent Robots and Systems*, Vol. 3, Osaka, Japan, pp. 1171–8.

Gat, E. (1992). Integrating planning and reacting in a heterogeneous asynchronous architecture for controlling real-world mobile robots. In *Proceedings of the Tenth National Conference on Artificial Intelligence*, San Jose, CA, USA, pp. 809–15.

Gat, E. (1997). On Three-Layer Architectures. In *Artificial Intelligence and Mobile Robots*, (D. Kortenkamp, R. P. Bonnasso and R. Murphy, eds). MIT/AAAI Press.

Gat, E., Slack, M., Miller, D.P. and Firby, R.J. (1990). Path Planning and Execution Monitoring for a Planetary Rover. In *Proceedings of the IEEE International Conference on Robotics and Automation*, Vol. 1, Cincinatti, OH, USA, pp. 20–25.

Harel, D. and Pnueli, A. (1985). On the development of reactive systems. In *Logics and Models of Concurrent Systems*, K.R. Apt (ed.). Springer-Verlag, pp. 477–98.

Humphrys, M. (1995). W-learning: Competition among selfish Q-learners. Technical Report 362 University of Cambridge Computer Laboratory. Available from *ftp://ftp.cl.cam.ac.uk/papers/reports/TR362-mh10006-w-learning.ps.gz.*

Jacobs, P. and Canny, J. (1989). Planning smooth paths for mobile robots. In *Proceedings of the IEEE International Conference on Robotics and Automation*, Scottsdale, AZ, USA, pp. 2–7.

Jacobs, P., Heinzinger, G., Canny, J. and Paden, B. (1989). Planning guaranteed near-time-optimal trajectories for a manipulator in a cluttered workspace. Research Report ESRC 89–20/RAMP 89–15. Engineering Systems Research Center, University of California, Berkeley, CA, USA.

Kanayama, Y., Kimura, Y., Myazaki, F. and Noguchi, T. (1991). A Stable Tracking Control Method for a Nonholonomic Mobile Robot. In *Proceedings of the IEEE-RSJ International Workshop on Intelligent Robots and Systems*, Vol. 2, Osaka, Japan, pp. 1236–41.

Kant, K. and Zucker, S. (1986). Toward efficient trajectory planning: the path-velocity decomposition. *International Journal of Robotics Research*, **5**(3), 72–89.

Kavraki, L., Svestka, P., Latombe, J,-C. and Overmars, M.H. (1996). Probabilistic roadmaps for path planning in high dimensional configuration spaces. *IEEE Transactions on Robotics and Automation*, **12**, 566–580.

Kyriakopoulos, K.J. and Saridis, G.N. (1991). Collision avoidance of mobile robots in non-stationary environments. In *Proceedings of the IEEE International Conference on Robotics and Automation*. Sacramento, CA, USA, pp. 904–9.

Latombe, J.-C. (1991). *Robot motion planning* Kluwer Academic Press.

Laugier, C. and Parent, M. (1999). Towards Motion Autonomy for Future vehicles. In *Proceedings of the International Symposium on Robotics Research*, Snowbird, USA, Invited paper.

Laugier, C., Fraichard, Th., Paromtchik, I.E. and Garnier, Ph. (1998). Sensor-Based Control Architecture for a Car-Like Vehicle. In *Proceedings of the IEEE-RSJ International Conference on Intelligent Robots and Systems*, Vol. 1, Victoria, BC, Canada, pp. 216–22.

Laumond, J.-P. (1986). Feasible trajectories for mobile robots with kinematic and environment constraints. In *Proceedings of the International Conference on Intelligent Autonomous Systems*. Amsterdam, The Netherlands, pp. 346–54.

Laumond, J.-P. (1987). Finding collision-free smooth trajectories for a nonholonomic mobile robot. In *Proceedings of the International Joint Conference on Artificial Intelligence*, Milan, Italy, pp. 1120–23.

Laumond, J.-P. (ed.) (1998). Robot motion planning and control. *Lecture Notes in Control and Information Science*, **229**, Springer.

Laumond, J.-P., Jacobs, P.E., Taïx, M. and Murray, R.M. (1994). A motion planner for non-holonomic mobile robots. *IEEE Trans! Robotics and Automation*, **10**(5), 577–93.

Laumond, J.-P., Sekhavat, S. and Lamiraux, F. (1998). Guidelines in nonholonomic motion planning for mobile robots. In *Robot motion planning and control*, J.-P. Laumond (ed.), *Lecture Notes in Control and Information Science*, **129**, pp. 1–53, Springer.

Laumond, J.-P., Siméon, T., Chatila, R. and Giralt, G. (1989). Trajectory planning and motion control for mobile robts. In *Geometry and Robotics*, J.-D. Boissonnat and J.-P. Laumond (eds). *Lecture Notes in Computer Science*, **391**, pp. 133–49, Springer.

Li, Z. and Canny, J.F. (eds) (1992). Nonholonomic Motion Planning. *The Kluwer International Series in Engineering and Computer Science*, **192** Kluwer Academic Press.

Lin, L.J. (1993). Reinforcement learning for robots using neural networks. Technical Report CMU-CS-93-103 Carnegie Mellon University.

Lozano-Perez, T. and Wesley, M.A. (1979a). An algorithm for planning collision-free paths among polyhedral obstacles. *Communications of the ACM* **22**(10), 560–70.

McFarland, D. (1987). *The Oxford Companion to Animal Behaviour*. Oxford University Press.

Moravec, H.P. (1983). The Stanford Cart and the CMU Rover. *Proceedings of the IEEE*, **71**(7), 872–84.

Murray, R.M. (1990). Robotic control and nonholonomic motion planning. Research Report UCB/ERL M90/117. Univ. of California at Berkeley, Berkeley, CA, USA.

Nakamura, Y. and Mukherjee, R. (1989). Nonholonomic path planning of space robots. In *Proceedings of the IEEE International Conference on Robotics and Automation*, Vol. 2, Scottsdale, AZ, USA, pp. 1050–55.

Nelson, W.L. (1989). Continuous curvature paths for autonomous vehicles. In *Proceedings of the IEEE International Conference on Robotics and Automation*, Vol. 3, Scottsdale, AZ, USA, pp. 1260–64.

Nilsson, N.J. (1984). Shakey The Robot. Technical note 323 AI Center, SRI International, Menlo Park, CA, USA.

Nilsson, N.J. (1980). *Principles of artificial intelligence.* Los Altos, CA, USA, Morgan Kaufmann.

Noreils, F. (1990). Integrating Error Recovery in a Mobile Robot Control System. In *Proceedings of the IEEE International Conference on Robotics and Automation*, Vol. 1, Cincinatti, OH, USA, pp. 396–401.

Ó'Dúnlaing, C. (1987). Motion planning with inertial constraints. *Algorithmica*, **2**, 431–75.

Paromtchik, I.E. and Laugier, C. (1996a). Autonomous Parallel Parking of a Nonholonomic Vehicle. In *Proceedings of the IEEE Symposium on Intelligent Vehicles*, Tokyo, Japan, pp. 13–18.

Paromtchik, I.E. and Laugier, C. (1996b). Motion Generation and Control for Parking an Autonomous Vehicle. In *Proceedings of the IEEE International Conference on Robotics and Automation*, Minneapolis, MN, USA, pp. 3117–22.

Paromtchik, I.E. and Laugier, Ch. (1997). Automatic Parallel Car Parking. In *Video-Proceedings of the IEEE International Conference on Robotics and Automation*, Albuquerque, NM, USA. Produced by Inst. Nat. de Recherche en Informatique et en Automatique-Unité de Communication et Information Scientifique (3 min.).

Payton, D.W. (1986). An Architecture for Reflexive Autonomous Vehicle Control. In *Proceedings of the IEEE International Conference on Robotics and Automation*. San Franciso, CA, USA, pp. 1838–45.

Payton, D.W., Rosenblatt, J.K. and Keirsey, D.M. (1990). Plan Guided Reaction. *IEEE Transactions on Systems, Man and Cybernetics*, **20**(6), 1370–82.

Pommier, E. (1991). Génération de trajectoires pour robot mobile nonholonome par gestion des centres de rotation. Thèse de doctorat Lab. d'Informatique, de Robotique et de Microélectronique Montpellier, France.

Reeds, J.A. and Shepp, L.A. (1990). Optimal paths for a car that goes both forwards and backwards. *Pacific Journal of Mathematics*, **145**(2), 367–93.

Reif, J. and Sharir, M. (1985). Motion planning in the presence of moving obstacles. In *Proceedings of the IEEE Symposium on the Foundations of Computer Science*, Portland, OR, USA, pp. 144–54.

Rich, E. and Knight, K. (1983). *Artificial Intelligence*. McGraw-Hill.

Rosenblatt, J.K. (1997). DAMN: A Distributed Architecture for Mobile Navigation. PhD thesis, Carnegie Mellon University.

Rosenblatt, J.K. and Payton, D.W. (1989). A Fine-Grained Alternative to the Subsumption Architecture for Mobile Robot Control. In *Proceedings of the International Joint Conference on Neural Networks*, Vol. 2, Washington, DC, pp. 317–23.

Sahar, G. and Hollerbach, J.H. (1985). Planning of minimum-time trajectories for robot arms. In *Proceedings of the IEEE International Conference on Robotics and Automation*. St Louis, MI, USA, pp. 751–8.

Scheuer, A. (1998). Planification de chemins à courbure continue pour robot mobile nonholonome. Thèse de doctorat Inst. Nat. Polytechnique de Grenoble, Grenoble, France.

Scheuer, A. and Laugier, Ch. (1998). Planning Sub-Optimal and Continuous-Curvature Paths for Car-Like Robots. In *Proceedings of the IEEE-RSJ International Conference on Intelligent Robots and Systems*, Vol. 1, Victoria, BC, Canada, pp. 25–31.

Shih, C.L., Lee, T.T. and Gruver, W.A. (1990). Motion planning with time-varying polyhedral obstacles based on graph search and mathematical programming. In *Proceedings of the IEEE International Conference on Robotics and Automation*. Cincinnatti, OH, USA, pp. 331–7.

Shiller, Z. and Chen, J.C. (1990). Optimal motion planning of autonomous vehicles in three-dimensional terrains. In *Proceedings of the IEEE International Conference on Robotics and Automation*. Cincinnatti, OH, USA, pp. 198–203.

Shiller, Z. and Lu, H.-H. (1990). Robust computation of path constrained time-optimal motion. In *Proceedings of the IEEE International Conference on Robotics and Automation*. Cincinnatti, OH, USA, pp. 144–9.

Shiller, Z. and Dubowsky, S. (1985). On the optimal control of robotic manipulators with actuator and end-effector constraints. In *Proceedings of the IEEE International Conference on Robotics and Automation*. St Louis, MI, USA, pp. 614–20.

Shiller, Z. and Dubowsky, S. (1988). Global time optimal motions of robotic manipulators in the presence of obstacles. In *Proceedings of the IEEE International Conference on Robotics and Automation*. Philadelphia, PA, USA, pp. 370–75.

Shiller, Z. and Dubowsky, S. (1989). Robot path planning with obstacles, actuator, gripper and payload constraints. *International Journal of Robotics Research*, **8**(6), 3–18.

Shin, K.G. and McKay, N.D. (1985). Minimum-time control of robotic manipulators with geometric path constraints. *IEEE Trans. Autom. Contr*, **30**, 531–41.

Simon, D., Espiau, B., Castillo, K. and Kapellos, K. (1993). Computer-Aided Design of a Generic Robot Controller Handling Reactivity and Real-Time Control Issues. *IEEE Transactions on Control Systems Technology*, **1**(4), 213–29.

Simmons, R.G. (1994). Structured Control for Autonomous Robots. *IEEE Transactions on Robotics and Automation*, **10**(1), 34–43.

Sussmann, H.J. and Tang, G. (1991). Shortest paths for the Reeds-Shepp car: a worked out example of the use of geometric techniques in nonlinear optimal control. Research Report Sycon-91–10. Rutgers University Center for Systems and Control.

Svestka, P. and Overmars, M.H. (1998). Probabilistic Path Planning. In *Robot motion planning and control*, J.-P. Laumond (ed.). *Lecture Notes in Control and Information Science*, **229**, pp. 255–304. Springer.

Udupa, S.M. (1977). Collision detection and avoidance in computer-controlled manipulators. In *Proceedings of the International Joint Conference on Artificial Intelligence*. Cambridge, MA, USA, pp. 737–48.

Wang, Y. (1996). Nonholonomic motion planning: a polynomial fitting approach. In *Proceedings of the IEEE International Conference on Robotics and Automation*, Vol. 3, Minneapolis, MN, USA, pp. 2956–61.

Wilfong, G. (1988). Motion planning for an autonomous vehicle. In *Proceedings of the IEEE International Conference on Robotics and Automation*. Philadelphia, PA, USA, pp. 529–33.

12

Brake modelling and control

Dragos B. Maciuca
BMW of North America, LLC.

This chapter presents an overview of hydraulic brakes for vehicles. It includes sections on brake dynamics modelling and suggests a form of nonlinear control. The control algorithm, although by no means the only possible one, can be used for automatic brake control for applications such as autonomous cruise control or fully automated platooning. The controller suggested is based on the variable structure control methodology and it was selected due to its ability to deal with the nonlinearities and uncertainties present in the brake system dynamics.

12.1 Brake modelling

Automatic brake actuation and control has generally lagged that of engine control. Although electronic fuel injection has been available for years, only recently dynamic stability systems, requiring independent brake control, have become available. And, although cruise control has been available for years, only recently has adaptive cruise control (ACC) been developed. Even in those cases, some manufacturers still offer braking capabilities only as much as the engine braking can supply. Therefore, historically there has been much less emphasis on modelling and control of automotive brake systems. However, as ACC becomes more popular and the future will see such systems as stop-and-go ACC, accident avoidance systems and even automated highway systems, it is essential to understand the dynamics of the automotive brake system, its limitations and methods for the proper actuation in order to achieve the desired results.

The first part of this chapter will develop a model for a typical hydraulic automotive brake system, while the second part will address the issue of brake control and actuation. Developing a model is essential in determining hardware requirements, controller development and simulation. The model is also used to determine the level of complexity necessary to achieve a good compromise between simplicity and accuracy. Actuation strategies can also be evaluated using this model. Such actuation strategies range from actuation at the brake pedal to independent wheel actuation. Test track time can also be reduced by extensive simulation, thus reducing the financial burden

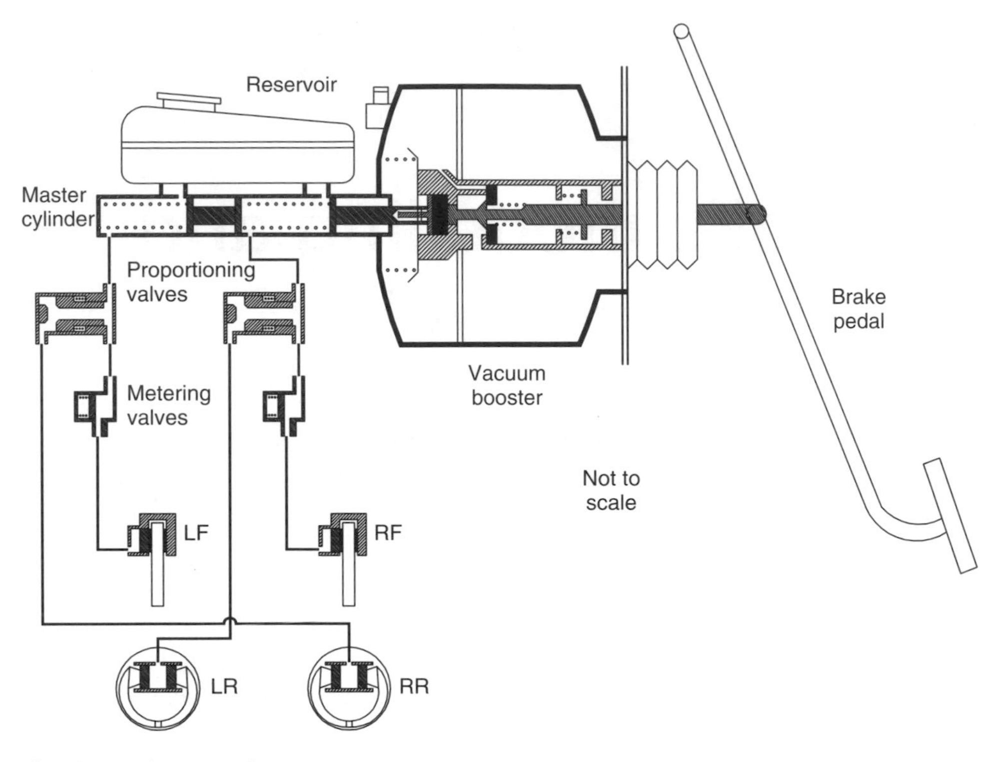

Fig. 12.1 Brake system diagram.

associated with such experiments. The model therefore needs to be accurate enough to capture the essential dynamics of the system yet simple enough to facilitate the development of model-based controllers. Most other models have concentrated on the design issues of the brake system. As such, these models are too complex (Fisher, 1970) for dynamic simulations. Khan *et al.* (1994) reduced this model to ten states which is still too complex for the demands at hand. Other models such as McMahon *et al.* (1990) considered a brake model that was a pure time delay followed by a first-order linear system. However, such a model fails to capture the nonlinearities of the brake system.

Therefore, this chapter will present a comprehensive brake system (Figure 12.1) model that is simple enough to facilitate the development of model-based controllers yet accurate enough to capture the essential dynamics. The vacuum booster model was developed by Maciuca *et al.* (1994) and Gerdes *et al.* (1993). The brake hydraulics model was developed by Gerdes and Hedrick (1996) and is consistent with Merritt (1967). The logical sequence is to follow the brake system, the same way that the forces follow it, from the applied force at the brake pedal to the resulting brake torque at the wheels.

12.1.2 Brake pedal

The brake pedal can be represented as a lever that amplifies the applied force.

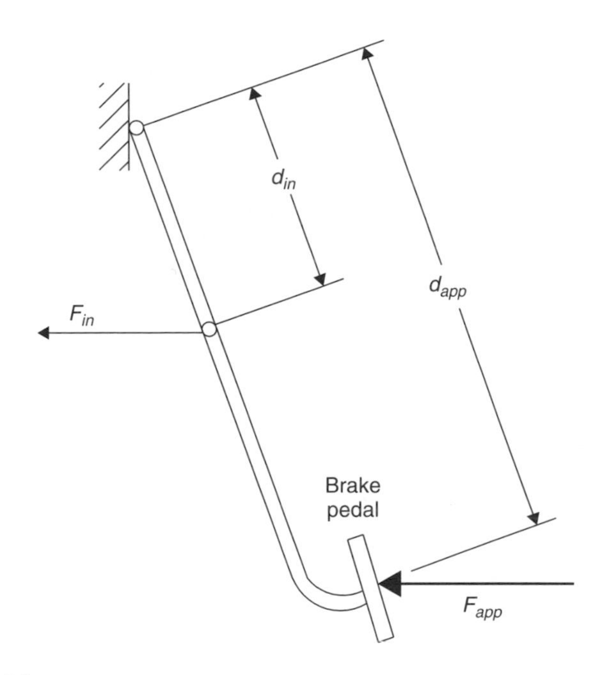

Fig. 12.2 Brake pedal.

Referring to Figure 12.2,

$$F_{in} = \frac{d_{app}}{d_{in}} F_{app} \tag{12.1}$$

where F_{app} is the force applied at the brake pedal and F_{in} is the input force to the vacuum booster.

12.1.3 Vacuum booster

Vacuum booster operation description

Disc brakes were first introduced in cars in the late 1940s. By the late 1970s they replaced the drum brakes on the front wheels of most vehicles. The major reasons for the change were better resistance to fade, better cooling, water and dirt resistance, less maintenance and greater surface area (Puhn, 1985). However, the servo action inherent to the drum brakes, explained on page 407, is lost so most cars with disc brakes require some form of power assist. In the vast majority of today's passenger cars this assist is provided by a vacuum booster (Figure 12.3).

The vacuum booster uses the negative pressure generated in the engine intake manifold to amplify the force produced at the pedal. This is achieved by the use of two air chambers separated by a diaphragm. One of the chambers (vacuum chamber) is constantly connected (through a check valve) to the intake manifold. The other chamber (apply chamber) can be connected (through a control valve) to the atmosphere or the vacuum chamber. The pressure difference between the two chambers applied over the surface of the diaphragm provides the amplified brake force (Robert Bosch GmbH,

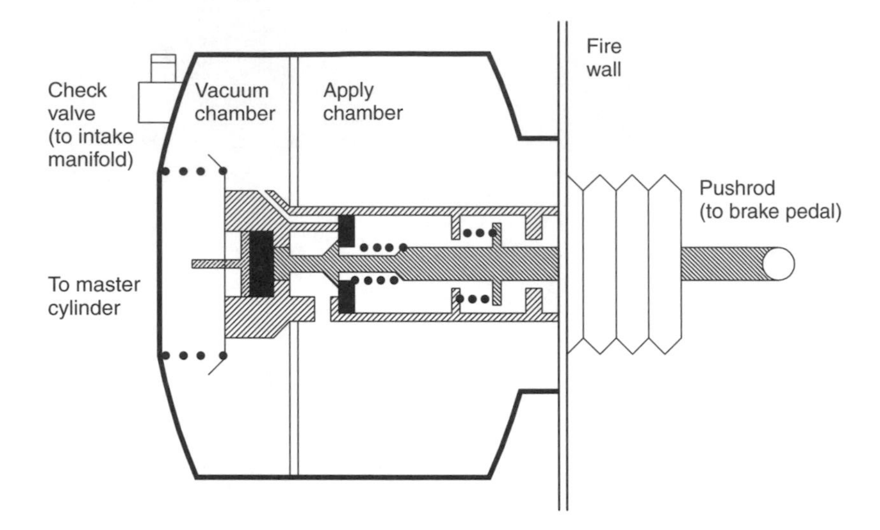

Fig. 12.3 Vacuum booster diagram.

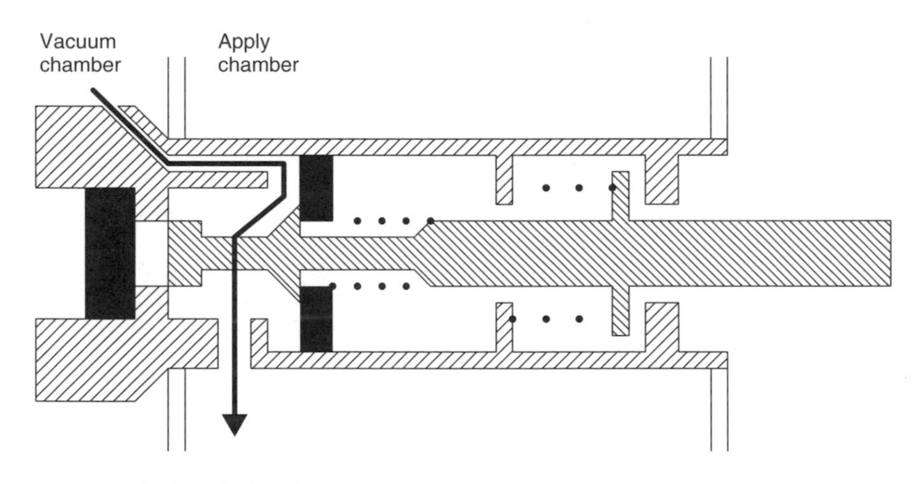

Fig. 12.4 Control valve – 'Release' stage.

1995). Under normal operation, the vacuum booster is in the 'release' stage. In this stage the two chambers are connected and therefore at the same pressure thus applying no force on the master cylinder push rod (Figure 12.4).

As the driver applies the brake pedal, the control valve seals the connection between the two chambers and opens the apply chamber to atmospheric pressure. As the air flows in the apply chamber the pressure difference created across the diaphragm moves it towards the vacuum chamber.

This motion serves a dual purpose. First, it starts applying force on the master cylinder push rod. Second, it starts closing the control valve opening to the atmosphere until it reaches the 'hold' stage.

This constitutes the internal feedback of the vacuum booster. At this stage a constant pressure difference across the diaphragm will be maintained. As the driver releases the brake pedal, the control valve again moves in the 'release' stage thus connecting the two chambers and sealing the apply chamber from the atmosphere. At this point there will be no brake force applied. The initial reasoning behind the study of the dynamics of the vacuum booster was the use of a retrofitted automatic brake system. The actuation in such a system is hydraulic and takes place at the brake pedal. There are several reasons why a retrofitted brake system was chosen over a system specifically designed for IVHS. Most importantly, such a system is easy to design and implement. The time frame and cost are considerably reduced. From a legal aspect, such a system presents less liability. Furthermore, such a system is more reliable since it relies heavily on the existing brake system. With this system, studies can also be made regarding the viability of retrofitting cars once the AHS system is operational. The vacuum booster though adds to the complexity of controlling the brakes by introducing a series of nonlinearities. Controlling the brakes using the vacuum booster is the topic of Section 12.2.2.

Control valve model

The vacuum booster dynamics have been divided into force balances and air dynamics.

The force balance across the control valve, shown in Figure 12.5, is as follows:

$$F_{in} - F_{VS} - rF_{mc} = 0 \tag{12.2}$$

$$F_{vs} + F_d - F_{rs} - (1 - r)F_{mc} = 0 \tag{12.3}$$

Here F_{in} is the input force from the brake pedal, F_{vs} is the force in the spring internal to the valve, F_{rs} is the return spring internal to the vacuum booster, F_{mc} is the force applied to the master cylinder and F_d is the force developed across the booster diaphragm. Since the pushrod of the control valve does not act over the entire area of the reaction washer, r represents the area of the reaction washer that is in contact with the control valve pushrod. The diaphragm force, F_d, is given by:

$$F_d = A_d(P_a - P_v) \tag{12.4}$$

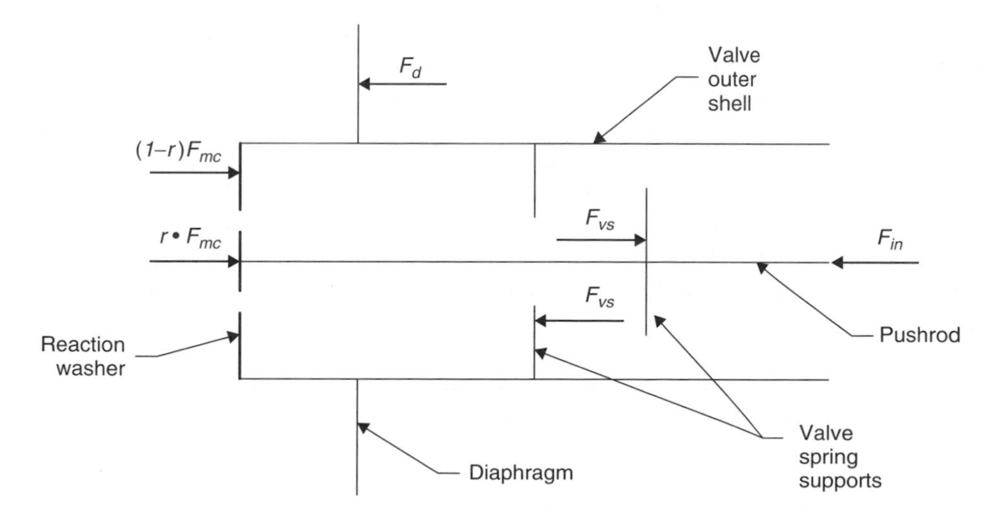

Fig. 12.5 Vacuum booster control valve force diagram.

where A_d is the area of the diaphragm, and P_a and P_v are the pressures in the apply and vacuum chambers, respectively.

The relative displacement of the valve pushrod, x_{pr}, and valve outer shell, x_{os}, is denoted as x_{rel}. Therefore the force in the valve spring is given by:

$$F_{vs} = F_{vs0} + K_{vs}x_{rel} \tag{12.5}$$

where F_{vs0} is the spring preload and K_{vs} is the spring constant. Similarly, the force in the booster return spring is related to the displacement of the outer shell, x_{os}, by:

$$F_{rs} = F_{rs0} + K_{rs}x_{os} \tag{12.6}$$

where F_{rs0} is the return spring preload and K_{rs} is the spring constant.

By summing Equations 12.2 and 12.3 the following force input to the master cylinder, F_{mc}, is obtained:

$$F_{mc} = \begin{cases} F_d + F_{in} - F_{rs} & F_d + F_{in} > F_{rs} \\ 0 & \text{otherwise} \end{cases} \tag{12.7}$$

Finally, the valve operation mode can be determined from x_{rel}:

$$\begin{aligned} 0 \le x_{rel} < x_h &\Rightarrow \text{'Release' stage} \\ x_h \le x_{rel} < x_a &\Rightarrow \text{'Hold' stage} \\ x_a \le x_{rel} < x_s &\Rightarrow \text{'Apply' stage} \end{aligned} \tag{12.8}$$

where x_h is the relative displacement at which the 'hold' stage begins, x_a is the relative displacement at which the 'apply' stage begins, and x_s is the maximum allowable relative displacement. The valve operation is summarized in Table 12.1

Air flow dynamics

The air flow into and out of each chamber determines the dynamics of the vacuum booster. A qualitative description of the dynamics follows the air flow from the 'apply' to 'hold' and finally to 'release' stage. As the brakes are applied, the air enters the apply chamber through the control valve increasing the pressure in the chamber. This results in the displacement of the diaphragm and thus the reduction in the volume of the vacuum chamber. The change in volume causes a pressure increase in the vacuum chamber which is followed by a pressure decay as the air flows into the intake manifold through the check valve. Throughout this process the pressure difference between the two chambers is kept constant by a slight leakage in the control valve.

As the brake pedal is released, the control valve allows the two chambers to be connected, thus the air from the apply chamber flows into the vacuum chamber causing

Table 12.1 Booster valve operation

Stage	Atmosphere to apply	Apply to vacuum
'Apply'	OPEN	CLOSED
'Hold'	CLOSED	CLOSED
'Release'	CLOSED	OPEN

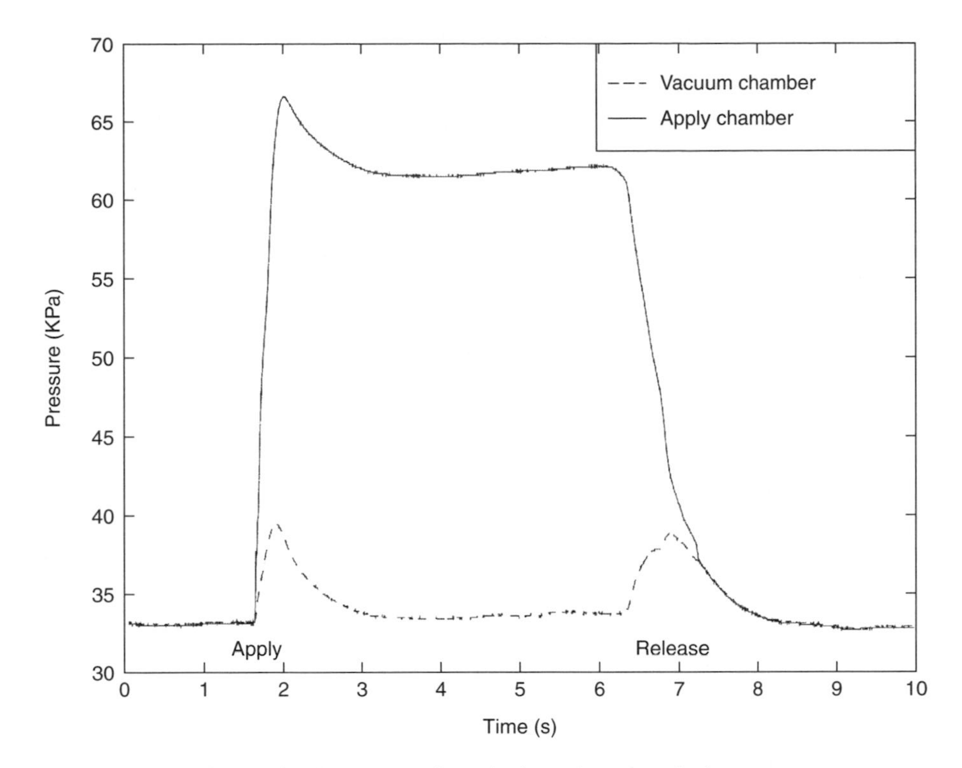

Fig. 12.6 Vacuum and apply chamber pressure during brake application and release.

an initial pressure increase in the latter one. However, since the vacuum chamber is connected to the intake manifold, air will flow out of the chamber at a rate allowed by the area of the check valve. Figure 12.6 shows a typical air pressure evolution in the two chambers during a typical brake application and release.

Quantitatively, the model assumes ideal gas behaviour and isothermal expansion. Under these assumptions, the pressures in the two chambers can be determined from the mass of air in each of them.

$$P_A = \frac{m_A RT}{V_{A0} + A_d x_{os}} \tag{12.9}$$

$$P_V = \frac{m_V RT}{V_{A0} + A_d x_{os}} \tag{12.10}$$

the two chambers. The mass flow into and out of the apply chamber then depends on the mode of operation:

$$\dot{m}_a = \begin{cases} 2.2 \times C_{aA}\rho_0 A_{aA} \times \sqrt{\left(\dfrac{P_{atm}}{P_A}\right)^{0.286} - 1} & \text{'apply' stage} \\ -2.2 \times C_{leak}\rho_0 A_{leak} \times \sqrt{\left(\dfrac{P_A}{P_V}\right)^{0.286} - 1} & \text{'hold' stage} \\ -2.2 \times C_{AV}\rho_0 A_{AV} \times \sqrt{\left(\frac{P_A}{P_V}\right)^{0.286} - 1} & \text{'release' stage} \end{cases} \tag{12.11}$$

Similarly the mass flow for the vacuum chamber is

$$\dot{m}_a = \begin{cases} -\dot{m}_{Vm} & \text{'apply' stage} \\ -\dot{m}_{Vm} - 2.2 \times C_{leak}\rho_0 A_{leak} \times \sqrt{\left(\dfrac{P_A}{P_V}\right)^{0.286} - 1} & \text{'hold' stage} \\ -\dot{m}_{Vm} - 2.2 \times C_{AV}\rho_0 A_{AV} \times \sqrt{\left(\dfrac{P_A}{P_V}\right)^{0.286} - 1} & \text{'release' stage} \end{cases} \quad (12.12)$$

Since this flow takes place through an idealized check valve, air can only flow out of it when $P_V \geq P_m$, and thus it can be modelled as:

$$\dot{m}_{Vm} = \begin{cases} 2.2 \times C_{Vm}\rho_0 A_{Vm} \times \sqrt{\left(\dfrac{P_V}{P_m}\right)^{0.286} - 1} & P_V > P_m \\ 0 & \text{otherwise} \end{cases} \quad (12.13)$$

In the previous equations, the following nomenclature was used:

P	pressure
m	mass
C	air flow coefficient
A	valve area
ρ	air density
subscript $_a$	'atmospheric conditions'
subscript $_A$	'apply chamber'
subscript $_V$	'vacuum chamber'
subscript $_m$	'intake manifold'

Although this model presents an accurate description of the dynamics of the vacuum booster, simplifications can be made to make it more compatible with control system analysis. Gerdes (1996), for example, suggests representing the mass air flow as proportional with the pressure difference and determining the flow constant experimentally. Depending on the application, each representation has its slight benefits over the other.

12.1.4 Brake hydraulics

A brief history of automotive braking systems reveals that early brakes were derived from horse-drawn wagons (Puhn, 1985). Next came the external drum brakes on the rear wheels only. This was due to the complexity of installing brakes on steered wheels and the fear of tipping the car on its nose. The internal drum brake was introduced by Renault in 1902. The 1920s saw the introduction of four-wheel braking and hydraulic actuation systems. In the late 1940s and early 1950s disc brakes were finally introduced. Beyond refinements and the recent popularity of ABS, the basic hydraulic brake system has not changed much since then. This section will analyse the most relevant components of a modern brake hydraulic system. We will follow the hydraulic circuit the way the fluid does, starting with the master cylinder.

Master cylinder

The braking process is initiated and controlled through the master cylinder. It has the main purpose of transferring and amplifying the force input from the brake pedal to the individual wheels (Puhn, 1985). Current safety regulations stipulate that cars be equipped with two separate brake circuits. This is satisfied by the use of a tandem master cylinder like the one depicted in Figure 12.7 (Robert Bosch GmbH, 1995).

The most common configuration is to use a diagonally split master cylinder in which one circuit serves the left front (LF) wheel and the right rear (RR) one and the other circuit serves the other two. This arrangement provides some braking and stability in case of failure in one of the circuits. Since the force transmission from the brake pedal to the wheels has to take place with the shortest amount of delay possible, the brake fluid used is incompressible. For this reason air should not be admitted in the system since air is compressible. However, when the brake pads wear out a larger amount of brake fluid needs to be displaced in order to get them in contact with the rotor. Also, if a small amount of leaks occur, the volume of the brake fluid is reduced. These problems are resolved by placing a brake fluid reservoir in contact with the brake system at the time the brakes are released. This guarantees that the master cylinder will compensate for the difference in brake fluid volume by accepting fluid from the reservoir rather than air. Since the flow from the reservoir into the master cylinder takes place when the brakes are released, it does not affect the dynamics of the hydraulic system. Therefore, the brake system is considered to consist of two sealed systems and the master cylinder transforms a force input into a pressure. The only additional detail is the return springs and the seals (Gerdes, 1996).

$$P_{mcp} = \frac{(F_{mc} - F_{csp} - F_{sfp})}{A_{mc}} \tag{12.14}$$

$$P_{mcs} = \frac{P_{mcp} - (F_{css} + F_{sfs})}{A_{mc}} \tag{12.15}$$

where F_{csp} and F_{css} are the return spring forces and F_{sfp} and F_{sfs} are the seal friction forces for the primary and secondary cylinders, respectively. The spring forces are given by:

$$F_{csp} = F_{csp0} + K_{csp}(x_{mcp} - x_{mcs}) \tag{12.16}$$

$$F_{css} = F_{css0} + K_{css}x_{mcs} \tag{12.17}$$

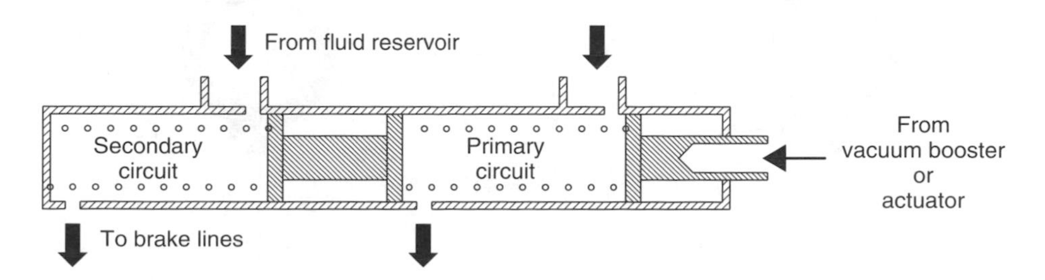

Fig. 12.7 Tandem master cylinder.

where x_{mcp} and x_{mcs} represent the displacements of the pistons. Since the vacuum booster's power piston is in contact with the primary piston of the master cylinder, x_{os}, used in Section 12.1.3, it is equal to x_{mcp} in this formulation. Furthermore, for the sake of simplicity, only one master cylinder pressure, P_{mc}, will be considered henceforth.

Proportioning valves

During braking a vehicle undergoes a shift in dynamic load from rear to front. Thus in a straight line deceleration on dry pavement the front wheels will experience a higher normal force while the rear wheels will experience an equal reduction in the normal force.

Referring to Figure 12.8, the normal force at the front and rear tyres are:

$$N_f = \frac{1}{2}\left(1 - \frac{D_{fw}}{D_w}\right)mg - \frac{mh_{cg}}{2D_w}a \tag{12.18}$$

$$N_r = \frac{D_{fw}}{2D_w}mg + \frac{mh_{cg}}{2D_w}a \tag{12.19}$$

where g is the gravitational acceleration, m is the mass of the vehicle and a is the vehicle acceleration.

Therefore, the braking force applied to the front wheels is greater than that at the rear. A similar phenomenon happens during cornering, the difference being that the weight transfers from one side to the other. The problem is that the braking force at each wheel should be proportional to the normal force at that wheel (Puleo, 1970).

$$\frac{F_{rf}}{N_{rf}} = \frac{F_{lf}}{N_{lf}} = \frac{F_{rr}}{N_{rr}} = \frac{F_{lr}}{N_{lr}} \tag{12.20}$$

where F is the braking force and N is the normal force at each wheel and the subscripts represent right-front (rf), left-front (lf), right-rear (rr) and left-rear (lr). If Equation 12.20 is violated at any of the wheels during hard deceleration, one of

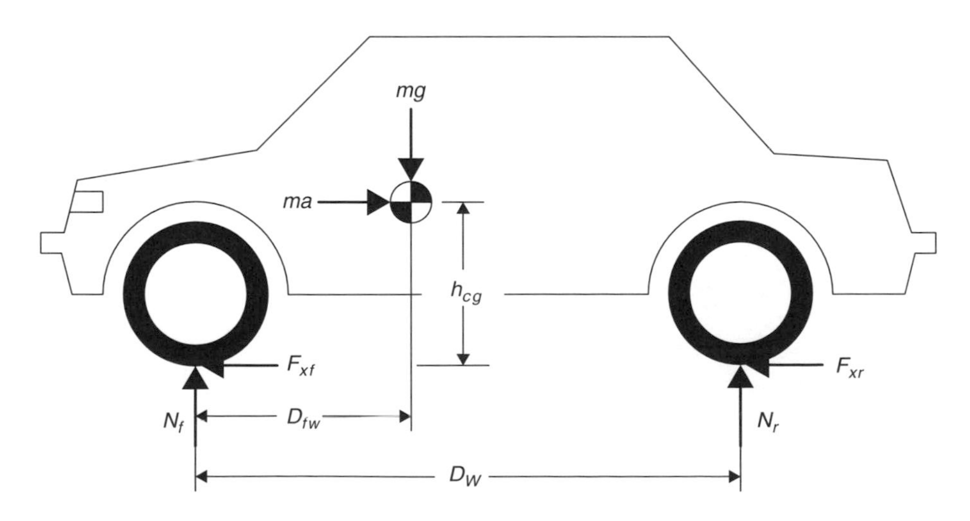

Fig. 12.8 Half car model.

two undesirable situations can arise. If the front tyres saturate (lock-up) before the rear tyres, steering control is lost. If the opposite happens (rear tyres lock-up before the front tyres) the vehicle becomes unstable. As an aside, the ratio in Equation 12.20 is known as 'peak friction coefficient' even though this is a misnomer since it is not really a friction coefficient but it is in the form of a friction coefficient definition. In order to achieve the desired force distribution most vehicle manufacturers include a brake proportioning valve in the brake hydraulic system (Limpert, 1992). The purpose of such a valve is to approximate the desired pressure distribution between the front and rear wheels. This is done by reducing the pressure increase to the rear wheels after a certain threshold known as knee pressure. That means that the pressure to the rear wheels will increase at a lower rate than the front wheels once the knee pressure is achieved. The pressure at the rear wheel, P_r, can be determined from the pressure at the master cylinder and the geometry of the valve:

$$P_r = \begin{cases} P_{mc} & P_{mc} \leq P_k \\ P_k + r_{rf}(P_{mc} - P_k) & \text{otherwise} \end{cases} \tag{12.21}$$

where P_k is the *knee pressure* of the proportioning valve. The ratio between the rear and front brake pressures above the changeover is denoted as r_{rf}.

The relationship between the front and rear brake pressures is shown in Figure 12.9 adapted from Puhn (1985).

One problem with this solution is that the changeover pressure and the ratio between the front and rear wheels is fixed in the factory and determined for an average weight distribution, with average tyres on an average road. Once any of these parameters

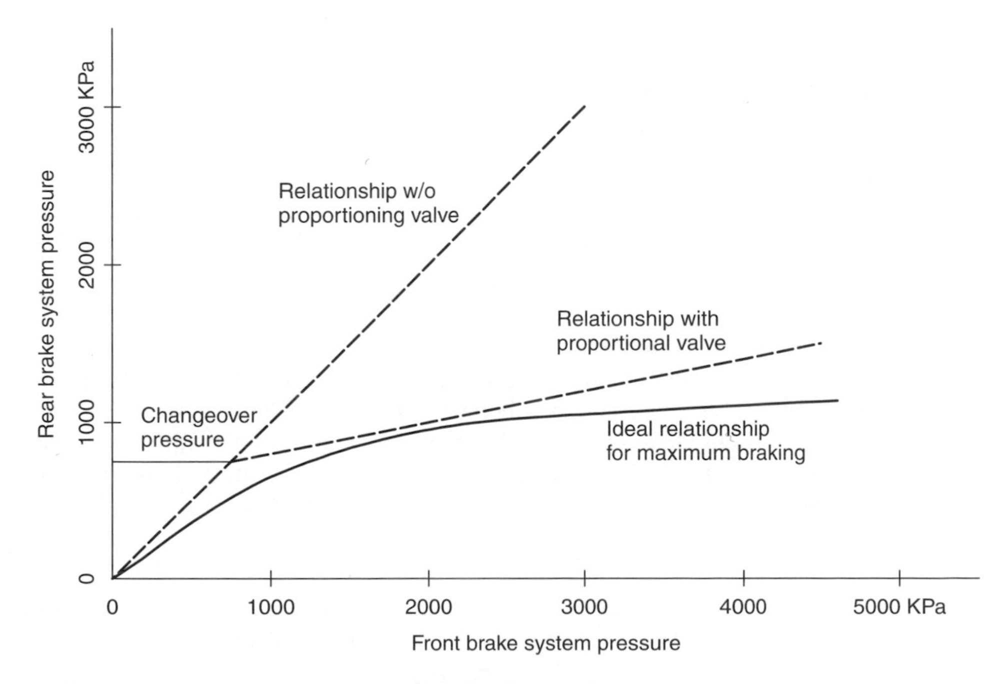

Fig. 12.9 Ideal and proportioning valve brake pressure distribution.

changes from its nominal value, the approximation of the brake distribution that the proportional valve provides, will not match the existing condition.

Several solutions exist to this problem such as driver selectable changeover pressure, load-sensitive regulating valves and deceleration-sensitive regulating valves. In one form or another they all attempt to control the vehicle dynamics such that Equation 12.20 is not violated.

Metering valves

Additionally, in cars that have front disc brakes and rear drum brakes a metering valve exists in the front brake hydraulic circuit. Its purpose is to prevent the application of the front brakes below a certain threshold. The reason for a metering valve is that it requires a lot more force and fluid movement to activate the drum brakes than the disc brakes. The metering valve thus allows the front and rear brakes to work more evenly at low pressures and it also prevents the lock-up of the front wheels when stopping on ice.

The valve itself operates like a residual-pressure valve (Puhn, 1985). When hydraulic pressure exerts enough force on the poppet, the spring preload is overcome and fluid is allowed to flow.

The pressure at the front wheels, P_f, can again be determined from the pressure at the master cylinder and the construction of the metering valve.

$$P_f = \begin{cases} 0 & P_{mc} \leq P_{tm} \\ P_{mc} - P_{tm} & \text{otherwise} \end{cases} \tag{12.22}$$

where P_{tm} represents the threshold of the metering valve. It can be calculated as:

$$P_m = \frac{F_{sp}}{A_{mv}} \tag{12.23}$$

where F_{sp} is the valve spring force and A_{mv} is the metering valve area. Typical values for P_{tm} vary from 500 to 1000 kpa.

Combination valves

In most modern cars the proportioning valve and the metering valve are built in one unit known as a combination valve (Limpert, 1992). Additionally, a differential pressure switch is included to warn of any hydraulic leaks.

Brake lines and slave cylinders

The brake hydraulic dynamics originally developed in Gerdes *et al.* (1993) were approximated by a linear first order system. Further work (Gerdes, 1996) showed that it is more accurate to represent the dynamics in the standard fashion as incompressible flow through an orifice. This is also consistent with Merritt (1967). The volume of fluid displaced depends on the displacement, x_{mc}, and area, A_{mc}, of the master cylinder:

$$V_b = x_{mc}A_{mc} \tag{12.24}$$

The brake lines and slave cylinders are assumed to have some capacitance. Therefore the pressure at each wheel, P_w, is a function of the volume displaced:

$$P_w = P_w(V_b) \tag{12.25}$$

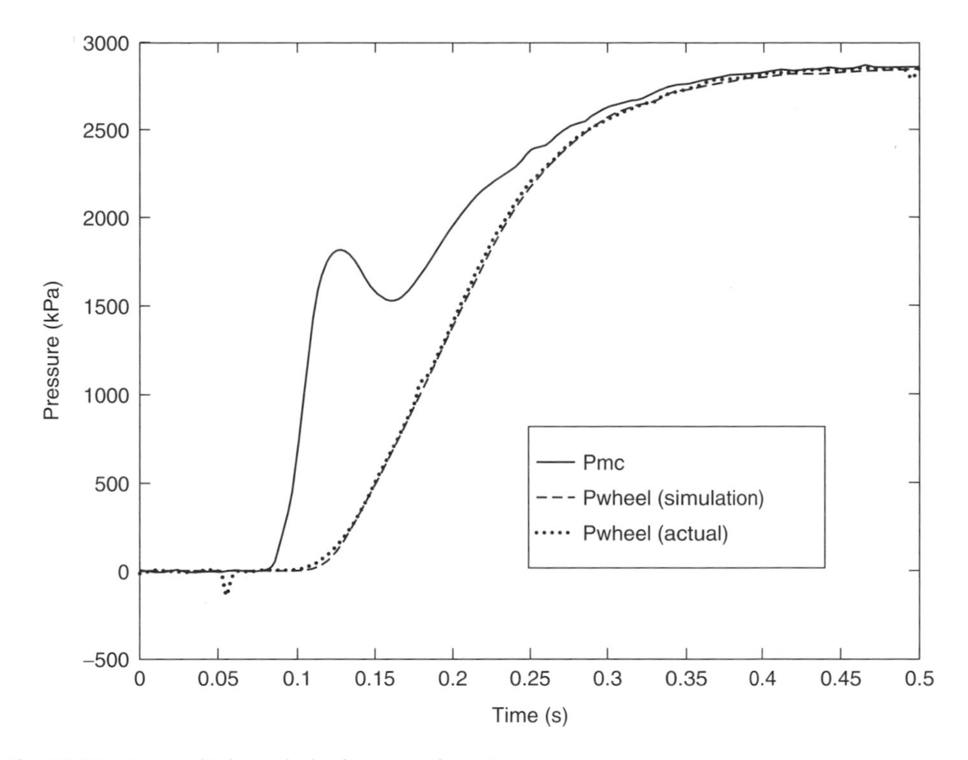

Fig. 12.10 Master cylinder and wheel pressure dynamics.

As can be seen, there is an initial fluid flow without pressure increase. This is due to brake line expansion. The capacitance of the brake lines may be approximated by a smooth function.

Finally the fluid flow to each wheel is represented as an incompressible flow through an orifice:

$$\dot{V}_b = \sigma_w C_w \sqrt{|P_{mc} - P_w|} \tag{12.26}$$

where $\sigma_w = \text{sign}(P_{mc} - P_w)$ and C_w is the fluid flow coefficient from the master cylinder to the wheels. Figure 12.10 shows the dynamics of the pressure at the wheel relative to those of the pressure at the master cylinder and verifies the model presented above.

12.1.5 Disc and drum brakes

The remaining part of the model is the relation between the brake pressure and the brake torque. The slave cylinders transform the brake pressure into a force that in various ways acts on the friction material to produce the braking torque. This is achieved through the use of disc brakes (Figure 12.11) and drum brakes (Figure 12.12).

Disc brakes

There are two basic types of disc brakes: fixed caliper (floating rotor) and floating caliper (fixed rotor). The difference is obvious upon visual inspection. A fixed caliper

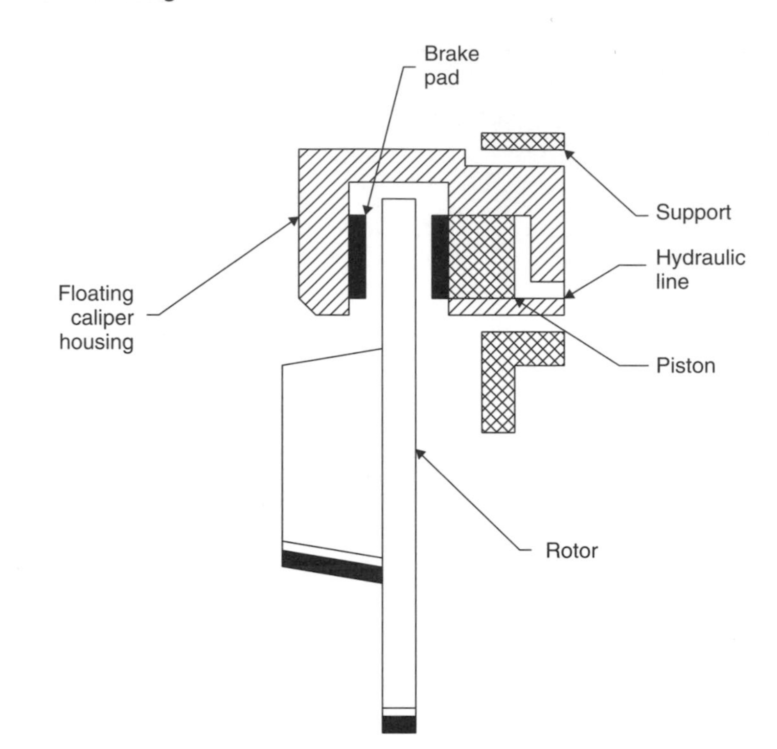

Fig. 12.11 Disc brake.

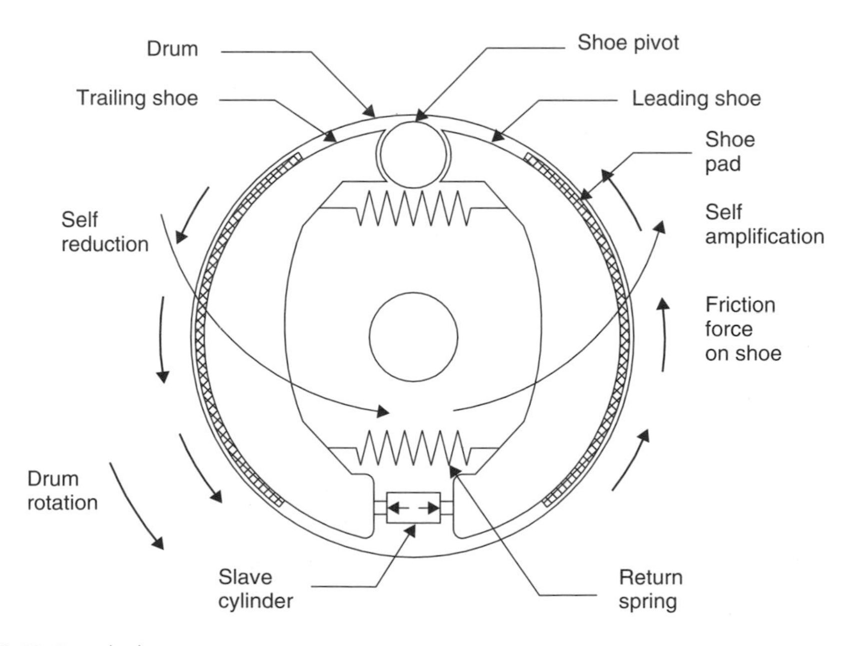

Fig. 12.12 Drum brake.

has one or two pistons on both sides of the rotor while a floating caliper has only one piston on one side of the rotor. In either case, the slave cylinder transforms the brake pressure into a force that acts normal to the rotor. Therefore the brake torque generated by the disc brake can be represented as:

$$T_{b,\text{disc}} = R_e \mu P_w A_{sc} \tag{12.27}$$

where R_e is the effective brake radius, μ is the friction coefficient between the brake pads and the rotor and A_{sc} is the area of the slave cylinder. This friction coefficient however varies greatly with external factors such as temperature, humidity, speed, friction material and wear of pads. Furthermore, the effective brake radius may vary with pad wear and pressure distribution across the pad. It is therefore convenient for future analysis to lump these uncertainties into a single parameter called the brake torque coefficient:

$$K_b = R_e \mu A_{sc} \tag{12.28}$$

Drum brakes

For the drum brakes the analysis gets a bit more involved due to the complexity of the mechanisms inside the drum brakes.

Referring to Figure 12.13, the input force to the drum mechanism, $F_{b,\text{drum}}$, is:

$$F_{b,\text{drum}} = P_w A_{wc} \tag{12.29}$$

where A_{wc} is the area of the drum slave cylinder, commonly known as the 'wheel cylinder'.

Due to the construction of the drum brake it has the interesting property of having a 'leading shoe' and a 'trailing shoe'. As it can be seen from the orientation of the friction

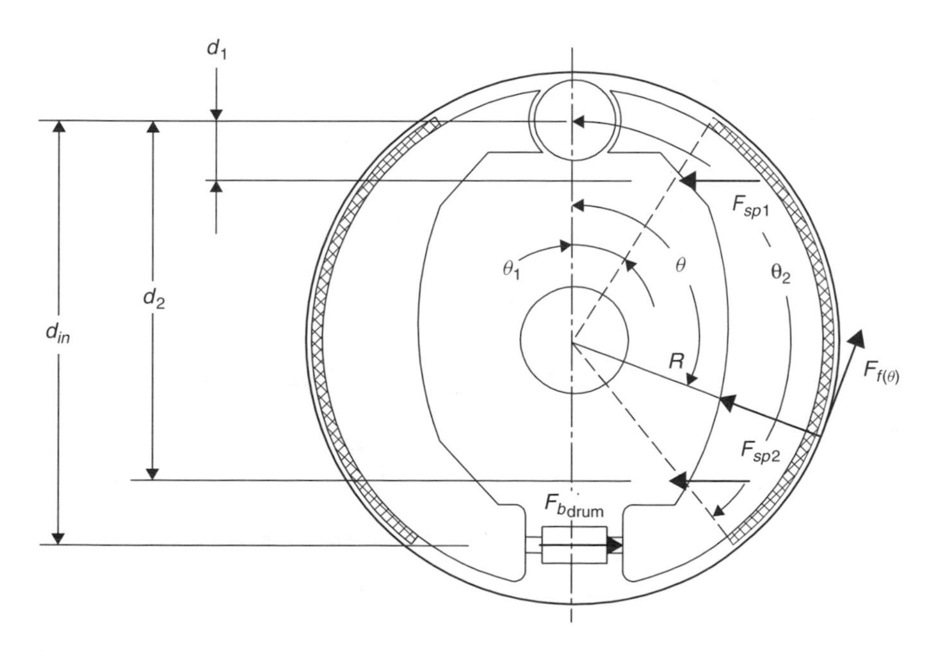

Fig. 12.13 Drum brake force diagram.

force on the shoes, the leading shoe will be attracted towards the drum providing even more friction force while the trailing shoe will be pushed away from the drum, providing less friction. Whether a shoe is leading or trailing depends on the direction of the rotation of the drum. Thus a leading shoe when the vehicle moves forward becomes a trailing shoe in reverse operation. Again, from Figure 12.13, the moment balance on the leading and trailing shoes yields respectively:

$$F_{b,\text{drum}}\, d_{in} - F_{sp1}d_1 - F_{sp2}d_2 + \int_{\Theta_1}^{\Theta_2} K_{f,l} d\Theta - \int_{\Theta_1}^{\Theta_2} K_{N,l} d\Theta = 0 \quad (12.30)$$

$$F_{b,\text{drum}}\, d_{in} - F_{sp1}d_1 - F_{sp2}d_2 - \int_{\Theta_1}^{\Theta_2} K_{f,t} d\Theta - \int_{\Theta_1}^{\Theta_2} K_{N,t} d\Theta = 0 \quad (12.31)$$

where $K_{f,l} = F_{f,l}(\Theta)\cos(\Theta - \alpha)R\dfrac{\sin\Theta}{\sin\alpha}$

$$K_{N,l} = N_l(\Theta)\sin(\Theta - \alpha)R\frac{\sin\Theta}{\sin\alpha}$$

$$K_{f,t} = F_{f,t}(\Theta)\cos(\Theta - \alpha)R\frac{\sin\Theta}{\sin\alpha}$$

$$K_{N,t} = N_t(\Theta)\sin(\Theta - \alpha)R\frac{\sin\Theta}{\sin\alpha}$$

and F_{sp1} and F_{sp2} are the forces of the return springs, F_f and N are the friction and normal forces, respectively, and R is the radius of the drum. The subscript l is used for the leading shoe and t for the trailing shoe. The friction force is related to the normal force by

$$F_{f,l}(\Theta) = \mu N_l(\Theta) \quad (12.32)$$

$$F_{f,t}(\Theta) = \mu N_t(\Theta) \quad (12.33)$$

And the brake torque generated by the brake drum is:

$$T_{b,\text{drum}} = R\left(\int_{\Theta_1}^{\Theta_2} F_{f,l}(\Theta)d\Theta + \int_{\Theta_1}^{\Theta_2} F_{f,t}(\Theta)d\Theta\right) \quad (12.34)$$

A few simplifications can now be made. Once again, from Figure 2.13:

$$\alpha = \tan^{-1}\left(R\frac{\sin\Theta}{1+\cos\Theta}\right) = \frac{1}{2}\Theta \quad (12.35)$$

Thus, substituting the above equation in Equations 12.30 and 12.31, the moment balance becomes:

$$F_{b,\text{drum}}d_{in} - F_{s1}d_1 - F_{s2}d_2 + \int_{\Theta_1}^{\Theta_2} N_l(\Theta) \times R[\mu\,(1+\cos\Theta) - \sin\Theta]\text{d}\Theta = 0 \quad (12.36)$$

$$F_{b,\text{drum}}d_{in} - F_{s1}d_1 - F_{s2}d_2 - \int_{\Theta_1}^{\Theta_2} N_t(\Theta) \times R[\mu\,(1+\cos\Theta) - \sin\Theta]\text{d}\Theta = 0 \quad (12.37)$$

If a certain normal force distribution is assumed, the problem can be further simplified. A reasonable distribution assumption is (Shigley and Mischke, 1989):

$$N(\Theta) = k \sin\Theta \left(1 - \frac{\Theta}{\pi}\right) \tag{12.38}$$

In this case one only needs to solve for k, a trivial task.

Again, it is important to note that μ can vary with temperature, speed, friction material, etc. It is therefore useful to consider the general 'gain' between the brake pressure at the wheel, P_w, and the lumped brake torque generated at all four wheels:

$$T_b = K_b P_w \tag{12.39}$$

where K_b lumps all the uncertainties in the model from brake pressure to brake torque.

12.1.6 Simplified model

Several simplifications can be made to the model for most of the applications considered. Some simplifications have already been made in the previous sections. The most important one comes from the fact that the total vehicle deceleration has more significance to control and vehicle dynamics than the individual wheel torques. This is true in the case of a single source actuation system. Individual wheel torques must be considered in the case of independent brake actuation systems.

Further simplifications come from the fact that under normal operation in automatic mode, the vehicle deceleration is in a range $(-2, 0)\,\text{m/s}^2$. At such low levels of deceleration the proportioning valve is not activated so both the rear and the front brakes experience the same brake pressure.

As such, the brake pressure at the master cylinder is obtained from the force input at the master cylinder pushrod, F_{mc}:

$$P_{mc} = \frac{(F_{mc} - F_{cs} - F_{sf})}{A_{mc}} \tag{12.40}$$

where A_{mc} is the area of the master cylinder and F_{cs} and F_{sf} are the force of the lumped master cylinder return springs and lumped seal friction forces, respectively.

From Equation 12.24, the displacement of the master cylinder is then related to the volume of fluid displaced by:

$$x_{mc} = \frac{V_b}{A_{mc}} \tag{12.41}$$

As stated in the section on brake lines and slave cylinders (page 404), the rate of volume displaced depends on the pressure difference between the master cylinder and the slave cylinders:

$$\dot{V}_b = \sigma_w C_w \sqrt{|P_{mc} - P_w|} \tag{12.42}$$

where C_w is the flow coefficient and $\sigma = sgn(P_{mc} - P_w)$. The pressure at the wheel, P_w, is a smooth function of the volume displaced:

$$P_w = P_w(V_b) \tag{12.43}$$

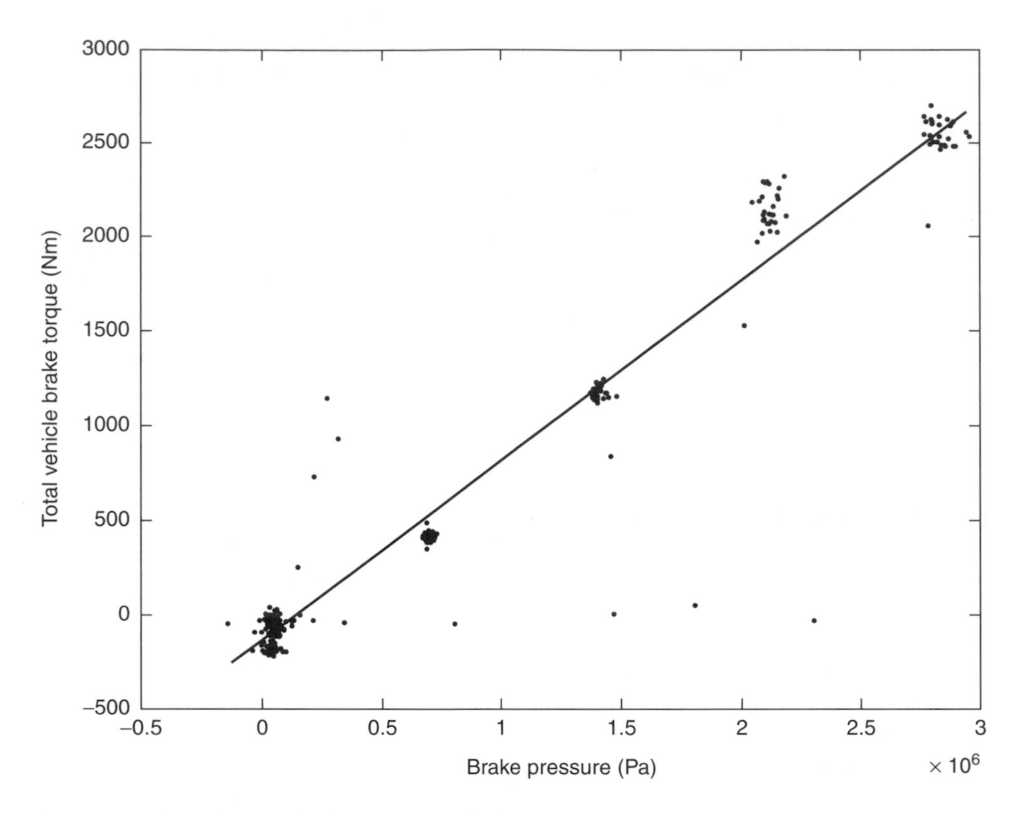

Fig. 12.14 Relation between brake pressure and brake torque.

Finally, the brake torque is considered to be proportional to the brake pressure at the wheel:

$$T_b = K_b P_w \tag{12.44}$$

where K_b is the lumped gain for the entire brake system.

Figure 12.14 shows the experimental data relating the brake pressure at the wheel and the brake torque justifying this linear relationship assumption.

12.1.7 Vehicle model

In order to simulate the behaviour of the brake system, a simplified vehicle model is developed with the dynamics of the brake system included. The torques developed by the brakes and engine affect the vehicle dynamics through the interaction between the road and the wheels. Figure 12.15, adapted from Gillespie (1992), shows the free-body diagram of a rear wheel drive vehicle.

The bicycle model force balance results in the following equation:

$$F_{x,f} + F_{x,r} - F_d = ma \tag{12.45}$$

where F_{xf} and F_{xr} are the traction forces and F_d is the drag force on the vehicle. Substituting the torque balances at each wheel in the above equation yields the following

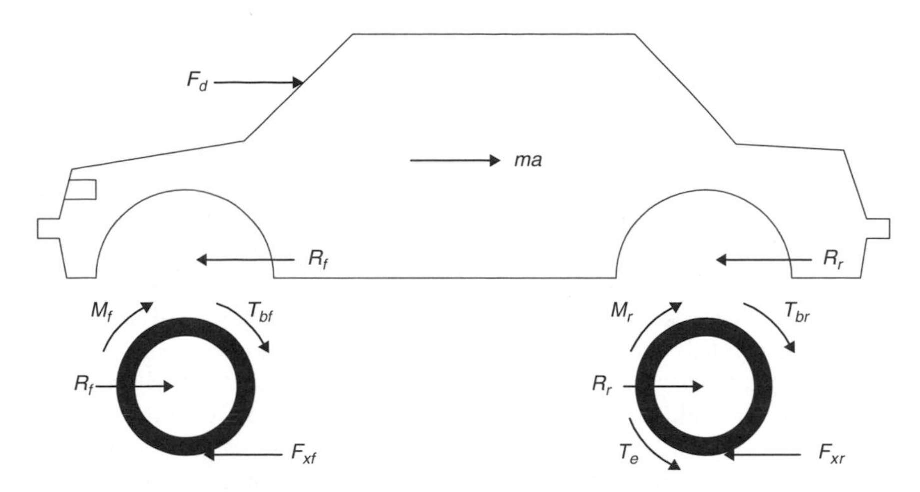

Fig. 12.15 Vehicle free-body diagram.

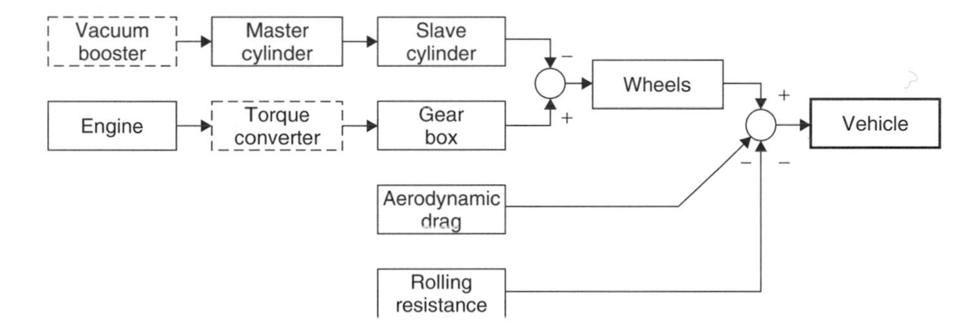

Fig. 12.16 Complete vehicle block diagram.

force balance:

$$\frac{[T_e - T_{b,r} - T_{b,f} - M_r - M_f - J_{wr,eff}\dot{\omega}_{wr} - J_{wf}\dot{\omega}_{wf}]}{r - F_d} = ma \tag{12.46}$$

where T_e is the engine torque, T_b is the brake torque, M is the rolling resistance momentum, J_w is the wheel inertia, $J_{w,eff}$ is the effective wheel inertia, r is the wheel radius and subscripts f and r stand for front and rear respectively.

Further simplifications and substitutions lead to the final longitudinal equation:

$$T_e - T_b - rR - rCv^2 = \beta a \tag{12.47}$$

where R is the rolling resistance, C is the drag coefficient, v is the vehicle speed and β is defined by the following equation:

$$\beta = \frac{1}{R_g^2 r}[J_e + R_g^2(J_{wr} + J_{wf} + mr^2)] \tag{12.48}$$

Figure 12.16 shows how all the components presented in this section come together in a vehicle.

12.2 Brake control

This section presents some of the issues associated with controlling the brakes in an automated system. The control algorithm is based on the models developed in Section 12.1. Two control actuation methods, and therefore control strategies, will be considered. The first one involves controlling the brakes through the vacuum booster. The second one considers brake actuation at the master cylinder.

The control through the vacuum booster is one of the most intuitive and energy efficient control methods. It is currently used by several automotive companies in their ACC systems. It is therefore presented here not only as a comparison for other methods available but as a viable control and actuation method. As it will be shown, it has the advantage of being easily implementable and operational with minimal energy consumption. The bandwidth is however too low for use in situations requiring a high level of deceleration resolution, particularly at low brake pressures, such as is the case of stop-and-go ACC. Another method is to control the brakes through actuation at the master cylinder, thus bypassing the vacuum booster. Such a system has been proved to better satisfy the stringent requirements of low speed, high accuracy control. The disadvantage is the energy consumption requirement and the present high cost. Another viable actuation scheme is using modified anti-lock braking systems (ABS) or traction control systems (TCS) to actuate the brakes near the slave cylinders. This system presents the advantages of reduced transport lag and the possibility of actuating the brake at each wheel independently. Current systems however have relatively slow actuation times and are intended for occasional, high pressure use rather than continuous application, high accuracy control and concern for passenger comfort.

12.2.1 Background

Some necessary mathematical preliminaries are presented below. This material is presented in detail in the book by Isidori (1995), Nijmeijer and Van der Schaft (1990) and Slotine and Li (1991).

We start with the definition of the Jacobian matrix.

Definition 12.1 (Jacobian matrix)

$$\nabla f = \begin{bmatrix} \frac{\partial f_1}{\partial x_1} & \cdots & \frac{\partial f_1}{\partial x_n} \\ \vdots & \ddots & \vdots \\ \frac{\partial f_n}{\partial x_1} & \cdots & \frac{\partial f_n}{\partial x_n} \end{bmatrix} \tag{12.49}$$

Another important mathematical operator on vector fields is the Lie derivative.

Definition 12.2 (Lie derivative)

Let $h : R^n \to R$ be a smooth scalar function, and $f : R^n \to R^n$ be a smooth vector field on R^n, then the Lie derivative of h with respect to f is a scalar function defined by $L_f h = \nabla h f$.

Thus, the Lie derivative $L_f h = \nabla h f$ is the directional derivative of h along the direction of the vector f. Repeated Lie derivatives can be defined as:

$$L_f^0 h = h$$
$$\vdots \qquad (12.50)$$
$$L_f^i h = L_f\left(L_f^{i-1}\right) = \nabla\left(L_f^{i-1} h\right) f$$

An associated operator is the Lie bracket.

Definition 12.3 (Lie bracket)
Let f and g be two vector fields on R^n. The Lie bracket of f and g is a third vector field defined by

$$[f, g] = \nabla g \times f - \nabla f \times g \qquad (12.51)$$

The Lie bracket $[f; g]$ is commonly written as ad_{fg} (where ad stands for 'adjoint'). Repeated Lie brackets can then be defined recursively by

$$ad^0_f g = g$$
$$\vdots \qquad (12.52)$$
$$ad^i_f g = [f, ad_f^{i-1} g]$$

In terms of the Lie derivatives, the *relative degree* of a system is defined as (Isidori, 1995):

Definition 12.4 (Relative degree)
The system

$$\dot{x} = f(x) + g(x)u$$
$$y = h(x)$$

is said to have relative degree r at a point x_0 if:

$$L_g L_f^k h(x) = 0 \quad \forall x \in nbhd(x_0), \forall k < r - 1$$
$$L_g L_f^{r-1} h(x_0) \neq 0$$

I/O linearization

While there are many control techniques to control linear systems, the same is not true for nonlinear ones. Therefore, if one could make a nonlinear system resemble a linear system, linear control techniques could be applied to the modified nonlinear system. One such idea is to linearize the relationship between the input and output of a system. The method, known as I/O linearization, considers the process of transforming the system

$$\dot{x} = f(x) + g(x)u$$
$$y = h(x) \qquad (12.53)$$

into the 'normal form' with $z_1 = y$ and

$$\begin{aligned}
\dot{z}_1 &= z_2 \\
\dot{z}_2 &= z_3 \\
&\vdots \\
\dot{z}_r &= v \\
\dot{z}_{r+1} &= q_{r+1}(z) \\
&\vdots \\
\dot{z}_n &= q_n(z)
\end{aligned} \tag{12.54}$$

through the state feedback

$$u = \alpha(x) + \beta(x)\nu \tag{12.55}$$

It can be easily seen that the relationship between the output and the new input, ν, is a chain of r integrators, where r is the relative degree of the system. From the definition of the relative degree, Equation 12.54 can be rewritten as:

$$\begin{aligned}
y = z_1 &= h(x) \\
\dot{y} = z_2 &= L_f h(x) \\
&\vdots \\
y^{(r-1)} = z_r &= L_f^{r-1} h(x) \\
y^{(r)} = \dot{z}_r &= L_f^r h(x) + L_g L_f^{r-1} h(x) u
\end{aligned} \tag{12.56}$$

Using the above equation, the system in Equation 12.53 can be transformed into the normal form of Equation 12.54 if the control input is:

$$u = \frac{1}{L_g L_f^{r-1} h(x)} (-L_f^r h(x) + v) \tag{12.57}$$

If $r = n$, this feedback law exactly linearizes the system such that the transformation $z = \Phi(x)$ is a local diffeomorphism. If, however, $r < n$, the normal form of the system can be written as:

$$x = \begin{bmatrix} r \times 1 \\ z \\ (n-r) \times 1 \\ \xi \end{bmatrix} \tag{12.58}$$

where ξ is known as the internal dynamics of the system. Its stability needs to be analysed, and Isidori (1995) and Slotine and Li (1991) present methods to do so. In this case the feedback law $u(x)$ has the form:

$$u(x) = \frac{1}{L_g L_f^{r-1} h(x)} \left(-L_f^r h(x) + y_{\text{des}}^{(r)} - \sum_{i=1}^{r} c_{i-1} \left(L_f^{i-1} h(x) - y_{\text{des}}^{(i-1)} \right) \right) \tag{12.59}$$

for some scalars $c_1, \ldots, c_{r-1} \in R$. Define the tracking error to be

$$e = y - y_{\text{des}} \tag{12.60}$$

This yields the following error dynamics

$$e^{(r)} + c_{r-1}e^{(r-1)} + \cdots + c_1\dot{e} + c_0 e = 0 \tag{12.61}$$

If c_i are chosen so that the polynomial is Hurwitz, the asymptotic tracking is achieved. This assumes that the desired trajectory is differentiable r times and that the internal dynamics are stable.

Although this method seems very powerful and straightforward, there are some inherent problems. First, there is no robustness guarantee in the presence of uncertainties. Second, the behaviour of the system is dependent on the initial conditions. The method of sliding control addresses some of these concerns.

Sliding control

Sliding control, a special case of variable structure control (VSC), evolved from work developed in the Soviet Union in the late 1950s and early 1960s. Some of the early developers were Filippov (1960), Emelyanov (1957) and Utkin (1977). It uses a discontinuous control structure to guarantee perfect tracking for a class of systems. It is particularly attractive to use this method in the control of automotive subsystems since most of them are highly nonlinear. Most methods of robust nonlinear control employ a Lyapunov synthesis approach where the controlled variable is chosen to make the time derivative of a Lyapunov function candidate negative definite. Under the terminology of variable structure control a switching surface representing some desirable dynamic behaviour is defined in the state space. Then, a discontinuous control law is developed the brings the state to the surface and forces the system to remain on that surface.

Consider the nonlinear system:

$$\begin{aligned}
\dot{x}_1 &= x_2 \\
\dot{x}_2 &= x_3 \\
&\vdots \\
\dot{x}_n &= f(x, t) + g(x, t)u
\end{aligned} \tag{12.62}$$

Suppose that u undergoes discontinuities on some plane $s = 0$, where

$$s = \sum_{i=1}^{n} c_i x_i \tag{12.63}$$

then the first derivative vector undergoes discontinuities in the same plane. If the trajectories are directed towards the plane, an $s = 0$ sliding mode will appear in this plane. The pair of inequalities

$$\lim_{s \to -0} \dot{s} > 0 \text{ and } \lim_{s \to +0} \dot{s} < 0 \tag{12.64}$$

are a sufficient condition for sliding mode to exist.

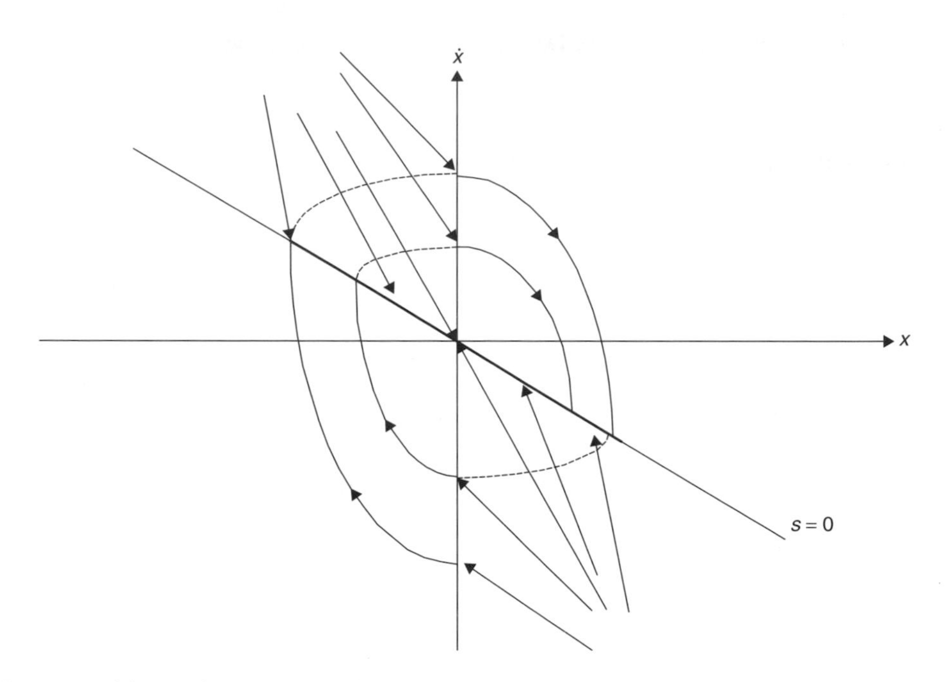

Fig. 12.17 Sliding mode in a second order VSC.

Now assume that the tracking objective is $x_1 \to x_{1,\text{des}}$ and $x_{1,\text{des}}$ is differentiable at least n times. First, for a notational simplification, define the error by:

$$e = x_1 - x_{1,\text{des}} \tag{12.65}$$

Then define a time-varying surface $S(t)$ in the state space R^n. A typical such surface is given by Slotine and Sastry (1983):

$$S(x, t) = \left(\frac{d}{dt} + \lambda\right)^{n-1} e \tag{12.66}$$

$S(x, t)$ can be expanded as:

$$S(x, t) = c_1 e + c_2 \dot{e} + \cdots + \frac{d^{n-1}e}{dt^{n-1}} \tag{12.67}$$

where $c_k = \dfrac{(n-1)!\lambda^{n-k}}{(n-k)!(k-1)!}$.

From Equations 12.62, 12.65 and 12.67, the error equation becomes:

$$\begin{aligned}
\dot{e} &= \dot{x}_1 - \dot{x}_{1,\text{des}} = x_2 - x_{2,\text{des}} = e_2 \\
&\vdots \\
\frac{d^{n-1}e}{dt^{n-1}} &= \dot{x}_{n-1} - \dot{x}_{n-1,\text{des}} = x_n - x_{n,\text{des}} = e_n \\
\frac{d^n e}{dt^n} &= f(x, t) + g(x, t)u - x_{\text{des}}^{(n)}
\end{aligned} \tag{12.68}$$

Then, the time derivative of S is:

$$\dot{S} = c_1 e_2 + c_2 e_3 + \cdots + c_{n-1} e_n + f(x,t) + g(x,t)u - x_{1,\text{des}}^{(n)} \tag{12.69}$$

In order to 'reach' the sliding surface, the condition $S\dot{S} \leq 0$ must be satisfied. This occurs for $\dot{S} = -\eta sgn(S)$ where $\eta > 0$. Therefore the control u

$$u = -\frac{1}{g}(c_1 e_2 + c_2 e_3 + \cdots + c_{n-1} e_n + f - x_{1,\text{des}}^{(n)} + \eta sgn(S)) \tag{12.70}$$

guarantees that $S \to 0$ in a time $t \leq \dfrac{|S(0)|}{\eta}$

12.2.2 Vacuum booster control

The basic idea in controlling the brakes through the vacuum booster is to modulate the pressure difference across the diaphragm. This can be accomplished by actuating the brake pedal which, in turn, controls the vacuum booster control valve, the same way a human operator would control the brakes. Figure 12.18 shows the basic details of this actuation method. Although cumbersome, it presents the advantage of being easily retrofittable.

Another method is to use a modified vacuum booster in which the flow of air in and out of the two chambers is controlled through two air valves. This actuation method

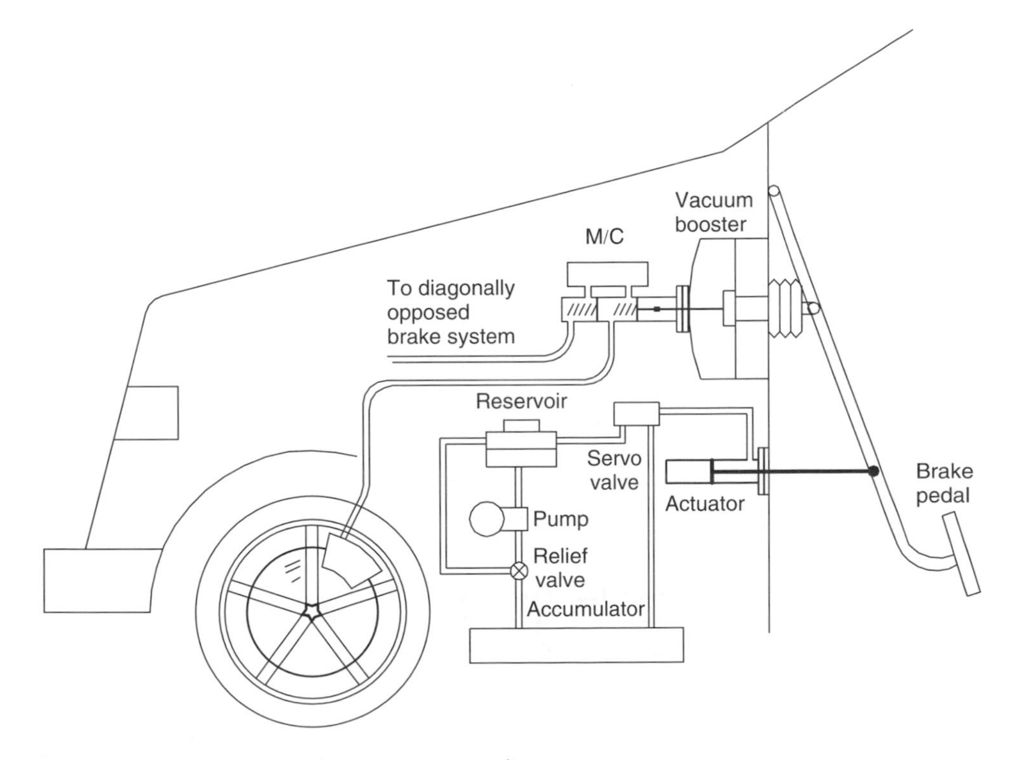

Fig. 12.18 Vacuum booster brake actuation system diagram.

was successfully used by Mitsubishi (Kishi *et al.*, 1993), Lucas (Martin, 1993) and Honda (Nakajima, 1994).

Controller development preliminaries

Based on the vacuum booster airflow dynamics presented on page 398, the pressure change across the diaphragm can be represented as:

$$\Delta\dot{P} = \dot{m}_{aA}\frac{RT}{V_A}u_1 - \dot{m}_{AV}RT\left(\frac{1}{V_A}+\frac{1}{V_V}\right)u_2 + \dot{m}_{Vm}\frac{RT}{V_V} \tag{12.71}$$

where the subscripts are as before and u_1 and u_2 are the inputs to Equation 12.71 and take the values 0 or 1 depending on the operation mode of the vacuum booster control valve. Table 12.2 details this operation. It is possible to assume that the vacuum booster control valve is either open or closed in one of the modes because the time constant of the valve is less than 1 ms. A simple modification can be made to include variable orifice opening (Gerdes, 1996). Furthermore, if a modified vacuum booster is used, Table 12.2 represents the desired status of the air valves.

Controller development

In this section it will be assumed that the desired brake torque is known. Obtaining the desired brake torque involves the powertrain model presented above. Furthermore, it will be assumed that the pressure in the vacuum chamber is low enough so that, in the 'apply' stage, the booster has the potential to provide the required force without additional input force from the actuator. In this case the actuator force needs only be large enough to open the booster valve to atmosphere. Since the actuator can only provide a 'pull' action, in release mode the actuator cannot help reduce the brake torque faster than the booster can purge the air to vacuum. The same is the case if air valves are used. Since the basic dynamic equation of the pressure difference across the diaphragm is a first order nonlinear type, a sliding mode controller is suggested. This method is relatively well established in the nonlinear control community. However, details can be found in Slotine and Li (1991). Using Definition 12.4 of that work, an error surface with relative degree one (the control input appears in the first time derivative) is defined:

$$S_{\Delta P} = \Delta P - \Delta P_{\mathrm{des}} \tag{12.72}$$

The desired error dynamics are defined such that $S_{\Delta P}$ approaches zero exponentially:

$$\dot{S}_{\Delta P} = -\lambda_{\Delta P} S_{\Delta P} \tag{12.73}$$

Table 12.2 Booster input

Stage	u_1	u_2
'apply'	1	0
'hold'	0	1
'release'	0	1

Table 12.3 Booster control

$S_{\Delta P}$	Stage	u_1	u_2	F_{vs}
$S_{\Delta P} < 0$	'apply'	1	0	F_{app}
$S_{\Delta P} = 0$	'hold'	0	0	F_{hold}
$S_{\Delta P} > 0$	'release'	0	1	F_{rel}

Since the surface has relative degree one, the control input is obtained by differentiating Equation 12.72 once:

$$\dot{S}_{\Delta P} = \Delta\dot{P} - \Delta\dot{P}_{\text{des}} \tag{12.74}$$

Substituting in the above equation the dynamics from Equation 12.71:

$$\dot{m}_{aA}\frac{RT}{V_A}u_1 - \dot{m}_{AV}RT\left(\frac{1}{V_A} + \frac{1}{V_V}\right)u_2 + \dot{m}_{Vm}\frac{RT}{V_V} - \Delta\dot{P}_{\text{des}} = -\lambda_{\Delta P}S_{\Delta P} \tag{12.75}$$

As can be seen, it would be very difficult to obtain values for u_1 and u_2 from the above equation. Furthermore, a large number of parameter uncertainties would be introduced since the flows are highly dependent on the orifice areas and chamber volumes which are very hard to measure accurately. However, since u_1 and u_2 can only take the values described previously, the problem simplifies significantly.

The objective is to have $S_{\Delta P}\dot{S}_{\Delta P} < 0$. Therefore when $S_{\Delta P} < 0$, $\dot{S}_{\Delta P} > 0$ which can only be achieved when $u_1 = 1$ and $u_2 = 0$ or apply mode. The opposite is true when $S_{\Delta P} > 0$. This statement is also physically intuitive. It says that if the pressure difference is too low, more pressure is needed in the apply chamber, and so the brakes should be applied. The significance of this statement is even more far reaching. It implies that this control scheme can be used with any vacuum booster that operates in this manner as long as the pressure difference measurement is available.

If the pressure difference is controlled through air valves, the analysis can stop here. If the booster is controlled by actuating the brake pedal, the operation mode of the control valve can be related to the force in the valve return spring as Table 12.3 shows. The advantage of this control scheme is that the force input to the master cylinder is obtained virtually free. The vacuum created by the engine already exists so the only energy consumed is the one needed to control the vacuum booster control valve or the air valves. It is using the basic idea under which most cruise controls operate. The drawback of such a system is its slow reaction time. Since only bandwidth in the range of 2–10 Hz is achievable, tracking errors of about 1–2 m should be expected at highway speeds. Although such errors are inadequate for implementation of platooning in an AHS environment, this system could be successfully used in an adaptive cruise control (ACC).

12.2.3 Master cylinder control

Controller development

One solution to improving the bandwidth of the brake actuation system is to have direct hydraulic actuation at the master cylinder as Figure 12.19 shows. Such a system

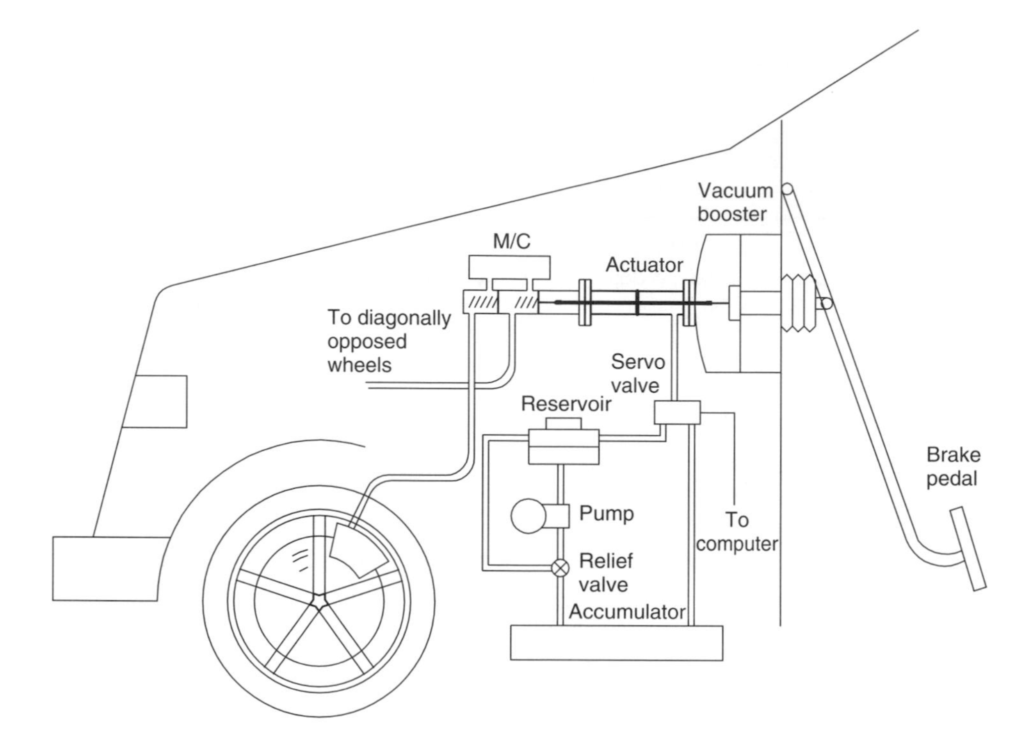

Fig. 12.19 Direct hydraulic brake actuation system diagram.

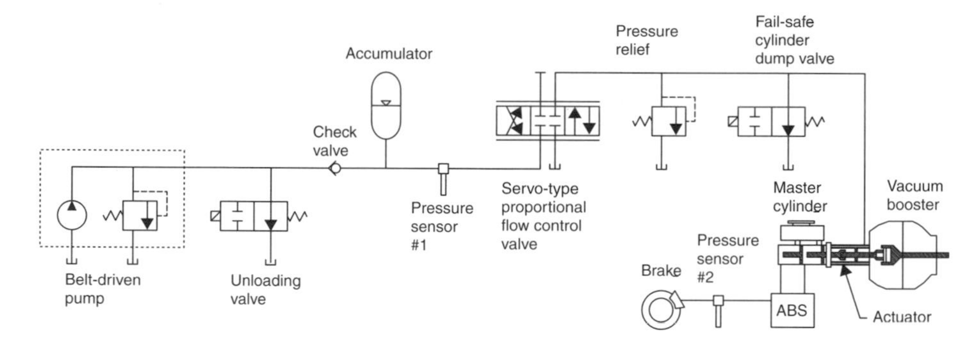

Fig. 12.20 Direct master cylinder actuation hydraulic diagram.

will be able to provide a bandwidth in the range of 20–50 Hz. Such bandwidth is high enough to be able to provide tracking with errors of less than 10 cm at speeds in excess of 100 km/h.

The pressure necessary for this hydraulic circuit could be produced by a second pump driven by the accessories belt. Pressurized fluid is stored in an accumulator whose maximum pressure is controlled by a relief valve. Pressure to the actuator is controlled through a servovalve. The valve is controlled by an analogue circuit with pressure feedback from the actuator. This setup provides for a fast actuator response time. Figure 12.20 shows the hydraulic diagram of this system.

As presented in Equation 12.44 there is a linear relationship between the desired brake torque, $T_{b,\text{des}}$, and the desired brake pressure at the wheel, $P_{w,\text{des}}$. Therefore, assuming that the desired brake torque is known, the desired brake pressure can be obtained from the aforementioned relation. The fluid dynamics were presented in Section 12.1.6. As described in Slotine and Li (1991) an error surface is defined:

$$S_w = P_w - P_{w,\text{des}} \tag{12.76}$$

The desired error dynamics are:

$$\dot{S}_w = -\lambda_w S_w \tag{12.77}$$

Differentiating Equation 12.76 once we obtain:

$$\dot{S}_w = \dot{P}_w - \dot{P}_{w,\text{des}} \tag{12.78}$$

Which can be rewritten as:

$$\dot{S}_w = \frac{\partial P_w}{\partial V_b}\dot{V}_b - \dot{P}_{w,\text{des}} \tag{12.79}$$

Substituting Equation 12.42 in the above equation yields:

$$\dot{S}_w = \frac{\partial P_w}{\partial V_b}\sigma_w C_w \sqrt{|P_{mc} - P_w|} - \dot{P}_{w,\text{des}} \tag{12.80}$$

Therefore, the desired master cylinder pressure is:

$$P_{mc,\text{des}} = P_w + \sigma_w \left(\frac{\dot{P}_{w,\text{des}} \quad \lambda_w S_w}{\dfrac{\partial P}{\partial V} C_w} \right)^2 \tag{12.81}$$

Now the desired force at the master cylinder, F_{mc}, can be obtained from Equation 12.40.

12.3 Conclusions

This chapter presented an automotive brake system model and several longitudinal control methods. The brake model was used to formulate brake control algorithms with input at the brake pedal, at the vacuum booster, at the master cylinder or at each wheel. Due to the implementation issues and bandwidth achievable (2–10 Hz), it is suggested that actuation at the brake pedal or vacuum booster can be used for adaptive cruise control (ACC) while actuation at the master cylinder or directly at the wheel, due to their higher bandwidth in the range of 20–50 Hz, is suggested for stop-and-go ACC or platooning in an AHS environment. However, current automatic brake actuation systems, such as ABS and TCS, are designed mostly for high pressure, occasional use situations only. It is therefore expected that for future automatic brake actuation application systems requiring low pressure and continuous use, novel hardware, designed specifically for the task at hand, will need to be developed.

References

Emelyanov, S.V. (1957). A Technique to Develop Complex Control Equations by Using Only the Error Signal of Control Variable and Its First Derivative. *Automatika i Telemekanika*, **10**, 873–85.

Filippov, A.F. (1960). Differential Equations With Discontinuous Right-Hand Sides. In: *Matematischeskii Sbornik.*, Vol. 51 (in Russian). Translated in English 1964.

Fisher, D.K. (1970). Brake System Component Dynamic Performance Measurement and Analysis. SAE Paper #700373.

Gerdes, J.C. and Hedrick, J.K. (1996). Vehicle Speed and Spacing Control Via Coordinated Throttle and Brake Actuation. *Proceedings of the 13th Triennal World Congress (IFAC)*. San Francisco, CA, pp. 183–8.

Gerdes, J.C., Maciuca, D.B., Devlin, P.E. and Hedrick, J.K. (1993). Brake System Modeling for IVHS Longitudinal Control. In: *Advances in Robust and Nonlinear Control Systems* (Eduardo A. Misawa, ed.). ASME. New Orleans. pp. 119–26.

Gillespie, T.D. (1992). *Fundamentals of Vehicle Dynamics*. Society of Automotive Engineers. Warrendale, PA.

Isidori, A. (1995). Nonlinear Control Systems. *Communications and Control Engineering*, (3rd edn). Springer-Verlag, New York.

Khan, Y., Kulkarni, P. and Youcef-Toumi, K. (1994). Modeling, Experimentation and Simulation of a Brake Apply System. *ASME Journal of Dynamic Systems, Measurement, and Control*, **116**, 111–22.

Kishi, M., Watanabe, T., Hayafune, K., Yamada, K. and Hayakawa, H. (1993). A Study on Safety Distance Control. *Proceedings of the 26th ISATA*. Aachen, Germany.

Limpert, R.(1992). *Brake Design and Safety*. Society of Automotive Engineers.Warrendale, PA.

Maciuca, D.B., Gerdes, J. and Hedrick, J.K. (1994). Automatic Braking Control for IVHS. *Proceedings of the International Symposium on Advanced Vehicle Control (AVEC)*. SAE. Tsukuba City, Japan. pp. 390–395.

Martin, P. (1993). Autonomous Intelligent Cruise Control Incorporating Automatic Braking. SAE Paper #930510.

McMahon, D.H., Hedrick, J.K. and Shladover, S.E. (1990). Vehicle Modelling and Control for Automated Highway Systems. *Proceedings of the 1990 American Control Conference*. San Diego, CA. pp. 297–303.

Merritt, H.E. (1967). *Hydraulic control systems*. Wiley, New York.

Nakajima, T. Chief Engineer (1994). Private communications. Honda R&D Co., Ltd.

Nijmeijer, H. and Van der Schaft, A.J. (1990). *Nonlinear Dynamical Control Systems*. Springer-Verlag, Berlin.

Puhn, F. (1985). *Brake Handbook*. HP Books, Los Angeles, CA.

Puleo, G. (1970). *Automatic Brake Proportioning Devices*. Technical Report 700375. SAE. Warrendale, PA.

Robert Bosch GmbH (1995). *Automotive Brake Systems (1st edn)*. SAE. Stuttgart, Germany.

Shigley, J.E. and Mischke, C.R. (1989). *Mechanical Engineering Design*. McGraw-Hill Publishing Co.

Slotine, J.-J.E., Hedrick, J.K. and Misawa, E.A. (1986). On Sliding Observers for Nonlinear Systems. *Proceedings of the 1986 American Control Conference*.

Slotine, J.-J.E. and Li, W. (1991). *Applied Nonlinear Control*. Prentice Hall, New Jersey.

Slotine, J.-J.E. and Sastry, S.S. (1983). Tracking control of nonlinear systems using sliding surfaces with applications to robot manipulators. *International Journal of Control*, **39**.

Utkin, V.I. (1977). Variable Structure Systems With Sliding Mode: A Survey. *Transactions on Automatic Control*, **2**. IEEE.

13

ACC systems – overview and examples

David Maurel
Technocentre Renault, France

and

Stéphane Donikian
IRISA-CNRS, France

Adaptive cruise control (ACC) techniques are now being put on the market as a safety and comfort feature. Further systems will perform longitudinal control in traffic jam situations and urban areas. These driving assistance features are described with highlights on control and sensor issues.

Detailed simulations try to assess the impact of such systems on traffic flow. Three different algorithms of ACC have been integrated in our simulator and are compared in terms of safety and comfort.

13.1 ACC overview

ACC stands for 'adaptive cruise control' and refers to extension of conventional cruise control to a higher level of sensors and control, including detection of vehicles in front of the equipped car, and distance regulation with the relevant targets. It first was called ICC (intelligent cruise control) or AICC (autonomous intelligent cruise control). The final acronym of ACC is more representative of the great number of highway situations the system can cope with.

The driver can select a cruising speed by means of buttons on the steering wheel. Then, if the driver releases his foot from the accelerator pedal, the car will automatically travel at the desired speed.

If a vehicle is detected by the range sensor in the same lane as the ACC vehicle (Figure 13.1), the car will slow down to the same speed and at a convenient distance from the vehicle in front. If the driver decides to overtake, the car will accelerate to the cruising speed as soon as the sensor no longer detects the preceding vehicle.

When the driver wants to go faster than the cruising speed for a short time, he can override the system by pushing the accelerator pedal. This overdrive phase will end when the driver releases his foot from the pedal.

The driver can also at any time take over the system by pushing the brake pedal. The system is then deactivated.

The sensor is mounted in front of the car and monitors the situation (it can be seen as the eyes of the system), calculating the distance and speed of preceding *moving* vehicles (Figure 13.1). A curve sensor helps to predict the course of the vehicle and to select the relevant preceding car. This information is then used to calculate the needed acceleration or deceleration of the car. Finally the adapted automatic action is commanded to the engine or the brakes. The driver receives information on the system status by means of a display.

The ACC function is only active at a speed of more than around 40 km/h. A lower speed limit is introduced for safety reasons to prohibit an automatic acceleration at low speed. This should avoid a dangerous approach to an obstacle that is outside of the lateral sensor range, but well inside the trajectory of the vehicle. This case is typical for crossing pedestrians in urban areas or nonaligned cars in traffic jams.

Above this speed of 40 km/h the system reacts on moving vehicles only (not the stopped ones). The driver retains his authority over and responsibility for the vehicle at all times. ACC is a typical driver assistance system, but not a collision system which prevents collision in all events.

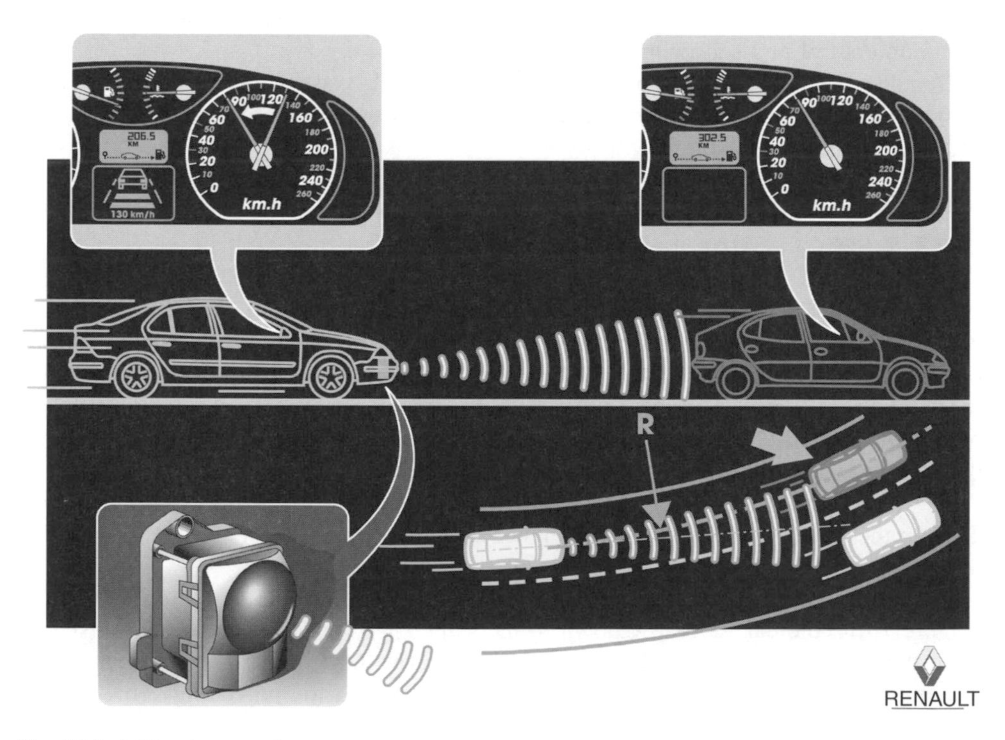

Fig. 13.1 ACC system overview.

13.1.1 Longitudinal control research

Longitudinal control has been under investigation since the late 1950s. First research works focused on the theoretical approach of the vehicle intelligent control, because of sensor development and high costs related to experimental vehicles.

The first steps were achieved by universities, and car manufacturers began to investigate in the early 1980s, with General Motors (Bender, 1982) and Daimler-Benz. Important research programmes like PROMETHEUS[1] in Europe, PATH[2] in the USA or PVS[3] in Japan enabled great co-operation between car manufacturers and sensor or actuator suppliers.

These projects experienced fully automated driving, including longitudinal and lateral control. ACC was then prototyped as a first step towards the automated highway.

Distance set-up

A vehicle with an ACC system will track any preceding vehicle at a desired distance. The policy which determines this tracking distance will have a great influence on system performance. For safety reasons, ACC system could use a very large tracking distance. However, vehicle flow rate on the highway is reduced as this distance is increased. Furthermore, the chance of a vehicle cut-in increases when the tracking distance is too large.

- **Constant distance**: the main advantage of the constant distance following policy is the increase in traffic flow rates. But distance sensors and control strategies cannot achieve enough efficiency at a reasonable cost. Moreover, the driver feels quite unsafe with the small delay left to him to take over the system in case of emergency.
- **Velocity dependent spacing (headway control)**: This is the usual computation of distance set-up, which entails smaller flow rates as constant spacing but improves driver feeling by imitation of his natural behaviour. There are upper and lower limits for the set time constant. The upper limit of 2 seconds is of no importance for the safety of the system, but it makes no sense extending this limit to higher values because very few drivers will select a larger time constant. The lower limit is far more relevant for the system safety. A minimum value must allow the driver to safely take over control of the vehicle in emergency situations. It must also be acceptable for the majority of drivers. Experiences made during the testing phase of ACC systems have shown that values of about 1.0 second are reasonable.
- **Other strategies**: Other strategies were also proposed with the use of square of the vehicle velocity (for example) and taking maximum acceleration and maximum jerk into account.

Control algorithms

Control is designed to maintain this desired distance with the preceding vehicle. Most of the algorithms use two layers:

- If no car is detected in front of equipped vehicle, a cruise control strategy is used to maintain the driver's desired speed.

[1] PROgraM for European Traffic with Highest Efficiency and Unprecedented Safety
[2] Program on Advanced Technology for the Highway
[3] Personal Vehicle System

- If a car is detected, control law determines a desired dynamics to track distance set-up.
- Regulation layer uses available actuators (like throttle, brake or transmission) to obtain this dynamics.
- To ensure safe actuation of throttle and brake systems, a switching logic between must be included.

A fuzzy logic controller was proposed to produce a desired acceleration as a function of relative distance and relative velocity. This solution was tested with success by Müller (1992) but fuzzy algorithms can be hard to tune even if adaptation to different kinds of engines seems easier.

Another approach consists of sliding mode control, mainly used in the PATH project. Use of these algorithms is motivated by the highly nonlinear behaviour of the engine. Experiments showed the possibility of ensuring safe distance control with a constant spacing of 4 metres with variations within 30 cm, while executing a driving cycle including moderate accelerations (0.1 g) and decelerations (0.05 g). These experiments related by Choi and Hendrick (1995) have been conducted using throttle control but no brake control, because related studies of braking actuation have shown the inadequate dynamic response of the brake actuator system.

Design of the control law is slightly linked with driver acceptance. It is then necessary to make reactions of the system as close as possible to driver normal reactions. Safety reactions should also occur slightly after the driver's normal reactions. Active braking must be restricted to moderate values to prevent the driver from getting into an unexpected situation. A usual value for limitation of the deceleration is $3\,\mathrm{m\,s^2}$. As a consequence, the ACC system is not able to produce emergency braking. For comfort reasons, the maximum value of positive acceleration is limited to a value generally chosen between 0.5 and $1\,\mathrm{m/s^2}$.

The regulation layer generally uses feedback and feedforward linearization. Problems of the load of the vehicle and inclination of the road should also be considered. That is why an estimation of mass of the vehicle and of the road grade is often implemented.

13.1.2 Sensor issues

Tracking strategies

Selection of the relevant preceding vehicle is the first and essential step in ACC. A tracking algorithm estimates the expected trajectory of the ACC-equipped car using the speed sensor and the yaw rate sensor. Then the distance sensor (radar or lidar) supplies the distance, relative speed and angle of multiple objects representing different preceding vehicles. Combination of expected trajectory and preceding vehicles' positions produces the most relevant target to be considered by ACC longitudinal control law.

In the simplest case, course prediction assumes movement in a straight line. Under this assumption, no dedicated curve sensor is needed.

In a more complex case, a constant course is assumed, meaning constant speed and curvature. This works well most of the time, since estimation errors usually only occur at the beginning and end of curves. Several ways of sensing the actual curvature are commonly used today: evaluation of the relative difference of the wheel revolution, using a steering angle sensor or measuring the lateral acceleration. A yaw rate sensor system can also be used.

A higher level of course prediction can be provided by using navigation systems (digital maps and course calculations). The limits of this technique depend on the agreement of the maps with the actual road and on the system's capability in determining the car's actual position. The prediction is faulty in construction areas and for new roads. With integration of DGPS on such systems, position resolution is widely improved, and navigation systems become useful sensors for future ACC.

New radar systems can also include road and lane prediction purely based on the radar information. Stationary objects like reflector posts and crash-barriers detected by the signal processor are used to reconstruct the road boundaries. Quality and robustness of such techniques are still improving.

Extracting road markers from video images is another way to predict the road course. Lane classification of the detected objects is possible by combining a video system with an angle resolving distance sensor. Current limitations are ambiguous lane markers which cannot be distinguished, and weather impact of fog and rain. Moreover, the prediction range is limited to about 100 m.

Technical limits of distance sensors

Sensor specifications also have an impact on control and driver acceptance. Main defaults are selection and deselection time of the targets. When the ACC vehicle changes lane, the driver expects the system to accelerate as soon as no vehicle is present in the expected trajectory (in manual driving, he accelerates before the lane change to prepare to overtake). But tracking algorithms use filtering and deselection of the target sometimes occurs when the ACC vehicle is in the second lane, with no more obstacle ahead, which is unacceptable for the driver.

Range is another issue in sensor design. Distance sensors have limited range, in longitudinal and lateral directions. Typical range values under ideal environment conditions are 150 m for longitudinal and 15° for lateral range. However, these values will be further reduced in two ways: either by weather conditions or by geometrical obstruction. Limitations due to geometrical obstructions are characteristic for all autonomous ranging sensors existing today. Tops of hills and bottoms of valleys naturally limit the longitudinal range. Crash barriers, walls and other side obstacles act in the same way in curves. Vehicles in the same or adjacent lanes can reduce both the longitudinal and the lateral range by hiding objects in front of them.

13.1.3 ACC products

First systems are now on the market, on the car manufacturers' more expensive models.

In Japan first, Mitsubishi and Nissan proposed ACC systems in 1997 and 1999, respectively with lidar and radar. These cars were equipped with other intelligent features like navigation and parking aid systems. This advance on other car manufacturers in the world is explained by the taste of Japan's customers for prototype-like cars.

In Europe, Mercedes S-Class comes with an ACC system called 'Distronic' as an option since 1999. Mercedes ACC is based on a radar. Other European car manufacturers (Jaguar, Audi, Renault) will follow between 2000 and 2002.

13.2 Systems based on ACC

13.2.1 Stop&Go

General principles

The Stop&Go function helps the driver to keep a safe distance with the preceding vehicle at low speed. This application will assist the driver in traffic jams on highways and in urban areas which will particularly benefit people in congested regions, reducing the need to accelerate and brake repeatedly.

As an extension of the adaptive cruise control systems (ACC), if slower vehicles are travelling ahead the speed is reduced in order to keep a safe distance. The Stop&Go is designed for low speed and dense traffic conditions where it assists the driver down to full stop behind a car and assists him for the go. The Stop&Go function is only active at a speed less than around 40 km/h. Under this speed the system reacts on moving or stopped vehicles. Thus, like ACC, the Stop&Go is a typical driver assistance system, but not a collision avoidance system which prevents collision in all events.

As a first step, Stop&Go could be limited to the highway for safety reasons: situations are easier to manage and control easier to design and make comfortable. No pedestrian or parked object is to be expected in these situations. In a second step, extension to urban areas is possible as far as sensors have been improved.

Table 13.1 shows the main differences between ACC and Stop&Go systems:

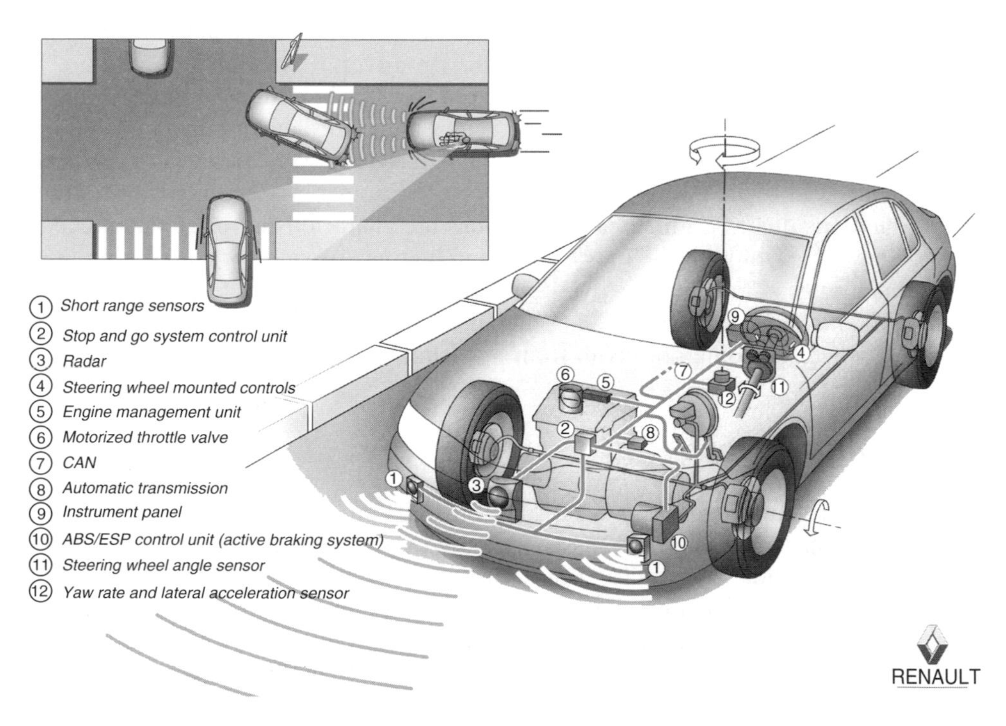

Fig. 13.2 Stop&Go system overview.

Table 13.1 Outline of the differences between ACC and Stop&Go systems

	ACC	Stop &Go
Conditions of operation	ACC performs driving assistance between 40 ± 10 km/h and 170 ± 10 km/h by action on accelerator and brakes	Stop &Go performs driving assistance between 0 km/h and 40 ± 10 km/h by action on accelerator and brakes
Preceding vehicle selection	Moving vehicles in the driving path of equipped vehicle	Moving and stopped vehicles in the driving path of equipped vehicle
Maximum braking capacity	Limited braking capacity (from -1.5 to $-3.5\,\text{m/s}^2$, depending on the systems)	Automatic braking capacity enhanced until $-4\,\text{m/s}^2$, or even $-5\,\text{m/s}^2$

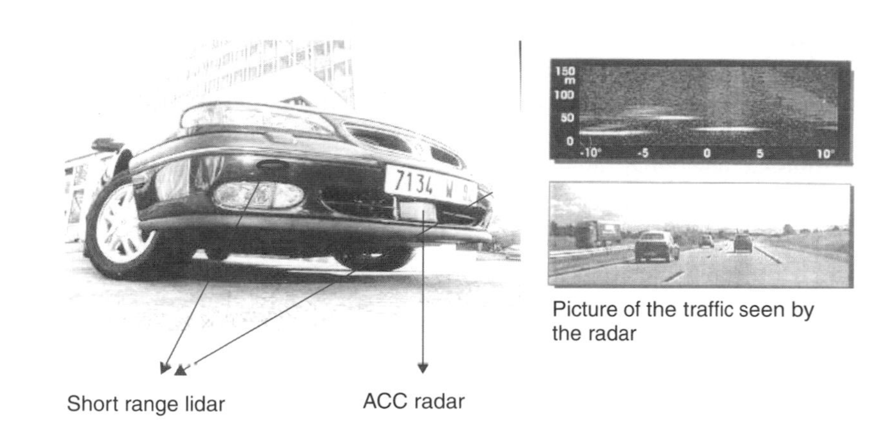

Fig. 13.3 Stop&Go sensor integration.

Tracking and control issues – examples

Tracking Traffic jams on highways and urban scenarios imply detection of a high number of targets and robust selection of the relevant one in front of the vehicle, from 0 to 50 metres and preferably to 100 meters. An ACC sensor with limited angle and blind zone in short range cannot be used as a Stop&Go sensor. Short range sensors based on optical or ultrasonic technologies can be added near headlights or fog lamps to detect any obstacle coming from the side of the vehicle trajectory.

A solution for the Stop&Go sensor is shown on Figure 13.3 with a combination of a radar and two lidars, which covers the whole scene in front of the car.

However, the driver is still responsible for detection of pedestrians suddenly coming between the front vehicle and the Stop&Go vehicle when the vehicle is stopped, for example. Another issue is to detect fixed obstacles without any false alarms in very complex scenarios (trees, sidewalk, traffic signs).

Control One of the control issues is to set a comfortable and safe 'stop' phase. Analysis of driver braking has been made, which shows three phases during an approach:

- At the beginning progressive increase of braking pressure.
- Constant braking until the vehicle stops.
- Just before complete stop, the driver releases the pedal to avoid discomfort.

Automatic braking strategy can be adapted from this behaviour and thus ensure driver acceptance. It has also been stated that in case of sensor false alarm, full braking is unbearable for the driver. A limited braking force between 0.4 g and 0.5 g is then necessary.

During automatic starting, another issue is to imitate driver behaviour without leaving enough space with the preceding vehicle to allow vehicles from adjacent lanes to cut-in. It can be useful to add relative acceleration in control law inputs to ensure better following of the target speed profile. But sensor noise level is a limiting factor for such a solution.

Examples Parent and Daviot (1994) developed a fully automated vehicle (longitudinal control and lateral guidance) based on electric powered urban vehicles. The sensor used was a linear camera combined with a co-operative target installed in the followed vehicle, which allowed full detection of the relative position of the two vehicles. Control was using Kalman filtering and a jerk command.

During the UDC European project,[4] Renault developed a Stop&Go controller, based on ACC algorithm with specific Stop&Go evolutions. Four main scenarios were considered for Stop&Go behaviour tuning:

- following of a vehicle driving at low speed in congested traffic,
- following of a vehicle decelerating to stop,
- following of a vehicle starting from stop,
- complete stop in front of a fixed obstacle.

Each of these scenarios implied evolutions of the control law and the acceleration loop. Attention focused on driving comfort in urban traffic, which greatly depends on distance set-up and acceleration levels. Computation of the distance set-up must be adapted to urban conditions. In ACC mode on highway, distance set-up is proportional with vehicle speed. But in congested traffic, the driver behaviour is appreciably different. During a starting phase, for example, the driver will intend to keep the distance with the preceding vehicle almost constant. The distance set-up for Stop&Go mode takes this behaviour into account. Distance following policy in Stop&Go mode can be compared with the driver keeping the highest distance from the preceding vehicle. This is mainly due to sensor delays and limited deceleration in automated control, which implies higher reaction time for the automated vehicle.

Gains of the control laws were also tuned to ensure comfort and safety in urban traffic. Transitions between throttle and brakes control were then adapted for new operating points of the engine.

A final experimentation took place in Turin, with infrastructure communication to perform stops at a red traffic light.

13.2.2 Anti-collision systems

System general principles

Natural evolution of driving assistance is a transition between comfort features and security features, like collision warning: based on distance sensor and time-to-collision

[4] UDC partners where TÜV, CRF, Jaguar, PSA, Renault, Turin city, CSST, Leica, Lucas, Mizar, Thomson-CSF, from 1996 to 1998.

Table 13.2 Comparison of anti-collision and ACC-like systems

	Anti-collision systems	ACC-like systems
Conditions of operation	Active *in emergency situations only*. During normal driving, the system is monitoring the situation through the distance sensor and vehicle sensors to detect potential crash, without action on actuators.	Activated by the driver during normal driving, then assists him by automatically accelerating and braking. But driver is still responsible for taking over in emergency situations.
Maximum braking level	Full braking capacity of the vehicle (around $-10\,\text{m/s}^2$)	Limited braking capacity (from -1.5 to $-3.5\,\text{m/s}^2$, depending on the systems)

analysis, this feature alerts the drivers through a visual or audio device. Warning systems can also use haptic feedback, via pedals, seat or steering wheel. Design of a sensor with acceptably low false alarm rates and sufficient reliability is hard to achieve. The danger of the situation to be evaluated by the system primarily depends on the future course of the vehicle: in most cases, a collision is avoided by a manoeuvre of the driver implying lateral motion of the car. However, anticipating this would require knowing the driver's actual and future intentions.

In a second step, the anti-collision system can become active and monitor the braking actuator to decelerate the car without driver intervention. Such a system needs perfect target selection because false alarms produce a maximum braking force, and thus collision with a potential following vehicle. Various intervention solutions are possible, from the basic solution of maximum braking force in case of danger to the solution of a controlled braking pressure to limit discomfort.

Table 13.2 shows main differences between anti collision and ACC like systems:

Control issues – examples

Collision warning design should take into account that:

- The driver needs time to react after the alarm.
- The alarm should be reliable, with no false alarm on bridges, etc.

Sensor performance needs to meet perfection: 100 per cent good detection and 0 per cent false alarms. Present distance sensors are designed for comfort features like distance regulation. Mainly, detection of fixed obstacles is an issue: current systems use speed of the different targets to perform target selection, but anti-collision systems need to detect fixed obstacle when driving at high speeds.

Collision warning and collision avoidance use similar algorithms. A critical distance is computed from target relative velocity and vehicle velocity. When the target relative distance is lower than this critical distance, a warning is given to the driver. A second distance can also be defined, as a braking critical distance, at which automatic braking will be applied.

ISO workgroups issued recommendations on maximal detection range for collision warning:

$$D_{\max} = V_{\max} \times T_{\max} + \frac{V_{\max}^2}{2 \times a_{\min}}$$

where: V_{max} = maximum speed of equipped vehicle as defined by car manufacturer.
T_{max} = maximum reaction delay of driver to brake after collision warning alarm.
a_{min} = minimum deceleration of equipped vehicle.

A value of 1.5 s is recommended for T_{max}.

Examples

The European project AC-ASSIST finished in 1999 and explored an anti-collision system based on the sensor fusion of the camera sensor (lane detection) and radar. Partners of the project were car manufacturers (Renault, Volvo, CRF, Jaguar, Rover), suppliers (Lucas, Matra, Magnetti Marelli), evaluators (TÜV, CSST, VTI), and for legal aspects (PVW&A). Renault, Volvo, CRF and Jaguar demonstrators were evaluated in Germany with 20 drivers, on normal roads for collision warning and on test tracks for anti-collision.

In AC-Assist final report, results are given concerning driver acceptance of such systems. The graphs in Figures 13.4 and 13.5 show test driver answers, for different car manufacturers (A,B,C,D) to questions concerning time of warning or scope for reaction after a collision warning. Both diagrams show that current prototyping of such systems are well accepted by usual drivers.

Vertical graduation show number of answers.

VORAD evaluated collision warning on 473 vehicles (mainly trucks and passenger cars) over 36 months. Accidents rate decreased from 1.61 accidents/million miles (1 accident/million km) for non-equipped vehicles to 0.38 (0.24 accidents/million km) for equipped vehicles.

Delphi is now studying a collision warning system based on fusion between camera and scanning radar.

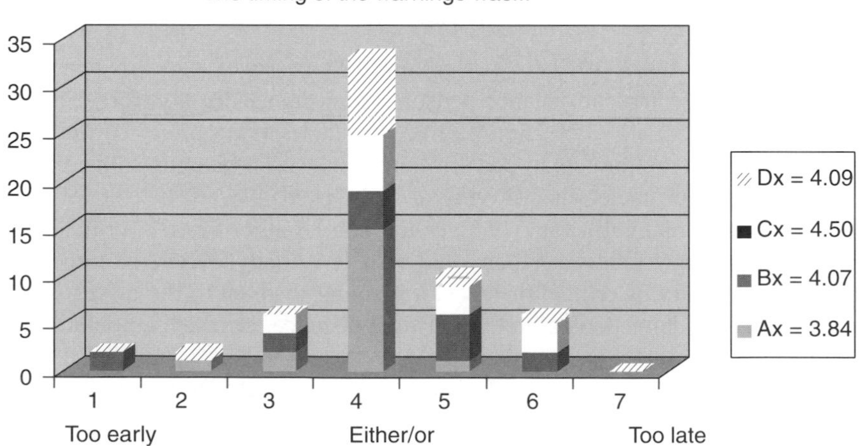

Fig. 13.4 AC-Assist evaluation – time of warnings.

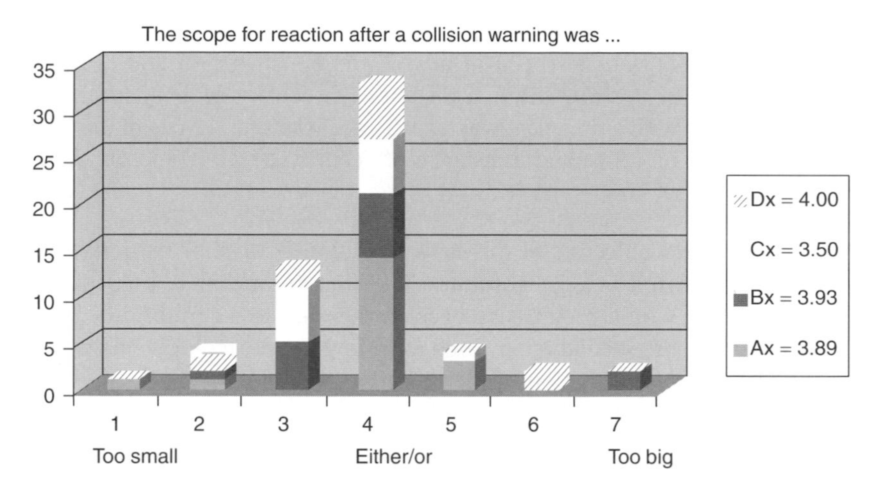

Fig. 13.5 AC-Assist evaluation – scope for reaction.

13.3 Impact of ACC on traffic and drivers

13.3.1 Traffic flow

Control law design

We have studied three different ACC algorithms. The first one, given by TRL (Transportation Research Laboratory, England), is based on the Jaguar algorithm and is used in the microscopic simulator SISTM (Hardman, 1996); we called it TRL ACC. The second algorithm is based on research performed in the European Project Prometheus and is used by TNO (TNO Inro, Traffic and Transportation Unit, The Netherlands) in their microscopic simulator MIXIC (van Arem, 1997); we called it TNO ACC. The third algorithm is a more complex algorithm developed and experimented at INRIA (French National Research Institute on Computer Science and Automation) in co-operation with a car manufacturer (Donikian, 1998a); we called it NEW ACC. Let us describe these three algorithms:

TRL ACC

$$\Gamma_{\mathrm{TRL\_ACC}} = \frac{[K_2(x_t - x_{t-1}/T + K_1(v \times t_0 - x_t)]}{\mathrm{M}}$$

$$-3\,\mathrm{m/s}^{-2} < \Gamma_{\mathrm{TRL\_ACC}} < 1\,\mathrm{m/s}^{-2}$$

where x_{t-1} and x_t are the inter-vehicle distances in metres in the previous and current epochs, v is the current vehicle speed in m/s, T the epoch length and t_0 the desired headway in seconds. M is the mass of the vehicle (1000 kg); K_1 and K_2 are damping constants ($-31.25\,\mathrm{kg/s}^{-2}$ and $500\,\mathrm{kg/s}^{-1}$).

TNO ACC

$$\Gamma_{\mathrm{TNO\_ACC}} = K_s \times (x_t - x_{t-1})/T + K_d(x_t - M - v \times t_0)$$

$$-5\,\mathrm{m/s}^{-2} < \Gamma_{\mathrm{TNO\_ACC}} < 3\,\mathrm{m/s}^{-2}$$

where x_{t-1} and x_t are the inter-vehicle distances in metres in the previous and current epochs, v is the current vehicle speed in m/s, T the epoch length and t_0 the desired headway in seconds. In this algorithm, a distance of security M is added to the time headway. To compare with a time headway of 1.5 s at 100 km/h, we use in this algorithm a desired headway t_0 of 1.14 s and a distance of security M of 10 metres. K_s and K_d are damping constants (3.0 s^{-1} and 0.2 s^{-2}).

NEW ACC The main objective of this new algorithm is to filter the jerk and to use both approaching and following algorithms. The first algorithm is used to approach the preceding vehicle with a progressive deceleration during a certain time and then a constant deceleration; the second one is used to follow the preceding vehicle, including a specific reaction to a deceleration of the preceding vehicle.

$$\Gamma_{NEW\_ACC} = MAX(\Gamma_{\text{follow}}, \Gamma_{\text{approach}})$$

$$-3\,\text{m/s}^{-2} < \Gamma_{\text{NEW\_ACC}} < 3\,\text{m/s}^{-2}$$

$$\text{with } \Gamma_{\text{follow}} = K_s S_r + K_d(D_r + [1.1 \times S_r]_{(sr<0)} - D_c) - [K](S_r^2(D_r + 1.1 \times S_r))]_{(sr<0)}$$

where D_r is the relative distance to the preceding vehicle, D_c the following *distance* $(M + H \times S_{\text{prec}})$. The calculation of the following distance differs because the speed taken into account is not the speed of the vehicle, as in the above algorithms, but is instead the speed of the preceding vehicle (S_{prec}). *The distance of security M* can take a value of 5 or 7 metres, while the headway time can vary between 0.8 and 1.32 s. The other constant are $K_s = 2.2\,\text{S}^{-1}$; $K_d = 0.2\,\text{s}^{-2}$; $K = 2$. The specificity is that specific terms (inside brackets) are added to the equation when the relative speed is negative ($S_f < 0$).

The approaching algorithm is more complex (see Figure 13.6). The objective is to impose, if possible, a constant value to the jerk which should be less than $0.05\,\text{m/s}^{-B}$. First, the duration of the progressive deceleration phase is calculated as a solution of the following third order polynomial:

$$\text{Jerk} \times T^3 + 6S_r \times T + 12D_r = 0$$

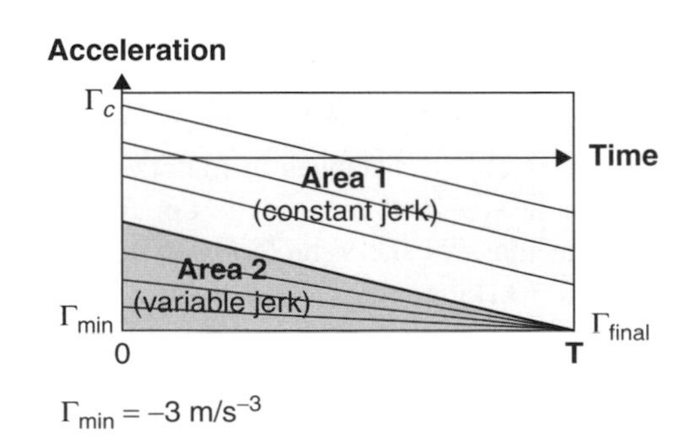

Fig. 13.6 Approaching algorithm.

Then the value of the acceleration at the beginning and at the end of the progressive deceleration phase are calculated as follows:

$$\Gamma_{\text{approach}} = S_r/T - \text{Jerk} \times T/2$$

$$\Gamma_{\text{final}} = S_r/T + \text{Jerk} \times T/2$$

If Γ_{final} is less than Γ_{min} ($-3\,\text{m/s}^{-2}$) then the variable jerk approach should be used. In that case, T is the solution of:

$$\Gamma_{\text{final}} \times T^2 + 2S_r \times T + 6D_r = 0$$

Then the acceleration is calculated as follows:

$$\Gamma_{\text{approach}} = 2 \times S_r/T - \Gamma_{\text{min}}$$

$$\Gamma_{\text{final}} = \Gamma_{\text{min}}$$

As we can see from these three algorithms, it would be much too simple to consider that an ACC implies a fixed headway and this headway is the only design criteria that we have to evaluate. There are many other design parameters which lead to various acceleration profiles and different vehicular distances. These design parameters have large influences on the comfort, safety and efficiency of the technique with regard to throughput. Our goal is to evaluate these factors.

13.3.2 Simulations

The model SSE (SIAMES Simulation Environment) was developed by the SIAMES team at IRISA as a very detailed simulator on a submicroscopic level. IRISA is not a specialist on transportation technology. Instead it specializes in software tools and control technology. The SSE simulator has been built with two goals. The first goal is to develop new software tools to program simulators which model the environment and the objects which move in this environment. The second objective is to use this simulator to test vehicle control technologies that are under development for the automotive industry. In particular, we wanted to test if the ACC control techniques developed in the automotive industry lead to safe and comfortable behaviour. Another objective was to estimate the effects of this technology on traffic flow. The objective of SSE is to reproduce a realistic multimodal traffic flow in urban environments and interurban road networks. The traffic is composed of autonomous entities (cars, bicycles, trams, pedestrians) evolving in the virtual environment. To perform modelling and simulation of virtual urban environments, models of the environment and of dynamic entities (mechanical models, motion control models, behavioural models, sensor models, geometric models) are needed.

SSE is composed of a set of modelling tools (see Figure 13.7). VUEMS is the acronym for Virtual Urban Environment Modelling System, and its main aim is to build a realistic representation of real road networks in which we would perform driving simulations (Donikian, 1997a). VUEMS produces two complementary outputs: the 3D geometric representation of the scene and its symbolic representation used by sensors and behavioural entities. In the realm of real-time behavioural simulation, it is

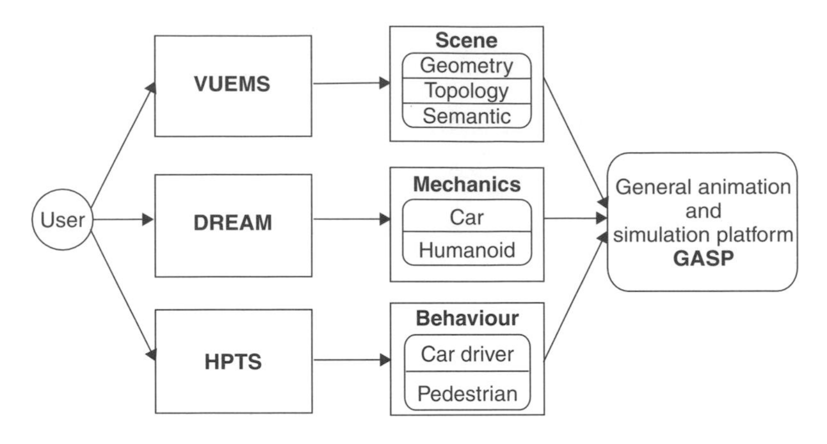

Fig. 13.7 SSE architecture.

impossible to completely simulate human vision and the building of a mental model of the environment. Therefore, the automatic driver gets a local view of its environment through a sensor which is in fact a filter of the whole environment database. Two different types of objects are taken into account in the sensor: static objects (buildings, road signals, traffic lights) and dynamic objects (cars, trucks, bicycles, trams, pedestrians). Objects that would be hidden by closer objects are eliminated thanks to a Z-buffer algorithm.

From our point of view and in accordance with some psychological studies, different paradigms are required to describe a behavioural model. This model should be both cognitive and reactive, treating flows of data to and from its environment. To describe realistic behaviours, we have defined a formal model based on Hierarchical Parallel Transition Systems (HPTS), including temporal characteristics like the reaction time (Moreau and Donikian, 1998). The car driver decisional model simultaneously performs three different activities: traffic handling (i.e. following the other cars, possibly overtaking them), road network following (i.e. following the road, changing lane, taking turns in crossroads) and traffic lights and road signals handling (i.e. adapting speed to the situation). The goal of the decisional model is to produce a target point and an output action with parameters for the low-level controller. These actions include a normal free driving mode at a desired speed, a following mode and different breaking modes. The goal of the low-level controller is to produce a guidance torque, an engine torque and a brake pedal pressure as inputs for the mechanical model. The mechanical aspect of the car is modelled with DREAM, our rigid and deformable bodies modelling system (Cozot, 1996). By means of Lagrange's equations, DREAM computes exact motion equations in a symbolic form for analysis and then generates numerical *C++* simulation code.

To integrate all these models the simulation platform GASP has been defined (Donikian *et al.*, 1998a). This platform takes into account real time synchronization and data communication between cooperative processes distributed on an heterogeneous network of workstations and parallel machines. SSE offers the simulation of technical details of the vehicle like motor characteristics, braking abilities, suspension etc. ACC algorithms have been integrated in the low level motion control model and have replaced (when active) the longitudinal control of the vehicle performed by the driver.

13.3.3 String stability

To compare ACC algorithms to the driver behaviour, we have reproduced in simulation some real case following sequences. For that, we have used data given by the Transportation Research Group (TRG) of the University of Southampton and collected on a French motorway near Lille. The TRG Instrumented Vehicle is one of the few fully instrumented units in Europe. The vehicle is equipped with a range of sensors allowing measurement of driver performance and how the motion of the vehicle relates to surrounding vehicles. Data provided by the instrumented vehicle are the time the data was recorded (every 0.1 seconds), the speed of the instrumented (lead) vehicle in km/hr × 10, the calculated acceleration of the lead vehicle, the inter-vehicular distance in centimetres and the relative velocity (negative = approaching) in cm/s and the speed of the rear vehicle. Figure 13.8 shows the results of one following sequence between two vehicles without any ACC system.

To study the string stability and to compare the different ACC algorithms we have forced a lead vehicle in the simulation to behave as the lead vehicle in the real situation, and we have generated following scenarios with the different ACC algorithms, each of them composed of five vehicles. The four following vehicles are equipped with an ACC system which has been activated (TRL, TNO or NEW ACC).

Figure 13.9 illustrates the simulation results. The lead car has been forced to accelerate as the lead car in the real case sequence. It is difficult to compare real case study of a following task with ACC algorithms, as for them the main task consists in maintaining the value of a target headway and we can see in the real case, that the real headway is very different, depending on the following phases, and varies between 0.4 and 1.4 s during an interval of time of 200 s. Simulation results show that the three algorithms are really different in their behaviour. The TRL algorithm is very smooth and slow to react while the TNO is very reactive and safe but to the detriment of the

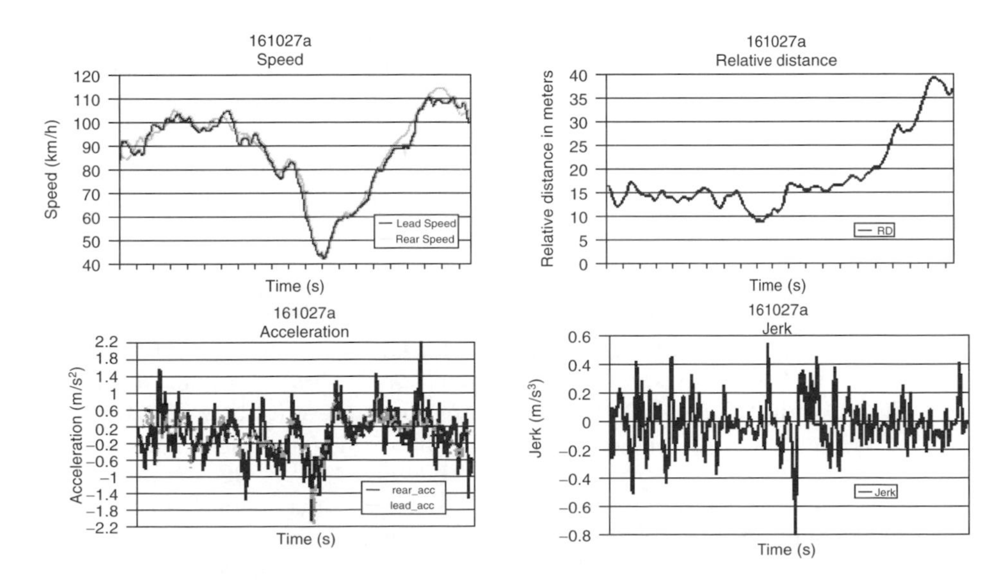

Fig. 13.8 Real following situation data.

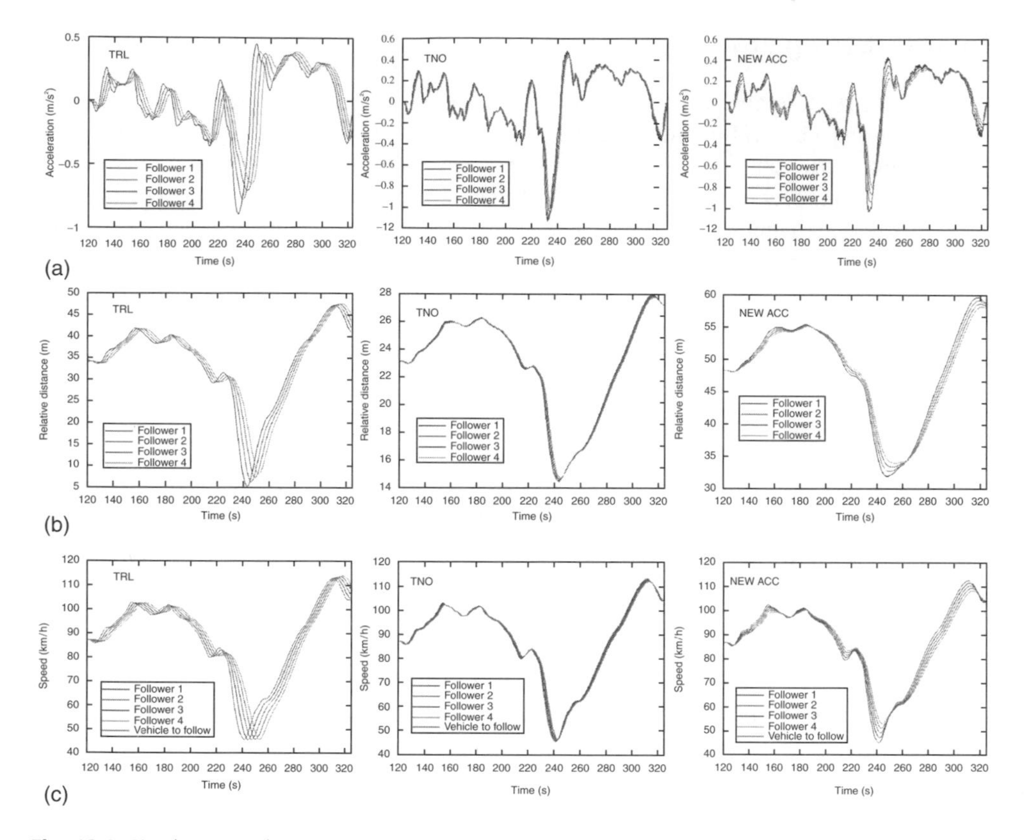

Fig. 13.9 Simulation results.

comfort of the driver. The new algorithm behaviour is between the two. The comfort of the driver has been improved in this New algorithm as the jerk (variation of the acceleration) value is taken into account: the algorithm controls the variation of the acceleration and tries to impose a constant value of the jerk during an acceleration or deceleration phase. The TNO ACC system can be viewed as an anti-collision system, while car manufacturers prefer to see ACC as a system able to assist the driver in car-following situations and to give a warning to the driver in case of emergency.

13.3.4 Impact on traffic flow in a merging situation

This second example concerns a comparison of ACC algorithms in a merging situation (Figure 13.10) In this scenario, 40 per cent of vehicles are equipped with one of the ACC algorithms. In this scenario, the input demand varies between 2000 veh/h and 4500 veh/h as shown in Figure 13.10. Four headway values have been used for each algorithm (0.8 s, 1.2 s, 1.38 s and 1.5 s). TNO and NEW ACC algorithms have good impacts on the traffic efficiency as they show a better journey time than the one obtained without ACC, while the TRL algorithm increases the journey time. For TNO and NEW ACC algorithms, the benefit in average speed is offset by an increase in low time to collision (TTC) values. This makes the system less acceptable

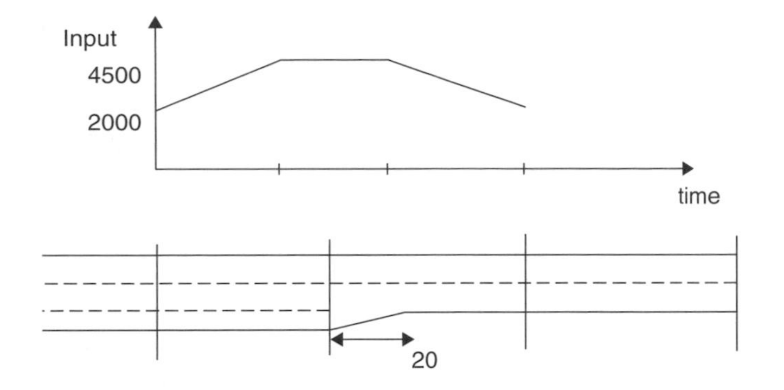

Fig. 13.10 A merging scenario.

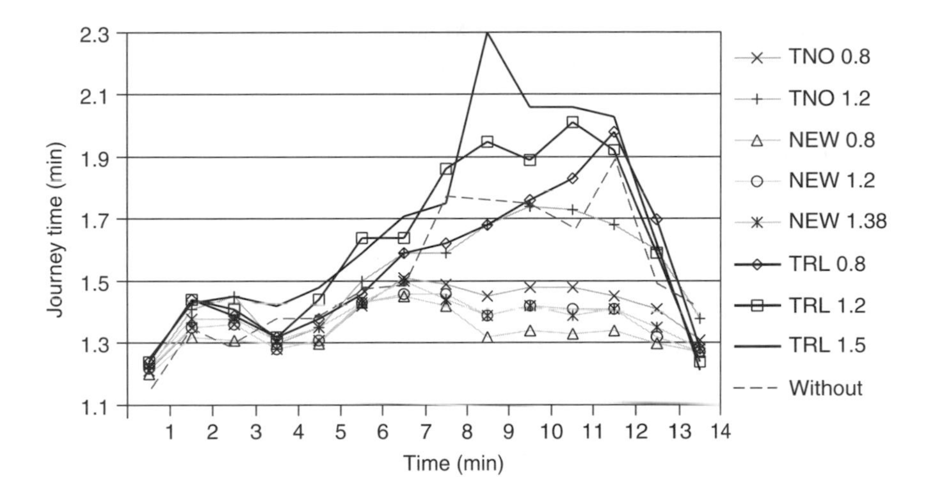

Fig. 13.11 ACC algorithms comparison.

to the user and less likely to be employed. The TRL algorithm which allows lower deceleration values under ACC, reduces the number of occurrences of low TTC in most cases. There appears to be some trade-off between driver comfort and traffic efficiency.

13.3.5 User acceptance

As the ACC system is just introduced now on the market, the understanding of how drivers will use this system and how drivers of non-ACC equipped vehicles will interact with the systems is crude. Up to now, in our simulations, we have considered that the user will fully accept the system when it is active. Depending on the breaking demand, if the system is capable of braking enough, the user will not deactivate it. When the situation makes that the breaking demand is higher than the maximal deceleration the system can produce, the driver, after a reaction time, de-activates the system by acting on the braking pedal.

There are only a few psychological studies on user acceptance. Ralf Risser (Risser, 1999), in a recent study has shown that drivers could be divided into three groups or types:

1. The active/dynamic drivers: this group drove fast and dynamically, but with quite good anticipation.
2. The risk-avoiding drivers: people of this group change lanes as soon as they discovered an obstacle, which meant that they drove with a better anticipation than the others.
3. The cool drivers that demonstrated their own very special driving style. This group illustrates people with awkward behaviours, especially in braking and checking situations. For them, the system was regarded as an obstruction for using their own driving style.

For his study, Risser had access to a prototype of an ACC system and 20 persons were tested on a route of 110 km length in Munich. A modified version of the Wiener Fahrprobe, an in-car-observation method, was used.

13.3.6 Conclusion

The results on the ACC algorithms comparison have shown that their behaviour is highly dependent on the baseline time headway. A short headway will improve the traffic flow, while a long headway will deteriorate it. However, short time headway will cause low time to collision values, which makes the system less acceptable to the user and less likely to be employed. There is some evidence to suggest that ACC systems should provide safety benefits to a driver in an emergency situation. However, there are also some concerns regarding the ability of the driver to make a good decision, as the ACC system could imply a loss of implication of the driver in the driving control loop.

References

van Arem, B. de Vos, A.P. and Vanderschuren, M.J.W.A. (1997). *The microscopic traffic simulation model MIXIC 1.3.* TNO-report INRO-VVG 1997–02b, Delft, January.

Bender, J.G. *et al.* (1982). System studies of automated highway systems. General Motors Transportation Systems Center report EP-81041A, 1981; FHWA Report No. FHWA/RD-82/003.

Choi, S. and Hedrick, J.K. (1995). Vehicle longitudinal control using an adaptive observer for automated highway systems. *Proceedings of the 1995 American Control Conference*, Seattle, WA. June.

Cozot, R. (1996). From multibody system modeling to distributed real-time simulation. *Proceedings of the 29th Annual Simulation Symposium*, IEEE-SCS.

Cremer, M., Demir, C., Donikian, S., Espié, S. and McDonald, M. (1998). Investigating the Impact of AICC Concepts on Traffic Flow Quality. *Proceedings of the 5th World Congress on Intelligent Transport Systems,* Seoul, South Korea, October.

Donikian, S. (1997a). VUEMS: a Virtual Urban Environment Modeling System. *Proceedings of the Computer Graphics International'97 Conference*, IEEE Computer Society Press, Hasselt-Diepenbeek, Belgium, June.

Donikian, S. (1997b). Driving Simulation in Virtual Urban Environments. *Proceedings of the Driving Simulation Conference DSC'97*, Lyon, France, September.

Donikian, S., Chauffaut, A., Duval, T. and Kulpa, R. (1998a). GASP: from Modular Programming to Distributed Execution. *Proceedings of the Computer Animation '98 Conference*, Philadelphia, June.

Donikian, S., Espié, S., Parent, M. and Rousseau, G. (1998b). Simulation Studies on the Impact of ACC. *Proceedings of the 5th World Congress on Intelligent Transport Systems*, Seoul, South Korea, October.

Donikian, S., Moreau, G. and Thomas, G. (1999). Multimodal Driving Simulation in Realistic Urban Environments. In *Progress in System and Robot Analysis and Control Design* (S. Tzafestas and G. Schmidt, eds), Lecture Notes in Control and Information Sciences (LNCIS 243), Springer Verlag.

Hardman, E.J. (1996). Motorway Speed Control Strategies Using SISTM. *Proceedings of the Eighth International Conference on Road Traffic Monitoring and Control*, 23–25 April 1996. Conference Publication No. 422, pp. 169–72. Institution of Electrical Engineers, Savoy Place, London, United Kingdom.

Hochstädter, A. and Cremer, M. (1997). Investigating the potential of convoy driving for congestion dispersion. *Proceedings of the 4th World Congress on Intelligent Transport Systems*, Berlin.

Moreau, G. and Donikian, S. (1998). From psychological and real-time interaction requirements to behavioural simulation. *Proceedings of the Eurographics Workshop on Computer Animation and Simulation*, Lisbon, Portugal, September.

Müller, R. and Nöcker, G. (1992). Intelligent Cruise Control with fuzzy logic. *Proceedings of the Intelligent Vehicles 92 Symposium, IEEE*, Detroit, MI, June-July, 173–8

Parent, M. and Daviet, P. (1994). Automatic driving in Stop and Go traffic. *Proceedings of the IEEE Conference on Intelligent Vehicles*. Paris, October.

Risser, R. (1999). Evaluation of an ACC System with the Help of Behaviour Observation. *Proceedings of the AEPSAT conference*, Angers, France, June.

Schladover, S.E. (1995). Review of the state of development of Advanced Vehicle Control Systems (AVCS). *Vehicle System Dynamics*, **24**, 551–95.

Yanakiev, D. and Kanellakopoulos, I. (1995). Variable time headway for string stability of automated heavy-duty vehicles. *Proceedings of the 34th conference on Decision and Control*.

McDonald, M., Marsden, G., Cremer, M., Dens, L. and Vaa, T. (1998). Deployment of Interurban ATT Test Scenarios (DIATS). Review and Evaluation. *Proceedings of the 5th World Congress on Intelligent Transport Systems*, Seoul, South Korea, October.

Part Four Case study

14

ARGO prototype vehicle

Alberto Broggi
Dipartimento di Informatica e Sistemistica, University of Pavia, Italy

and

Massimo Bertozzi, Gianni Conte, and Alessandra Fascioli
Dipartimento di Ingegneria dell'Informazione,
University of Parma, Italy

This chapter presents the experience of the ARGO project. It started in 1996 at the University of Parma, based on the previous experience within the European PROMETHEUS Project. In 1997 the ARGO prototype vehicle (described in Section 14.3) was set up with sensors and actuators, and the first version of the GOLD software system – able to locate one lane marking and generic obstacles on the vehicle's path – was installed. In June 1998 the vehicle underwent a major test (the *MilleMiglia in Automatico*, a 2000 km tour on Italian highways discussed in Section 14.4) in order to test the complete equipment. The analysis of this test enabled the improvement of the system. Section 14.2 presents the current implementation of the GOLD system, featured by enhanced lane detection abilities and extended obstacle detection abilities, such as the detection of leading vehicles and pedestrians.

14.1 Introduction: the ARGO project

The main target of the ARGO project is the development of an active safety system with the ability to act also as an automatic pilot for a standard road vehicle.

In order to achieve autonomous driving capabilities on the existing road network with no need for specific infrastructures, a robust perception of the environment is essential. Although very efficient in some fields of application, active sensors – besides polluting the environment – feature some specific problems in automotive applications due to inter-vehicle interference amongst the same type of sensors, and due to the wide variation in reflection ratios caused by many different reasons, such as obstacles' shape or material. Moreover, the maximum signal level must comply with safety rules and must be lower than a safety threshold. For this reason in the implementation of the ARGO vehicle only the use of passive sensors, namely *cameras*, has been considered.

A second design choice was to keep the system costs low. These costs include both production costs (which must be minimized to allow a widespread use of these devices) and operative costs, which must not exceed a certain threshold in order not to interfere with the vehicle performance. Therefore low-cost devices have been preferred, both for the image acquisition and the processing: the prototype installed on ARGO is based on *cheap cameras* and a *commercial PC*.

Section 14.2 presents the main functionalities integrated on the ARGO vehicle:

- lane detection and tracking
- obstacle detection
- vehicle detection and tracking.

A further functionality, pedestrian detection – needed for urban driving – is currently under development.

14.2 The GOLD system

GOLD is the acronym used to refer to the software that provides ARGO with autonomous capabilities. It stands for Generic Obstacles and Lane Detection since these were the two functionalities originally developed. Currently it integrates two other functionalities: vehicle detection and the new pedestrian detection, which is under development.

14.2.1 The inverse perspective mapping

The lane detection and obstacle detection functionalities share the same underlying approach: the removal of the perspective effect obtained through the inverse perspective mapping (IPM) (Bertozzi and Broggi, 1998; Broggi *et al.*, 1999).

The IPM is a well-established technique that allows the perspective effect to be removed when the acquisition parameters (camera position, orientation, optics etc.) are completely known and when a piece of information about the road is given, such as a *flat road hypothesis*. The procedure aimed at removing the perspective effect resamples the incoming image, remapping each pixel toward a different position and producing a new two-dimensional array of pixels. The so-obtained remapped image represents a top view of the road region in front of the vehicle, as it were observed from a significant height.

Figures 14.1(a) and (b) show an image acquired by ARGO's vision system and the corresponding remapped image respectively.

14.2.2 Lane detection

Lane detection functionality is divided in two parts: a lower-level part, which, starting from iconic representations of the incoming images, produces new transformed representations using the same data structure (array of pixels), and a higher-level one, which analyses the outcome of the preceding step and produces a symbolic representation of the scene.

Low- and medium-level processing for lane detection

Lane detection is performed assuming that a road marking in the remapped image is represented by a quasi-vertical bright line of constant width on a darker background

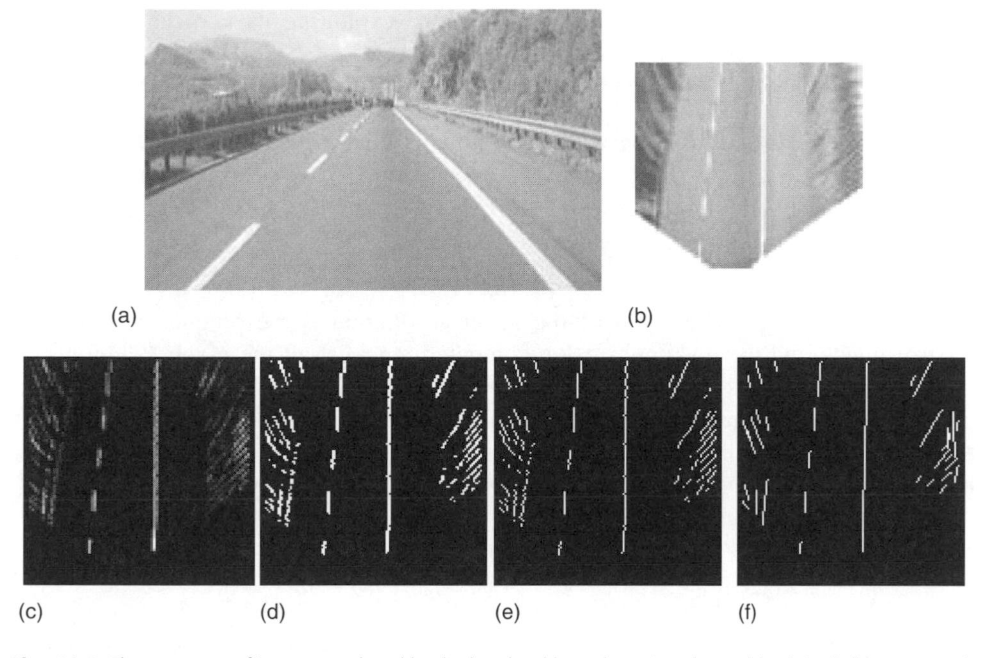

Fig. 14.1 The sequence of images produced by the low-level lane detection phase: (a) original; (b) remapped; (c) filtered; (d) enhanced; (*e*) binarized; (*f*) polylines.

(the road). Consequently, the pixels belonging to a road marking feature a higher brightness value than their left and right neighbours at a given horizontal distance.

The first phase of road markings detection is therefore based on a line-wise determination of horizontal black-white-black transitions, while the following medium-level process is aimed at extracting information and reconstructing road geometry.

Feature extraction The brightness value $b(x, y)$ of a generic pixel belonging to the remapped image is compared to its horizontal left and right neighbours at a distance m: $b(x, y-m)$ and $b(x, y+m)$, with $m \geq 1$.

A new image, whose values $r(x, y)$ encode the presence of a road marking, is then computed according to the following expression:

$$r(x, y) = \begin{cases} d_{+m}(x, y) + d_{-m}(x, y) & \text{if } (d_{+m}(x, y) > 0) \wedge (d_{-m}(x, y) > 0) \\ 0 & \text{otherwise} \end{cases} \tag{14.1}$$

where

$$\begin{cases} d_{+m}(x, y) = b(x, y) - b(x, y+m) \\ d_{-m}(x, y) = b(x, y) - b(x, y-m) \end{cases} \tag{14.2}$$

represent the horizontal brightness gradient. The m parameter is computed according to generic road markings width, image acquisition process, and parameters used in the remapping phase. The resulting filtered image is shown in Figure 14.1(c).

Due to different light conditions (e.g. in presence of shadows), pixels representing road markings may have different brightness, yet maintain their superiority relationship with their horizontal neighbours. Consequently, since a simple threshold seldom gives a satisfactory binarization, the image is enhanced exploiting its vertical correlation; then an adaptive binarization is performed.

The enhancement of the filtered image is performed through a few iterations of a *geodesic morphological dilation* with the following binary structuring element $\begin{smallmatrix} & \bullet & \\ \bullet & \bullet & \bullet \\ & \bullet & \end{smallmatrix}$, where

$$c(x, y) = \begin{cases} 1, & \text{if } r(x, y) \neq 0 \\ 0, & \text{otherwise} \end{cases} \tag{14.3}$$

is the *control image*. The result of the geodesic dilation is the product between the control image and the maximum value computed amongst all the pixels belonging to the neighbourhood described by the structuring element. The iterative application of the geodesic dilation corresponds to the widening of the neighbourhood, except towards the directions in which the control values are 0. Figure 14.2 shows the results of three iterations on a portion of an image that represents a lane marking.

Moreover, since

$$r(x, y) \neq 0 \Longrightarrow \begin{cases} r(x, y - m) = 0 \\ r(x, y + m) = 0, \end{cases} \tag{14.4}$$

the filtered image's pixels at a distance m from a road marking assume a zero value; according to Equation 14.3 their control value is 0 and as a result they form a barrier to the propagation of the maximum value. The enhanced image (after eight iterations) is shown in Figure 14.1(d).

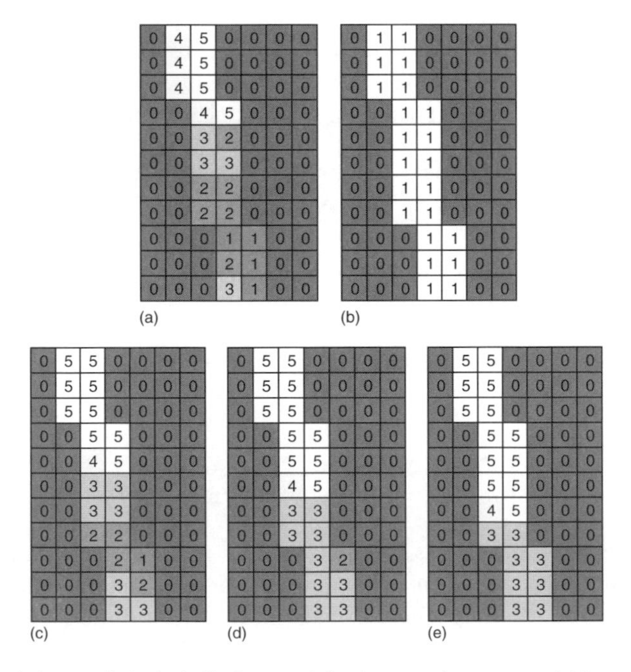

Fig. 14.2 The geodesic morphological dilation used for image enhancement: (*a*) input image, (*b*) control image, and (*c–e*) results of the first three iterations.

Finally, the binarization is performed by means of an adaptive threshold:

$$t(x, y) = \begin{cases} 1, & \text{if } e(x, y) \geq \dfrac{m(x, y)}{k}, \\ 0, & \text{otherwise} \end{cases} \quad (14.5)$$

where $e(x, y)$ represents the enhanced image, $m(x, y)$ the maximum value computed in a given $c \times c$ neighbourhood, and k is a constant. The result of the binarization of Figure 14.1(d) (considering $k = 2$ and $c = 7$) is presented in Figure 14.1(e).

Generation of higher-level data structures The binary image is scanned row by row in order to build chains of eight connected non-zero pixels. Following the choice of $m = 2$ at the previous step, every non-zero pixel can have at the most one adjacent non-zero pixel in the horizontal direction; as a result for each pair of horizontally adjacent pixels the one farthest to the right is chosen so that chains are built with a thickness of one pixel.

Each chain is approximated with a *polyline* made of one or more segments, by means of an iterative process. Firstly, the two extrema (A and B) of the polyline are determined averaging the position of the last few pixels at either end of the chain, in order to cope with possible ripples. The segment that joins these two extrema is considered. A first approximate segment ($\overline{AB}$), built using these two extrema, is considered and the horizontal distance between its mid point M and the chain is used to determine the quality of the approximation. In case it is larger than a threshold, two segments ($\overline{AM'}$ and $\overline{M'B}$) sharing an extremum are considered for the approximation of the chain, where M' is the intersection between the chain and the horizontal line that passes through M (see Figure 14.3). The process is iterated on the approximate segments $\overline{AM'}$ and $\overline{M'B}$; at the end of the processing all chains will be approximated by polylines (see Figure 14.1(f)).

High-level processing for lane detection

After the first low-level stage, in which the main features are localized, and after the second stage, in which the main features are extracted, the new data structure (a list of polylines) is now processed in order to semantically group homologous features and to produce a high-level description of the scene.

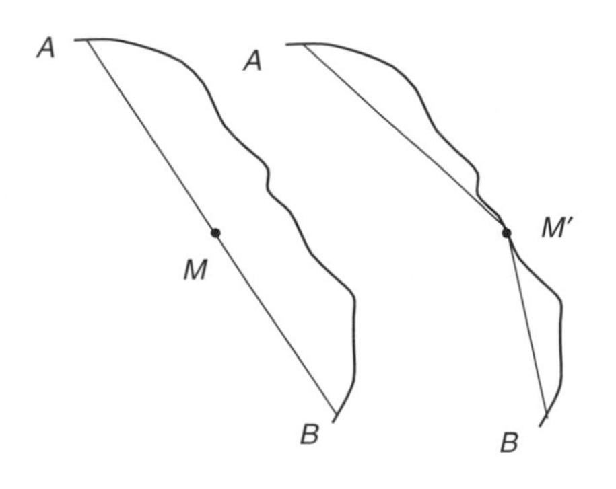

Fig. 14.3 Approximation of chains with polylines.

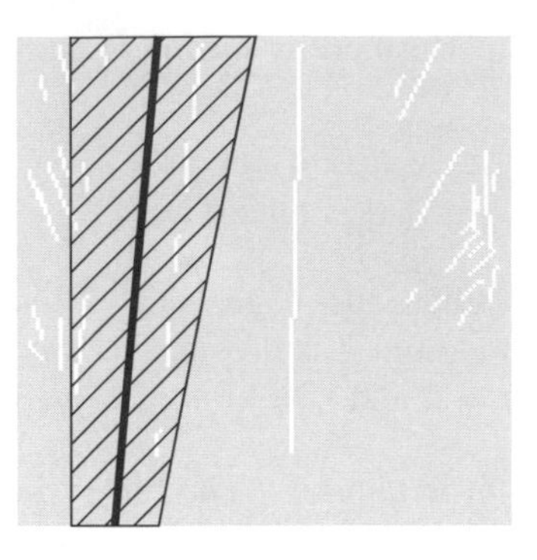

Fig. 14.4 Selection of polylines almost matching the previous left result.

This process in divided into: filtering of noisy features and selection of the feature that most likely belongs to the line marking; joining of different segments in order to fill gaps caused by occlusions, dashed lines, or even worn lines; selection of the best representative and reconstruction of the information that may have been lost, on the basis of continuity constraints; then the result is kept for reference in the next frames and displayed onto the original image.

Feature filtering and selection Each polyline is compared against the result of the previous frame, since continuity constraints provide a strong and robust selection procedure. The distance between the previous result and each extremum of the considered polyline is computed: if all the polyline extrema lay within a stripe centred onto the previous result then the polyline is marked as useful for the following process. This stripe is shaped so that it is small at the bottom of the image (namely close to the vehicle, therefore short movements are allowed) and larger at the top of the image (far from the vehicle, where also curves that appear quickly must be tracked). This process is repeated for both left and right lane markings.

Figure 14.4 shows the previous result with a heavy solid line and the search space with a gridded pattern; it refers to the left lane marking.

Polylines joining Once the polylines have been selected, all the possibilities are checked for their joining. In order to be joined, two polylines must have similar direction; must not be too distant; their projections on the vertical axis must not overlap; the higher polyline in the image must have its starting point within an elliptical portion of the image; in case the gap is large also the direction of the connecting segment is checked for uniform behaviour. Figure 14.5 shows that polyline A cannot be connected to: B due to high difference of orientation; C due to high distance (does not lay within the ellipse); D due to the overlapping of their vertical projections; E since their connecting segment would have a strongly mismatching orientation. It can only be connected to F.

Selection of the best representative All the new polylines, formed by concatenations of the original ones, are then evaluated. Starting from a maximum score, each of the following rules provides a penalty. First each polyline is segmented; in case the polyline does not cover the whole image, a penalty is given. Then, the polyline length

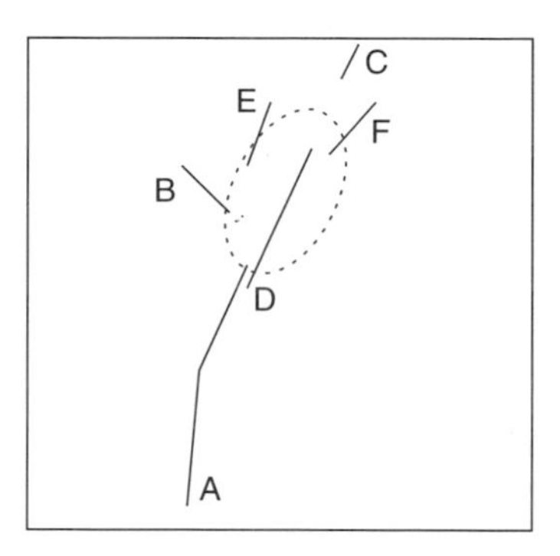

Fig. 14.5 Joining of similar polylines.

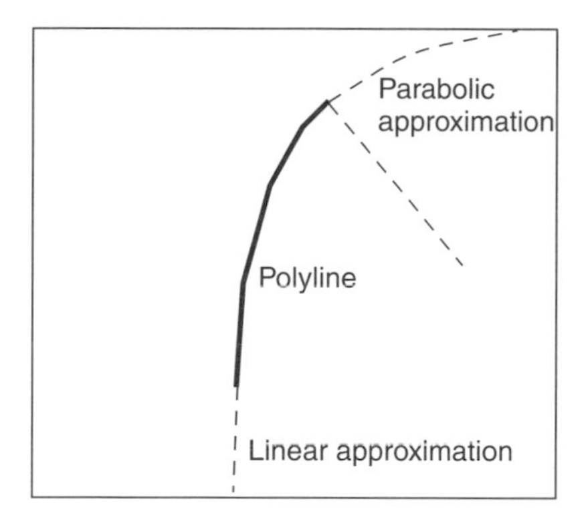

Fig. 14.6 Continuation of short polylines.

is computed and a proportional penalty is given to short ones, as well as to polylines with extremely varying angular coefficients. Finally, the polyline with the highest score is selected as the best representative of the lane marking.

Reconstruction of lost information The polyline that has been selected at the previous step may not be long enough to cover the whole image; therefore a further step is necessary to extend the polyline. In order to take into account road curves, a parabolic model has been selected to be used in the prolongation of the polyline in the area far from the vehicle. In the nearby area, a linear approximation suffices. Figure 14.6 shows the parabolic and linear prolongation.

Model fitting The two reconstructed polylines (one representing the left and one the right lane markings) are now matched against a model that encodes some more knowledge about the absolute and relative positions of both lane markings on a standard road. A model of a pair of parallel lines at a given distance (the assumed lane width) and in a specific position is initialized at the beginning of the process; a specific learning phase allows to adapt the model to errors in camera calibration (lines may be non-perfectly parallel). Furthermore, this model can be slowly changed during the processing to adapt to new road conditions (lane width and lane position), thanks to a learning process running in the background.

The model is kept for reference: the two resulting polylines are fitted to this model and the final result is obtained as follows. First the two polylines are checked for non-parallel behaviour; a small deviation is allowed since it may derive from vehicle movements or deviations from the flat road assumption, that cause the calibration to be temporarily incorrect (diverging of converging lane markings). Then the quality of the two polylines, as computed in the previous steps, is matched: the final result will be attracted toward the polyline with the highest quality with a higher strength.

In this way, polylines with equal or similar quality will equally contribute to the final result; on the other hand, in case one polyline has been heavily reconstructed, or is far from the original model, or is even missing, the other polyline will be used to generate the final result. The weights for the left and right polylines are computed; then, each horizontal line of the two polylines is used to compute the final results (see Figure 14.7).

Finally, Figure 14.8 presents the resulting images referring to the example presented in Figure 14.1. It shows the results of the selection, joining, and matching phases for the left (upper row) and for the right (bottom row) lane markings.

Figure 14.9 presents the final result of the process.

Results of lane detection

This subsection presents a few results of lane detection in different conditions (see Figure 14.10) ranging from ideal situations to road works, patches of non-painted roads, and entries and exits from tunnels. Both highway and extra-urban scenes are provided for comparison; the system proves to be robust with respect to different illumination situations, missing road signs, and overtaking vehicles which occlude the visibility of the left lane marking. If two lines are present – a dashed and a continuous one – the system selects the continuous one.

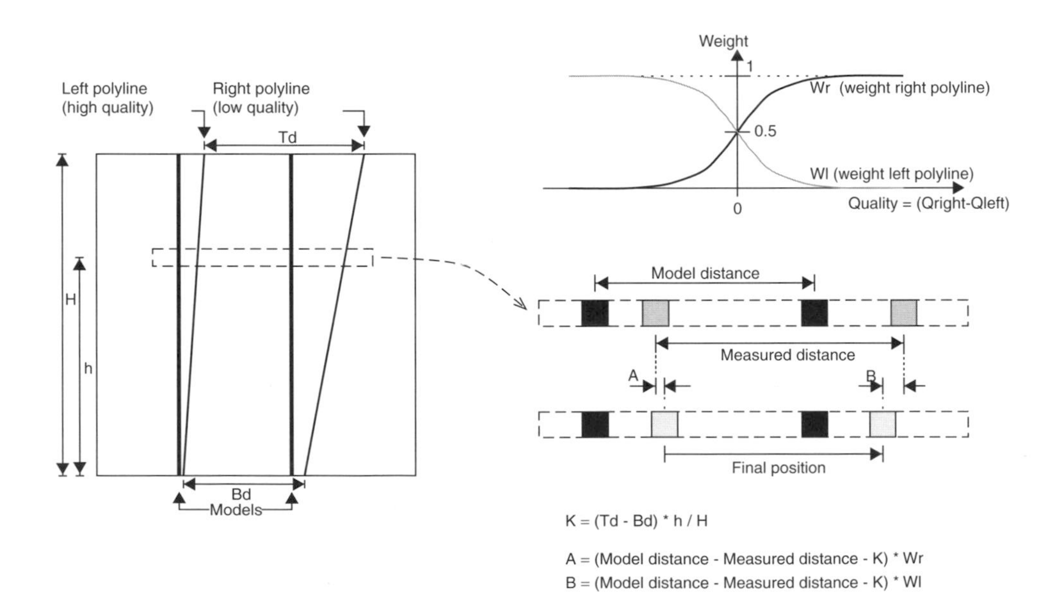

Fig. 14.7 Generation of the final result.

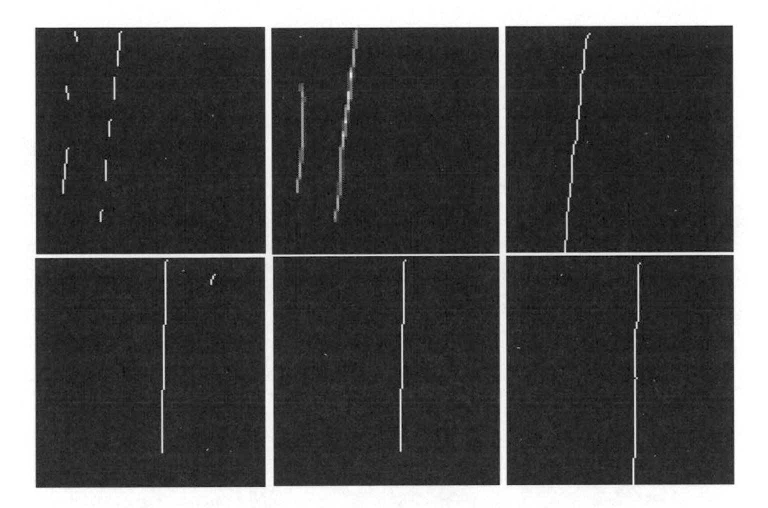

Fig. 14.8 Filtered polylines, joined polylines, and model fitting for the left (upper row) and right (bottom row) lane markings.

Fig. 14.9 The result of lane detection: black markers represent actual road markings while white markers represent interpolations between them.

14.2.3 Obstacle detection

The obstacle detection functionality is aimed at the *localization* of generic objects that can obstruct the vehicle's path, without their complete *identification* or *recognition*. For this purpose a complete 3D reconstruction is not required and a matching with a given model is sufficient: the model represents the environment without obstacles, and any deviation from the model detects a potential obstacle. In this case the application of IPM to stereo images (Bertozzi *et al.*, 1998b), in conjunction with *a priori* knowledge on the road shape, plays a strategic role.

Low-level processing for obstacle detection

Assuming a *flat road* hypothesis, IPM is performed on both stereo images. The flat road model is checked through a pixel-wise difference between the two remapped images: in correspondence to a *generic obstacle* in front of the vehicle, namely anything rising up from the road surface, the difference image features sufficiently large clusters of

Fig. 14.10 Some results of lane detection in different conditions.

non-zero pixels that possess a particular shape. Due to the stereo cameras' different angles of view, an ideal homogeneous square obstacle produces two clusters of pixels with a triangular shape in the difference image, in correspondence to its vertical edges (Broggi *et al.*, 1999). This behaviour is shown in Figure 14.11, which depicts the left and right views of an ideal white square obstacle on a gridded dark background, the two corresponding remapped images, and the thresholded difference also showing the overlapping between the two viewing areas.

Unfortunately due to the texture, irregular shape, and non-homogeneous brightness of generic obstacles, in real cases the detection of the triangles becomes difficult. Nevertheless, in the difference image some clusters of pixels with a quasi-triangular shape are anyway recognizable, even if they are not clearly disjointed. Moreover, if two or more obstacles are present in the scene at the same time, more than two triangles appear in the difference image. A further problem is caused by partially visible obstacles which produce a single triangle.

The low-level portion of the process, detailed in Figure 14.12, is consequently reduced to the computation of difference between the two remapped images, a threshold, and a morphological opening aimed at removing small-sized details in the thresholded image.

Medium- and high-level processing for obstacle detection

The following process is based on the localization of pairs of triangles in the difference image by means of a quantitative measurement of their shape and position (Fascioli, 2000). It is divided into: computing a polar histogram for the detection of triangles, finding and joining the polar histogram's peaks to determine the angle of view under which obstacles are seen, and estimating the obstacle distance.

Polar histogram A *polar histogram* is used for the detection of triangles: it is computed scanning the difference image with respect to a point called *focus* and counting the number of overthreshold pixels for every straight line originating from the focus. The polar histogram's values are then normalized using the polar histogram obtained by scanning an image where all pixels are set (*reference image*). Furthermore, a low-pass filter is applied in order to decrease the influence of noise (see Figure 14.12(f) and (g)).

The polar histogram's focus is placed in the middle point between the projection of the two cameras onto the road plane; in this case the polar histogram presents an appreciable peak corresponding to each triangle (Broggi, 1999). Since the presence of an obstacle produces two disjointed triangles (corresponding to its edges) in the difference image, obstacle detection is limited to the search for pairs of adjacent peaks. The position of a peak in fact determines the angle of view under which the obstacle edge is seen (Figure 14.13).

Peaks may have different characteristics, such as amplitude, sharpness, or width. This depends on the obstacle distance, angle of view, and difference of brightness and texture between the background and the obstacle itself (see Figure 14.15).

Peaks joining Two or more peaks can be joined according to different criteria, such as similar amplitude, closeness, or sharpness. The analysis of a large number of different situations made possible the determination of a parameter embedding all of the above quantities. According to the notations of Figure 14.14, R is defined as the ratio between areas A_1 and A_2. If R is greater than a threshold, two adjacent peaks are considered as generated by the same obstacle, and then joined; otherwise, when the two peaks are far apart or the valley is too deep they are left alone (not joined). Figure 14.15 shows some examples of peak joining.

Obviously, a partially visible obstacle produces a single peak that cannot be joined to any other. The amplitude and width of peaks, as well as the interval between joined peaks, are used to determine the angle of view under which the whole obstacle is seen.

Estimation of obstacle distance The difference image can also be used to estimate the obstacle distance. For each peak of the polar histogram a *radial histogram* is computed scanning a specific sector of the difference image. The width α_i of the sector is determined as the width of the polar histogram peak in correspondence to 80 per cent of the peak maximum amplitude h_i. The number of overthreshold pixels is computed and the result is normalized. The radial histogram is analysed to detect the corners of triangles, which represent the contact points between obstacles and road plane, therefore allowing the determination of the obstacle distance through a simple threshold (see Figure 14.16).

Results of obstacle detection

Figure 14.17 shows the results obtained in a number of different situations. The result is displayed with black markings superimposed on a brighter version of the left image; they encode both the obstacle's distance and width.

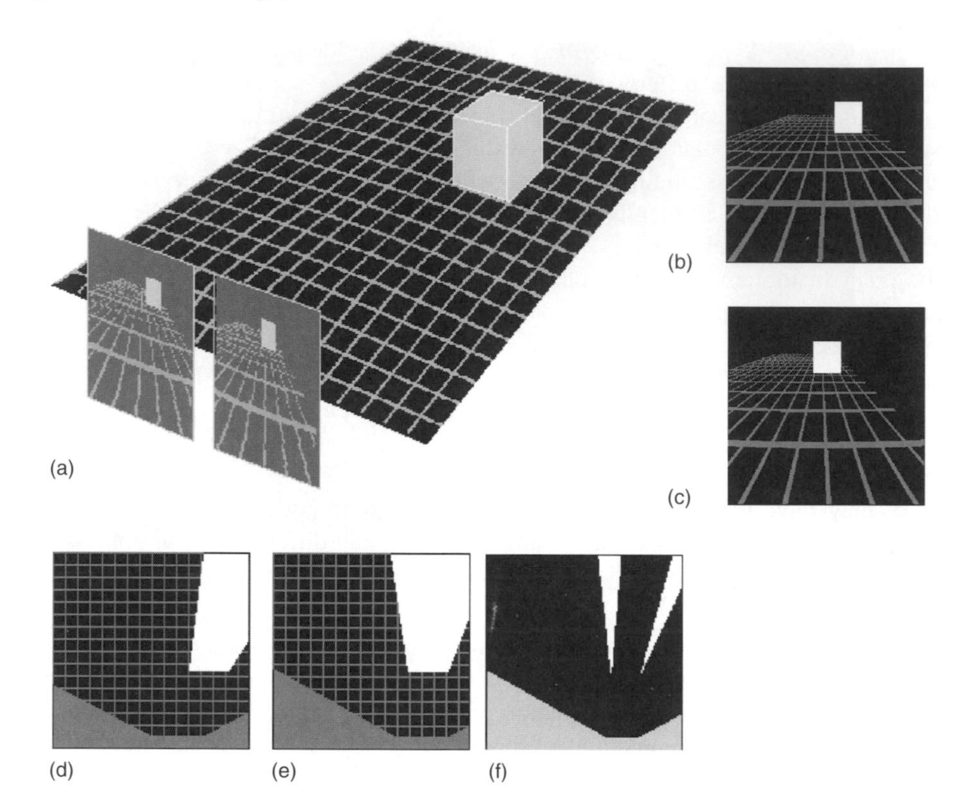

Fig. 14.11 The acquisition of an ideal homogeneous square obstacle: (*a*) 3D representation; (*b*) left image; (*c*) right image; (*d*) left remapped image; (*e*) right remapped image; (*f*) difference image in which the grey area represents the region of the road unseen by both cameras.

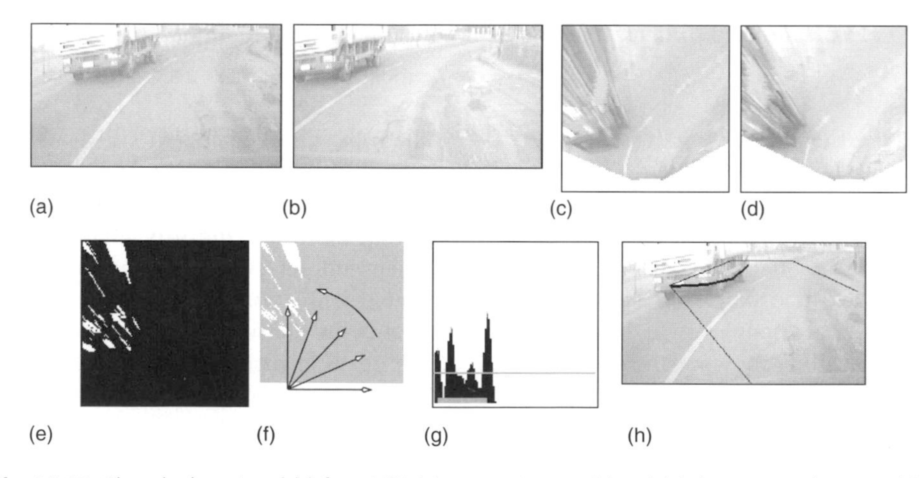

Fig. 14.12 Obstacle detection: (*a*) left and (*b*) right stereo images, (*c*) and (*d*) the remapped images, (*e*) the difference image, (*f*) the angles of view overlapped with the difference image, (*g*) the polar histogram, and (*h*) the result of obstacle detection using a black marking superimposed on the acquired left image; the thin black line highlights the road region visible from both cameras.

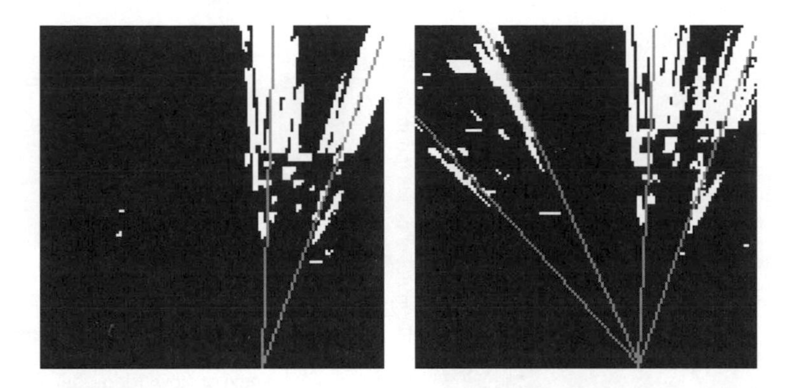

Fig. 14.13 Correspondence between triangles and directions pointed out by peaks detected in the polar histogram.

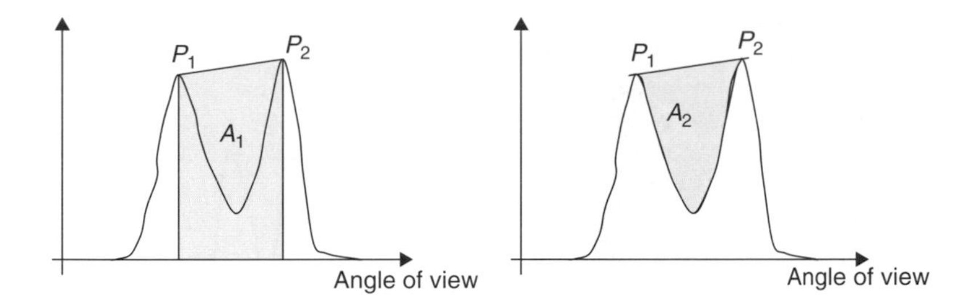

Fig. 14.14 If the ratio between areas A_1 and A_2 is greater than a threshold, the two peaks are joined.

14.2.4 Vehicle detection

The platooning task is based on the detection of the distance, speed, and heading of the preceding vehicle. Since obstacle detection does not generate sufficiently reliable results – in particular regarding obstacle distance – a new functionality, vehicle detection, has been considered; the vehicle is localized and tracked using a single monocular image sequence.

The vehicle detection algorithm is based on the following considerations: a vehicle is generally symmetric, characterized by a rectangular bounding box which satisfies specific aspect ratio constraints, and placed in a specific region of the image. These features are used to identify vehicles in the image in the following way: first an area of interest is identified on the basis of road position and perspective constraints. This area is searched for possible vertical symmetries; not only grey level symmetries are considered, but vertical and horizontal edges symmetries as well, in order to increase the detection robustness. Once the symmetry position and width has been detected, a new search begins, which is aimed at the detection of the two bottom corners of a rectangular bounding box. Finally, the top horizontal limit of the vehicle is searched for, and the preceding vehicle localized.

The tracking phase is performed through the maximization of the correlation between the portion of the image contained into the bounding box of the previous frame (partially stretched and reduced to take into account small size variations due to the increment and reduction of the relative distance) and the new frame.

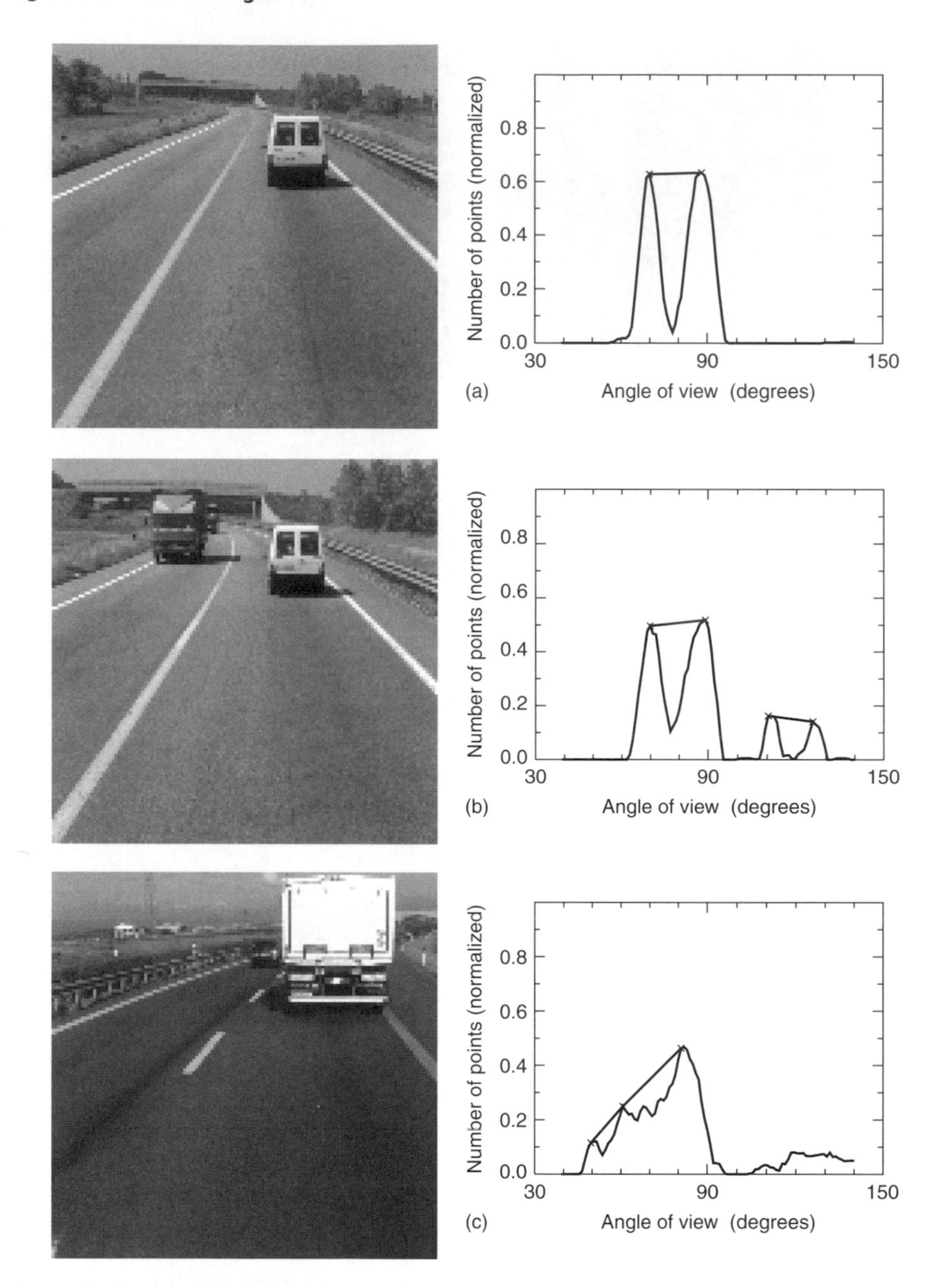

Fig. 14.15 Some examples of peaks join: (*a*) one obstacle, (*b*) two obstacles and (*c*) a large obstacle.

Symmetry detection

In order to search for symmetrical features, the analysis of grey level images is not sufficient. Figure 14.18 shows that strong reflections cause irregularities in vehicle symmetry, while uniform areas and background patterns present highly correlated symmetries. In order to get rid of these problems, also symmetries in other domains are computed.

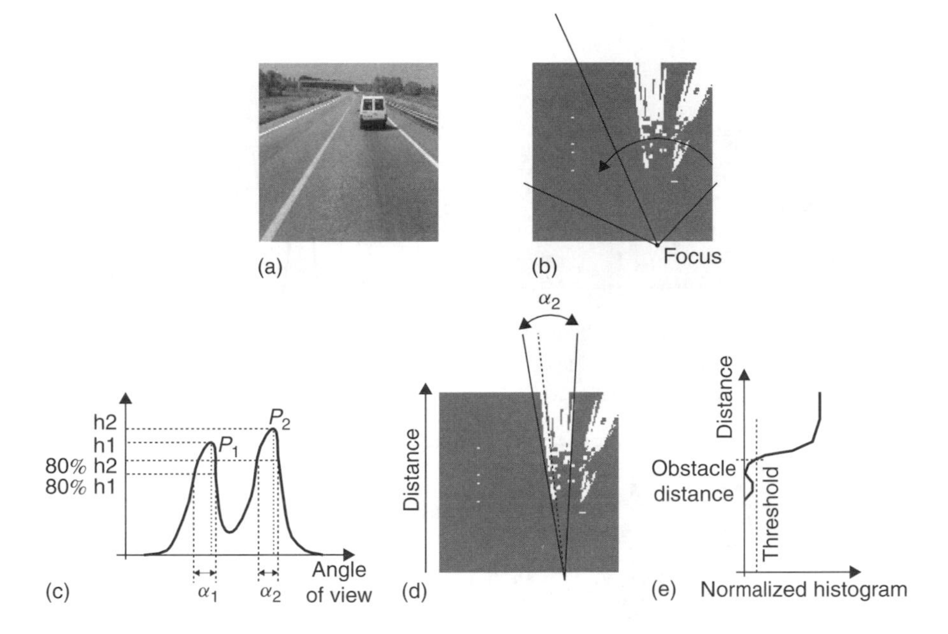

Fig. 14.16 Steps involved during the computation of radial histogram for peak P_2: (*a*) original image; (*b*) binary difference image; (*c*) polar histogram; (*d*) sector used for the computation of the radial histogram; (*e*) radial histogram.

Fig. 14.17 Obstacle detection: the result is shown with a black marking superimposed onto a brighter version of the image captured by the left camera; a black thin line limits the portion of the road seen by both cameras.

Fig. 14.18 Typical road scenes: in the leftmost image a strong sun reflection reduces the vehicle grey level symmetry; in the centre image a uniform area can be regarded as a highly symmetrical region; the rightmost image shows background symmetrical patterns.

(a) (b)

Fig. 14.19 Edges enforce the detection of real symmetries: strong reflections have lower effects while uniform areas are discarded since they do not present edges.

Fig. 14.20 Grey level symmetries: symmetry maps encoding high symmetries with bright points are shown in an enlarged version.

To get rid of reflections and uniform areas, vertical and horizontal edges are extracted and thresholded, and symmetries are computed into these domains as well. Figure 14.19 shows that although a strong reflection is present on the left side of the vehicle, edges are anyway visible and can be used to extract symmetries; moreover, in uniform areas no edges are extracted and therefore no symmetries are detected. Figure 14.20 shows two examples in which grey level symmetries can be successful for vehicle detection, while Figure 14.21 shows the result of edge symmetry.

For each image, the search area is shown in dark grey and the resulting vertical axis is superimposed. For each image its symmetry map is also depicted both in its original size and – on the right – zoomed for better viewing. Bright points encode

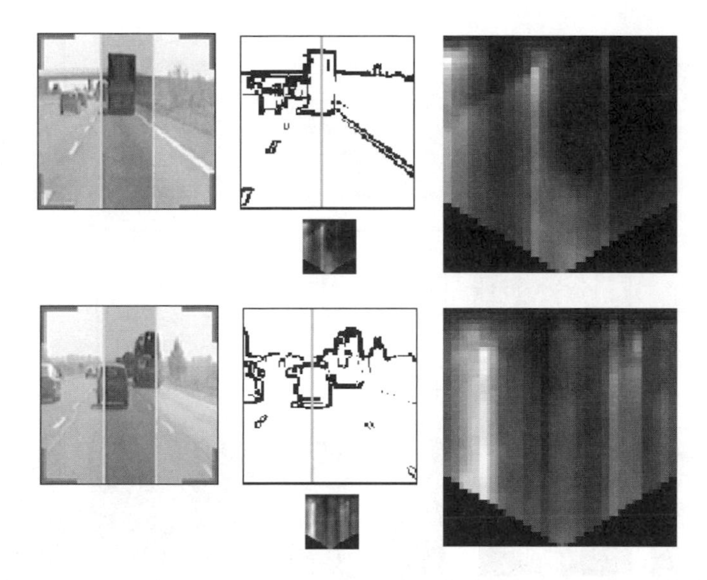

Fig. 14.21 Edge symmetries: the symmetries are computed on the binary images obtained after thresholding the gradient image.

the presence of high symmetries. The 2D symmetry maps are computed by varying the axis' horizontal position within the grey area (shown in the original image) and the symmetry horizontal size. The lower triangular shape is due to the limitation in scanning large horizontal windows for peripheral vertical axes.

Similarly, the analysis of symmetries of horizontal and vertical edges produces other symmetry maps, which – with specific coefficients detected experimentally – can be combined with the previous ones to form a single symmetry map. Figure 14.22 shows all symmetry maps and the final one, that allows to detect the vehicle.

Bounding box detection

After the localization of the symmetry, the width of the symmetrical region is checked for the presence of two corners representing the bottom of the bounding box around the vehicle. Perspective constraints as well as size constraints are used to reduce the search. Figure 14.23 shows possible and impossible bottom parts of the bounding box, while Figure 14.24 presents the results of the lower corners detection.

This process is followed by the detection of the top part of the bounding box, which is looked for in a specific region whose location is again determined by perspective and size constraints. Figure 14.25 shows the search area.

Backtracking

Sometimes it may happen that in correspondence to the symmetry maximum no correct bounding boxes exist. Therefore, a backtracking approach is used: the symmetry map is again scanned for the next local maximum and a new search for a bounding box is performed. Figure 14.26 shows a situation in which the first symmetry maximum, generated by a building, does not lead to a correct bounding box; on the other hand, the second maximum leads to the correct detection of the vehicle.

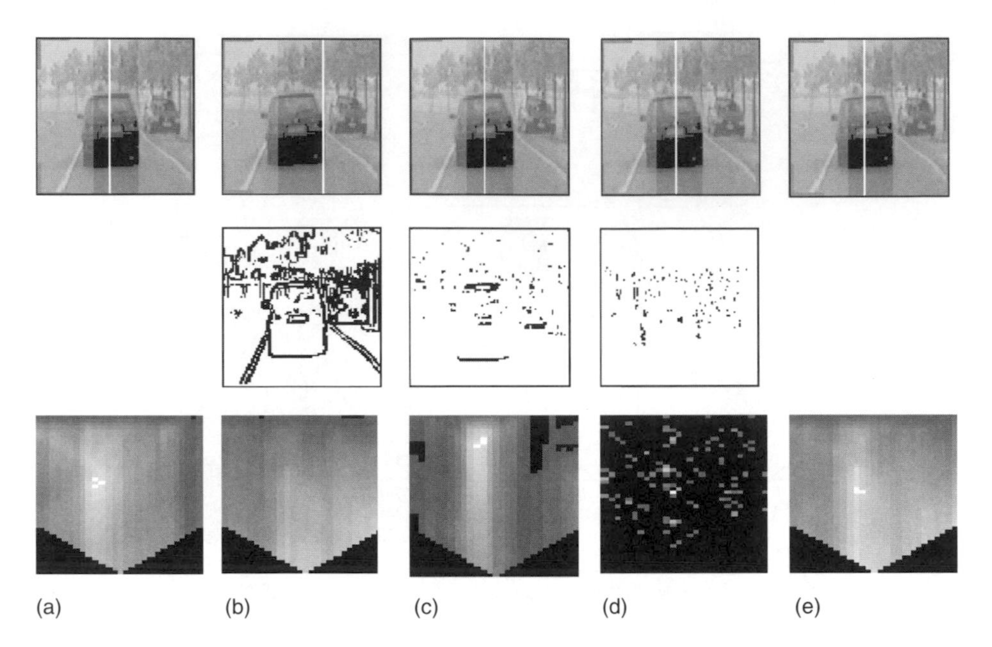

Fig. 14.22 Computing the resulting symmetry: (*a*) grey-level symmetry; (*b*) edge symmetry; (*c*) horizontal edges symmetry; (*d*) vertical edges symmetry; (*e*) total symmetry. For each column the resulting symmetry axis is superimposed onto the top original image.

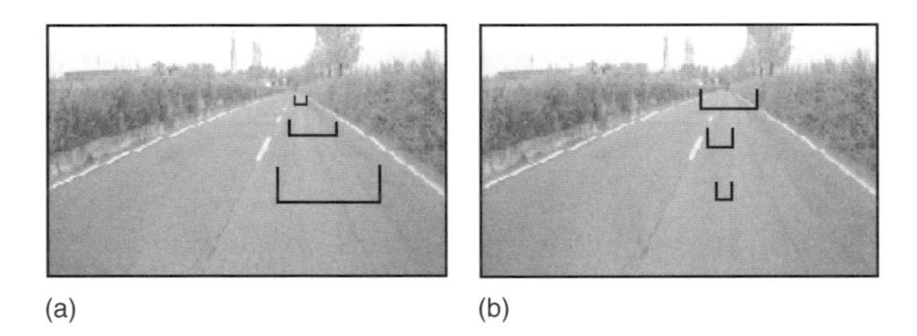

Fig. 14.23 Detection of the lower part of the bounding box: (*a*) correct position and size, taking into consideration perspective constraints and knowledge on the acquisition system setup, as well as typical vehicles' size; (*b*) incorrect bounding boxes.

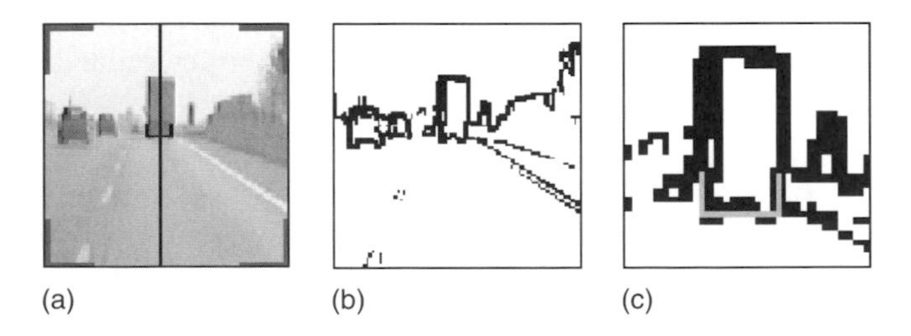

Fig. 14.24 Detection of the lower part of the bounding box: (*a*) original image with superimposed results; (*b*) edges; (*c*) localization of the two lower corners.

Fig. 14.25 The search area for the upper part of the bounding box is shown in dark grey. It takes into account knowledge about the typical vehicle's aspect ratio.

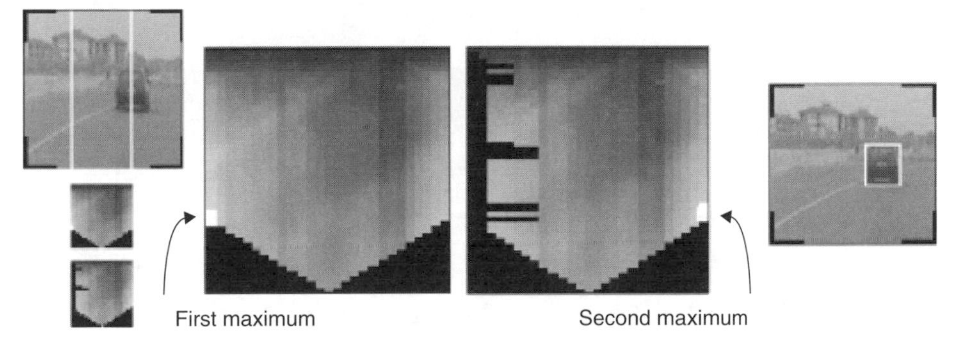

Fig. 14.26 A case in which the total background symmetry is higher than the vehicle symmetry. The search for the bounding box reveals that the left symmetry axis does not correspond to a valid vehicle. The backtracking approach removes the peak near the maximum from the symmetry map and spawns a new bounding box search.

Results of vehicle detection

Figure 14.27 shows some results of vehicle detection in different situations.

14.2.5 Pedestrian detection

A new functionality, still under development and test, allows detection of pedestrians in a way similar to what happens for vehicle detection.

Namely, a pedestrian is defined as a symmetrical object with a specific aspect ratio. With very few modifications to the vehicle detection algorithm presented in the previous section, the system is able to locate and track single pedestrians.

Moreover a new phase, based on stereo vision, aimed at the precise computation of pedestrian distance is currently under study.

Figure 14.28 shows a few examples of correct detection.

Fig. 14.27 Results of vehicle detection in different road scenes.

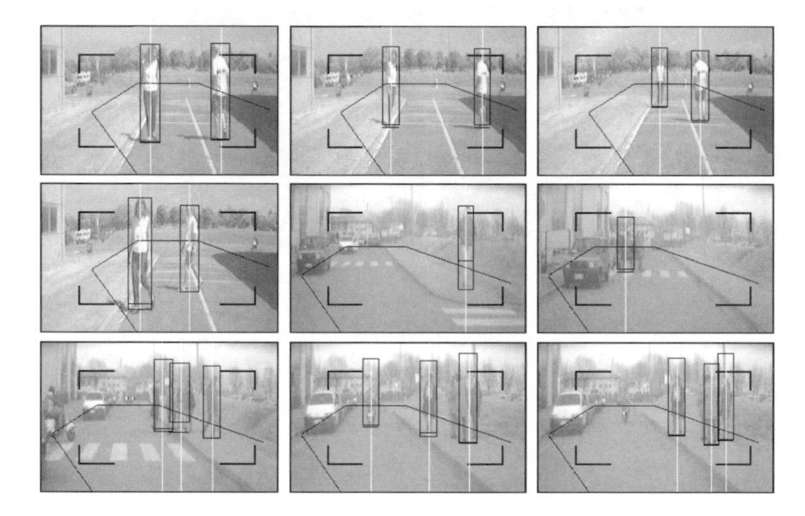

Fig. 14.28 Some preliminary results of pedestrian detection.

14.2.6 The software system's architecture

GOLD is a complex software system made of several interacting modules, all written in C language for a total amount of more than 10 000 code lines.

The program flow chart is sketched in Figure 14.29. Different processing stages can be recognized: after a beginning phase devoted to preliminary actions such as parameter configuration, storage allocation, variables and devices initialization, the main processing cycle is entered and repeatedly executed until termination is requested. Some concluding operations, such as storage deallocation and device closing, are performed before termination.

At each cycle data are first received from different sources: environmental data are acquired from the sensors (cameras and speedometer) into the computer memory, and

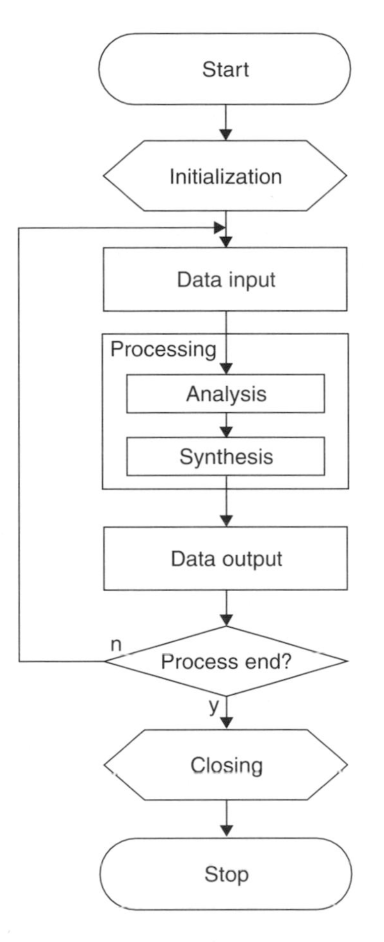

Fig. 14.29 The program flow chart.

commands are received from the user interface. The subsequent processing phase is made of two steps: the analysis of the input data, aimed at the extraction of information concerning the environment, and the exploitation of this knowledge in the decision on the actions to carry on regarding the warning and driving strategies. The data output, therefore, consists in controlling the steering wheel and issuing acoustical and optical warnings. For debugging purposes a visual output can also be supplied to the user and intermediate results can be recorded on disk.

Figure 14.30 sketches the organization and relations among the different program modules from the data flow point of view. As shown, data input and output is performed by means of a modular hierarchical structure: the multiple modules in charge of acquisition or output rely on lower-level modules which interact with the devices' drivers or supply system services. This division between the front end and the back end both in the input and in the output eases the integration of new boards and peripheral devices.

Depending on the configuration parameters, images can be acquired from the cameras, by means of the frame grabber module, when the program is run on board

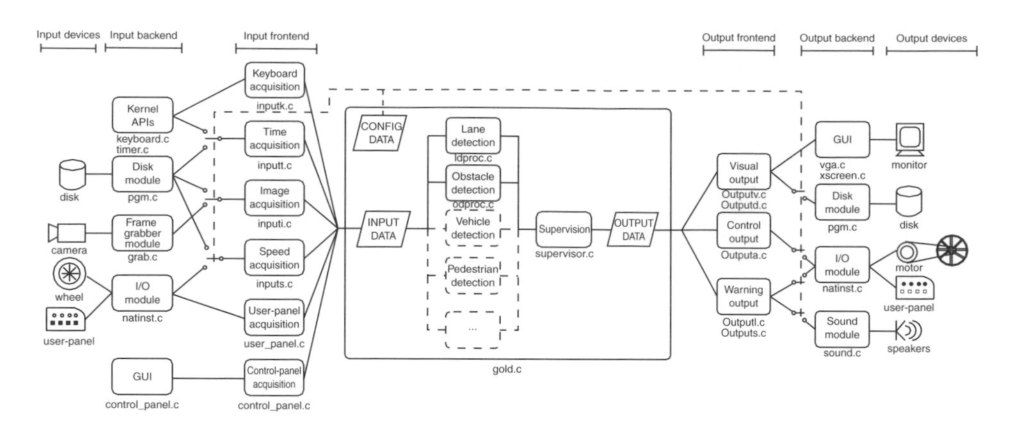

Fig. 14.30 The GOLD system's software architecture. For each module the names of the source files implementing it are displayed.

of the vehicle, or from mass storage (hard or compact disks), through disk managing routines, when it is being tested in the laboratory on pre-recorded image sequences. In the former case the vehicle's speed and the time elapsed between consecutive acquisitions, which are used to assess the spatial distance between frames, are obtained from the speedometer and from the system clock by means of I/O routines and interaction with the kernel APIs, respectively. Conversely, in the latter case they are read from a specific position in the image file where they have been previously encoded during acquisition.

User commands can be entered through the keyboard exploiting the kernel APIs, through the mouse thanks to a software control-panel realized by means of graphical routines, or through a hardware button-based user-panel integrated in the vehicle, managed by the acquisition module through I/O routines.

Several independent modules realize the different functionalities related to the environmental information extraction. The lane detection and obstacle detection functionalities have been initially developed, a module for the localization of the leading vehicle for platooning is currently under development, and, anyway, thanks to the modular structure of GOLD, other modules could be easily integrated, such as the detection of pedestrians or the recognition of road signs.

A supervisor module collects the results from all these processing modules and, by applying strategical reasoning and decisional capabilities, produces the data to output.

Data to control the vehicle trajectory are issued to the motor which drives the steering wheel through I/O modules, which are also used as an interface towards the warning LEDS integrated in the on-board user panel.

The visual output for debugging is produced by means of the routines of the graphical user interface and, depending on the available graphic environment, can be displayed under the X windowing system or can be directly sent to the VGA.

The full modularity and hierarchical structure of the GOLD software architecture allows expanding the system by adding new functionalities, guarantees portability, hardware independence and reconfigurability by allowing the use of different devices, and eases the debugging and testing of the algorithms.

14.2.7 Computational performance

Table 14.1 shows the timing performance obtained on the computing system currently installed on ARGO (see Section 14.3.3); since obstacle and lane detection share the removal of the perspective effect, the timings for IPM are separated from the others. In addition, due to the different computational burden of vehicle detection when looking for a vehicle or tracking an already found one, two distinct timings for vehicle detection and tracking are shown.

The acquisition adapter installed on the ARGO system is able to continuously capture images into a circular buffer in main memory, therefore not requiring a synchronization with the processing (see Figure 14.31).

When all three functionalities are on, the system can work up to a 45 Hz rate.

Table 14.1 Timings of processing steps

	Pentium 200MMX	Pentium II 450MMX	Speedup
IPM	9.9 ms	3.4 ms	2.9
Lane detection	14.4 ms	3.5 ms	4.1
Obstacle detection	17.5 ms	6.3 ms	2.7
Vehicle tracking	24.8 ms	8.8 ms	2.8
Vehicle detection	47.6 ms	19.9 ms	2.4

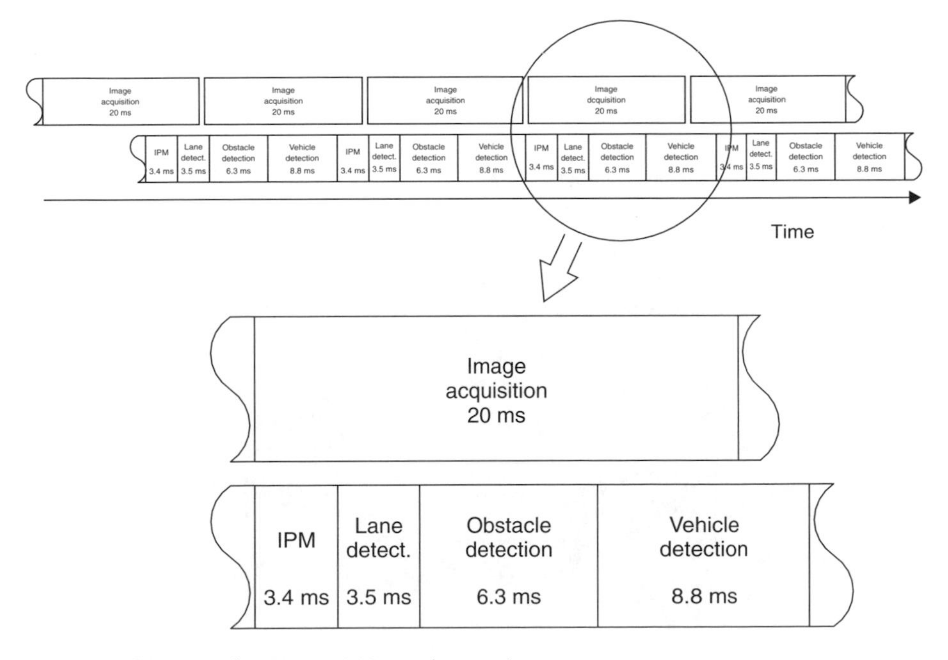

Fig. 14.31 Performance of image acquisition and processing.

14.3 The ARGO prototype vehicle

ARGO, shown in Figure 14.32, is an experimental autonomous vehicle equipped with vision systems and an automatic steering capability.

It is able to determine its position with respect to the lane, to compute the road geometry, to detect generic obstacles on the path, and to localize a leading vehicle. The images acquired by a stereo rig placed inside the windscreen are analysed in real time by a computing system located into the boot. The results of the processing are used to drive an actuator mounted onto the steering wheel and other driving assistance devices.

The system was initially conceived as a safety enhancement unit: in particular it is able to supervise the driver behaviour and issue both optic and acoustic warnings or even take control of the vehicle when dangerous situations are detected. Further developments have extended the system functionalities to automatic driving capabilities fully.

14.3.1 Functionalities

Thanks to the control panel the driver can select the level of system intervention. The following three driving modes are integrated.

- **Manual driving:** the system simply monitors and logs the driver's activity.
- **Supervised driving:** in case of danger, the system warns the driver with acoustic and optical signals.

Fig. 14.32 The ARGO prototype vehicle.

- **Automatic driving:** the system maintains the full control of the vehicle's trajectory, and the two following functionalities can be selected:
  - **Road following:** consisting of the automatic movement of the vehicle inside the lane. It is based on: *lane detection* (which includes the localization of the road, the determination of the relative position between the vehicle and the road, and the analysis of the vehicle's heading direction) and *obstacle detection* (which is mainly based on localizing possible generic obstacles on the vehicle's path).
  - **Platooning:** namely the automatic following of the preceding vehicle, that requires the localization and tracking of a target vehicle (*vehicle detection and tracking*), and relies on the recognition of specific vehicle's characteristics.

14.3.2 The data acquisition system

Only passive sensors (two cameras and a speedometer) are used on ARGO to sense the surrounding environment: although vision is extremely computationally demanding, it offers the possibility to acquire data in a non-invasive way, namely without altering and polluting the environment. In addition, a button-based control panel has been installed enabling the driver to modify a few driving parameters, select the system functionality, issue commands, and interact with the system.

The vision system

The ARGO vehicle is equipped with a stereoscopic vision system consisting of two synchronized cameras able to acquire pairs of grey level images simultaneously. The installed devices are small (3.2 cm × 3.2 cm) low-cost cameras featuring a 6.0 mm focal length and 360 lines resolution, and can receive the synchronism from an external signal.

The cameras lie inside the vehicle at the top corners of the windscreen, so that the longitudinal distance between the two cameras is maximum. This allows the detection of the third dimension at long distances. The optical axes are parallel and, in order to also handle non-flat roads, a small part of the scene over the horizon is also captured, even if the framing of a portion of the sky can be critical due to abrupt changes in image brightness: in case of high contrast, the sensor may acquire over-saturated images.

The images are acquired by a PCI board, which is able to grab three 768 × 576 pixel images simultaneously. The images are directly stored into the main memory of the host computer thanks to the use of DMA. The acquisition can be performed in real time, at 25 Hz when using full frames or at 50 Hz in the case of single field acquisition.

System calibration

Since the process is based on stereo vision, camera calibration plays a fundamental role in the success of the approach. The calibration process is divided into two steps:

- **Supervised calibration:** the first part of the calibration process is an interactive step: a grid with a known size, shown in Figure 14.33, has been painted onto the ground and two stereo images are captured and used for the calibration. Thanks to an X Windows-based graphical interface a user selects the intersections of the grid lines using a mouse; these intersections represent a small set of points whose world coordinates are known to the system; this mapping is used to compute the

Fig. 14.33 View of the calibration grid painted on a reference road segment.

calibration parameters. This set of homologous points is used to minimize different cost functions, such as the distance between each point and its neighbours as well as line parallelism.

This first step is intended to be performed only once when the orientation of the cameras or the vehicle trim has changed. Since the homologous points are few and their coordinates may be affected by human imprecision, this calibration represents only a rough guess of the parameters, and a further process is required.

- **Automatic parameters tuning:** after the supervised phase, the computed calibration parameters have to be refined. Moreover, small changes in the vision system setup or in the vehicle trim require a periodic tuning of the calibration. For this purpose, an automatic procedure has been developed (Bertozzi *et al.*, 1998b). Since this step is only a refinement, a structured environment, such as the grid, is no longer required and a mere flat road in front of the vision system is sufficient. The parameters' tuning consists of an iterative procedure, based on the application of the IPM transform to stereo images, which takes about 20 seconds: iteratively small deviations from the coarse parameters computed during the previous step are used to remap the captured images; the aim is to get the remapped images as similar as possible. All the pixels of the remapped images are used to test the correctness of the calibration parameters through a least square difference approach.

The speed sensor

The vehicle is also equipped with a speedometer to detect its velocity. A Hall effect-based device has been chosen due to its simplicity and its reliability and has been interfaced to the computing system via a digital I/O board with event counting facilities. The resolution of the measuring system is about 9 cm/s.

The user interface

Finally, a set of buttons on a control panel allows the user to interact with the system: the driver can select the functionality, modify some driving parameters, and – in this first stage – can use it as an interface for a basic debugging tool. The control panel is shown in Figure 14.34.

The keyboard

The on-board PC's keyboard is used as a set of auxiliary buttons to trigger on or off specific debugging facilities and modify the internal driving parameters that are hidden from the driver. It is therefore only used to debug and tune the system's parameters.

Fig. 14.34 The control panel, which includes the two sets of LEDs and a series of buttons.

14.3.3 The processing system

Two different architectural solutions were considered and evaluated: a special-purpose and a general-purpose processing system. Although the dedicated architecture presents the advantages of an *ad hoc* solution, it requires an expensive and complex design, whilst, on the other hand, the standard architectural solutions offer nowadays a sufficient computational power even for hard real-time constrained applications as this one.

For these reasons, the architectural solution which was originally installed and tested on the ARGO vehicle was based on a standard 200 MHz MMX Pentium processor. Software performance was boosted exploiting the SIMD MMX enhancements of the traditional instruction set, which permits processing in parallel multiple data elements using a single instruction flow. The processing system currently installed on ARGO is based on a standard 450 MHz Pentium II processor. Figure 14.35 shows the ARGO boot in which the PC is visible.

14.3.4 The output system

Four different output devices have been installed on ARGO: some of them are actuators, some were integrated in order to test their usefulness, and a few were included just for debugging purposes. Figure 14.40 on shows an internal view of the driving cabin.

The acoustical devices

A pair of stereo loudspeakers are used to issue warnings to the driver in case dangerous conditions are detected, e.g. when the distance from the leading vehicle is under a safety threshold or when the vehicle position within the lane is unsafe. The possibility of using

Fig. 14.35 Internal view of ARGO's boot: both the PC and inverter are visible.

the loudspeakers individually is exploited in order to enhance the effectiveness of the acoustical messages. The driver can choose amongst several kinds of acoustic signals: chimes only, vocal messages in different languages (Italian or English) and with a different information content.

The optical devices

Two different optical devices, both shown in Figure 14.36, have been installed: a LED-based control panel and a colour monitor. The former is used to display the current functionality (bottom row LEDs) and a visual information about the relative position of the vehicle compared to the lane, namely the vehicle offset with respect to the road centre line (top row LEDs).

The latter is mainly used as a debugging tool, since its integration on a commercial vehicle would distract the driver's attention: a visual feedback is supplied to the driver by displaying the results of the process on a 6-inch on-board colour monitor (see Figure 14.37). The monitor presents the acquired left image with markings highlighting lane markings as well as the position of obstacles and preceding vehicles. It also presents on the right side a bi-dimensional reconstruction of the road scene, showing the position of obstacles, vehicles and lane markings, and, on the bottom, the vehicle's speed, a software version of the control panel's LEDs, and a measure of the vehicle offset with respect to the road centre line. As a different option it can display images representing intermediate processing results and the value of some basic parameters for debugging purposes.

Figure 14.38 shows the debugging tool, running under the X11 windowing system; it allows tuning of the GOLD parameters and computing of processing statistics.

Fig. 14.36 The control panel and on-board monitor.

The mechanical devices

A single mechanical device, shown in Figure 14.39, has been integrated on ARGO to provide autonomous steering capabilities. It is composed of an electric engine linked to the steering column by means of a belt.

This device can operate in three different ways:

- as a *warning* device, it is used to vibrate the steering wheel for a short while in order to warn the driver when an unsafe manoeuvre is undertaken or in correspondence to a momentary reduction in driving capabilities (e.g. drowsiness, wrong security distance, etc.);
- as an *actuating* device, it is used during humanly operated vehicle driving to restore a safe lateral position of the vehicle in case of dangerous situations: it steers the

Fig. 14.37 The output of the GOLD system on the on-board monitor.

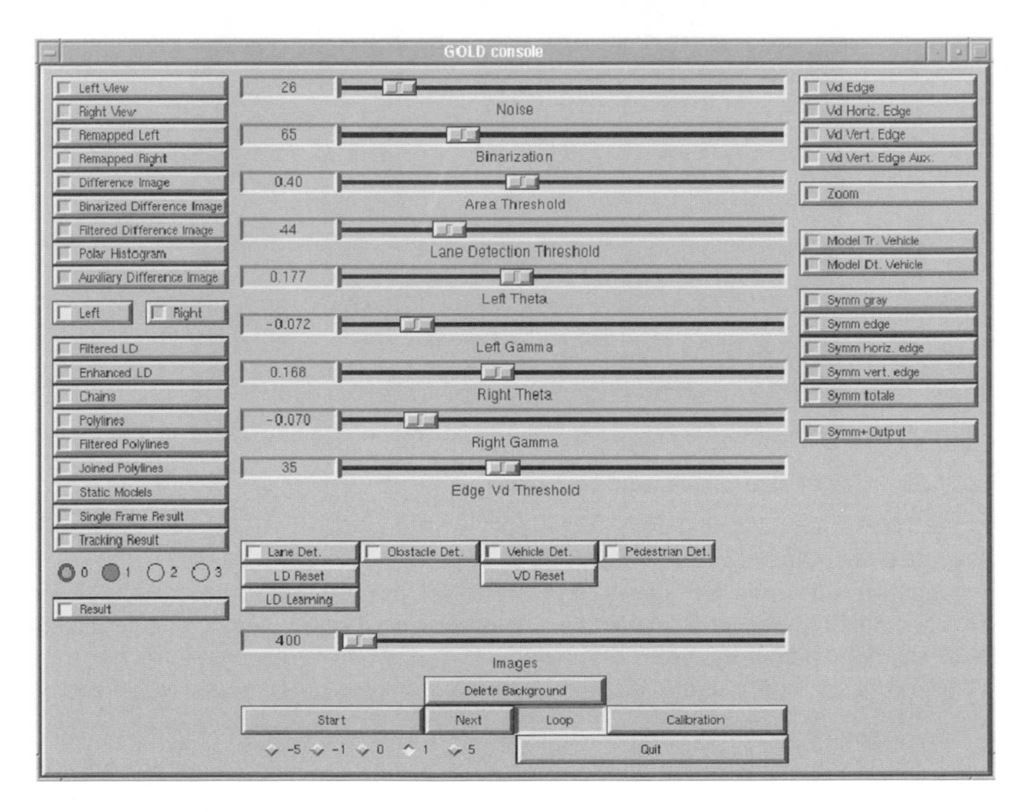

Fig. 14.38 The software control panel used for debugging.

Fig. 14.39 View of the electric engine installed on the steering column.

vehicle and releases the control of the steering wheel to the driver only when the vehicle is in a safe position within the lane;

- again as an *actuating* device, to drive the vehicle autonomously: the output provided by the GOLD vision system, i.e. the lane centre at a given distance in front of the vehicle for road following or the position of the leading vehicle for platooning, are used to steer the wheel so to direct the vehicle towards that point.

14.3.5 The control system

The controller currently adopted for the ARGO vehicle was initially designed and optimized for the road following task. Minor changes have been introduced to implement also the platooning functionality.

Once the road geometry ahead of the vehicle has been recovered, a software module is used to compute signals to be issued to the steering wheel actuator. More precisely, starting from the knowledge regarding road geometry and vehicle speed, this module determines the vehicle yaw and computes the position that the vehicle will assume in a given time interval (in the current implementation it is 1.5 s). In case the future vehicle position is not compatible with the requirements (the vehicle should keep central with respect to the driving lane), a rotation is imposed on the steering wheel.

To keep the control system simple, the steering wheel rotation is computed as the result of a temporal average amongst a set of values proportional to the offsets between the future position and the lane centre in successive time instants. Despite the delay caused by temporal average, this low-pass filter has the specific advantage of reducing noise caused by possible incorrect results of the vision system, due – for example – to calibration drifts induced by vehicle movements.

Since the processing rate is high (25 frames per second or more), the refinements of the steering wheel positions occur at a high rate; this eases the filtering of incorrect, vibrating movements caused by noise.

Moreover, an emergency feature has been added to the control system to avoid sudden and therefore dangerous movements of the steering wheel. Each new position – computed as mentioned above – is compared to the current position and processed through a limiting filter, whose threshold depends on vehicle speed. In other words, when the vehicle is moving at a low speed, two consecutive positions of the steering wheel can be quite different; on the other hand, when driving at a high speed, only small steering wheel movements are permitted.

When the road following functionality is active, by means of specific buttons in the control panel, left or right lane change manoeuvres can be triggered off. In order to change the lane, the offset between the vehicle's position and the lane marking that the system is tracking is progressively varied until the vehicle reaches the central position in the new lane. This implies the knowledge of the lane width. Then the system locks onto the new lane marking (delimiting the new lane), which is easily localized lying parallel to the previous one at a known distance.

The control strategy adopted for platooning takes advantage of the previously defined control scheme. The main difference with respect to the path-following functionality is on the estimation of the offset error. When the platooning functionality is activated, the target point is centred on the preceding vehicle so that the target look-ahead distance is neither constant nor the most appropriate for the current velocity. Obviously, the use of this look-ahead distance and the corresponding offset error could degrade the performance of the functionality. The efficiency of the platooning control algorithm is recovered by scaling the tracking error measured at that distance to an estimated offset error given by a *virtual* target point placed at the appropriate look-ahead distance.

14.3.6 Other vehicle equipments and emergency features

Besides the main devices for input, output and data processing, some additional equipments have been installed on the vehicle.

- Since all internal instrumentation works at 220 V @ 50 Hz, an inverter has been included to provide the correct power supply.

- The vehicle has been connected to the Internet thanks to two GSM cellular channels. The connection is set up on two serial links with the PPP (Point to Point) protocol and the packets are switched onto the two lines for load balancing reasons. This connection permits the transfer of images and statistical data from the vehicle to Internet.
- Finally, the ARGO vehicle has been equipped with several emergency devices to be activated by hand in case of system failures. The availability of these devices has been necessary to ensure safety during the tests and to face insurance requirements.

Besides acting on the control panel to deactivate automatic driving features, the passenger can intervene at different levels on a number of emergency devices (see Figures 14.40 and 14.41).

- **Emergency joystick:** a joystick is used to overcome the commands issued by the system. The passenger, after pushing one of the two joystick buttons, can take control of the steering wheel by moving the joystick. This emergency feature is intended only to refine the commands issued by the system. When the button is released, the system keeps working automatically.
- **Emergency pedal:** located near the clutch, it is used to set the *manual driving* functionality. As soon as the pedal is pressed, the control software switches to manual driving and the passenger can take control of the vehicle.
- **Emergency button:** an emergency button, located near the manual brake, performs via hardware the same action of the emergency pedal, detaching the electric engine control signal.
- **Power supply switch:** this last device can be used in case of the hardware or software system's failure, and allows the release of the power supply for the steering wheel electric engine.

14.4 The MilleMiglia in Automatico test

14.4.1 Description

In order to extensively test the vehicle under different traffic situations, road environments, and weather conditions, a 2000 km journey was carried out in June 1998. Other prototypes were tested on public roads with long journeys (CMU's Navlab *No Hands Across America*, and a tour from Munich to Odense organized by the Universität der Bundeswehr, Germany) whose main differences were that the former was relying also on non-visual information (therefore handling occlusions in a different way) and that the latter was equipped with complex computing engines.

The *MilleMiglia in Automatico* test was carried out in 1998, and the system was much more primitive than it is currently. Only lane detection and obstacle detection were tested: lane detection was based on the localization of a single line, while the detection of the preceding vehicle was performed by the obstacle detection module; no tracking was done and only the road following functionality was available.

In the following subsections the test is described, and a critical discussion is presented. The main bottlenecks of the system, as well as some considerations which led to the current implementation (described in the previous sections) are highlighted.

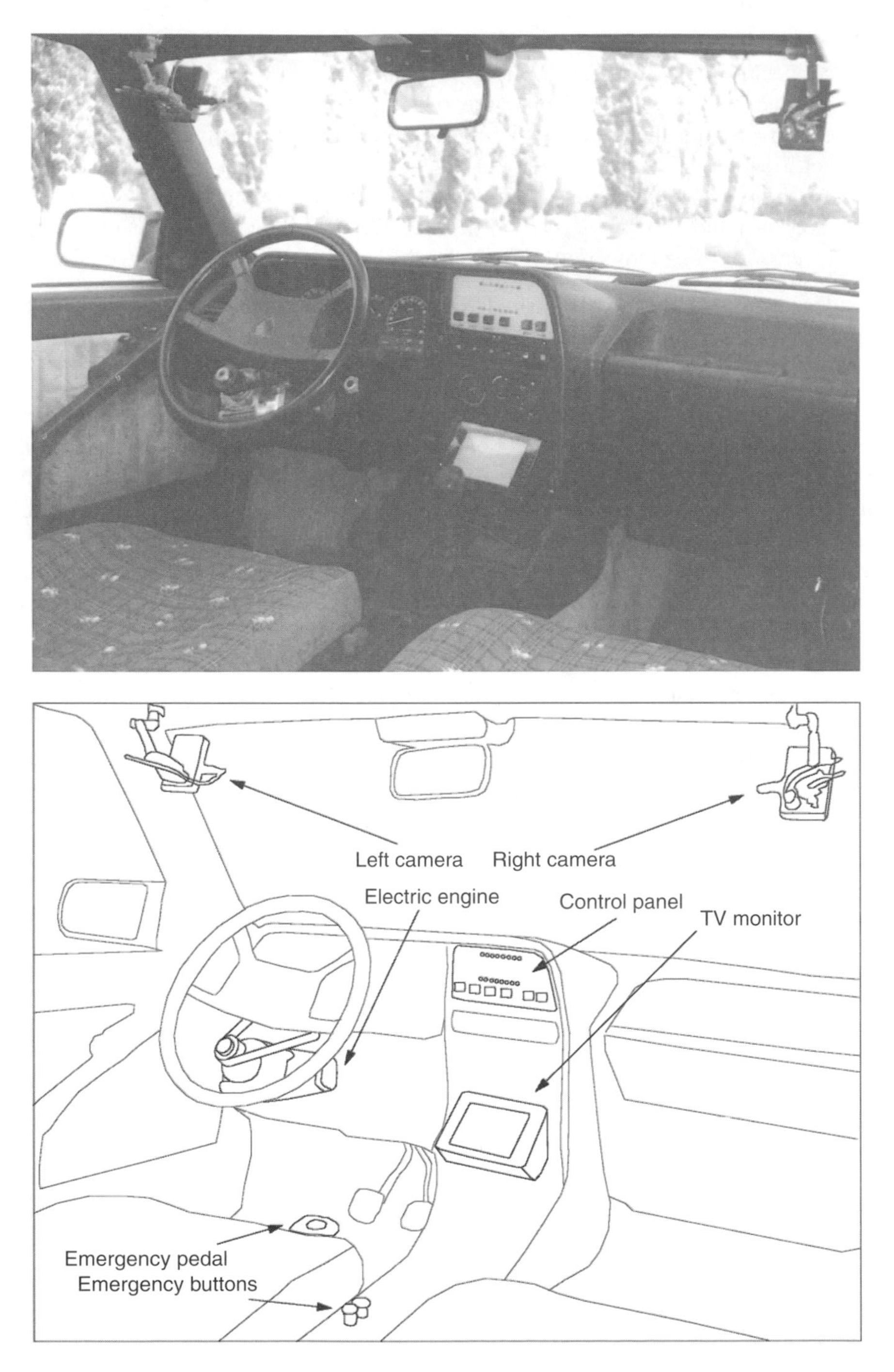

Fig. 14.40 Internal view of the ARGO vehicle.

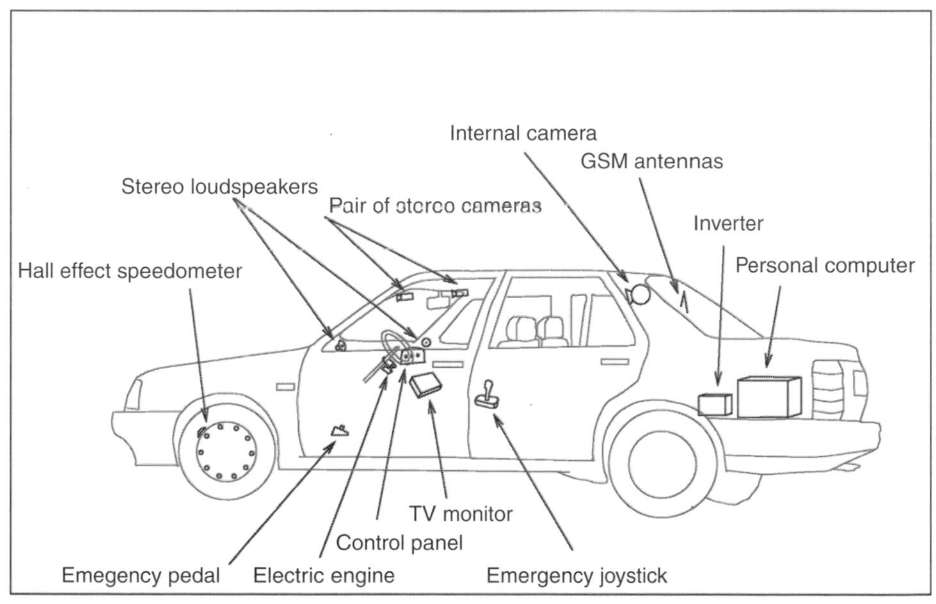

Fig. 14.41 Equipment of the ARGO vehicle.

Test schedule

During this test, ARGO drove itself autonomously along the Italian highway network, passing through flat areas and hilly regions including viaducts and tunnels. The Italian road network is particularly suited for such an extensive test since it is characterized by quickly varying road scenarios which included changing weather conditions and a

Fig. 14.42 Automatic driving during the *MilleMiglia in Automatico* tour.

generally good amount of traffic. The tour took place on highways and freeways, but the system proved to work also on sufficiently structured rural extra-urban roads with no intersections.

The *MilleMiglia in Automatico* tour took place between 1 and 6 June, 1998: it was subdivided into seven stages of about 250 km each, and carried out within six days.

- Monday, 1 June 1998
  9.00–12.00 Parma–Turin .... 245 km
  Mainly flat.

- Tuesday, 2 June 1998
  9.00–11.00 Turin–Milan–Pavia .... 175 km
  Flat; the Milan by-pass taken.
  16.00–20.00 Pavia–Milan–Ferrara .... 340 km
  Flat; again through the Milan by-pass.

- Wednesday, 3 June 1998
  12.00–15.30 Ferrara–Bologna–Ancona .... 260 km
  The Bologna by-pass taken; flat until Rimini, then hilly crossing the Appennine region towards Ancona.

- Tuesday, 4 June 1998
  8.30–13.00 Ancona–Pescara–Rome .... 365 km
  Hilly until Rome.

- Friday, 5 June 1998
  9.00–12.30 Rome–Florence .... 280 km
  Flat and hilly. The GRA (*Grande Raccordo Anulare*) taken.

- Saturday, 6 June 1998
  8.30–12.00 Florence–Bologna–Parma.................................. 195 km
  Mountainous: crossing the Appennine region as far as Bologna, then flat until Parma.

Data logging

During the journey, besides the normal tasks of data acquisition and processing for automatic driving, the system logged the most significant data, such as speed, position of the steering wheel, lane changes, user interventions and commands, and dumped the whole status of the system (images included) in correspondence to situations in which the system had difficulties in detecting reliably the road lane.

This data has been processed off-line after the end of the tour in order to compute the overall system performance, such as the percentage of automatic driving, and to analyse unexpected situations. The possibility of re-processing the same images, starting from a given system status, allows the reproduction of conditions in which the fault was detected and a way of solving it is found. At the end of the tour, the system log contained more than 1200 Mbyte of raw data.

Live broadcasting of the event via the Internet

During the tour, the ARGO vehicle broadcasted a live video stream on the Internet: two GSM cellular modems were connected to the Vision Laboratory of the Dipartimento di Ingegneria dell'Informazione in Parma, and used to transfer both up-to-date news on the test's progression and images acquired by a camera installed in the driving cabin, shown in Figure 14.43, demonstrating automatic driving.

The set-up of a permanent data link from a moving platform implies the use of mobile telecommunications facilities, such as GSM modems. The use of GSM modems for live video streaming faces the following constraints:

- low bit-rate for transmission (usually 9600 bps) and
- high bandwidth variability during movements.

In order to increase the throughput of the link, two modems – and therefore two channels – working simultaneously have been installed on ARGO. The transmission hardware is composed of a PC running Linux with two serial ports and a colour camera connected to the parallel port. The network traffic is split across the two serial channels thanks to the EQL (EQualizer Load-balancer) protocol, able to split the network traffic across multiple links. Moreover, the communication software was designed in order to dynamically adapt the network traffic to the throughput.

To prove the scientific community, mass media and general public's high interest, the web site received more than 350 000 contacts during the tour and more than 3000 Mbyte of information were transferred, with a peak of 16 000 contacts per hour during the first day of the tour.

14.4.2 System performance

This section deals with various issues related to system performance. It covers the following: hardware issues regarding the vision system, considerations made on the image processing algorithms and processing speed, issues related to the vehicle control

Fig. 14.43 The internal camera framing the driving cabin; a GSM antenna is also visible in the background.

system and in particular to the visual feedback used in the vehicle control loop, the experience regarding the use of the man-machine interface installed on ARGO, and a discussion on how environmental conditions influence the whole system. This chapter ends with an overview of some of the problems encountered during the tour and a detailed analysis of one hour of automatic driving.

Vision system

Since one of the goals of the whole project is the development of a sufficiently low-cost system so as to ease its integration in a large number of vehicles, the use of low-cost acquisition devices was a clear starting point: in particular videophone cameras (small sized 3.5 cm × 3.5 cm sensors at an average cost of US$100 each) were installed.

Although having a high sensitivity even in low light conditions (such as during the night), a quick change in the illumination of the scene causes a degradation in the image quality (for example at the entrance or exit from a tunnel). In particular, having a slow automatic gain control (they have been designed for applications characterized by constant illumination such as video-telephony), in correspondence to the exit from a tunnel for a period of time of about 100 ÷ 200 ms, the acquired images become completely saturated and their analysis is impossible. This problem also happens when the vehicle crosses an area that contains patches of new (black) and old (light grey or even white) asphalt. A possible solution to this problem could be to use cameras with a higher performance rate and, in particular, a faster automatic gain control and higher dynamics. CMOS-based sensors are also now being evaluated: their logarithmic response should allow higher dynamics, and their intrinsic slowness due to the specific pixel-by-pixel addressing mode, should pose no additional problems since the early stages of processing (IPM) are based on a strong subsampling of the incoming image. Moreover, their use should speed-up the acquisition process since a complete scanning and transfer of the whole image is no longer required.

Processing system

The processing system proved to be powerful enough for the automatic driving of the vehicle. Moreover, current technology provides processing systems with characteristics that are definitely more powerful than the one installed on ARGO during the test: a commercial PC with a 200 MHz Pentium processor and 32 megabytes of memory.

On such a system, enhanced by a video frame grabber able to acquire simultaneously two stereo images (with 768 × 576 pixel resolution), the GOLD system processes up to 25 pairs of stereo frames per second and provides the control signals for autonomous steering every 40 ms.[1] Obviously the processing speed influences the maximum vehicle speed: the higher the processing speed, the higher the maximum vehicle speed.

Visual processing

The approach used for both obstacle and lane detection, based on the inverse perspective mapping transform, proved to be effective for the whole trip. Even if on Italian highways the flat road assumption (required by the IPM transform) is not always valid, apart from exceptions, the approximation of the road surface with a planar surface was acceptable. The calibration of the vision system, in fact, should be tuned to reflect the modifications of road slope ahead of the vehicle, since a wrong calibration generates a lateral offset in the computation of the vehicle trajectory. Nevertheless, since the highway lanes' width is sufficiently large, this offset has never caused serious problems, but it is for this reason that an enhancement to the IPM transform is currently under development allowing to get rid of the flat-road assumption and to also handle generic roads (Bertozzi, 1998a).

The only drawback which exists, due to the use of the IPM transform, is that the vehicle movements (pitch and roll) do not allow a reliable detection of obstacles at distances further than 50 metres.

During navigation the system locked onto the right lane marking (as already mentioned, lane detection was based on the localization of a single line), hence overtaking vehicles did not occlude its visibility. This choice, however, was critical during the complex situations of highway exits, where two lines are present, a continuous one for the exit and a dashed one for the lane. In such cases the user

[1] This is equivalent to one refinement on the steering wheel position for every metre when the vehicle drives at 100 km/h.

could select whether to stay on the road or to exit it. Moreover, occlusions by other vehicles, even if much less frequent as they could have been for the left lane marking, could represent a problem. For this reason a more robust version of the lane detection algorithm (discussed in Section 14.2.2) has been developed, which can handle both lines. This choice enforces the system's reliability also in case either of the two lines is worn or missing.

On the contrary, the resolution of these devices turned out to be satisfactory and the framing of the scene (no horizon to reduce strong light conditions and direct sunlight) correct for the type of roads considered (sloping gently).

During the whole tour the system processed about 1 500 000 images[2] totalling about 330 Gbyte of input data.

Control system

The control system has been designed and developed focusing on its simplicity. Regarding the *mechanical* part, an electric stepping motor allows the rotation of the steering wheel with a high resolution and a reduced power consumption.

On the other hand, the simplicity of the *logical* part of the control system has the main advantage of keeping the entire system robust: only a temporal average and a simple measurement are used to compute the steering wheel angles.

With this kind of control, for speeds reaching around $90 \div 95$ km/h there is no noticeable difference in comparison to a human driver, while for higher speeds (during the test a peak of 123 km/h was reached) the vehicle tends to demonstrate slightly unstable behaviour, oscillating inside the lane.

A more sophisticated control system is currently under study also including a strong road model; this should allow reaching higher speeds with more stability. This new control mechanism is undergoing tests on a system simulator that permits the emulation of vehicle behaviour on curved roads with more than one lane. It has also been used to test lane change manoeuvre.

Man–machine interface

The automatic driving system is managed by a control panel by means of six buttons and eight LEDs which indicate the currently selected driving functionality. Since this is still in an early development stage, it is possible to also modify the value of a number of driving parameters, which should normally be hardcoded and not adjustable by the final user. Nevertheless, this possibility does not increase the interface complexity. For security reasons the user is required to press two buttons simultaneously for each command to be executed. The modifications in the system's state are then notified to the user through vocal messages which confirm the reception and execution of the command. Moreover, for some commands whose execution is not instantaneous (such as the lane change command) an acoustic signal is also generated upon completion of the manoeuvre.

The emergency systems allows the driver to take the control of the vehicle during emergency situations in three different ways: tuning driving parameters (through the control panel buttons), temporarily taking over the system (by means of a joystick), completely taking over the system (by pressing a pedal close to the clutch, or the control panel buttons, or an emergency button close to the handbrake).

---

[2] Each with size 768×288 pixel.

Environmental conditions

During the six-day test, the system's behaviour was evaluated in various environmental conditions. The route was chosen in order to include areas with different morphological characteristics: from flat areas (Parma, Piacenza, Turin, Milan, Verona, Padova, Ferrara, Bologna), to sloping territories of the Appennines region (Pescara, L'Aquila, Rome, Florence, Bologna), and heavy traffic zones (Rome, Turin and Milan's by-passes), inevitably encountering stretches of highway with road works, absent or worn horizontal road signs, and diversions.

Moreover, the different weather conditions (in particular the light conditions) demonstrated the robustness of the image processing algorithms. In fact, the system was always able to extract the information for the navigation task even in critical light conditions, with the sun in front of the cameras, high or low on the horizon, during the evening as well as during the day, with high or low contrast.

The second leg of the tour ended in the late evening; at night-time the system's behaviour is improved by the absence of sunlight reflections and shadows, whilst the area of interest is constantly illuminated by the vehicle headlights.

One of the problems, which is now being solved through shadowing devices and a light polarizing filter, is the light's reflection within the internal surface of the windscreen, as shown in Figure 14.44. Figure 14.45 shows some of the typical effects of this reflection.

Finally, the system proved to be surprisingly robust despite the high temperatures measured during the tour: in some cases the external temperature reached 35°C and the system continued to work reliably even with no air conditioning.

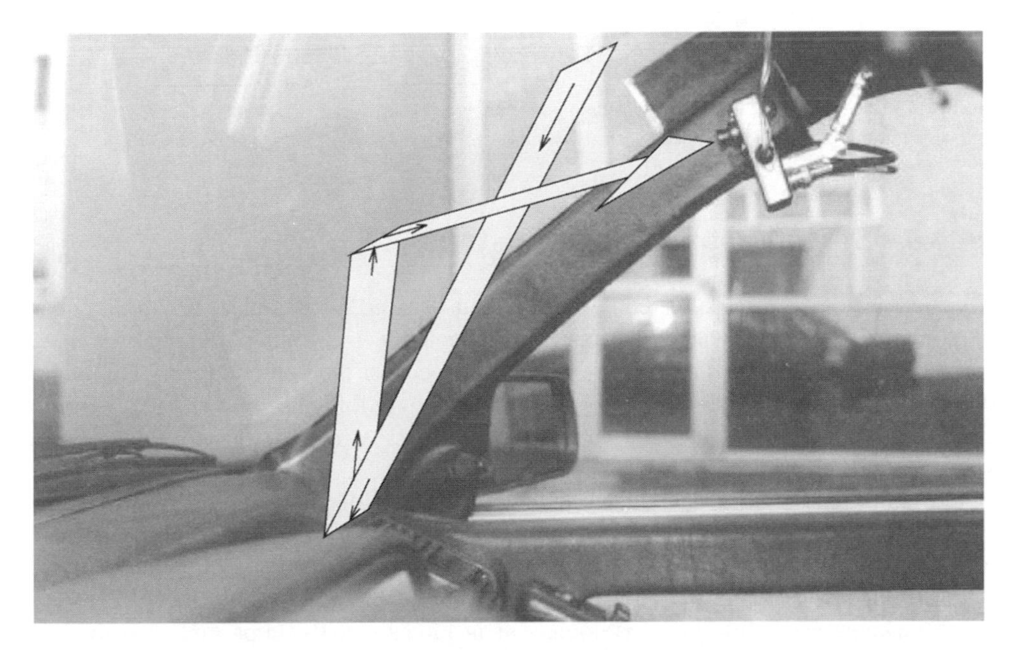

Fig. 14.44 The light's reflection within the internal surface of the windscreen; this causes the images to be over-saturated.

Fig. 14.45 The reflection problem: (*a*) image with no reflections; (*b*) image with a strong reflection; (*c*) oversaturated image due to reflections on the windscreen.

14.4.3 Statistical analysis of the tour

The analysis of the data collected during the tour allowed the computation of a number of statistics regarding system performance (see Table 14.2). In particular, for each stage of the tour the average and the maximum speed of the vehicle during automatic driving were computed. The average speed was strongly influenced by the heavy traffic conditions (especially on Turin's and Milan's by-passes) and by the presence of toll stations, junctions, and road works.

The automatic driving percentage and the maximum distance automatically driven show high values despite the presence of many tunnels (particularly during the Appennines routes Ancona–Rome and Florence–Bologna) and of several stretches of road with absent or worn lane markings (Ferrara–Ancona and Ancona–Rome) or even no lane markings at all (Florence–Bologna). It is fundamentally important also to note that some stages included passing through toll stations and transiting in by-passes with heavy traffic and frequent queues during which the system had to be switched off.

Table 14.2 Statistical data regarding the system performance during the tour

Stage	Date	Departure	Arrival	Kilometres	Average speed (km/h)	Maximum speed (km/h)	Percentage of automatic driving	Maximum distance in automatic (km)
1	1 June 1998	Parma	Turin	245	86.6	109	93.3	23.4
2	2 June 1998	Turin	Pavia	175	80.2	95	85.1	42.2
3	2 June 1998	Pavia	Ferrara	340	89.8	115	86.4	54.3
4	3 June 1998	Ferrara	Ancona	260	89.8	111	91.1	15.1
5	4 June 1998	Ancona	Rome	365	88.4	108	91.1	20.8
6	5 June 1998	Rome	Florence	280	87.5	110	95.4	30.6
7	6 June 1998	Florence	Parma	195	89.0	123	95.1	25.9

14.4.4 Detailed analysis of one hour of automatic driving

Figure 14.46 shows the system behaviour during a one hour period of time (from 7 pm to 8 pm on 2 June 1998) of the tour's third stage (approximately 90 km from Verona to Ferrara). Figure 14.46 shows the following from top to bottom:

- The selected functionality: during this period of time the system always worked in automatic mode except in four situations indicated with A, B, C and D, in which:
  (A) ARGO passed through two tunnels (near Vicenza);
  (B) the system at the end of an overtaking manoeuvre did not recognize the lane markings due to an occlusion;
  (C) and (D) the vehicle passed two working areas with no lane markings.
  In the first two cases, the system switched back to manual driving automatically and warned the driver; whilst in the other two cases the system was switched off manually. It is important to note that in correspondence to these four situations the logged data (apart from speed) is meaningless. The images dumped on disk on these occasions are shown in Figure 14.47.
- The vehicle position on the lane: ARGO made a number of lane changes, moving from the rightmost lane to the central lane and vice versa (ARGO was running on a three-lane highway).
- The distance between the right wheel and the right lane markings (in centimetres). It can be noted that during lane changes the distance alters abruptly; when the vehicle position is again central with respect to the new lane, the systems locks onto the new lane and the distance returns to the nominal value. The selected value of the distance is 80 cm.
- The steering wheel angle automatically chosen by the system. It is possible to note that this angle is proportional to the distance shown by the previous graph, with the only exception of the peaks due to lane changes.
- Finally, the speed of the vehicle. It is possible to note the low speed in the situations identified by C and E respectively, due to road works and driving on a junction (between A4 and A13, in which is also visible an evident movement of the steering wheel).

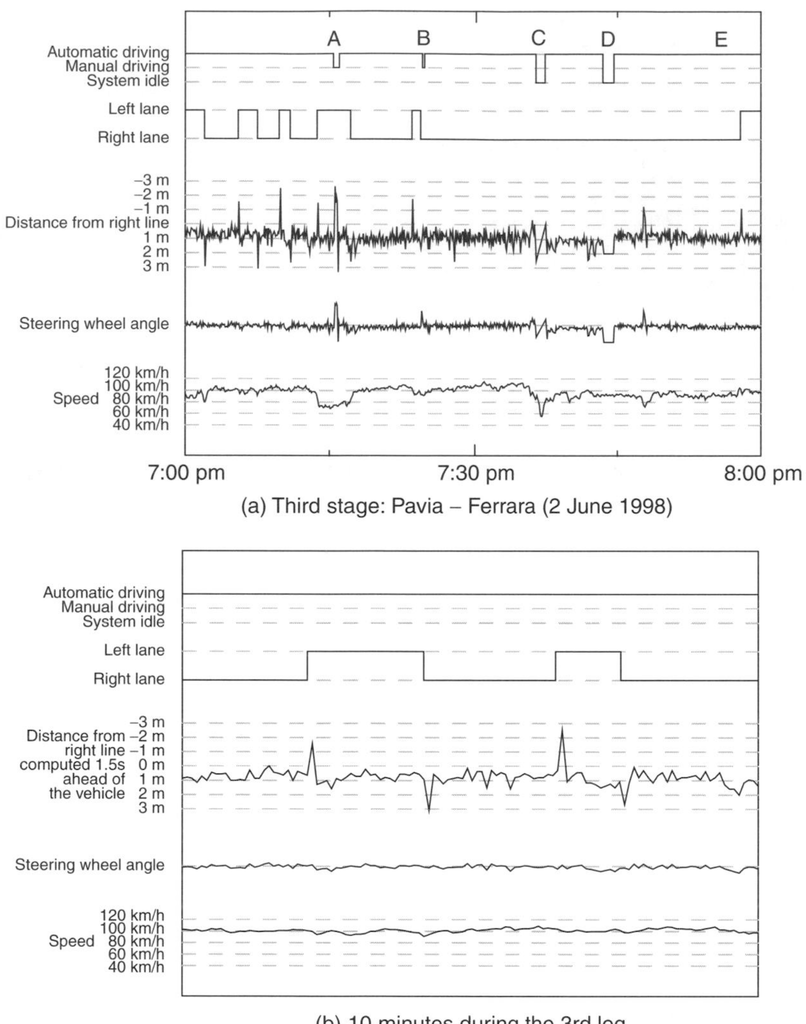

Fig. 14.46 System behaviour during (*a*) one hour and (*b*) 10 minutes of automatic driving.

14.4.5 Discussion and current enhancements

This section summarizes the main problems that lowered the system performance. As mentioned above, the presence of tunnels and long bridges, when combined with strong sunshine, produces very contrasted shadows and sun spots. In order to adjust the value of the automatic gain control, the cameras compute the average brightness of the whole frame. This produces poor results when the scene is highly contrasted as shown in Figure 14.48.

Highly contrasted scenes were mainly found in the situations explained above (tunnels and bridges), but also when passing on different patches of asphalt (usually a new dark grey and an old light grey one), as shown in Figure 14.49. In the same

Fig. 14.47 Images dumped in the anomalous situations A, B, C, and D respectively.

Fig. 14.48 The problem of high contrast: (*a*) exit from a tunnel; (*b*) long bridge; (*c*) short bridge.

figure a strong reflection within the internal windscreen surface – which produces a high image saturation – is also visible.

Figure 14.50 shows a situation in which the system had a major fault: a vehicle occluded the visibility of the right lane marking in correspondence to a highway exit. Since the only visible lane marking was the continuous line representing the exit, the system followed that line, therefore requiring a user intervention. Since the GOLD system does not include the possibility of automatically choosing whether to stay on the highway or to exit, a user intervention was always requested at each exit. This explains why the percentage of automatic driving was high even when the maximum distance driven in automatic was not that long (see Table 14.2). Other user interventions were required in correspondence to toll stations and road works.

Fig. 14.49 The saturation problem: two different asphalt patches produce a highly contrasted scene which, along with a strong reflection on the windscreen, saturates the image.

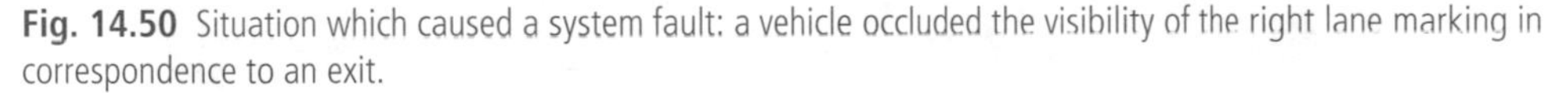

Fig. 14.50 Situation which caused a system fault: a vehicle occluded the visibility of the right lane marking in correspondence to an exit.

Another serious problem happened in a stretch of road just after the conclusion of road works: on the ground were still present the lines that were painted to delimit the diversion and, as a result, the vehicle followed those lines.

The system demonstrated high robustness with respect to the following: horizontal road signs (Figure 14.51(a)) and fog markings (Figure 14.51(b)); in presence of forks, junctions, and exits, as shown in Figure 14.51(c); heavy traffic conditions that were encountered, mainly on the Turin, Milan, and Rome by-passes, as shown in Figure 14.51(d); and the presence of the guard-rail in Figure 14.51(e).

Moreover, also high temperatures, different light conditions (the third stage of the tour terminated during the night), and high speeds (up to 120 km/h) did not influence the stability and robustness of the whole system, both for hardware and software. The vehicle only experienced a slightly oscillating behaviour in correspondence to sloping stretches of road.

Finally, a few minor problems experienced during the tour were: a power failure due to an overloading of the power supply, interference of an on-board cellular phone with the cameras and acquisition board, and some human errors in the use of the control panel.

Fig. 14.51 Situations in which the system demonstrated good reliability: (*a*) horizontal road signs; (*b*) fog markings; (*c*) exits; (*d*) heavy traffic; (*e*) guard-rail.

These problems have been deeply analysed and the following solutions have been considered and implemented (as discussed in the previous sections).

- Lane detection based on two lines: when the visibility of one lane marker is low or missing, the system relies on the other one, reconstructing the missing information on the basis of historical information; the system can learn from past experience.
- New cameras (with higher dynamics and faster automatic gain control) have been considered and will be tested in the future.
- The new vehicle detection procedure has been developed, together with vehicle tracking.
- A new functionality – platooning – has been demonstrated, which allows ARGO to automatically follow any preceding vehicle.

References

Bertozzi, M. and Broggi, A. (1998). GOLD: a Parallel Real-Time Stereo Vision System for Generic Obstacle and Lane Detection. *IEEE Trans. on Image Processing*, **7**(1), 62–81, January.

Bertozzi, M., Broggi, A. and Fascioli, A. (1998a). An extension to the Inverse Perspective Mapping to handle non-flat roads. *Proceedings of the IEEE Intelligent Vehicles Symposium '98*, pp. 305–10, Stuttgart, Germany, October.

Bertozzi, M., Broggi, A. and Fascioli, A. (1998b). Stereo Inverse Perspective Mapping: Theory and Applications. *Image and Vision Computing Journal*, **8**(16), pp. 585–90.

Broggi, A., Bertozzi, M., Fascioli, A. and Conte, G. (1999). *Automatic Vehicle Guidance: the Experience of the ARGO Vehicle*. World Scientific, April. ISBN 981–02–3720–0.

Fascioli, A. (2000). Vision-based Automatic Vehicle Guidance: Development and Test of a Prototype. PhD thesis, Dipartimento di Ingegneria dell'Informazione, Università di Parma, Italy, January.

Index